T0142916

Lecture Notes in Computer Science

Lecture Notes in Artificial Intelligence **13937**

The series Lecture Notes in Artificial Intelligence (LNAI) was established in 1988 as a topical subseries of LNCS devoted to artificial intelligence.

The series publishes state-of-the-art research results at a high level. As with the LNCS mother series, the mission of the series is to serve the international R & D community by providing an invaluable service, mainly focused on the publication of conference and workshop proceedings and postproceedings.

Hisashi Kashima · Tsuyoshi Ide · Wen-Chih Peng
Editors

Advances in Knowledge Discovery and Data Mining

27th Pacific-Asia Conference
on Knowledge Discovery and Data Mining, PAKDD 2023
Osaka, Japan, May 25–28, 2023
Proceedings, Part III

Editors
Hisashi Kashima
Kyoto University
Kyoto, Japan

Tsuyoshi Ide
IBM Research, Thomas J. Watson Research Center
Yorktown Heights, NY, USA

Wen-Chih Peng
National Chiao Tung University
Hsinchu, Taiwan

ISSN 0302-9743 ISSN 1611-3349 (electronic)
Lecture Notes in Artificial Intelligence
ISBN 978-3-031-33379-8 ISBN 978-3-031-33380-4 (eBook)
https://doi.org/10.1007/978-3-031-33380-4

LNCS Sublibrary: SL7 – Artificial Intelligence

This Springer imprint is published by the registered company Springer Nature Switzerland AG
The registered company address is: Gewerbestrasse 11, 6330 Cham, Switzerland

General Chairs' Preface

On behalf of the Organizing Committee, we were delighted to welcome attendees to the 27th Pacific-Asia Conference on Knowledge Discovery and Data Mining (PAKDD 2023), held in Osaka, Japan, on May 25–28, 2023. Since its inception in 1997, PAKDD has long established itself as one of the leading international conferences on data mining and knowledge discovery. PAKDD provides an international forum for researchers and industry practitioners to share their new ideas, original research results, and practical development experiences across all areas of Knowledge Discovery and Data Mining (KDD). PAKDD 2023 was held as a hybrid conference for both online and on-site attendees.

We extend our sincere gratitude to the researchers who submitted their work to the PAKDD 2023 main conference, high-quality tutorials, and workshops on cutting-edge topics. We would like to deliver our sincere thanks for their efforts in research, as well as in preparing high-quality presentations. We also express our appreciation to all the collaborators and sponsors for their trust and cooperation.

We were honored to have three distinguished keynote speakers joining the conference: Edward Y. Chang (Ailly Corp), Takashi Washio (Osaka University), and Wei Wang (University of California, Los Angeles, USA), each with high reputations in their respective areas. We enjoyed their participation and talks, which made the conference one of the best academic platforms for knowledge discovery and data mining. We would like to express our sincere gratitude for the contributions of the Steering Committee members, Organizing Committee members, Program Committee members, and anonymous reviewers, led by Program Committee Co-chairs: Hisashi Kashima (Kyoto University), Wen-Chih Peng (National Chiao Tung University), and Tsuyoshi Ide (IBM Thomas J. Watson Research Center, USA). We feel beholden to the PAKDD Steering Committees for their constant guidance and sponsorship of manuscripts.

Finally, our sincere thanks go to all the participants and volunteers. We hope all of you enjoyed PAKDD 2023 and your time in Osaka, Japan.

April 2023

Naonori Ueda
Yasushi Sakurai

PC Chairs' Preface

It is our great pleasure to present the 27th Pacific-Asia Conference on Knowledge Discovery and Data Mining (PAKDD 2023) as the Program Committee Chairs. PAKDD is one of the longest-established and leading international conferences in the areas of data mining and knowledge discovery. It provides an international forum for researchers and industry practitioners to share their new ideas, original research results, and practical development experiences from all KDD-related areas, including data mining, data warehousing, machine learning, artificial intelligence, databases, statistics, knowledge engineering, big data technologies, and foundations.

This year, PAKDD received a record number of 869 submissions, among which 56 submissions were rejected at a preliminary stage due to policy violations. There were 318 Program Committee members and 42 Senior Program Committee members involved in the reviewing process. More than 90% of the submissions were reviewed by at least three different reviewers. As a result of the highly competitive selection process, 143 submissions were accepted and recommended to be published, resulting in an acceptance rate of 16.5%. Out of these, 85 papers were primarily about methods and algorithms and 58 were about applications. We would like to thank all PC members and reviewers, whose diligence produced a high-quality program for PAKDD 2023. The conference program featured keynote speeches from distinguished researchers in the community, most influential paper talks, cutting-edge workshops, and comprehensive tutorials.

We wish to sincerely thank all PC members and reviewers for their invaluable efforts in ensuring a timely, fair, and highly effective PAKDD 2023 program.

April 2023

Hisashi Kashima
Wen-Chih Peng
Tsuyoshi Ide

Organization

General Co-chairs

Naonori Ueda	NTT and RIKEN Center for AIP, Japan
Yasushi Sakurai	Osaka University, Japan

Program Committee Co-chairs

Hisashi Kashima	Kyoto University, Japan
Wen-Chih Peng	National Chiao Tung University, Taiwan
Tsuyoshi Ide	IBM Thomas J. Watson Research Center, USA

Workshop Co-chairs

Yukino Baba	University of Tokyo, Japan
Jill-Jênn Vie	Inria, France

Tutorial Co-chairs

Koji Maruhashi	Fujitsu, Japan
Bin Cui	Peking University, China

Local Arrangement Co-chairs

Yasue Kishino	NTT, Japan
Koh Takeuchi	Kyoto University, Japan
Tasuku Kimura	Osaka University, Japan

Publicity Co-chairs

Hiromi Arai	RIKEN Center for AIP, Japan
Miao Xu	University of Queensland, Australia
Ulrich Aivodji	ÉTS Montréal, Canada

Proceedings Co-chairs

Yasuo Tabei	RIKEN Center for AIP, Japan
Rossano Venturini	University of Pisa, Italy

Web and Content Chair

Marie Katsurai	Doshisha University, Japan

Registration Co-chairs

Machiko Toyoda	NTT, Japan
Yasutoshi Ida	NTT, Japan

Treasury Committee

Akihiro Tanabe	Osaka University, Japan
Aya Imura	Osaka University, Japan

Steering Committee

Vincent S. Tseng	National Yang Ming Chiao Tung University, Taiwan
Longbing Cao	University of Technology Sydney, Australia
Ramesh Agrawal	Jawaharlal Nehru University, India
Ming-Syan Chen	National Taiwan University, Taiwan
David Cheung	University of Hong Kong, China
Gill Dobbie	University of Auckland, New Zealand
Joao Gama	University of Porto, Portugal
Zhiguo Gong	University of Macau, Macau
Tu Bao Ho	Japan Advanced Institute of Science and Technology, Japan
Joshua Z. Huang	Shenzhen Institutes of Advanced Technology, Chinese Academy of Sciences, China
Masaru Kitsuregawa	University of Tokyo, Japan
Rao Kotagiri	University of Melbourne, Australia
Jae-Gil Lee	Korea Advanced Institute of Science & Technology, Korea

Contents – Part III

Big Data

Financial Data

Information Retrieval and Search

Internet of Things

Medical and Biological Data

Multimedia and Multimodal Data

Recommender Systems

Big Data

Toward Explainable Recommendation via Counterfactual Reasoning

Haiyang Xia[1], Qian Li[2], Zhichao Wang[3], and Gang Li[4(✉)]

[1] Australian National University, Canberra, ACT 2601, Australia
[2] Curtin University, Perth, WA 6102, Australia
[3] Insurance Australia Group, Sydney, NSW 2000, Australia
[4] Strategic Research Center for Cyber Resilience and Trust, Deakin University, Melbourne, VIC 3126, Australia
gang.li@deakin.edu.au

Abstract. Recently, counterfactual explanation models have shown impressive performance in adding explanations to recommendation systems. Despite their effectiveness, most of these models neglect the fact that not all aspects are equally important when users decide to purchase different items. As a result, the explanations generated may not reflect the users' actual preferences. Furthermore, these models typically rely on external tools to extract aspect-level representations, making the model's explainability and recommendation performance are highly dependent on external tools. This study addresses these research gaps by proposing a co-attention-based fine-grained counterfactual explanation model that uses co-attention and aspect representation learning to directly capture user preferences toward different items for recommendation and explanation. The superiority of the proposed model is demonstrated through extensive experiments.

Keywords: Recommendation System · Explainability · Aspect · Counterfactual Reasoning · Co-attention

1 Introduction

Explainability as an important factor of recommendation systems has attracted extensive research interest in recent years [15]. Classical explainable recommendation models are usually equipped with inherent or post-hoc explanations. Inherent explanation models rely on transparent models to generate explanations [1]. Post-hoc explanation models aim to provide explanations for existing recommendation models in the form of feature importance or natural language [6]. Different from classical explanation models that generate explanations by observing the correlations between recommendation inputs and outputs, counterfactual explanation models are proposed to generate explanations by identifying minimal changes that could alter the recommendation results and have demonstrated superior performance [12].

H. Kashima et al. (Eds.): PAKDD 2023, LNAI 13937, pp. 3–15, 2023.
https://doi.org/10.1007/978-3-031-33380-4_1

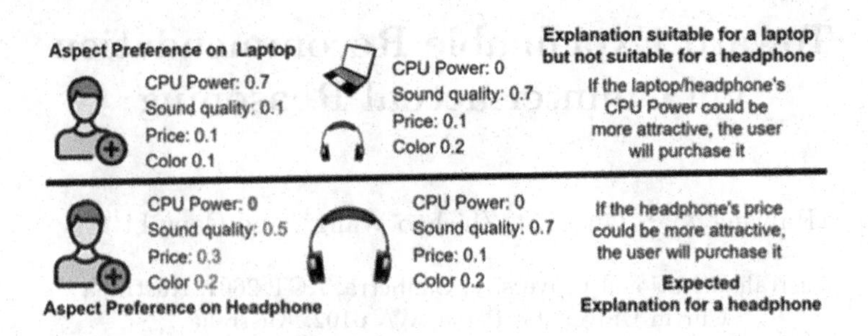

Fig. 1. Inappropriate explanations caused by ignoring the user's aspect preferences toward different items. The correct explanation for not purchasing that headphone is on its price aspect rather than CPU power because CPU power is not considered by `user A` when purchasing headphones.

Despite the impressive performance achieved by these explanation models, there are still two issues to be resolved. To begin with, most of these models explore user reviews to generate explanations via external sentiment analysis tools. As a result, the models' explainability is highly tied to external sentiment analysis tools [3]. In addition, these models neglect the fact that aspects are not equally important when users decide to purchase different items [3]. Leading to the explanations may not reflect the users' actual preferences. Consider the scenario shown in Fig. 1, assume that `user A` mentioned four aspects in his review - `CPU power`, `sound quality`, `price`, and `color`. When buying a laptop, `user A` cares more about the power of `CPU` than other aspects. By contrast, when buying a headphone, `user A` focus the `sound quality` and `price` aspects rather than `CPU power`. As overlooking the aspect importance difference between items, explanations generated by traditional counterfactual models for an item with aspect importance 0, 0.7, 0.1, 0.2 could be "`if the item could improve its attractiveness on the CPU power aspect, user A who has not yer bought the item, will purchase it.`" This is an intuitive explanation for a laptop but not for a headphone, because `CPU power` is not an aspect of considered when `user A` purchases headphones.

To alleviate these issues, this paper proposed a co-attention-based Fine-Grained Counterfactual Explainable Recommendation (`FGCR`) model. Specifically, in `FGCR` an embedding generation subcomponent and an aspect representation learning subcomponent are designed to learn the aspect-level user and item representations directly from review data; and a co-attention-based importance estimation subcomponent is designed to capture users' aspect preferences toward different items for recommendation and explanation.

The main contributions of this paper are:

- For the first time, we propose an end-to-end counterfactual explainable recommendation model, which can extract aspect representations directly from review data for recommendation and explanation.

- We introduce the technique of co-attention to model the fine-grained aspect-level user preferences toward different items for further improving the explainability of existing counterfactual recommendation models.
- We perform extensive experiments on several benchmark datasets to demonstrate the effectiveness of our proposed model.

2 Related Work

2.1 Classical Explainable Recommendations

Classical explainable recommendation models can be categorized into inherent and post-hoc explanation models. Popular inherent explanation models include EFM (*Explicit Factor Model*) [13], A2CF (*Attribute-Aware Collaborative Filtering*) [1], and MATM (*Multi-modal Aspect-Aware Topic Model*) [2]. EFM generates explanations by aligning the latent dimensions of explicit features with item aspects extracted from user reviews [13]. A2CF generates explanations by conducting attribute-level comparisons between user-visited items and corresponding alternatives [1]. MATM generates explanations by tracing the prediction results back to the training data [2].

Different from inherent explanation models that generate explanations during the recommendation process, post-hoc explanation models generate explanations for the recommendation results [10]. For instance, Peake and Wang [6] proposed an association rule-based post-hoc explanation model that generates explanations by analyzing the association rules between recommendation inputs and outputs. Ribeiro et al. [7] proposed LIME (*Local Interpretable Model-agnostic Explanation*) that generates post-hoc explanations by using interpretable linear models to approximate the nonlinear classifiers. Wang et al. [11] proposed a reinforcement learning-based post-hoc explanation model that generates sentence-level explanations by integrating the explanation generator into a personalized attention neural network.

2.2 Counterfactual Explainable Recommendations

Different from traditional explanation models that utilize correlations between inputs and outputs for explanations, counterfactual recommendation models capture the causality of recommendation behavior for explanation generation [4,8]. The natural explainability of causality provides higher explanation performance for counterfactual explanation models [9]. Early counterfactual explanation models focus on providing explanations for HIN (*Heterogeneous Information Network*) and ratings-based recommendation models [5,9]. Examples of such explanation models include PRINCE (*Provider-side Interpretability with Counterfactual Evidence*) [5] and ACCENT (*Action-based Counterfactual Explanations for Neural Recommenders for Tangibility*). PRINCE generates explanations [9] by identifying a minimal set of counterfactual actions on user-specific HINs [5]. ACCENT provides counterfactual explanations for neural collaborative filtering-based recommenders that leverage user ratings [9].

In recent years, the counterfactual explanation research has shifted to providing explanations for aspect-level review-based recommendation models. For example, Zhou et al. [14] proposed CNR which relies on BPR loss and counterfactual reasoning to explain why a user prefers a certain aspect of an item over other items. Xiong et al. [12] used counterfactual reasoning as a data-augmentation method that could not only improve the recommendation models' accuracy but also provide pair-wise aspect-level explanations for users. Tan et al. [8] introduced the concepts of explanation strength and explanation complexity to promote counterfactual explanation models generate simple and efficient aspect-level explanations.

Although existing aspect-level review-based counterfactual explanation models have achieved improved explainability, they typically model user and item importance by static user-aspect and item-aspect matrices [3]. However, in reality, users' aspect preferences change dynamically across items, and items attract users to different aspects. Capturing these dynamic and fine-grained aspect interactions to improve the explanation performance of existing models constitutes the main motivation of this study.

3 Problem Formulation

Across the paper, all the individual elements are denoted as lowercase letters, such as user u, item v, and aspect a. All sets of elements are denoted as calligraphic uppercase letters, such as user set $\mathcal{U}$, item set $\mathcal{V}$, and aspect set $\mathcal{A}$. We conclude the meaning of key notations in Table 1.

Table 1. Notation table

Notations	Meaning
$\mathcal{U}$	The set of users
$\mathcal{V}$	The set of items
$\mathcal{A}$	The set of aspects
$\mathcal{D}_u$	User review corpus (all reviews of user u)
$\mathcal{D}_v$	Item review corpus (all reviews of item v)
e_a	The embedding vector of aspect a
$p_{u,a}$	The latent representation of user u on aspect a
$q_{v,a}$	The latent representation of item v on aspect a
$\lambda_{u,a}$	The importance of aspect a for user u
$\lambda_{v,a}$	The importance of aspect a for item v
$s_{u,v}$	User u's actual preference sore on item v
$\hat{s}_{u,v}$	User u's predicted preference sore on item v
$M_{u,a}$	Aspect level document embedding for user u on aspect a
Δ	The changes in item aspect importance scores

Based on the notations defined, suppose we have a recommendation model $f(\mathcal{A}|\mathcal{D}_u, \mathcal{D}_v)$ that aims to recommend top-k items $v \in V$ for user $u \in U$ by ranking the preference score $s_{u,v}$. Specifically, $s_{u,v}$ is estimated by synthesizing learned aspect-level user representation $p_{u,a}$, item representation $q_{v,a}$, user aspect importance score $\lambda_{u,a}$ and item aspect importance score $\lambda_{v,a}$.

Problem 1 (Aspect-level counterfactual explanation). Assume an item v was not been recommended to user u in recommendation model $f(\mathcal{A}|\mathcal{D}_u, \mathcal{D}_v)$ based on her/his history reviews. Aspect-level counterfactual explanation aims to find an "optimal" perturbation vector $\triangle = \{\delta_1, \delta_2, ..., \delta_r\}$ on item v's aspect importance score $\lambda_{v,a}$ so that v will be recommended to user u.

Intuitively, the generated explanation is: *If item v could be slightly improved on its attractiveness (i.e., item importance score) in aspects* $[\delta_1, \delta_i, \delta_{i+1}...]$, *it will be recommended to user u who previously disliked it.*

4 Methodology

Figure 2 shows an overview of `FGCR`, which consists of an `aspect-Level recommendation component` and a `counterfactual explanation component`. The `aspect-level recommendation component` aims to learn the aspect-level user/item representations and corresponding aspect importance for recommendations. The `counterfactual explanation component` aims to generate explanations based on the learned aspect representations and importance.

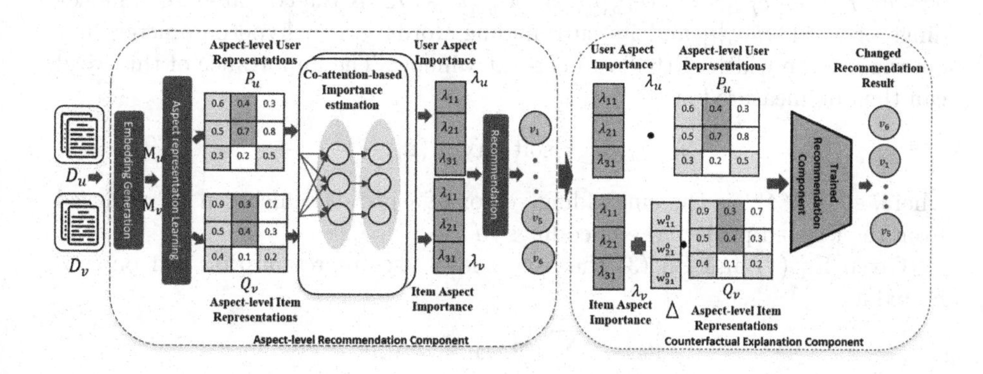

Fig. 2. The overall framework of FGCR.

4.1 Aspect-Level Recommendation Component

The aspect-level recommendation component has four subcomponents: embedding generation, aspect representation learning, co-attention-based importance estimation, and recommendation. Among them, the embedding generation subcomponent aims to retrieve embeddings of the words in $\mathcal{D}_u$ and $\mathcal{D}_v$ from pre-trained embeddings, and transfer $\mathcal{D}_u$ and $\mathcal{D}_v$ into corresponding matrix $M_u \in \mathbb{R}^{n \times d}$ and $M_v \in \mathbb{R}^{n \times d}$ for subsequent analysis. Where n represents the number of words in the user/item document, d represents the dimension of embeddings. The `Word2Vec`[1] embeddings are used in this paper to initialize embeddings. The aspect representation learning subcomponent aims to learn a set of aspects $\mathcal{A} \in \mathbb{R}^K$ and corresponding user representation $P_u = \{p_{u,a} | a \in \mathcal{A}\}$ and item representation $Q_v = \{q_{v,a} | a \in \mathcal{A}\}$ over this set of aspects. Since the sentiment of the same word can be different for different aspects, a learnable word projection matrix $W_a \in \mathbb{R}^{d \times h_1}$ is introduced to add variations to `Word2Vec` embeddings w.r.t. different aspects. More specifically, for the i-th word $M_u[i]$ in user u's review corpus $\mathcal{D}_u$, its projected aspect-specific embedding $M_{u,a}[i]$ is measured as:

$$M_{u,a}[i] = M_u[i] W_a \tag{1}$$

Accordingly, the projected aspect-specific embedding on aspect a can be represented as $M_{u,a} \in \mathbb{R}^{n \times h_1}$, where h_1 is the hyperparameter that controls the number of latent factors used in aspect-level representations. Besides, due to the fact that words in different aspects are not equally important and the importance of each word is influenced by its surrounding words [3], we then adopted a local context window $z_{u,a}[i]$ (in which the i-th word is located at the center) to estimate the importance of the word w.r.t different aspects:

$$z_{u,a}[i] = (M_{u,a}[i - c/2], \cdots, M_{u,a}[i], \cdots, M_{u,a}[i + c/2]) \tag{2}$$

where $M_{u,a}[i - c/2], \cdots, M_{u,a}[i], \cdots, M_{u,a}[i + c/2]$ is the concatenated embeddings of word $M_{u,a}[i]$ and its surrounding words. c is a hyperparameter that determines the width of the local context window. The importance of this word can then be measured as:

$$imp_{u,a}[i] = \text{softmax}(e_a (z_{u,a}[i])^T) \tag{3}$$

where, $e_a \in \mathbb{R}^{c \times h_1}$ is the embedding vector of aspect a, which can be retrieved from the learned projected embeddings $M_{u,a}$.

Given Eq. (1) and Eq. (3) the aspect-level user representation can be estimated as:

$$p_{u,a} = \sum_{n}^{i=1} imp_{u,a}[i] \cdot M_{u,a}[i]. \tag{4}$$

Similarly, the aspect-level item representation can be estimated as:

$$q_{v,a} = \sum_{n}^{i=1} imp_{v,a}[i] \cdot M_{v,a}[i] \tag{5}$$

The co-attention-based importance estimation subcomponent aims to model users' aspect preferences toward different items via co-attention. Following [3],

[1] https://code.google.com/archive/p/word2vec/.

an affinity matrix $M_S \in K \times K$ that indicates the affinity between user and item at the aspect-level is calculated first according to the following formula:

$$M_S = \phi(P_u W_s Q_v^T) \tag{6}$$

where $P_u \in \mathbb{R}^{K \times h_1}$ is the set of aspect-level user representations $p_{u,a}$ (measured by Eq. (4)), $Q_v \in \mathbb{R}^{K \times h_1}$ is the set of aspect-level item representations $q_{v,a}$ (measured by Eq. (5)), $W_s \in \mathbb{R}^{h_1 \times h_1}$ is a learnable weight matrix, and ϕ is the ReLU function. After obtaining the affinity matrix M_S, the aspect-level user and item importance can be estimated as:

$$\lambda_u = \text{softmax}(P_u W_x + S^T (Q_v W_y) v_x) \tag{7}$$

$$\lambda_v = \text{softmax}(Q_v W_y + S^T (P_u W_x) v_y) \tag{8}$$

where $v_x, v_y \in \mathbb{R}^{h2}$, $W_x, W_y \in \mathbb{R}^{h_1 \times h_2}$ are learnable parameters. Intuitively, both the aspect-level user representation P_u and item representation P_v are considered when estimating the aspect-level user and item importance.

The recommendation subcomponent aims to predict the user and item preference score $\hat{s}_{u,v}$, according to the learned aspect-level representations and importance for recommendations. Formally,

$$\hat{s}_{u,v} = \sum_{a \in \mathbb{A}} (\lambda_u \cdot \lambda_v) \cdot (p_{u,a} \cdot q_{v,a}) + b_u + b_v + b_g \tag{9}$$

where b_u, b_v, and b_g are user, item, and global biases.

The training of `FGCR` mainly involves train the `aspect-level recommendation component`, which is achieved by gradient descent on mean squared error loss between estimated and actual user-item preference score, i.e., $\hat{s}_{u,v}$ and $s_{u,v}$.

4.2 Counterfactual Explanation Component

The `counterfactual explanation component` aims to identify item aspects on which minimal changes in importance (Eq. (8)) will flip the recommendation result for explanations. Let vector $\triangle$ denote the modifications that counterfactual explanation component applied to the item aspect importance. The counterfactual explanation task can then be formulated as following optimization problem:

$$\begin{aligned} &\min_{\triangle} \parallel \triangle \parallel_2^2 + \gamma \parallel \triangle \parallel_0 \\ &\text{s.t.,}\ \hat{s}_{u,v_\triangle} - \hat{s}_{u,v} \geq \epsilon \end{aligned} \tag{10}$$

where the first item of the optimization function $\parallel \triangle \parallel_2^2$ represents the total changes that have been applied to item aspect importance. The second item of the optimization function $\parallel \triangle \parallel_0$ represents the number of aspects whose importance has been changed. γ is a weight used to balance these two criteria.

$\hat{s}_{u,v_\triangle}$ is the estimated preference score after $\triangle$ is applied to item v' aspect importance. More formally,

$$\hat{s}_{u,v_\triangle} = \sum_{a \in \mathbb{A}} (\lambda_u \cdot (\lambda_v + \triangle)) \cdot (p_{u,a} \cdot q_{v,a}) + b_u + b_v + b_g \tag{11}$$

ϵ is a threshold to ensure the changes of the preference score caused by adding $\triangle$ could hit an unrecommended item v into user u's recommendation list. Suppose the predicted preference score of the k-th item in the original recommendation list is $\hat{s}_{u,v_k}$, ϵ can then be calculated as the margin of the preference score of user u on the $k-1$-th item and item v itself. Formally:

$$\epsilon = \hat{s}_{u,v_{k-1}} - \hat{s}_{u,v}. \tag{12}$$

Thereby, Eq. (10) can be simplified as:

$$\begin{aligned} &\min_{\triangle} \| \triangle \|_2^2 + \gamma \| \triangle \|_0 \\ &\text{s.t.}, \ \hat{s}_{u,v_\triangle} \geq \hat{s}_{u,v_{k-1}} \end{aligned} \tag{13}$$

Solving the optimization problem in Eq. (13) is challenging because both the objective function and constraint are not differentiable. Following [8], to make the objective convex, we relax the $\| \triangle \|_0$ to l_1-norm $\| \triangle \|_1$ and $\hat{s}_{u,v_\triangle} \geq \hat{s}_{u,v_{k-1}}$ to hinge loss:

$$\mathcal{L}(\hat{s}_{u,v_\triangle}, \hat{s}_{u,v_{k-1}}) = \max(0,\ \alpha + \hat{s}_{u,v_k - 1} - \hat{s}_{u,v_\triangle}) \tag{14}$$

Adding hinge loss to the total objective, we get the final optimization function for generating counterfactual explanations as below,

$$\min_{\triangle} \| \triangle \|_2^2 + \gamma \| \triangle \|_1 + \lambda \mathcal{L}(\hat{s}_{u,v_\triangle}, \hat{s}_{u,v_{k-1}}) \tag{15}$$

where λ is the balance weight. Because Eq. (15) is convex, it can then be solved by Adam optimizer.

5 Experiments

5.1 Experimental Setup

Datasets and Baselines. Five of the publicly available *Amazon*[2] datasets - `Cell_Phone`, `Software`, `CDs_and_Vinyl`, `Magazine_Subscriptions` and `Kindle_Store` were used in our experiments. Similar to previous studies, we remove the users with less than 10 reviews for dense datasets `Cell_Phone`, `Software` and `CDs_and_Vinyl`; and users with less than 5 reviews for relatively sparse datasets `Magazine_Subscriptions` and `Kindle_Store`. We compare our model with two state-of-the-art aspect-based explainable recommendation models `CountER` [8], `A2CF` [1], and `Random` selection - a frequently used baseline to demonstrate the difficulty of explainable recommendation [8].

[2] https://nijianmo.github.io/amazon/.

Evaluation Metrics. Two sets of recently proposed evaluation metrics [8] are introduced in this paper to measure the explainability of the proposed model. The first is user-side-evaluation metrics - `Precision`, `Recall` and `F1` that aim to measure whether the generated explanations match the user's actual preferences. The second is model-side-evaluation metrics - the probability of necessity (`PN`), the probability of sufficiency (`PS`), and F_{NS} that aim to measure whether the generated explanations could justify the model's behavior. The higher the values of these metrics, the better the model.

Implementation. Our proposed model is implemented by PyTorch framework. The reviews are tokenized using NLTK (https://www.nltk.org/) toolkit. the top 10 000 frequently used words are selected as the corpus for each dataset. The user and item document length $|D_u|$ and $|D_v|$ are set to 500. The ration of training and testing samples is set to 8 : 2. The pre-trained `300-d Google News`[3] was used in our experiments. The number of aspects n for recommendation is set to 5, other recommendation hyperparameters such as the size of the convolutional filter are set to the same values as in [3]. The counterfactual reasoning hyperparameter λ, α, and ρ are set to 250, 0.2 and 1, respectively. The ablation study of parameters n and λ are presented in Sect. 5.3.

5.2 Experimental Results

We first report the percentage of items that our model can successfully generate explanations in Fig. 3. We can see that in the five datasets our model can fully generate explanations for `Magazine_Subscriptions` and `Cell_Phone` datasets. For datasets `CDs_and_Vinyl`, `Kindle_Store` and `Software`, our model can generate explanations for more than 98% of the recommended items. These high percentages of successful explanations demonstrate the effectiveness of our model in generating explanations.

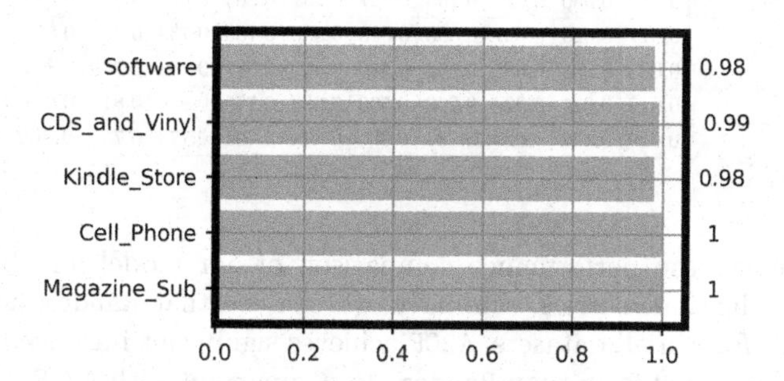

Fig. 3. The percentage of items that successfully generate explanations

[3] https://drive.google.com/file/d/0B7XkCwpI5KDYNlNUTTlSS21pQmM/edit?resourcekey=0-wjGZdNAUop6WykTtMip30g.

Table 2 presents the explainability comparison between our model and baselines in user-side-evaluation metrics. Random achieves the lowest Precision, Recall and F1 in all datasets. This not only demonstrates the difficulty of explainable recommendation problems but also confirms the effectiveness of A2CF, CountER, and FGCR in generating meaningful explanations for recommendation models. CountER shows better performance than A2CF in all datasets in terms of Precision, Recall and F1, its average Precision, Recall and F1 improvement reach 33%, 15%, and 25% respectively. This demonstrates the benefit of introducing counterfactual reasoning in improving the explainability of recommendation models. FGCR achieves the highest Precision, Recall and F1 in all datasets, its Precision, Recall and F1 improvement compared to CountER reach to 53.4%, 18.5%, and 30.4% respectively. This demonstrates the benefits of capturing detailed interactions of review data in improving the recommendation model's ability to explain user preferences.

Table 2. Performance comparison in user-side-evaluation metrics

Datasets	Random			A2CF			CountER			FGCR		
	Precision	Recall	F1	Precision	Recall	F1	Precision	Recall	F1	Precision	Recall	F1
Magazine_Subscriptions	0.021	0.042	0.028	0.167	0.500	0.250	0.745	0.423	0.540	1.000	0.565	0.704
Cell_Phone	0.014	0.032	0.019	0.083	0.167	0.111	0.381	0.468	0.420	1.000	0.575	0.708
Kindle_Store	0.033	0.019	0.024	0.167	0.500	0.250	0.574	0.477	0.521	1.000	0.670	0.785
CDs_and_Vinyl	0.011	0.027	0.016	0.114	0.103	0.108	0.246	0.326	0.280	1.000	0.640	0.762
Software	0.047	0.023	0.031	0.150	0.097	0.118	0.383	0.421	0.401	1.000	0.591	0.720
Average	0.025	0.029	0.024	0.136	0.273	0.182	0.466	0.423	0.432	**1.000**	**0.608**	**0.736**

Table 3. Performance comparison in model-side-evaluation metrics

Datasets	Random			A2CF			CountER			FGCR		
	PN	PS	F_{NS}	PN	PS	F_{NS}	PN	PS	F_{NS}	PN	PS	F_{NS}
Magazine_Subscriptions	0.031	0.043	0.036	0.500	0.500	0.500	0.456	0.572	0.507	0.638	0.965	0.768
Cell_Phone	0.022	0.030	0.025	0.133	0.867	0.231	0.864	0.821	0.842	0.869	0.891	0.880
Kindle_Store	0.017	0.029	0.021	0.500	0.500	0.500	0.739	0.739	0.739	0.816	0.959	0.882
CDs_and_Vinyl	0.035	0.035	0.035	0.054	0.946	0.102	0.687	0.883	0.773	0.807	0.953	0.874
Software	0.042	0.021	0.028	0.250	0.500	0.333	0.775	0.850	0.811	0.904	0.928	0.916
Average	0.029	0.032	0.029	0.287	0.663	0.401	0.704	0.773	0.734	**0.807**	**0.939**	**0.864**

Table 3 shows the performance comparison of our model and baselines in model-side-evaluation metrics. Similarly, we can see that Random achieves the lowest PN, PS, F_{NS} in all datasets. A2CF achieves significant improvement on PN, PS and F_{NS} in all datasets than Random. In comparison with A2CF, the average PN, PS, F_{NS} improvement of CountER reach 41.7%, 11.0%, and 33.3% respectively. FGCR achieves the highest PN, PS, F_{NS} in all datasets, its performance improved by 10.3%, 16.6%, and 13.0% when compared to CountER. This demonstrates the effectiveness of FGCR in justifying recommendation model's behavior.

5.3 Hyperparameter Sensitivity

Number of Aspects. Figure. 4 shows the performance of our model with varying the number of aspects between 3 and 11. From Fig. 4a, Fig. 4b and Fig. 4c, we can see that `FGCR` achieves slightly higher `PN`, `PS` and `F_NS` when the number of aspects equals 9. However, Fig. 4d and Fig. 4e demonstrate that `FGCR` achieves higher `Recall` and `F1` when the number of aspects equals 5. This indicates that the optimal number of aspects is determined by whether the explanation focuses on users' preferences or models' behavior. We set the number of aspects to 5 across our experiments as we focus more on explaining users' preferences.

Balance Weight $\boldsymbol{\lambda}$. The balance weight λ in Eq. (15) controls the extent to which the explanation model prefers explanations that have minimal changes on the aspect importance weights, or explanations that can effectively change the recommendation results. Namely, explanation complexity ($\| \triangle \|_2^2 + \gamma \| \triangle \|_1$) or explanation strength ($\mathcal{L}(\hat{s}_{u,v_\triangle}, \hat{s}_{u,v_{k-1}})$) [8] Small λ indicates the explanation model emphasizes more on the former, large λ indicates the explanation model emphasizes more on the later.

Figure 5a shows the changes in explanation complexity with varying λ between 100 and 300. We can see that the explanation complexity of all datasets increases with increasing λ. Figure 5b shows the change in `F1` with varying λ between 100 and 300. It is clear that `FGCR` achieves better $F1$ when λ equals 250. As such, we set λ to 250 across all experiments.

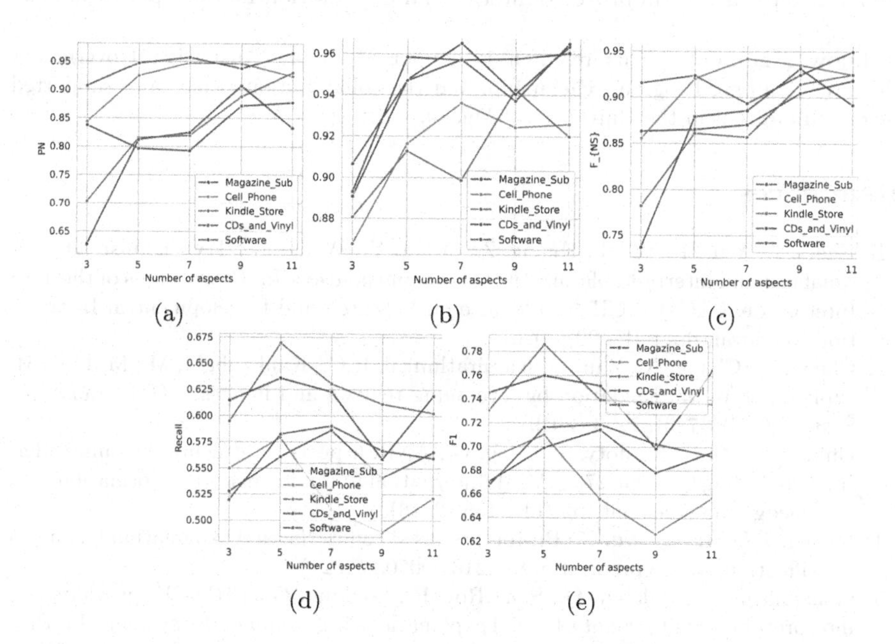

Fig. 4. The influence of the number of aspects on model and user side evaluations

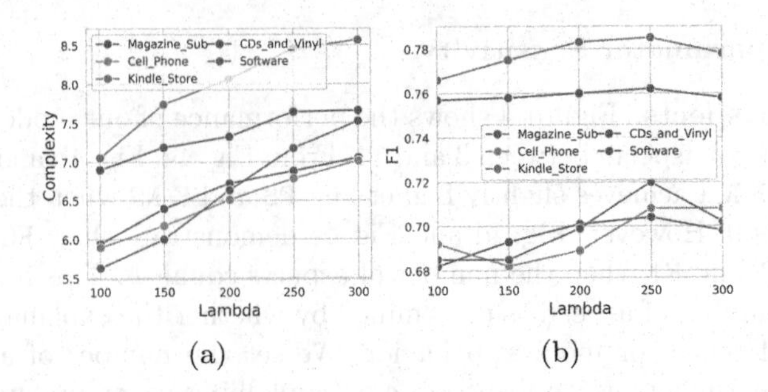

Fig. 5. The influence of the number of aspects on model and user side evaluations

6 Conclusions

In this paper, we proposed an end-to-end fine-grained counterfactual explanation model FGCR. It improves the explainability of traditional counterfactual explanation models by using co-attention technique to capture users' aspect-level preferences toward different items. Extensive experiments on publicly available datasets demonstrated that FGCR leads to state-of-the-art explanation performance in terms of several user-side and model-side evaluations. Our proposed FGCR is not without limitations, in reality, it is not infrequent that the same user's preferences toward the same item may change with the lipase of time. Future studies could explore the possibility of integrating the time dimension of user preference to further improve counterfactual explanation models' performance.

Acknowledgements. This research is supported by an Australian Government Research Training Program scholarship, and the collaboration is partially supported by one SRG grant in the University of Macau.

References

1. Chen, T., Yin, H., Ye, G., Huang, Z., Wang, Y., Wang, M.: Try this instead: personalized and interpretable substitute recommendation. In: Proceedings of the 43rd International ACM SIGIR Conference on Research and Development in Information Retrieval, pp. 891–900 (2020)
2. Cheng, Z., Chang, X., Zhu, L., Kanjirathinkal, R.C., Kankanhalli, M.: MMALFM: explainable recommendation by leveraging reviews and images. ACM Trans. Ins. Syst. (TOIS) **37**(2), 1–28 (2019)
3. Chin, J.Y., Zhao, K., Joty, S., Cong, G.: ANR: aspect-based neural recommender. In: Proceedings of the 27th ACM International Conference on Information and Knowledge Management, pp. 147–156 (2018)
4. Duong, T.D., Li, Q., Xu, G.: Prototype-based counterfactual explanation for causal classification. arXiv preprint arXiv:2105.00703 (2021)
5. Ghazimatin, A., Balalau, O., Saha Roy, R., Weikum, G.: PRINCE: provider-side interpretability with counterfactual explanations in recommender systems. In: Proceedings of the 13th International Conference on Web Search and Data Mining, pp. 196–204 (2020)

6. Peake, G., Wang, J.: Explanation mining: post hoc interpretability of latent factor models for recommendation systems. In: Proceedings of the 24th ACM SIGKDD International Conference on Knowledge Discovery & Data Mining, pp. 2060–2069 (2018)
7. Ribeiro, M.T., Singh, S., Guestrin, C.: "Why Should I Trust You?" Explaining the Predictions of any Classifier. In: Proceedings of the 22nd ACM SIGKDD International Conference on Knowledge Discovery and Data Mining, pp. 1135–1144 (2016)
8. Tan, J., Xu, S., Ge, Y., Li, Y., Chen, X., Zhang, Y.: Counterfactual explainable recommendation. In: Proceedings of the 30th ACM International Conference on Information & Knowledge Management, pp. 1784–1793. CIKM 2021, Association for Computing Machinery, New York, NY, USA (2021). https://doi.org/10.1145/3459637.3482420
9. Tran, K.H., Ghazimatin, A., Saha Roy, R.: Counterfactual explanations for neural recommenders. In: Proceedings of the 44th International ACM SIGIR Conference on Research and Development in Information Retrieval, pp. 1627–1631 (2021)
10. Verma, S., Dickerson, J., Hines, K.: Counterfactual explanations for machine learning: a review. arXiv preprint arXiv:2010.10596 (2020)
11. Wang, X., Chen, Y., Yang, J., Wu, L., Wu, Z., Xie, X.: A reinforcement learning framework for explainable recommendation. In: 2018 IEEE International Conference on Data Mining (ICDM), pp. 587–596. IEEE (2018)
12. Xiong, K., et al.: Counterfactual review-based recommendation. In: Proceedings of the 30th ACM International Conference on Information & Knowledge Management, pp. 2231–2240 (2021)
13. Zhang, Y., Lai, G., Zhang, M., Zhang, Y., Liu, Y., Ma, S.: Explicit factor models for explainable recommendation based on phrase-level sentiment analysis. In: Proceedings of the 37th International ACM SIGIR Conference on Research & Development in Information Retrieval, pp. 83–92 (2014)
14. Zhou, Y., Wang, H., He, J., Wang, H.: From intrinsic to counterfactual: on the explainability of contextualized recommender systems. arXiv preprint arXiv:2110.14844 (2021)
15. Zhou, Y., et al.: Explainable hyperbolic temporal point process for user-item interaction sequence generation. ACM Trans. Inf. Syst. **41**, 1–26 (2022)

Online Volume Optimization for Notifications via Long Short-Term Value Modeling

Yuchen Zhang[1], Mingjun Zhao[2], Chenglin Li[2], Weiyu Tou[1], Haolan Chen[1], Di Niu[2], Cunxiang Yin[1(✉)], Yancheng He[1], and Fei Guo[1]

[1] Tencent Inc., Shenzhen, China
{ericyczhang,raintou,haolanchen,jasonyin,collinhe,richardfguo}@tencent.com
[2] University of Alberta, Edmonton, Canada
{zhao2,ch11,dniu}@ualberta.ca

Abstract. App push notifications are an essential tool for app developers to engage with their users actively and provide them with timely and relevant information about the app. However, determining the proper volume of notifications sent to each user is a key challenge for improving user experience, particularly for new users whose preferences on push notifications are unknown. In this paper, we address the problem of app notification volume optimization for newly onboarded users and propose a systematic approach to solve this problem. We incorporate a multi-task learning technique to accurately modeling both the short-term and long-term effects of different volumes of push notifications, and utilize online linear programming to achieve real-time notification allocation with volume constraints. We have conducted both offline and online experiments to evaluate the effectiveness of our method, and the results demonstrate that our approach dramatically improves multiple core metrics of the user experience, such as daily active users.

Keywords: Notification Volume Optimization · Long-Term Value Model · Online Linear Program

1 Introduction

Nowadays, the competition among different apps has become increasingly fierce with the rapid development of the mobile Internet. Attracting users with information of their interest into persistent app users has become a fundamental challenge. Push notifications are one of the most important channels because they allow apps actively push content to users and guide them to consume and experience the app's product, with the goal of improving business metrics such as daily active users (DAU) and monthly active users (MAU).

One of the biggest challenge in a practical notification system is to determine the optimal number of notifications to send to each user. On the one hand, sending more notifications can increase user engagement with the app. However,

H. Kashima et al. (Eds.): PAKDD 2023, LNAI 13937, pp. 16–28, 2023.
https://doi.org/10.1007/978-3-031-33380-4_2

some users, especially those just onboard with no attachment to the app, may become annoyed by excessive notifications and either turn them off or uninstall the app. On the other hand, many cellphone manufacturers have begun setting limits on the daily volume of push notifications. In this context, developing intelligent algorithms to optimize notification volume for different users to maximize business targets becomes a pressing need. Existing studies on the notification volume optimization problem often focus on short-term metrics such as click-through rate (CTR) or other business targets using traditional machine learning models [2,7,11,16], or on long-term business targets using reinforcement learning [3,4,10,12,15].

However, the problem of notifications for new users has not been adequately addressed in previous work. Many existing models are inefficient or even inapplicable due to the lack of information about new users, such as their interests. In this paper, we propose a systematic approach to solve the notification volume optimization problem for new users. In order to fully exploit the potential of push notifications and avoid excessive usage, we design a multi-task network to accurately model both the short-term and long-term rewards of different volume of push notifications. Additionally, we incorporate an online linear programming algorithm to achieve near real-time optimization of the allocation of the limited notification volumes to each user. Our proposed method has been deployed in the Tencent Mobile QQ Browser (QB), involving over 100 million active users, and currently serves as the main notification strategy. The results of offline experiments and online A/B tests demonstrate our method greatly improves the business target via optimizing the strategy of sending notifications to new users.

The contributions of this work are summarized as follows:

- To the best of our knowledge, this is the first attempt to consider long short-term values when solving notification volumes optimization, and we proposed a volumes-specified multi-task model called MMOE-ATT.
- We incorporate online linear programming to solve decision policy in real-time and achieve the optimal total reward under the limited volume constraint.
- The proposed push notification optimization method has been applied to the Tencent Mobile QQ Browser user onboarding strategy, generating business benefits.

2 Related Work

Predicting positive responses to push notifications is a well-defined problem and is most relevant to our work. As senders, we can build models to predict CTR for each user on push notifications and leverage them to decide the volume. If users like to interact with push notifications, we can send them more frequently. But there are still several challenging issues to address, as described in [14]. First, a proper objective function is essential. Second, increasing or decreasing volume typically has a complicated return on the objective function, so the framework must be able to leverage models to capture this effect. Third, the optimization algorithm must be very efficient and scalable to handle many users. Recently, LinkedIn proposed the problem of email volume optimization for large-scale online services [2]. This paper presents the strategy of optimizing notifications to balance various utilities, such as engagement and send volume, by formulating the problem using constrained optimization. To guarantee the freshness of push notifications, they implemented the solution in a stream computing system in near real-time. Zhao proposed a novel machine learning approach that has been deployed to production at Pinterest [16]. This novel approach computes weekly notification volume for each user such that long-term user engagement is optimized and a global volume constraint is met. This significantly reduced notification volume and improved CTR of notifications and site engagement metrics compared with the previous machine learning approach. An adaptive mobile notification scheduling, which detects opportune timings based on mobile sensing and machine learning, has been proposed to alleviate users' limited attentional resources [11]. As we will explain in the following sections, there are a few limitations of these approaches, which we aim to improve in our new framework. Besides the difficulty of push notification volume optimization, We can never predict the requests generated by the pacer in production. The existing approach for new users, which is the focus of this paper, increases uncertainty. Previous work fails to address this challenge, so we introduce the OVOLS framework to address this weakness.

3 Preliminaries

This section discusses the definition and meaning of the variables used in the following work. Table 1 shows the variables used in this paper and their corresponding interpretations.

Table 1. Variables used in this paper and corresponding explanations

Variables	Explanations
I	the set of users, described in this paper is the set of new users in QB
i	$user_i$, $i \in I$
t	time, this paper indicates the hour
$\mu_{i,t}$	click-through rate for $user_i$ at time t
n_t	the number of push notifications before time t
$n_{i,t}$	the number of push notifications for $user_i$ before time t
$l(i, n_{i,t})$	long-term value for $user_i$ on condition of $n_{i,t}$
α, β, γ	parameters to control the importance of each objective in reward r
$r_{i,t}$	reward function for $user_i$ at time t
B	constraint on the absolute number of push notifications
b_t	the total number of push notifications before time t
π	decision policy

We depict the definition of the problem proposed in this paper as follows. Given the total number of push notifications constraint B, the objective is to find a decision policy π that maximizes the total rewards, which can be expressed formally as Eq. 1. Decision policy will be applied to each user's push delivery at each time, only containing 0 or 1, that is, sending notifications or not, and the aggregate decision for a user is the result of push notifications volume optimization.

$$\begin{aligned} \max_{\pi(i,t)} \quad & \sum_i \sum_t \pi(i,t) * r_{i,t} \\ \text{s.t.} \quad & \sum_i \sum_t \pi(i,t) \leq B \\ \text{where} \quad & \pi(i,t) \in \{0,1\}, \forall i,t. \end{aligned} \tag{1}$$

4 Methods

In this section, we introduce the (Online Volume Optimization for Notifications via Long Short-Term Value Modeling) OVOLS method proposed in this paper to generate the volume of push notifications for new users through online linear programming and long-term value modeling. The challenge of optimizing the volume of push notifications for new users lies in balancing exploration and exploitation, as well as matching long-term value goals. We propose an optimization framework for production that has been applied to the new users at QB and has produced stable improvement. Figure 1 shows the general flow of the framework, which consists of three main components, long-term value model, short-term value, and online linear programming. The following subsections explain each component in detail.

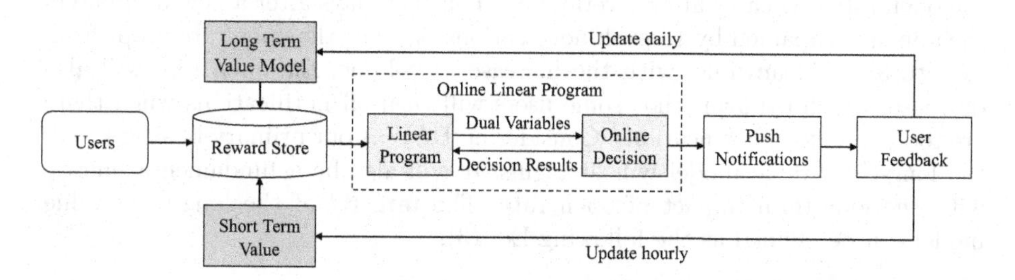

Fig. 1. OVOLS framework in Tencent Mobile QQ browser.

4.1 Short-Term User Value Modeling

New users reveal fewer feedback features and less interaction in the push system than existing users, such as their interests. In this case, some recommendation or decision systems relying on feedback features may bring varying performance across new users. For example, the system will converge quickly on new users with abundant feedback features. In contrast, users with fewer feedback features will perform poorly or even not get traffic or volume. Generally speaking, the industry gives new users a fixed amount of traffic for exploration during the cold start period in case the system discriminates against them, i.e., a fixed volume is given to new users in our system, which is the previous practice. The mentioned problem of new users is a kind of Exploration-Exploitation (EE) problem. An imbalance between exploration and exploitation will lead the system into two extremes. How to balance the two directions has been widely studied. We follow Upper Confidence Bound (UCB) [6] approach to solving the problem to meet short-term goals, and the short-term value of $r_{i,t}^{short}$ can be described as the following Eq. 2

$$r_{i,t}^{short} = \mu_{i,t} + \sqrt{\frac{2ln(n_t)}{n_{i,t}}} \tag{2}$$

4.2 Long-Term User Value Modeling

Our long-term user value model is a multi-task model, which contains two tasks that reflect the long-term value of users, open rate, and logins(number of days users are active in a period, and this paper uses seven days as the period). Many push systems tend to greedily push content to users to improve short-term targets such as CTR and DAU. For example, increasing the number of push notifications may increase DAU. However, some users prefer limited or fewer notifications and might turn off notifications, which greatly jeopardizes the long-term goals and reduces the opportunities to engage with users, especially for new users.

Figure 3 shows the correlation between the volume of push notifications and the open rate for new users. Regardless of the number of push notifications sent, the open rate decreases after several days. Fourteen days after a new user enters the app, accompanied by 15 push notifications per day, the open rate drops from 97% to 91%. Meanwhile, with the increase in volume, the open rate will also decrease, which confirms that some users will turn off notifications when there are too many push notifications. Considering DAU as our primary business goal, the long-term value model will fit logins. It will also be a fundamental metric with the long-term impact of open rate. The target l of the long-term value model can be defined as the following Eq. (3):

$$l(i, n_{i,t}) = logins(i, n_{i,t}) * openRate(i, n_{i,t})^c, \tag{3}$$

where $logins(i, n_{i,t})$ denotes the number of active days on condition of receiving $n_{i,t}$ push notifications within seven days, and $openRate(i, n_{i,t})$ indicates whether the user turns on the push notification after seven days. Both targets are combined by a control parameter c to obtain the final predicted value of the long-term value model. Long-term value contains two different tasks, so long-term value models can also be considered multi-task learning.

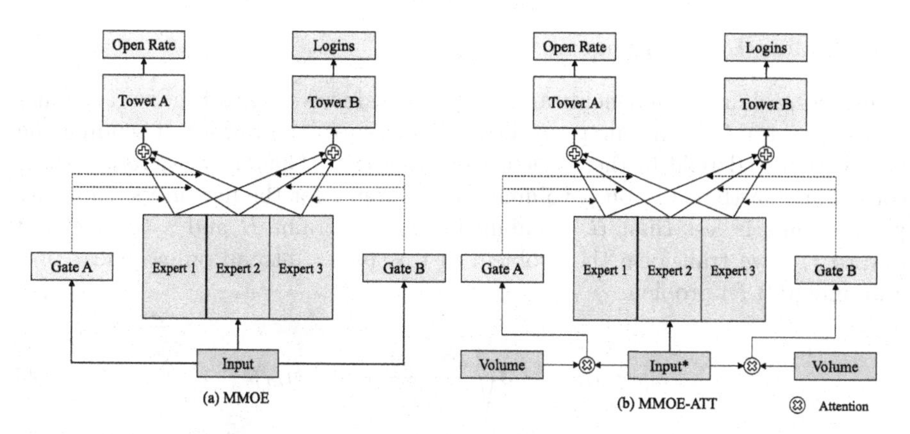

Fig. 2. MMoE and MMoE-ATT structure.

Figure 2(a) shows the structure of the MMoE, a multi-task model, which contains a shared expert network, a gating structure, and a multi-task specified tower. As described in Fig. 3, the volume significantly impacts the long-term value. Existing multi-task models do not consider the impact of certain features on the task but associate all inputs with the task and obtain the gating parameters. To capture the impact of certain features on different tasks and to strengthen the role played by these features in the model. We propose a feature-specified MMOE-ATT model, as described in Fig. 2(b). Comparing the Input of the MMOE, we remove the volume feature to get Input*, then take the volume to make scaled dot-product attention with Input* to get the gating parameters for subsequent.

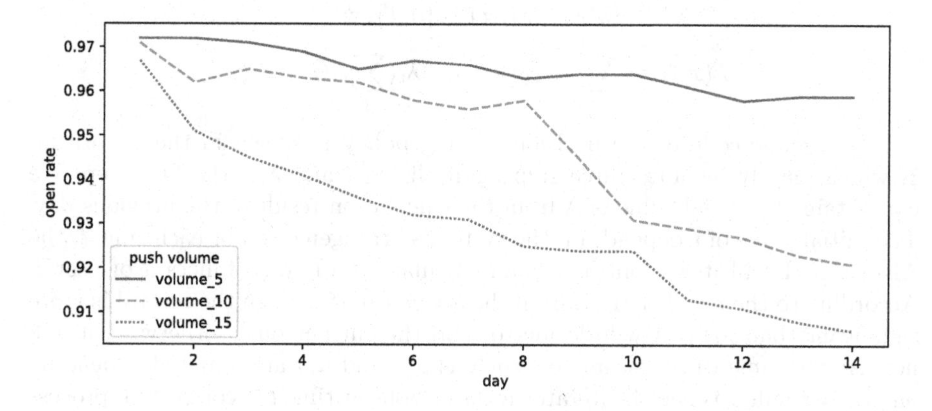

Fig. 3. The open rate of notifications for new users under different volume.

4.3 Online Linear Program

Considering short-term benefits and long-term value, we defined the reward function $r_{i,t}$ as Eq. 1. From our perspective, it measures the reward of sending the notification and reflects the tendency of sending it. Generally, there are some constraints on the daily push volume. For example, for the new users, the daily push volume is less than B. Combined with constraint B and a time-related reward $r_{i,t}$, we transform the problem in this paper into an online linear programming(OLP) problem

$$r_{i,t} = \alpha\mu_{i,t} + \beta\sqrt{\frac{2ln(n_t)}{n_{i,t}}} + \gamma l(i, n_{i,t}) \tag{4}$$

We adopt the Near-Optimal Algorithm for the OLP problem proposed by [1]. Let λ_t be the dual variable for each time t, the following allocation principle is easily derived

$$\pi(i,t) = \begin{cases} 1, & if(r_{i,t} \geq \lambda_{t-1}) \\ 0, & otherwise \end{cases} \tag{5}$$

Once we know the allocation rules, when the online request arrives, a push notification is sent to the user if the reward $r_{i,t}$ exceeds the dual price λ_{t-1}. The calculation of λ_t is a small linear program, as shown below from Eq. (6) to (7), which the solver calculates, and the calculation process is defined as $\lambda_t = g(t, b_t)$

$$\begin{aligned} \max_{\pi(i,t)} \; & \sum_i \pi(i,t) * r_{i,t} \\ \text{s.t.} \; & \sum_i \pi(i,t) \leq b_t \\ \text{where} \; & \pi(i,t) \in \{0,1\}, \forall i. \end{aligned} \tag{6}$$

$$L(x, \lambda) = \sum_i \pi(i,t) * r_{i,t} - \lambda_t(\sum_i \pi(i,t) - b_t) \tag{7}$$

The pseudo-code of the decision-making policy is shown in the Algorithm 1, which mainly includes three steps. Initializing t and λ is the first step. We can obtain the initial value of λ from the calculation result of the previous day. The initial value of t depends on the system's arrangement. For each time t, the Algorithm 1 will iterate all users and calculate the reward function of user i. According to the result derived from the above Eq. (5), when the reward is more significant than λ_{t-1}, it will decide to send the current push; otherwise, it will not. At the same time, parameters such as $\mu_{i,t}$ and $n_{i,t}$ are updated depending on policy results. When the iterator for user polls at time t is completed, process g solves the λ_t, which is a standard linear programming process that can be solved by an APIs such as SciPy.

Algorithm 1. Decision-making policy of push notifications

```
t ← t_0; λ_t ← λ_0
n_i ← 0, ∀i; n_{i,t} ← 0, ∀i, j; μ_{i,t} ← α_i, ∀i, j
for t = t_1, t_2, t_3... do
  for i ∈ I do
    if r ≥ λ_{t-1} then
      π(i, t) = 1
    else
      π(i, t) = 0
    end if
    n_{i,t+1} ← n_{i,t} + π(i, t); n_i ← n_i + π(i, t)
  end for
  t ← t + 1; λ_t ← g(t, b_t)
end for
```

5 Experiments

Datasets. The long-term value model relies on random (the daily number of push notifications) experimental data for training, which can reflect the user's long-term value under different volume. However, the current push system's decision policy on the quantity is a pre-defined result, which does not satisfy the randomness. Before the experiment, we generated a random volume of push notifications to some new users every day for a while. The specified number was randomly determined when the user entered the app for the first time. By accumulating these data, we finally received a dataset containing over 20k samples and the corresponding 27 features, with an average open rate of 0.918 and average logins is 2.241 for the samples.

Measurements. For the offline part of the experiments, the long-term value described in this paper contains *logins* and *openRate*, and their model performance is evaluated by MSE and AUC, respectively. As for the online part of the experiments, the business target DAU is the first to be considered, open rate indicates the future reward, and CTR is an indirect metric to illustrate the impact of the strategy on user feedback behavior.

Implementation Details. Implement our long-term value model through the TensorFlow framework and scipy module solver for the LP problem. We followed the best setup from their original paper for the multi-task learning model [8,9,13] compared.

5.1 Compared Methods

Considering our contributions in this paper, we conducted offline and online experiments and compared the methods described below.

- **Baseline** Baseline method considering the fewer features and volume support for new users, the mentioned above constraint B is divided equally among users in our push system.
- **Offline LP** Suppose we set t to the day level. In that case, the solution becomes an offline linear programming (offline LP), and the optimal daily volume can be generated directly by the solver when the long-term value and the fixed short-term reward are given in advance.We can convert $r_{i,t}$ to r_{i,n_i}, and the variables in the function r become i and n_i instead of t.
- **PID controller** Proportional-integral-derivative controller (PID) [5] achieves the effect of control traffic by fitting the error, given the constraint B and the number of push notifications in real-time; the error is calculated by the control function and then multiplied by $r_{i,t}$ to make decision.
- **OVOLS-short, OVOLS-long, OVOLS** OVOLS is the method proposed, we remove the short-term component from $r_{i,t}$ to construct OVOLS-long, and OVOLS-short is constructed by eliminating the long-term part.
- **Multi-task Model** Co-optimization of logins and open rate through a multi-task model is a reasonable way. We compare three efficient multi-task models in offline experiments, including ESMM, PLE, and MMOE.

5.2 Offline Experiments Results

A time-related short-term value is difficult to simulate by offline experiments, while modeling long-term value can be evaluated by AUC and MSE with its corresponding ground truth. Therefore, for offline experiments, we mainly compare the performance of the multi-task models on the dataset.

For multi-task models, we report various models' AUC and MSE metrics, which are the average over 10 runs. Table 2 shows the results of offline experiments. A seesaw phenomenon appears in models like ESMM, making it outperform the baseline on some tasks but underperform on *logins* due to inter-task correlation during training. Gating mechanisms alleviate this phenomenon by adjusting the correlation across several tasks and features with a trainable gating parameter. MMOE and PLE exceeded the baseline by 2.043% and 2.947% on two tasks by such mechanisms. As described above, for the MMOE-ATT model, we introduced the volume feature and its corresponding attention score with other features as gating parameters. Attention score resulted in the optimal effect we obtained on the target of *openRate*, which exceeded the baseline of 2.286%; at the same time, there was no decrease in *logins* (compared to PLE). Offline results and Fig. 3 illustrate the effect of the push volume on the push-related target, such as DAU and CTR. The optimal performance demonstrated by the MMOE-ATT model also shows the impact of our proposed enhanced gating parameter through the attention mechanism.

5.3 Online A/B Test Results

We tested several of the compared methods described above and our proposed method under different parameters through 7 days of online experiments. Every

Table 2. Offline experiments performance on multi-task models

Models	AUC	MSE	Multi-task Gain	
	openRate	*logins*	%Incr. in AUC	%Incr. in MSE
Two DNNs	0.8225	0.02172	–	–
ESMM	0.8303	0.02243	+0.948	−3.269
MMOE	0.8393	0.02112	+2.043	+2.762
PLE	0.8359	**0.02108**	+1.629	**+2.947**
MMOE-ATT	**0.8413**	**0.02108**	**+2.286**	+2.946

day, over 100k users activate the QB for the first time. Once they complete this motion, our push system will group them into different experimental strategies, including our proposed method and various compared methods.

The offline A/B tests are carried out in the production environment of the QB. All the data in this section, such as DAU and CTR, are expressed as relative increments. Table 3 compares allocation algorithms and ablation studies for the method proposed in this paper. The comparison metrics specifically include DAU, CTR, and open rate. Another factor to be considered for the multi-day push policy for new users is the stability of the above algorithms. Therefore, metrics in *day*1, *day*7, and *average* are all in our experimental data to illustrate the performance of algorithm for new users.

Compared with several methods as shown in Table 3, the OVOLS proposed in this paper gets state-of-the-art in both key metrics DAU and Open-Rate. Although there are better methods in CTR, it is still optimal for our crucial business metrics. Among these compared methods, PID outperforms the Offline LP method, which may come from the time sensitivity of the PID. However, it does not solve the volume as a linear programming problem. If users and content keep changing throughout the day, greedy decisions and nearly satisfying the constraint are very robust practical approaches. Removing some components from the reward function makes it possible to study the impact of different parts on the method. An ablation experiment's results are consistent with our assumptions about the functionality of each component. OVOLS-short performs very strongly on the short-term metric (CTR) and poorly on the long-term metric (open rate), or OVOLS-long has the opposite performance. The performance trend of these two methods on DAU can also illustrate the difference between each component. OVOLS-long is inferior on *day*1 but exceeds OVOLS-short on *day*7, and an enormous difference may be observed through a more extended online experiment.

We further tested the parameter sensitivity of the OVOLS, containing the performance of the three parameters α, β, and γ on two key metrics. The most significant variability in Fig. 4 is *openRate*, especially for γ, where γ controls the importance of long-term value in rewards. *openRate* abrupt change with γ illustrates the impact of our modeling of long-term. At the same time, the fact that *openRate* is not sensitive to α and β also indicates that short-term values

Table 3. Online A/B test results for several baseline methods and ablation study

Algorithm	%Incr. in DAU			%Incr. in CTR			%Incr. in Open-Rate
	*day*1	*day*7	*avg*	*day*1	*day*7	*avg*	*day*7
Baseline	–	–	–	–	–	–	–
Offline LP	+0.70	+0.47	+0.66	+5.19	+4.48	+4.78	+0.9321
PID control	+0.89	**+0.91**	+0.90	**+11.77**	+10.16	+10.45	+0.9748
OVOLS-short	+0.64	+0.64	+0.71	+9.51	**+10.53**	**+10.75**	-0.3813
OVOLS-long	+0.31	+0.78	+0.68	+2.60	+1.76	+2.42	+0.9639
OVOLS	**+1.17**	**+0.91**	**+0.99**	+9.43	+9.03	+10.12	**+1.0407**

have difficulty influencing long-term values. For DAU, the individual control parameters appear to be coupled with each other, and the optimal parameters need to be generated through online a/b tests. In our QB push system practice, we usually constrain the lower limit of DAU increment and optimize *openRate* on this basis, which will derive OVOLS with parameters $\alpha = 0.2$, $\beta = 0.2$, and $\gamma = 0.6$ is our optimal strategy applied to our push system.

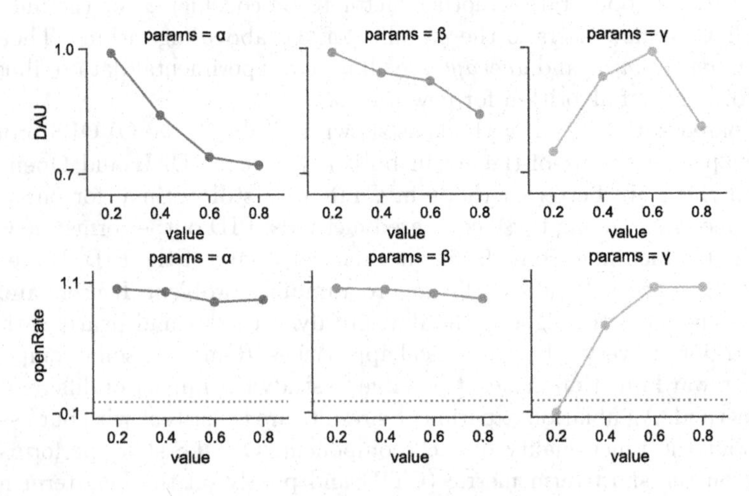

Fig. 4. Experimental results of control parameters(α, β, γ) in reward function r on DAU and open rate

6 Conclusions

We introduced a novel framework optimizing push notification volumes for new users. The proposed method alleviates the lack of feedback features for new users and further optimizes long-term values. OVOLS constructs a long-term value training component and a dual variables calculation part to form an online decision system applied to production. Furthermore, offline experiments, online a/b tests, and long-term observations in QB have shown that OVOLS significantly outperforms other baseline methods.

References

1. Agrawal, S., Wang, Z., Ye, Y.: A dynamic near-optimal algorithm for online linear programming. Oper. Res. **62**(4), 876–890 (2014)
2. Gao, Y., et al.: Near real-time optimization of activity-based notifications. In: Proceedings of the 24th ACM SIGKDD International Conference on Knowledge Discovery & Data Mining, pp. 283–292 (2018)
3. Gupta, R., Liang, G., Tseng, H.P., Holur Vijay, R.K., Chen, X., Rosales, R.: Email volume optimization at Linkedin. In: Proceedings of the 22nd ACM SIGKDD International Conference on Knowledge Discovery and Data Mining, pp. 97–106 (2016)
4. Ho, B.J., Balaji, B., Koseoglu, M., Srivastava, M.: Nurture: notifying users at the right time using reinforcement learning. In: Proceedings of the 2018 ACM International Joint Conference and 2018 International Symposium on Pervasive and Ubiquitous Computing and Wearable Computers, pp. 1194–1201 (2018)
5. Johnson, M.A., Moradi, M.H.: PID Control. Springer, London (2005). https://doi.org/10.1007/1-84628-148-2
6. Jouini, W., Ernst, D., Moy, C., Palicot, J.: Upper confidence bound based decision making strategies and dynamic spectrum access. In: 2010 IEEE International Conference on Communications, pp. 1–5. IEEE (2010)
7. Liu, Y., et al.: Personalized execution time optimization for the scheduled jobs. arXiv preprint arXiv:2203.06158 (2022)
8. Ma, J., Zhao, Z., Yi, X., Chen, J., Hong, L., Chi, E.H.: Modeling task relationships in multi-task learning with multi-gate mixture-of-experts. In: Proceedings of the 24th ACM SIGKDD International Conference on Knowledge Discovery & Data Mining, pp. 1930–1939 (2018)
9. Ma, X., et al.: Entire space multi-task model: an effective approach for estimating post-click conversion rate. In: The 41st International ACM SIGIR Conference on Research & Development in Information Retrieval, pp. 1137–1140 (2018)
10. O'Brien, C., Wu, H., Zhai, S., Guo, D., Shi, W., Hunt, J.J.: Should i send this notification? Optimizing push notifications decision making by modeling the future. arXiv preprint arXiv:2202.08812 (2022)
11. Okoshi, T., Tsubouchi, K., Tokuda, H.: Real-world product deployment of adaptive push notification scheduling on smartphones. In: Proceedings of the 25th ACM SIGKDD International Conference on Knowledge Discovery & Data Mining, pp. 2792–2800 (2019)
12. Sutton, R., Fraser, K., Conlan, O.: A reinforcement learning and synthetic data approach to mobile notification management. In: Proceedings of the 17th International Conference on Advances in Mobile Computing & Multimedia, pp. 155–164 (2019)
13. Tang, H., Liu, J., Zhao, M., Gong, X.: Progressive layered extraction (PLE): a novel multi-task learning (MTL) model for personalized recommendations. In: Fourteenth ACM Conference on Recommender Systems, pp. 269–278 (2020)
14. Visuri, A., van Berkel, N., Okoshi, T., Goncalves, J., Kostakos, V.: Understanding smartphone notifications' user interactions and content importance. Int. J. Hum. Comput. Stud. **128**, 72–85 (2019)

15. Yuan, Y., Muralidharan, A., Nandy, P., Cheng, M., Prabhakar, P.: Offline reinforcement learning for mobile notifications. arXiv preprint arXiv:2202.03867 (2022)
16. Zhao, B., Narita, K., Orten, B., Egan, J.: Notification volume control and optimization system at Pinterest. In: Proceedings of the 24th ACM SIGKDD International Conference on Knowledge Discovery & Data Mining, pp. 1012–1020 (2018)

Discovering Geo-referenced Frequent Patterns in Uncertain Geo-referenced Transactional Databases

Palla Likhitha[1,3](✉), Pamalla Veena[2], Uday Kiran Rage[1,3], and Koji Zettsu[1]

[1] National Institute of Information and Communications Technology, Tokyo, Japan
uday.rage@gmail.com, zettsu@nict.go.jp
[2] Sri Balaji PG College, Ananthapur, AP, India
rage.vinny@gmail.com
[3] The University of Aizu, Fukushima, Japan
likhithapalla7@gmail.com

Abstract. An uncertain geo-referenced transactional database represents the probabilistic data produced by stationary spatial objects observing a particular phenomenon over time. Useful patterns that can empower the users to achieve socio-economic development lie hidden in this database. Finding these patterns is challenging as the existing frequent pattern mining studies ignore the spatial information of the items in a database. This paper proposes a generic model of Geo-referenced Frequent Patterns (GFPs) that may exist in an uncertain geo-referenced transactional database. This paper also introduces two new upper-bound constraints, namely "*neighborhood-aware prefix item camp*" and "*neighborhood-aware expected support*", to effectively reduce the search space and the computational cost of finding the desired patterns. An efficient neighborhood-aware pattern-growth algorithm has also been presented in this paper to find all GFPs in a database. Experimental results demonstrate that our algorithm is efficient.

Keywords: frequent patterns · uncertain data · geo-referenced series

1 Introduction

An uncertain geo-referenced transactional database is a basic form of a spatiotemporal database. It represents the probabilistic data generated by stationary spatial items observing a particular phenomenon over time. Many applications naturally produce this data. Examples include air pollution data gathered by ground monitoring stations, raster data produced by satellites, and traffic congestion data collected by sensors located at specified fixed locations in a road

R. U. Kiran—First three authors have equally contributed to the paper.
This research was partially funded by JSPS Kakenhi 21K12034.

H. Kashima et al. (Eds.): PAKDD 2023, LNAI 13937, pp. 29–41, 2023.
https://doi.org/10.1007/978-3-031-33380-4_3

network. However, crucial information that can facilitate the users to achieve socio-economic development lies hidden in this data. Previous works [3] explored clustering and classification techniques to discover interesting information in this data. This paper focuses on finding valuable patterns hidden in this data.

Frequent pattern mining [2,9] involves finding all frequently occurring patterns in a transactional database. Several algorithms [1,7,8] have been described in the literature to find these patterns effectively in an uncertain transactional database. However, these studies have found limited practicality in spatial informatics due to the following limitation: "*The basic model of frequent pattern implicitly assumes that the spatial information of the items, if any, will not determine the interestingness of a pattern in the data. However, this assumption is too restrictive and often does not reflect reality, as the items' spatial information typically influences the interestingness of a pattern.*" With this motivation, this paper focuses on finding a class of user interest-based patterns, called geo-referenced frequent patterns (GFPs), that may exist in an uncertain geo-referenced transactional database.

A GFP represents a set of neighboring items (or an area) in which a particular phenomenon was frequently observed in a database. Thus, there exists value in finding these patterns in real-world databases. However, finding these patterns is a non-trivial and challenging task due to the following reasons:

1. The itemset lattice represents the search space of pattern mining. Thus, the size of this search space is $2^n - 1$, where n represents the total number of items within the database. We must explore new upper-bound measures to find all GFPs in an itemset lattice effectively. The reason is existing upper-bound measures [1] for frequent patterning, say *tubeP* [7] and *prefix item cap* [8], ignore the spatial information of the items.
2. Existing studies [1,9] find frequent patterns by performing the conventional breadth or depth-first search on the itemset lattice. Unfortunately, these conventional search approaches were found to be inadequate for finding GFPs effectively. Thus, we need to explore alternative search techniques to reduce the computational cost of finding the GFPs.

The contributions of this paper are as follows. First, this paper proposes a generic model of GFPs that may exist in an uncertain (binary) geo-referential transactional database. Our model is generic as it facilitates us to capture spatial items of heterogeneous shapes, such as pixels, points, lines, and polygons. Second, we introduce two novel upper-bound constraints, namely '*neighborhood-aware prefix item cap*' and '*neighborhood-aware expected support*,' to reduce the search space and the computational cost of finding the desired patterns. Third, we explore a new search technique, namely '*neighborhood-aware pattern-growth technique*,' and present an efficient depth-first search algorithm to find the complete set of GFPs in a database. We call our algorithm Geo-referential Frequent Pattern-growth (GFP-growth). Experimental results demonstrate that our algorithm is both memory and runtime efficient.

The rest of the paper is organized as follows. Section 2 describes related work on finding frequent patterns in uncertain databases and spatiotemporal pattern analysis. Section 3 describes a proposed model of GFPs in a database. Section 4 presents the GFP-growth algorithm. Section 5 reports the experimental results. Finally, in Sect. 6, we conclude and discuss future research.

2 Related Work

Agrawal et al. [2] described frequent pattern mining as a key intermediary step to discover association rules in a transactional database. Several algorithms [9] have been described to find these patterns in certain transactional databases. Chui et al. [4] extended the frequent pattern model to discover regularities in an uncertain transactional database. Since then, several algorithms (e.g., PUF-growth [8], CUFP-growth, TubeP, TubeS [7], DISC, and BLIMP [1]) have been described to find these patterns in uncertain transactional databases. As GFPs are a subset of frequent patterns, we can develop a naïve geo-referential frequent pattern mining algorithm by extending anyone of the existing frequent pattern mining algorithm. The naïve algorithm involves finding all frequent patterns in a geo-referenced transactional database and selecting a subset of frequent patterns representing GFPs. Unfortunately, this naïve algorithm was found to be highly inefficient due to its increased search space and mining costs.

Identifying spatiotemporal association rules in spatiotemporal databases has received considerable attention from researchers [3,6]. These algorithms typically segment the data over space and time using a clustering algorithm and apply traditional association rule mining algorithms on each cluster to find interesting associations between them. The limitations of these studies are as follows: (*i*) These studies suffer from an open problem of determining the total number of clusters. (*ii*) Clustering itself is a computationally expensive process. (*iii*) Association rules across the multiple clusters (or subsets of the data) will be completely missed, and (*iv*) Too many input parameters must be specified by the user to carry out both clustering and association rule mining algorithms. The proposed model does not rely on the clustering of spatial items. Consequently, our model does not suffer from any of the above-mentioned limitations.

Veena et al. [10] recently studied the problem of finding fuzzy geo-referenced periodic-frequent patterns in a certain quantitative geo-referenced time series database. It has to be noted that this work ignores the uncertain nature of the geo-referenced transactional (or time series) data. Overall, the proposed model of GPFs is novel and distinct from existing studies.

3 Proposed Model

Let $I = \{i_1, i_2, \cdots, i_n\}$, $n \geq 1$, be a set of items. A **location database**, denoted as LD, is a collection of items and their coordinates. That is, $LD = \cup_{i_j \in I}(i_j, Coor_{i_j})$, where $Coor_{i_j}$ represent the set of coordinates of an item $i_j \in I$. Please note that the coordinates of an item can represent a point, a line, or a polygon. Let $X \subseteq I$ be an itemset (or a pattern). A pattern containing k number of items is called a k-pattern. A uncertain transaction, t_{tid}, consists of

Table 1. Uncertain database

Tid	Transaction
1	q(0.1) r(0.8) s(0.9)
2	p(0.7) r(0.7) s(0.1)
3	p(0.8) q(0.6) r(0.4)
4	r(0.3) s(0.4) t(0.9)
5	q(0.3) r(0.5) s(0.6) t(0.4)
6	p(0.6) q(0.5) r(0.7) s(0.9) t(0.9)
7	p(0.5) q(0.3) r(0.4) s(1.0)
8	p(0.9) r(0.4) s(0.7) t(0.5)
9	p(0.2) q(0.3) r(0.4) s(0.5)
10	p(0.4) q(0.8) r(0.9) s(0.8) t(0.9)

Table 2. Locations

Item	Point
p	(9, 4)
q	(5, 1)
r	(2, 8)
s	(4, 1)
t	(5, 3)

Table 3. Neighbors

item	Neighbors
p	–
q	s, t
r	–
s	q, t
t	q, s

Table 4. Uncertain frequent patterns and geo-referenced frequent patterns generated from Table 1. The terms 'FP' and 'GFP' denote frequent pattern and geo-referenced frequent pattern, respectively

Pattern	Support	FP	GFP	Pattern	Support	FP	GFP
p	4.1	✓	✓	ps	2.16	✓	✗
q	2.9	✓	✓	qr	1.78	✓	✗
r	5.5	✓	✓	qs	1.81	✓	✓
s	5.8	✓	✓	rs	3.44	✓	✗
t	3.6	✓	✓	rt	2.11	✓	✗
pr	2.23	✓	✗	st	2.48	✓	✓

a transaction identifier (tid) and a pattern Y. That is, $t_{tid} = (tid, Y)$. More importantly, each item $i_k \in Y$ is also associated with an existential probability value $P(i_k, t_{tid}) \in (0, 1)$, which represents the likelihood of the presence of i_k in t_{tid}. A uncertain transactional database, $UTDB = \{t_1, t_2, \cdots, t_m\}$, $m \geq 1$.

Example 1. Let $I = \{p, q, r, s, t\}$ be a set of fixed sensors (or items). A hypothetical uncertain transactional database constituting these items is shown in Table 1. The location database storing the spatial information of these items is shown in Table 2. The set of items p and r, i.e., $\{p, r\}$ (or pr, in short) is a pattern. This is a 2-pattern as it contains only two items.

Definition 1 ***(Expected support of pattern*** X ***in a transaction).*** *The existential probability of* X *in* t_{tid}*, denoted as* $P(X, t_{tid})$*, represents the product of corresponding existential probability values of all items in* X *when these items are independent. That is,* $P(X, t_{tid}) = \prod_{\forall i_j \in X} P(i_j, t_{tid})$*. The expected support of* X *in* $UTDB$*, denoted as* $expSup(X) = \sum_{tid=1}^{m} P(X, t_{tid})$.

Example 2. The pattern *pr* occurs in transactions with *tids* of 2, 3, 6, 7, 8, 9, and 10. The existential probability of *pr* in the second transaction, i.e., $P(pr, t_2) = P(p, t_2) \times P(r, t_2) = 0.7 \times 0.7 = 0.49$. Similarly, $P(pr, t_3) = 0.32$, $P(pr, t_6) = 0.42$, $P(pr, t_7) = 0.2$, $P(pr, t_8) = 0.36$, $P(pr, t_9) = 0.08$, $P(pr, t_{10}) = 0.36$. The expected support of *pr* in the entire database, i.e., $expSup(pr) = 0.49 + 0.32 + 0.42 + 0.2 + 0.36 + 0.08 + 0.36 = 2.23$.

Definition 2 ***(Frequent pattern X).*** *A pattern X is said to be frequent if* $expSup(X) \geq minSup$, *where minSup represents the user-specified minimum support value.*

Example 3. If the user-specified $minSup = 1.6$, we consider *pr* a frequent pattern because $expSup(pr) \geq minSup$. Similarly, *qs* will also be considered as a frequent pattern because $expSup(qs) \geq minSup$. The basic frequent pattern model implicitly assumes that both the patterns, i.e., *pr* and *qs*, are equally interesting irrespective of their items' spatial locations. However, the user may consider *pr* an uninteresting pattern because both items are far apart in a coordinate system (see Table 2.) In the next definition, we try to prune such uninteresting frequent patterns whose items are far apart.

Definition 3 ***(Geo-referenced frequent pattern X).*** *The frequent pattern X is said to be a GFP if the maximum distance between any two of its items is less than or equal to the user-specified* maximum distance *(maxDist) value. That is, X is a geo-referential frequent pattern if* $max(dist(i_p, i_q)|\forall i_p, i_q \in X) \leq maxDist$, *where dist() is a distance function that satisfies the commutative property. The Euclidean and Geodesic are some of the popular distance-measuring functions that satisfy the commutative property.*

Example 4. If Euclidean is considered the distance function, then the distance between the items *p* and *r*, i.e., $dist(p, r) = 8.06$. If the user-specified $maxDist = 5$, then we consider the frequent pattern *pr* as not a geo-referential frequent pattern as $max(dist(p, r)) > maxDist$. In contrast, the pattern *qs* is considered as a geo-referential frequent pattern as $max(dist(q, s)) \leq maxDist$. The complete set of frequent patterns and geo-referential frequent patterns discovered from Table 1 are shown in Table 4. It can be observed that several frequent patterns fail to be GFPs if we consider the spatial information of the items.

Definition 4 ***(Problem definition).*** *Given an uncertain transactional database (UTDB), location database (LD), the user-specified* minimum support *(minSup) and* maximum distance *(maxDist) constraints, find the complete set of GFPs that satisfy the maxDist and minSup values. Please note that minSup can also be specified in the percentage of UTDB.*

4 Proposed Algorithm

4.1 Basic Idea: Potential Geo-referenced Frequent Patterns

The search space size to find GFPs in a database is $2^n - 1$, where n represents the total number of items in a database. We explore the concept of "neighbors" and introduce two new tighter upper-bound constraints to reduce this search space effectively.

Definition 5 ***(Neighbors of an item i_k).*** *The neighbors of an item $i_k \in I$, denoted as N_{i_k}, represent the items whose distance is within the maxDist value. That is, $N_{i_k} \subseteq I$ such that $\forall i_x \in N_{i_k}$, $dist(i_x, i_k) \leq maxDist$.*

Example 5. The neighbors for the item s in Table 1 are q and t. It is because $dist(q, s) \leq maxDist$ and $dist(t, s) \leq maxDist$. Thus, $N_s = \{q, t\}$. The neighbors for all items in Table 1 are shown in Table 3.

Definition 6 ***(Neighborhood-aware prefix item cap).*** *Let $t_{tid}.Y = \{i_1, i_2, \cdots, i_k, \cdots, i_l\}$, $1 \leq k \leq l \leq n$ be a transaction in a database. The neighborhood-aware prefix item cap of an item i_k, denoted as $NPI^{cap}(i_k, t_{tid})$, represents the product of $P(i_k, t_{tid})$ and the highest existential probability value among all its neighboring items from i_1 to i_{k-1} in t_{tid}. That is, $NPI^{cap}(i_k, t_{tid}) = P(i_k, t_{tid}) \times max(P(i_y, t_{tid}) | \forall i_y \in \{i_1, i_2, \cdots, i_{k-1}\} \cap N_{i_k}))$.*

Example 6. Consider the first transaction shown in Table 1. The neighborhood-aware prefix item cap of s in this transaction, i.e., $NPI^{cap}(s, t_1) = P(s, t_1) \times max(P(i_y, t_1) | \forall i_y \in \{q, r\} \cap \{q, t\}) = P(s, t_1) \times P(q, t_1) = 0.9 \times 0.1 = 0.09$. The set of items $\{q, r\}$ in the above equation represents the items that have appeared before s in the first transaction. The set of items $\{q, t\}$ in the above equation represents the neighboring items of s in the entire database.

Definition 7 ***(Neighborhood-aware expected support cap of X).*** *The neighborhood-aware expected support cap of X, denoted as $NexpSup^{cap}(X)$, is defined as the sum of all neighborhood-aware prefix item caps of i_k in all the transactions containing X, i.e., $NexpSup^{cap}(X) = \sum_{j=1}^{m} \{NPI^{cap}(i_k, t_j) | X \subseteq t_j\}$.*

Example 7. In the pattern qs, s is the last (or k^{th}) item. In Table 1, this pattern appears in the transactions whose *tids* are 1, 5, 6, 7, 9 and 10. The item cap of s in t_1, i.e., $NPI^{cap}(s, t_1) = 0.09$ (see Example 6). Similarly, $NPI^{cap}(s, t_5) = 0.18$, $NPI^{cap}(s, t_6) = 0.45$, $NPI^{cap}(s, t_7) = 0.3$, $NPI^{cap}(s, t_9) = 0.15$ and $NPI^{cap}(s, t_10) = 0.64$. Thus, the neighborhood-aware expected support cap of qs in the entire database, i.e., $NexpSup^{cap}(qs) = 0.09 + 0.18 + 0.45 + 0.3 + 0.15 + 0.72 = 1.89$. Since the $NexpSup^{cap}(qs) \geq minSup$, we consider qs as a potential GFP whose supersets may also be GFPs.

The *neighborhood-aware expected support cap* (see Property 1) satisfies the downward closure property. Hence, we employ this measure to reduce the search space effectively.

Definition 8 ***(Potential geo-referenced frequent pattern X).*** *The pattern X is said to be a potential geo-referenced uncertain frequent pattern if $NexpSup^{cap}(X) \geq minSup$ and $dist(X) \leq maxDist$.*

Example 8. Continuing with the previous example, qs is a potential geo-referenced frequent pattern because $NexpSup^{cap}(qs) \geq minSup$ and q is a neighbor of s.

Since potential geo-referenced frequent patterns satisfy the *downward closure property*, we can employ a conventional breadth-first search or depth-first search technique to find all GFPs. However, these conventional search techniques were inefficient as they do not effectively utilize the neighborhood information of the items. In this context, we propose an alternative neighborhood-aware depth-first search technique (or neighborhood-aware pattern-growth technique) to find all GFPs effectively.

Property 1 (**The downward closure property** [2,8]). If Y is a potential geo-referenced frequent pattern, then $\forall X \subset Y$ and $X \neq \emptyset$, X is also a potential geo-referenced uncertain frequent pattern. That is, if $NexpSup^{cap}(Y) \geq minSup$ and $\forall X \subset Y$ and $X \neq \emptyset$, $NexpSup^{cap}(X) \geq minSup$.

Definition 9 ***(Neighborhood-aware pattern-growth technique).*** *Let Tr be a tree generated from an uncertain transactional database. The prefix paths of an item i_j in Tr represents the conditional pattern base. Existing pattern mining algorithms construct conditional pattern bases for an item i_j by considering all other items in its prefix paths. This approach makes GPF-mining ineffective. To tackle this problem, we propose a neighborhood-aware pattern-growth technique that involves constructing conditional pattern bases for an item i_j by considering only its neighboring items in its prefix paths. (This topic is further discussed in the latter parts of this paper.)*

4.2 GFP-growth

The proposed algorithm involves the following three steps: (*i*) compress the given uncertain transactional database into a Geo-referenced Frequent Pattern tree (GFP-tree), (*ii*) find all potential GFPs by recursively mining the GFP-tree using $NexpSup^{cap}$ constraint, and (*iii*) discover all GFPs from potential GFPs by scanning the database. Before we describe these three steps, we describe the structure of the GFP-tree.

Step 1: Construction of GFP-tree. Since GFPs satisfy the downward closure property, geo-referenced frequent items (or 1-length patterns) plays a key role in the efficient discovery of GFPs in an uncertain database. These 1-length GFPs is generated when scanning the database for the first time and stored in descending order of their support, respectively. Let *GFP-list* denote the sorted set of 1-length GFPs.

Next, we perform a second scan on the database and construct GFP-tree as given in Algorithm 1. Figure 1 illustrates the step-by-step process of constructing a GFP-tree by scanning the database. Please note that the node-links are maintained between the items in GFP-tree for tree-traversal. In this paper, we are not showing these links for brevity.

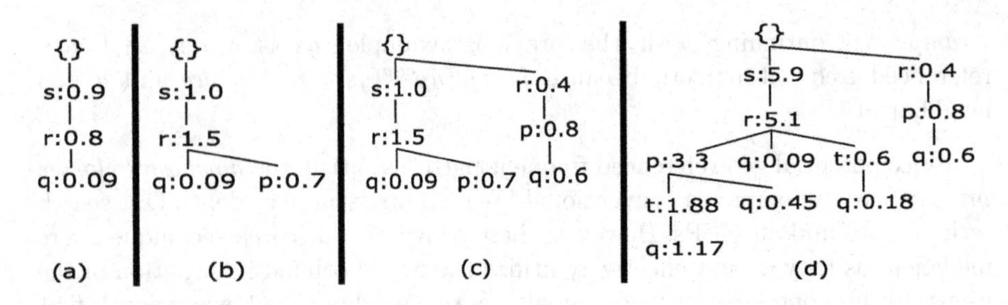

Fig. 1. Constructing the GFP-tree by scanning the uncertain database. (a) After scanning the first transaction. (b) After scanning the second transaction. (c) After scanning the third transaction. (d) GFP-tree after scanning all the transactions in the uncertain database.

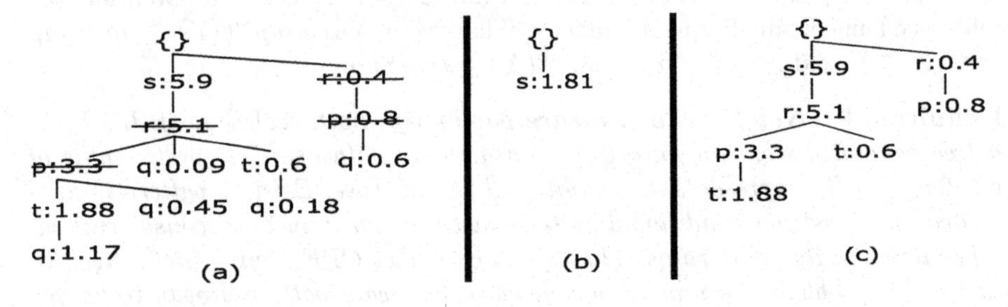

Fig. 2. Mining GFP-tree. (a) branches containing item q. (b) conditionalGFP-tree of q. (c) GFP-tree after pruning item q

Step 2: Finding Potential Geo-Referenced Frequent k-patterns. Potential geo-referenced frequent k-patterns, $k \geq 2$, are generated by recursively mining the GFP-tree using bottom-up search as shown in Algorithm 3. Consider item q, which is the bottom-most item in the GFP-list. The branches containing q in GFP-tree are shown in Fig. 2(a). Considering q as the suffix item, we construct its neighborhood-aware conditional pattern base, say CPB_q, by considering only its neighboring items. The resultant CPB_q is shown in 2(b). The neighborhood expected support cap of r in CPB_q is 1.89 $(= 1.17+0.45+0.09+0.18)$. But the r is not a neighbor of q, So r is removed from conditional GFP tree. The neighborhood expected support cap of s is 1.89 $(= 1.17+0.45+0.09+0.18)$. As s is a neighbor of q, we declare qs as GFP and s is added to CPB_q. The neighborhood expected support cap of qs in CPB_q is 1.89. As $NexpSup^{cap}(qs) \geq minSup$ and $dist(qs) \leq maxDist$, we consider qs as a potential GFP. Similarly, the Neighbourhood expected support cap of t is 1.35 $(= 1.17 + 0.18)$. As t is a neighbor of q, but the neighborhood expected support cap of qt in CPB_q is 1.35. As $NexpSup^{cap}(qt) < minSup$, we do not consider qt as a potential GFP. The conditional GFP-tree after removing the non-neighbors from CPB_q is shown in 2(b). Once we complete the mining process of q, we prune it from the original GFP-tree as shown in Fig. 2(c). Similar process is repeated for the remaining items in the original GFP-tree to find all potential GFPs. This bottom-up neighborhood-aware pattern-growth technique is efficient as it reduces the search space dramatically.

Algorithm 1. GFP-Tree ($UTDB$, GFP-list (1-length frequent patterns))

1: Create the root node in GFP-tree, $Tree$, and label it as "*null*".
2: **for** each transaction $t \in UTDB$ **do**
3: Select the frequent items in t and sort them in L order. Let the sorted list be $[e|E]$, where e is the first item with its existential probability value and E is the remaining list. Call $insert\_tree([e|E], ts_{cur}, Tree)$.
4: call GFP-growth ($Tree$, $null$);

Algorithm 2. insert_tree($[e|E]$, ts_{cur}, T)

1: **while** E is non-empty **do**
2: **if** T has a child N such that $e.itemName \neq N.itemName$ **then**
3: Create a new node N. Set $N.itemName = e.itemName$ and $N.expSup^{cap} = NPI^{cap}(e.itemName, ts_{cur})$. Let its parent link be linked to T. Let its node-link be linked to nodes with the same $itemName$ via the node-link structure. Remove e from E.
4: **else**
5: update $N.expSup^{cap} += NPI^{cap}(e.itemName, ts_{cur})$;

Algorithm 3. GFP-growth ($Tree$, α, $neighborList$)

1: **while** item i_j in the header of $Tree$ **do**
2: Generate pattern $\beta = i_j \cup \alpha$. Traverse $Tree$ using the node-links of β, and construct an array, TS^{β}, which represents the existential probability. Construct β's conditional pattern base by checking if item is in $neighborList[i_j]$ and β's conditional GFP-tree $Tree_{\beta}$ if $NexpSup$ is greater than or equal to $minSup$.
3: **if** $Tree_{\beta} \neq \emptyset$ **then**
4: call GFP-growth ($Tree_{\beta}$, β, $neighborList$);
5: Remove i_j from the $Tree$.

5 Experimental Results

This section first shows that GFPs are an order of value smaller than the Frequent Patterns (FPs) found in uncertain transactional databases, especially at a low $minSup$ value. Next, we show that GFP-growth is efficient in memory and runtime compared to state-of-the-art frequent pattern mining algorithms, such as PUF-growth, TubeP, and TubeS. Since PUF-growth outperformed the TubeP and TubeS algorithm in most cases, we confined our experiments to GFP-growth and PUF-growth algorithms for brevity.

Since no algorithm exists to find GFPs in an uncertain geo-referenced transactional database, we show that GFP-growth is efficient by comparing it against the PUF-growth [8]. GFP-growth and PUF-growth algorithms were written in Python 3.7 and executed on a DELL 2U Tower server machine containing two Intel(R) Xeon(R) Gold 6140 CPUs running at 2.30 GHz. This server machine has 32 GB of memory and runs on CentOS 7. The experiments have been conducted on synthetic (**T10I4D100K**) and real-world (**Pollution** and **Congestion**) databases.

The **T10I4D100K** is a sparse synthetic database generated using the procedure described in [2]. The uncertain values were randomly assigned for every item in a transaction. This database contains 1,00,000 transactions and 870 items. The *minimum*, *average*, and *maximum* transaction lengths of this database are 1, 10, and 29, respectively. The **Pollution** [10] is a dense real-world database containing 30 d of hourly pollution recordings of 1200 ground stations. Thus, this database contained 720 transactions and 1200 items. This database's *minimum*, *average*, and *maximum* transaction lengths are 11, 460, and 971, respectively. The **Congestion** is a high-dimensional sparse database provided by an anonymous company for Kobe, Japan, for July 2015. It contains 8,928 transactions and 1,414 items (or road segments). The *minimum*, *average*, and *maximum* transaction lengths of this database are 1, 58, and 337, respectively. It is to be noted that in all of the above databases, uncertainty values were randomly set between 0 to 1. The runtime is calculated in seconds, while the memory is calculated in kilobytes. The code, databases, and other experimental results were provided in [5] to verify the correctness of our experiments.

Figure 3(1)–Fig. 3(3) shows the number of FPs and GFPs generated in various databases at distinct *minSup* values. In this experiment, the *maxDist* values in the T10I4D100K, Congestion, and Pollution databases have been set at 60, 1, and 5, respectively. The following observations can be drawn from these figures: (*i*) Increase in *minSup* has decreased the number of UFPs and GFPs. It is because many patterns have failed to satisfy the increased *minSup* value. (*ii*) The total number of GFPs generated were relatively smaller than the total number of UFPs found at any *minSup* value. (*iii*) PUF-growth was unable to generate the patterns at low *minSup* values in congestion and pollution databases. In contrast, GFP-growth could generate the desired patterns even at low *minSup* values. Some of the interesting patterns discovered in congestion database are shown in Fig. 4. These patterns represents the neighboring road segments in which people have frequently faced traffic congestion. When such information is visualized along with other data sources, say rainfall data of a typhoon interpolated along with real world database the produced information may found to be useful to the users for various purposes, such as monitoring the traffic and developing smart navigation systems.

Figure 3(4)–Fig. 3(6) shows the runtime requirements of PUF-growth and GFP-growth algorithms over different databases at distinct *minSup* values. It can be observed that GFP-growth is significantly faster than the PUF-growth. It is because our pruning technique (see Definition 9) effectively reduced the search space by avoiding all the extensions of patterns which are not at close distance.

Figure 3(7)–Fig. 3(9) shows the memory consumption of both PUF-growth and GFP-growth algorithms over different databases at distinct *minSup* values. The subsequent observation drawn from these figures is as follows: (*i*) Increase in *minSup* has resulted in a decrease in the memory requirements of both PUF-

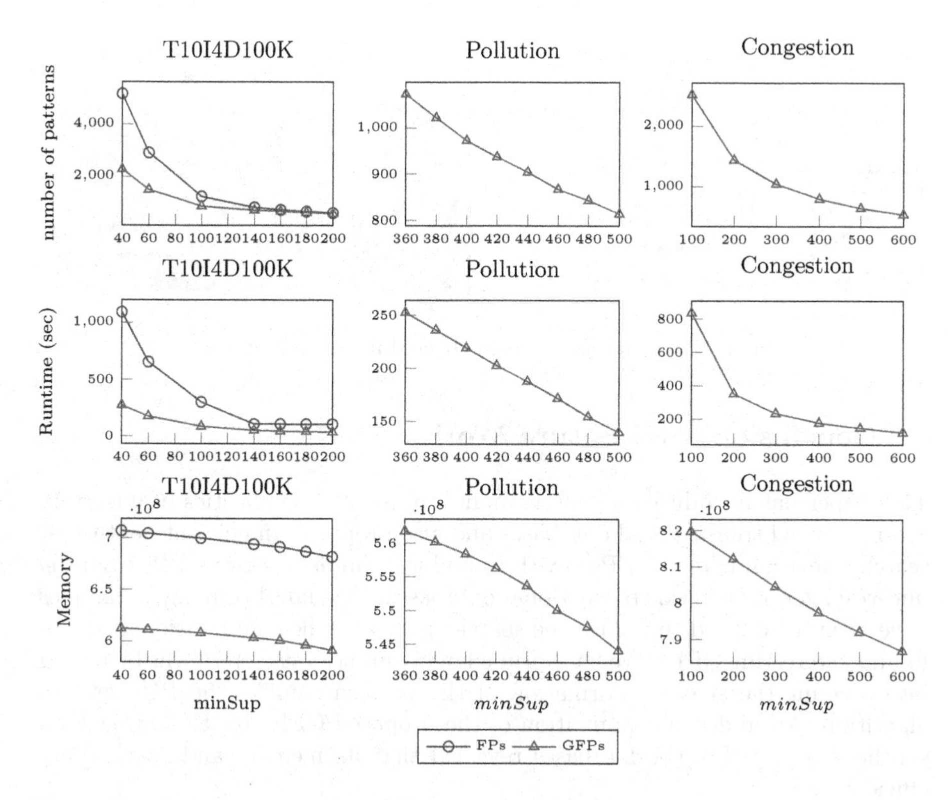

Fig. 3. Number of patterns, Runtime, memory evaluation of algorithms PUF-growth (*FPs*) and GFP-growth (*GFPs*) by varying *minSup*

growth and GFP-growth algorithms. It is because both algorithms have to spend fewer resources to generate lesser number of FPs and GFPs. (*ii*) The memory requirements of GPF-growth were smaller than that of PUF-growth as our algorithm effectively prunes the search space.

At a fixed *minSup*, an increase in *maxDist* increases the number of GFPs, while FPs remain the same. As a result, the runtime and memory requirements of the GFP-growth algorithm increase with the increase in *minSup*. At a very large *maxDist* value, both algorithms consume similar memory and runtime as every GFP will also be a frequent pattern. Unfortunately, we were unable to present the results with varying *maxDist* due to page limitations.

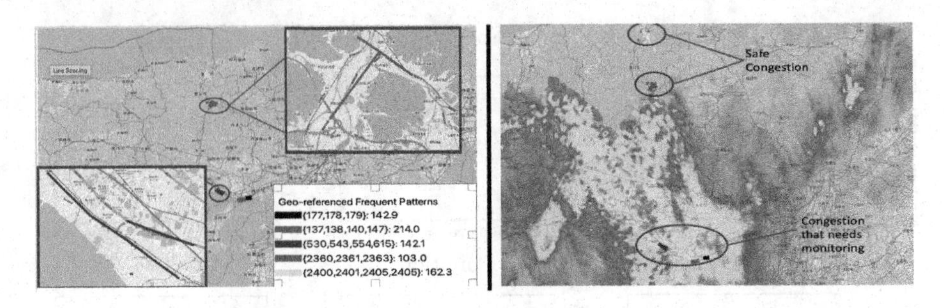

Fig. 4. Interesting patterns discovered in Congestion data

6 Conclusions and Future Work

This paper has introduced a generic model to discover regularities in uncertain geo-referenced transactional databases and also proposed an efficient depth-first search algorithm named GFP-growth to find the complete set of GFPs from the uncertain geo-referenced transactional databases. Two novel pruning techniques have been introduced to reduce the search space and the computational cost of finding interesting GFPs. We have pruned many uninteresting patterns by taking into account the spatial information of the patterns, unlike the PUF-growth algorithm. An in-depth examination of the proposed GFP-growth approach on synthetic and real-world databases revealed that its memory and runtime are efficient.

As part of future work, we would like to focus on reducing the search space and extending it into uncertain data streams.

References

1. Aggarwal, C.C., Yu, P.S.: A survey of uncertain data algorithms and applications. IEEE Trans. Knowl. Data Eng. **21**(5), 609–623 (2009)
2. Agrawal, R., Imieliński, T., Swami, A.: Mining association rules between sets of items in large databases. In: SIGMOD, pp. 207–216 (1993)
3. Ansari, M., Ahmad, A., Khan, S., Bhushan, G., Siddique, M.: Spatiotemporal clustering: a review. Artif. Intell. Rev. **53**, 2381–2423 (2020). https://doi.org/10.1007/s10462-019-09736-1
4. Chui, C.-K., Kao, B., Hung, E.: Mining frequent itemsets from uncertain data. In: Zhou, Z.-H., Li, H., Yang, Q. (eds.) PAKDD 2007. LNCS (LNAI), vol. 4426, pp. 47–58. Springer, Heidelberg (2007). https://doi.org/10.1007/978-3-540-71701-0_8
5. GFP: Geo-referenced Frequent Pattern (GFP) (2022). https://github.com/Likhitha-palla/GeoReferencedFrequentPatterns.git. Accessed 8 Dec 2022
6. Kiran, R.U., Shrivastava, S., Fournier-Viger, P., Zettsu, K., Toyoda, M., Kitsuregawa, M.: Discovering frequent spatial patterns in very large spatiotemporal databases. In: SIGSPATIAL, pp. 445–448 (2020)
7. Leung, C.K., MacKinnon, R.K., Tanbeer, S.K.: Fast algorithms for frequent itemset mining from uncertain data. In: ICDM, pp. 893–898 (2014)

8. Leung, C.K.S., Tanbeer, S.K.: PUF-Tree: a compact tree structure for frequent pattern mining of uncertain data. In: PAKDD, pp. 13–25 (2013)
9. Luna, J.M., Fournier-Viger, P., Ventura, S.: Frequent itemset mining: a 25 years review. Wiley Interdiscip. Rev. Data Min. Knowl. Discov. **9**(6), e1329 (2019)
10. Veena, P., et al.: Discovering fuzzy geo-referenced periodic-frequent patterns in geo-referenced time series databases. In: FUZZ-IEEE, pp. 1–8 (2022)

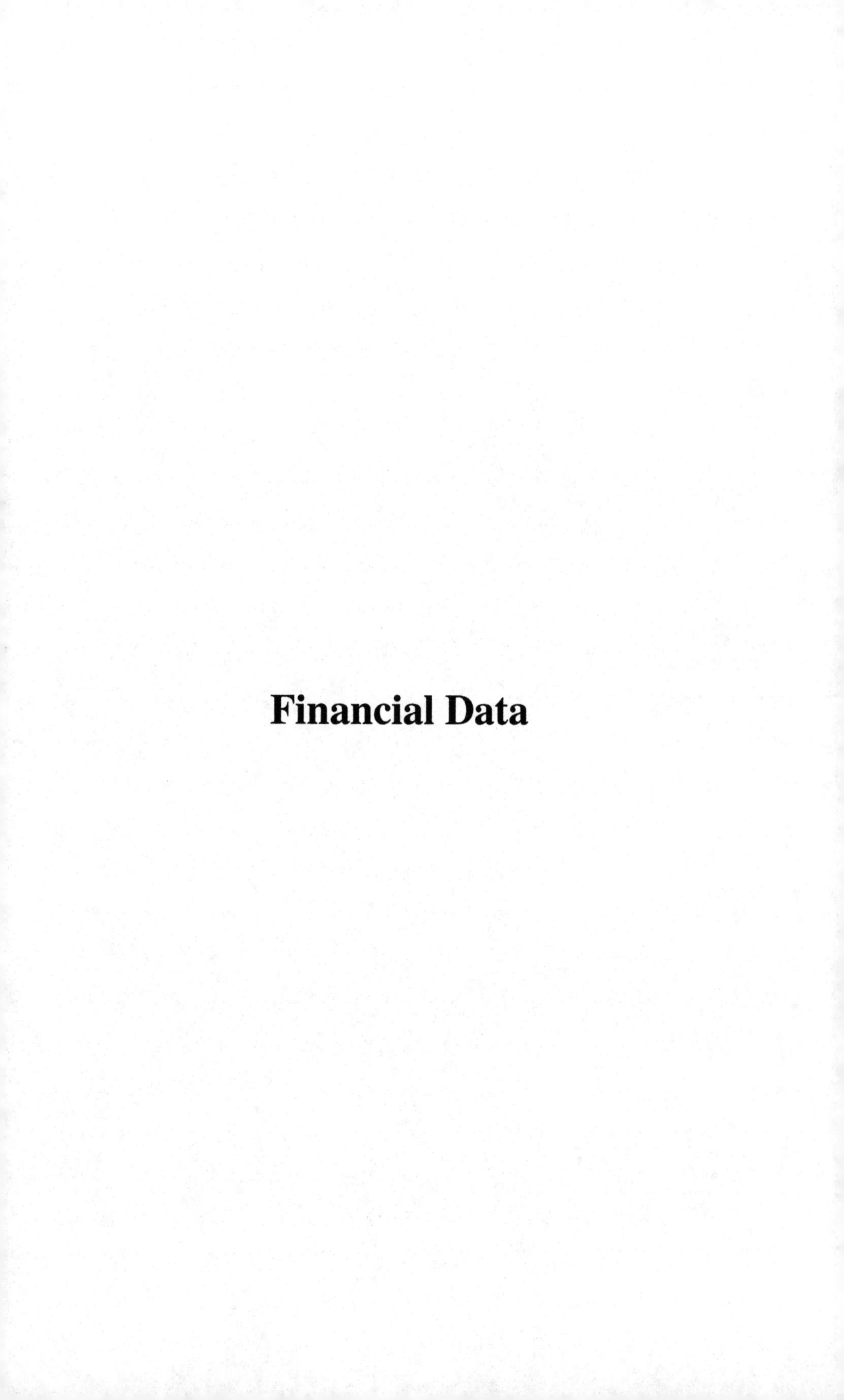

Financial Data

Joint Latent Topic Discovery and Expectation Modeling for Financial Markets

Lili Wang[1(✉)], Chenghan Huang[2], Chongyang Gao[3], Weicheng Ma[1], and Soroush Vosoughi[1]

[1] Dartmouth College, Hanover, NH 03755, USA
{lili.wang.gr,soroush}@dartmouth.edu
[2] Millennium Management, LLC, New York, NY 10022, USA
[3] Northwestern University, Evanston, IL 60208, USA

Abstract. In the pursuit of accurate and scalable quantitative methods for financial market analysis, the focus has shifted from individual stock models to those capturing interrelations between companies and their stocks. However, current relational stock methods are limited by their reliance on predefined stock relationships and the exclusive consideration of immediate effects. To address these limitations, we present a groundbreaking framework for financial market analysis. This approach, to our knowledge, is the first to jointly model investor expectations and automatically mine latent stock relationships. Comprehensive experiments conducted on China's CSI 300, one of the world's largest markets, demonstrate that our model consistently achieves an annual return exceeding 10%. This performance surpasses existing benchmarks, setting a new state-of-the-art standard in stock return prediction and multiyear trading simulations (i.e., backtesting).

Keywords: stock trend prediction · trading simulation · expectation modeling

1 Introduction

The efficient-market hypothesis in traditional finance posits that stock prices reflect all available market information, with current prices consistently trading at their fair value [5]. Consequently, predicting future stock prices is challenging without access to new information. However, markets are often less efficient in reality [11], with stock market fluctuations driven by behavioral factors such as expectations, confidence, panic, euphoria, or herding behavior. These inefficiencies enable the use of machine learning to predict future stock movements based on historical trends.

Stock-affecting behavioral factors can be categorized into short- and long-term factors. Factors like panic, euphoria, or herding behavior are typically short-term, while subjective expectations and confidence tend to be long-term factors,

H. Kashima et al. (Eds.): PAKDD 2023, LNAI 13937, pp. 45–57, 2023.
https://doi.org/10.1007/978-3-031-33380-4_4

only influencing stock prices imperceptibly over extended periods. These factors do not solely impact individual stocks; their effects often spread to topically related stocks, which share similarities across various explicit or latent dimensions. Recent stock prediction works [14,15,21] utilize topic stocks to improve prediction capabilities. However, most of these methods exhibit two key limitations:

(1) Topics are typically assumed to be static and known beforehand. However, real-world topics can change and new topics may emerge. For example, during the COVID-19 pandemic, pharmaceutical companies investing in COVID vaccines (e.g., Pfizer[1] and Moderna[2]) experienced stock price fluctuations under the new COVID topic.

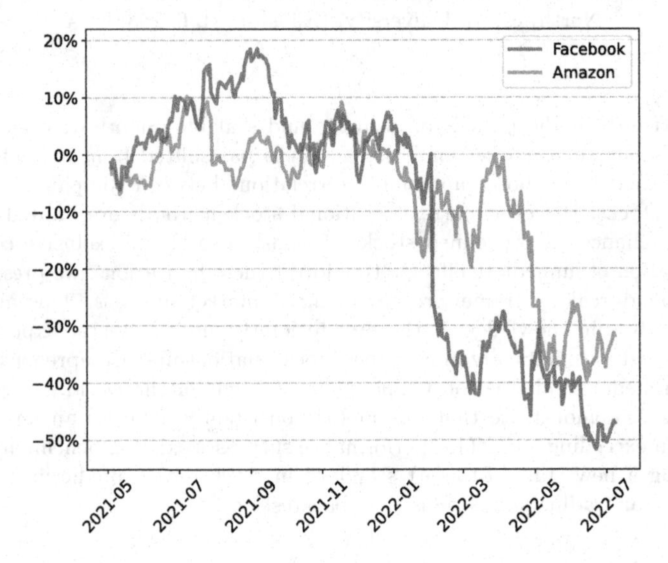

Fig. 1. The return of Amazon and Facebook (Meta) stocks from 2021-05-03 to 2022-07-08 with respect to their stock prices at 2021-05-03.

(2) Only the *short-term impact* between stocks is considered, neglecting the *long-term subjective expectations.* Unlike analyst expectations, subjective expectations are based on human psychology and behavior and can be irrational. Figure 1 illustrates that Amazon and Facebook stock prices often correlate, and previous methods might reason that a significant drop in Facebook's price would also lead to plummeting Amazon stocks. However, in the second half of 2021, Amazon's return was lower than Facebook's, lowering investor expectations for Amazon. Thus, when Amazon released an unremarkable financial report[3] on February 3, 2022, its stock rose 13.5

[1] https://investors.pfizer.com/Investors/Stock-Info/default.aspx.

[2] https://investors.modernatx.com/Stock-Info/default.aspx.

[3] https://s2.q4cdn.com/299287126/files/doc_financials/2021/q4/business_and_financial_update.pdf.

In this paper, we introduce a novel attention-based framework for stock trend prediction that simultaneously discovers topical relations between stocks and models both the *short-term impact* and *long-term subjective expectations* of topically similar stocks. To the best of our knowledge, our framework is the first to:

- Model the influence of investors' subjective expectations on stock prices.
- Automatically identify dynamic topics between stocks without making assumptions or requiring additional knowledge.

Through comprehensive experiments against 16 well-established baselines, we demonstrate that our method achieves the current state-of-the-art on the Qlib [22] quantitative investment platform.

2 Related Work

The stock price prediction and stock selection problems can be easily formed as a time series forecasting problem. Therefore, traditional and deep-learning-based machine learning (ML) methods, especially those for sequence learning, have been directly applied to these tasks are widely used by investment institutions. Specifically, Qlib [22], a popular quantitative investment platform, benchmarks models based on the following ML methods: multi-layer perceptron (MLP); TabNet [1]; TCN [2]; gradient boosting models: CatBoost [12], LightGBM [8]; Recurrent Neural Network (RNN) based models: long short-term memory (LSTM) [6], gated recurrent unit (GRU) [3], DA-RNN [13], AdaRNN [4]; and attention-based models: Transformer [18], and Localformer [7]. To model the co-movement and relations among stocks, some research, such as MAN-SF [15] and STHAN-SR [14]), also adopted graph neural network methods like GCN [10] and GATs [19] to mine the correlation between different stocks.

More recent models include those specifically designed for stock trading. DoubleEnsemble [23] is an ensemble model which utilizes learning-trajectory-based sample reweighting and shuffling-based feature selection for stock prediction. ADD [16] attempt to extract clean information from noisy data to improve prediction performances. Specifically, they proposed a method for separating the inferential features from the noisy raw data to a certain degree using disentanglement, dynamic self-distillation, and data augmentation. Xu et al. assume that inter-dependencies may exist among different stocks at different time series and propose a method called IGMTF [20] to mine these relations. In their other work, they propose HIST [21], a three-step framework to mine the concept-oriented shared information and individual features among stocks.

We use most of the above-mentioned methods as baselines in our experiments.

3 Framework

3.1 Problem Definition

We formulate the stock trend prediction problem as a regression problem. Let $stock_1$, $stock_2$, ..., $stock_n$ denote n different stocks. For each stock $stock_j$ on date

i, the closing price is $price_j^i$. Given the historical information before date i, our task is to predict the one-day return $r_j^i = \frac{price_j^i - price_j^{i-1}}{price_j^{i-1}}$ for each stock j on date i. In the rest of this paper, we use r^i to denote $(r_1^i, r_2^i, ..., r_n^i)$.

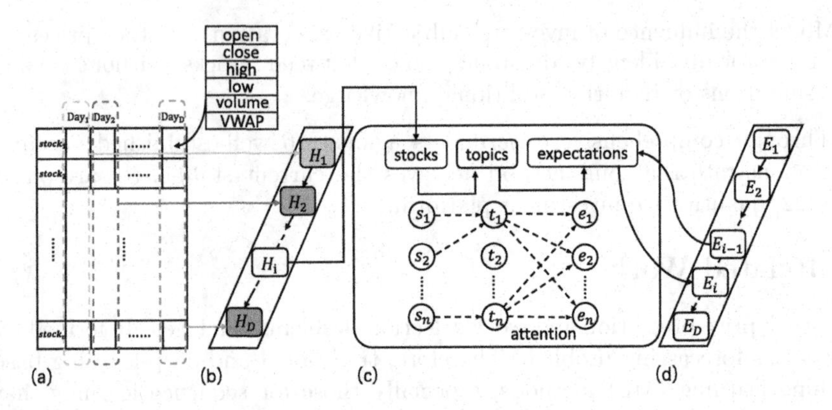

Fig. 2. Our model's framework consists of: (a) Extracting Alpha360 features from raw data: For each stock on a given day, we combine the opening price, closing price, highest price, lowest price, trading volume, and volume-weighted average price (VWAP) into a 6-D feature vector. We then concatenate this vector with similar 6-D vectors from the preceding 59 days to form a 360-D feature vector. (b) The LSTM module processes the extracted Alpha360 features to learn temporal representations. (c) The left half of this section represents the topic module, which uses stock embeddings as input to extract latent topics. The right half illustrates the expectations module, which takes the E_{i-1} output from the expectation LSTM in part (d) as an initial embedding, employs attention with topics to update it to $\hat{E}_{i-1}$, and feeds it back into the expectation LSTM as input for day i. (d) The second LSTM module models the evolution of each stock's expectation.

3.2 Overview of the Framework

The architecture of our model is shown in Fig. 2. The model consists of three jointly optimized modules: temporal stock representation (which aims to extract temporal stock features), topic module (aims to discover the dynamic topics based on the extracted features), and expectation module (aims to model the subjective expectations for each stock). Below we describe each module in detail.

3.3 Temporal Stock Representation

The first step of our learning framework is to extract the Alpha360 features [22] from the raw data. The Alpha360 is a 360-D feature vector that is widely used in the quantitative investment domain. As shown in Fig. 2(a), for each stock on each day, we combine the opening price, closing price, highest price, lowest price, trading volume and volume-weighted average price (VWAP) as a 6-D feature vector and concatenate it with similar 6-D vectors from the past 59 days to get a 360-D feature vector.

To extract the temporal representation of stocks, we adopt an LSTM layer shown in Fig. 2(b)). Our framework is trained recursively by date: for each trading day i, the input is the Alpha360 features H_i of $stock_1$, $stock_2$, ..., $stock_n$ for that day and the output of the LSTM layer is S_i, which is comprised of s_1^i, s_2^i, ..., s_n^i, denoting the embeddings of each stock.

3.4 Topic Module

As mentioned before, the relations among stocks may evolve overtime, so our framework needs to be able to capture the evolution of topics and discover new topics each day. Figure 2 (c) shows the topic and expectation modules of our framework.

First, for each day i, we initialize the n topic embeddings $T_i = (t_1^i, t_2^i, ..., t_n^i)$ using the n stock embeddings $S_i = (s_1^i, s_2^i, ..., s_n^i)$. Then, we compute the Tanimoto coefficient ($\mathcal{T}$) [17] between all pairs of $t_{j_1}^i$ (topic j_1 in day i) and $s_{j_2}^i$ (stock j_2 in day i), for $\forall j_1, j_2 \in [1, n]$ with the following equation:

$$\mathcal{T}(t_{j_1}^i, s_{j_2}^i) = \frac{t_{j_1}^i s_{j_2}^i}{\|t_{j_1}^i\|^2 + \|s_{j_2}^i\|^2 - t_{j_1}^i s_{j_2}^i} \tag{1}$$

We define a function $\phi^i(s_{j_2}^i)$ that for each stock embedding, $s_{j_2}^i$, returns the most similar topic index j_1, except for its own topic (i.e., $j_1 \neq j_2$) in date i, based on the Tanimoto coefficient:

$$\phi^i(s_{j_2}^i) = \arg\max_{j_1} \left(\mathcal{T}(t_{j_1}^i, s_{j_2}^i), j_1 \neq j_2\right) \tag{2}$$

In the example shown in Fig. 2(c) (with the dashed lines) $\phi^i(s_1^i) = 1$, $\phi^i(s_2^i) = 1$, $\phi^i(s_n^i) = n$.

We further construct a set $valid^i$ that contains "valid" topics for each day i, i.e., those that are the most related to at least one stock:

$$valid^i = \{x | \exists j, x = \phi^i(s_j^i)\} \tag{3}$$

This set denotes the topics we discovered for each day. Only if a topic $t_{j_1}^i$ is the most similar topic to at least one stock, it will be include in this set, other topics (e.g., t_2^i in Fig. 2(c)) will be excluded from the following calculations.

To update each topic embedding $t_{j_1}^i$ ($j_1 \in valid^i$), we train the fully connected layer with weight matrix W_t, bias matrix b_t and activation function $tanh$ to aggregate the stock embeddings using the Tanimoto coefficient:

$$t_{j_1}^i = tanh\left(W_t\left(\sum_{\phi^i(s_{j_2}^i)=j_1} \mathcal{T}(t_{j_1}^i, s_{j_2}^i) s_{j_2}^i\right) + b_t\right) \tag{4}$$

3.5 Expectation Module

The expectations of investors change over time and our framework needs to take that into consideration. As shown in Fig. 2(d), we adopt an LSTM to model the evolving expectations of each stock. Each E_i consists of n expectation embeddings e_1^i, e_2^i, ..., e_n^i, we assume that at the first timestamp, the investor's expectations are all decided by the stocks themselves, so the initial embedding $E_1 = (e_1^1, e_2^1, ..., e_n^1)$ are initialized as the n stock embeddings $S_1 = (s_1^1, s_2^1, ..., s_n^1)$.

The expectation for one stock can also be affected by the performance of other stocks under related topics. So for day i, we take the output $E_{i-1} = (e_1^{i-1}, e_2^{i-1}, ..., e_n^{i-1})$ of the LSTM and adopt an attention mechanism to learn the importance of each topic j_1 to the expectations:

$$\alpha(t_{j_1}^i, e_{j_2}^{i-1}) = \frac{\exp\left(\mathcal{T}\left(t_{j_1}^i, e_{j_2}^{i-1}\right)\right)}{\sum_{j \in valid^i} \exp\left(\mathcal{T}\left(t_j^i, e_{j_2}^{i-1}\right)\right)} \tag{5}$$

$$\hat{e}_{j_2}^{i-1} = tanh\left(W_e^1 e_{j_2}^{i-1} + W_e^2 \left(\sum_{j_1 \in valid^i} \alpha(t_{j_1}^i, e_{j_2}^{i-1}) t_{j_1}^i\right) + b_e\right) \tag{6}$$

where $\alpha(t_{j_1}^i, e_{j_2}^{i-1})$ measures the importance of topic j_1 to the expectation of stock j_2, and the updated $\hat{E}_{i-1} = (\hat{e}_1^{i-1}, \hat{e}_2^{i-1}, ..., \hat{e}_n^{i-1})$ then feed back to the LSTM (d) as the input of day i.

3.6 Loss Function

The objective of our model is to predict the one-day return r of each stock. The objective relies on three components: r_{stock}, r_{topic}, and $r_{expectation}$.

The r_{stock} and $r_{expectation}$ are learnt from the temporal stock embeddings and the expectation embeddings, respectively:

$$r_{stock}^i = tanh\left(W_{stock} S_i + b_{stock}\right) \tag{7}$$

$$r_{expectation}^i = tanh\left(W_{expectation} E_i + b_{expectation}\right) \tag{8}$$

To learn r_{topic}, we first learn the importance of each topic to the stocks using a similar attention mechanism as the expectation module:

$$\beta(t_{j_1}^i, s_{j_2}^i) = \frac{\exp\left(\mathcal{T}\left(t_{j_1}^i, s_{j_2}^i\right)\right)}{\sum_{j \in valid^i} \exp\left(\mathcal{T}\left(t_j^i, s_{j_2}^i\right)\right)} \tag{9}$$

$$o_{j_2}^i = tanh\left(W_s \left(\sum_{j_1 \in valid^i} \beta(t_{j_1}^i, s_{j_2}^i) t_{j_1}^i\right) + b_s\right) \tag{10}$$

where $\beta(t_{j_1}^i, s_{j_2}^i)$ measures the importance of topic j_1 to the expectation of stock j_2 on day i. Note that different from the expectation module which includes the term $W_e^1 e_{j_2}^{i-1}$, here $o_{j_2}^i$ measures the impact of all the topics on the stock j_2 on day i, without considering s_i. This is because s_i is already included in r_{stock}. We use O^i to denote $(o_1^i, o_2^i, ..., o_n^i)$; r_{topic} is learnt as:

$$r_{topic}^i = tanh\left(W_{topic}O^i + b_{topic}\right) \tag{11}$$

The predicted return $\hat{r}$ is learnt by combining these three components:

$$\hat{r}^i = tanh\left(W_{\hat{r}} r_{stock}^i + W_{\hat{r}} r_{topic}^i + W_{\hat{r}} r_{expectation}^i + b_{\hat{r}}\right) \tag{12}$$

The loss function of our model is defined as the mean squared error between $\hat{r}$ and r:

$$\mathcal{L} = \frac{\sum_{i\in[1,D]} \left(r^i - \hat{r}^i\right)^\top \left(r^i - \hat{r}^i\right)}{D \cdot n} \tag{13}$$

where D corresponds to the number of trading days. Algorithm 1 shows the pseudocode of our method.

Algorithm 1. Training pseudo-code

Input: $H = \{H_1, H_2, \ldots, H_{|D|}\}$: the Alpha360 features for each trading day

Parameters:
Θ : the initialized model parameters, *epochs* : the number of training epochs, η : learning rate

Output: The predicted return $\hat{r}$

1: **for** *epoch* $\leftarrow \{1, \ldots, epochs\}$ **do**
2:   **for** $i \leftarrow \{1, \ldots, D\}$ **do**
3:     $S_i \leftarrow LSTM_b(H_i)$
4:     **if** $t == 1$ **then**
5:       $T_i \leftarrow S_i$
6:       $E_i \leftarrow S_i$
7:     **end if**
8:     $\mathcal{T} \leftarrow$ Calculate Tanimoto coefficient (Eq. 1)
9:     $valid^i \leftarrow$ Calculate the valid topic set according to $\mathcal{T}$ (Eq. 3)
10:     $T_i \leftarrow$ Aggregate information from S_i according to $\mathcal{T}$ (Eq. 4)
11:     $\alpha \leftarrow$ Calculate the attention weight (Eq. 5)
12:     $\hat{E}_i \leftarrow$ Aggregate information from T_i according to α (Eq. 6)
13:     $E_{i+1} \leftarrow LSTM_d(\hat{E}_i)$
14:   **end for**
15:   Compute the stochastic gradients of Θ (Eq. 13)
16:   Update model parameters Θ according to learning rate η and gradients.
17: **end for**
18: **return** the predicted return $\hat{r}$

3.7 Model Training

Our model is optimized by minimizing the global loss $\mathcal{L}$. This was done using the Adam optimizer [9]. The hyper-parameters are set as follows: the embedding size is set to 128, the learning rate is set to 0.001, the training epoch is set to 300, the dropout rate is set to 0.1. All experiments are run on a Lambda Deep Learning 2-GPU Workstation (RTX 2080) with 24 GB of memory, and the random seed is set to 0 at the beginning of each experiment.

4 Experiments

4.1 Datasets

We run comprehensive evaluations of our framework on the China's CSI 300 financial markets, from 2008 to 2022. We use the data from 01/01/2008 to 12/31/2014 as the training set, the data from 01/01/2015 to 12/31/2016 as the validation set for hyper-parameter fine-tuning, and the data from 01/01/2017 to 07/10/2022 as the test set.

4.2 Baselines

We compare our framework with a comprehensive list of 16 well-known methods which are widely used in the financial sector. These methods span six different categories and are:

- **Classic Models -** MLP, TCN [2], GATs [19]
- **Tabular Learning -** TabNet
- **Gradient Boosting Models -** CatBoost [12], LightGBM [8]
- **RNN-Based Methods -** LSTM [6], GRU [3], DA-RNN [13], AdaRNN [4]
- **Attention-Based Methods -** Transformer [18], Localformer [7]
- **Financial Prediction Methods -** DoubleEnsemble [23], ADD [16], HIST [21], IGMTF [20]

Note that although our method can mine the latent topics among stocks, the tasks in our experiments only assume access to price and volume features (opening price, closing price, highest price, lowest price, VWAP). Several recently proposed methods require additional information such as company relations [14] or social media text [15], thus these methods cannot be included as baselines.

4.3 Results

We use **stock trend prediction** and **trading simulation** for our experiments.

Table 1. The results of stock trend prediction on the CSI300 market from 01/01/2017 to 07/10/2022. All the results are averaged after 10 runs, and the standard deviations are shown. * corresponds to statistically significant differences between a baseline and our method ($p < 0.05$ using t-test).

Model Name	IC	ICIR	Rank IC	Rank ICIR
Transformer	0.0143±0.0024 *	0.0910±0.0180 *	0.0317±0.0024 *	0.2192±0.0190 *
TabNet	0.0286±0.0000 *	0.1975±0.0000 *	0.0367±0.0000 *	0.2798±0.0000 *
MLP	0.0267±0.0017 *	0.1845±0.0154 *	0.0362±0.0018 *	0.2681±0.0157 *
Localformer	0.0358±0.0036 *	0.2633±0.0334 *	0.0477±0.0019 *	0.3643±0.0218 *
CatBoost	0.0326±0.0000 *	0.2328±0.0000 *	0.0394±0.0000 *	0.2998±0.0000 *
DoubleEnsemble	0.0362±0.0005 *	0.2725±0.0036 *	0.0444±0.0004 *	0.3450±0.0038 *
LightGBM	0.0347±0.0000 *	0.2648±0.0000 *	0.0443±0.0000 *	0.3520±0.0000 *
TCN	0.0384±0.0015 *	0.2834±0.0164 *	0.0455±0.0012 *	0.3546±0.0077 *
ALSTM	0.0413±0.0034 *	0.3166±0.0329 *	0.0504±0.0032 *	0.3974±0.0280 *
LSTM	0.0402±0.0030 *	0.3194±0.0271 *	0.0496±0.0027 *	0.4040±0.0212 *
ADD	0.0370±0.0025 *	0.2669±0.0254 *	0.0511±0.0018 *	0.3756±0.0235 *
GRU	0.0417±0.0029 *	0.3284±0.0367 *	0.0510±0.0014 *	0.4137±0.0224 *
AdaRNN	0.0380±0.0117 *	0.2999±0.1022 *	0.0472±0.0095 *	0.3744±0.0974 *
GATs	0.0430±0.0010 *	0.3221±0.0096 *	0.0543±0.0012 *	0.4217±0.0099 *
IGMTF	0.0419±0.0004 *	0.3152±0.0055 *	0.0538±0.0014 *	0.4213±0.0171 *
HIST	0.0437±0.0012 *	0.2952±0.0108 *	0.0581±0.0013 *	0.3912±0.0096 *
Our Method	**0.0489±0.0026**	**0.3593±0.0143**	**0.0605±0.0023**	**0.4514±0.0225**

Stock Trend Prediction. This task aims to evaluate the ability of models to predict the future stock price trend. For each trading day i, we calculate the 1-day return $\hat{r}^i$ of each stock based on its historical information before date i. For the results, we report the averaged information coefficient (IC), ranked information coefficient (Rank IC), information ratio of IC (ICIR), and information ratio of Rank IC (Rank ICIR). IC^i is the daily IC that measures the Pearson correlation between the predicted ratio $\hat{r}^i$ and the ground-truth ratio r^i:

$$\mathrm{IC^i} = \frac{(\hat{r}^i - \mathrm{mean}(\hat{r}^i))^\top (r^i - \mathrm{mean}(r^i))}{n \cdot \mathrm{std}(\hat{r^i}) \cdot \mathrm{std}(r^i)} \tag{14}$$

The IC is calculated for the average of each trading day:

$$\mathrm{IC} = \frac{\sum_{i \in [1,D]} \mathrm{IC^i}}{D} \tag{15}$$

The ICIR is used to show the stability of IC, which is calculated by dividing IC by its standard deviation:

$$\mathrm{ICIR} = \frac{\mathrm{IC}}{\mathrm{std(IC)}} \tag{16}$$

Table 2. The results of trading simulation on the CSI300 market from 01/01/2017 to 07/10/2022. All the results are averaged after 10 runs, and the standard deviations are shown. * corresponds to statistically significant differences between a baseline and our method ($p < 0.05$ using t-test).

Model Name	Annualized Return	Max Drawdown	Information Ratio
Transformer	0.0069±0.0181 *	-0.2131±0.0868 *	0.0753±0.2138 *
TabNet	0.0719±0.0000 *	-0.1139±0.0000	0.8155±0.0000 *
MLP	0.0441±0.0153 *	-0.1512±0.0375 *	0.5163±0.1882 *
Localformer	0.0498±0.0228 *	-0.1268±0.0235	0.6194±0.2843 *
CatBoost	0.0585±0.0013 *	-0.1364±0.0051	0.7270±0.0162 *
DoubleEnsemble	0.0642±0.0112 *	-0.0900±0.0103 *	0.8234±0.1398 *
LightGBM	0.0707±0.0000 *	**-0.0835±0.0000** *	0.9487±0.0000 *
TCN	0.0781±0.0203 *	-0.0849±0.0151 *	1.0205±0.2350 *
ALSTM	0.0777±0.0220 *	-0.1031±0.0204	1.0226±0.2859 *
LSTM	0.0826±0.0242 *	-0.0908±0.0132 *	1.0706±0.2771 *
ADD	0.0759±0.0178 *	-0.0939±0.0237	0.9471±0.2101 *
GRU	0.0815±0.0258 *	-0.0917±0.0270 *	1.0826±0.3671
AdaRNN	0.0619±0.0589 *	-0.1392±0.1622	0.8439±0.7172
GATs	0.0886±0.0115 *	-0.1022±0.0184	1.1524±0.1469 *
IGMTF	0.0903±0.0095 *	-0.0986±0.0174	1.1825±0.1035
HIST	0.0854±0.0119 *	-0.0919±0.0152 *	1.0879±0.1504 *
Our Method	**0.1063±0.0187**	-0.1191±0.0301	**1.3315±0.2169**

For the calculation of Rank IC^{i}, we first use $R^i = rank(r^i)$, and $\hat{R}^i = rank(\hat{r}^i)$ to denote the ranks of the ground-truth and the predicted ratios, respectively:

$$\text{Rank IC}^{\text{i}} = \frac{(\hat{R}^i - \text{mean}(\hat{R}^i))^\top (R^i - \text{mean}(R^i))}{n \cdot \text{std}(\hat{R}^i) \cdot \text{std}(R^i)} \tag{17}$$

The Rank IC and Rank ICIR are calculated similarly as before:

$$\text{Rank IC} = \frac{\sum_{i\in[1,D]} \text{Rank IC}^{\text{i}}}{D} \tag{18}$$

$$\text{Rank ICIR} = \frac{\text{Rank IC}}{\text{std(Rank IC)}} \tag{19}$$

The results of the stock trend prediction task on the test set of the China CSI300 market (01/01/2017 to 07/10/2022) are shown in Table 1. Our method significantly outperforms all the 16 baselines across all four metrics (IC, ICIR, Rank IC, and Rank ICIR) with around 10% enhancement over the second-place model for each metric. These results indicate the importance of modeling expectations and dynamic topics in financial market analysis. It is also interesting to note that the traditional RNN-based methods (such as GRU and LSTM) achieve similar or even better results compared to the models specifically designed for financial analysis (such as ADD, IGMTF, and DoubleEnsemble). This may be attributed to the low signal-to-noise ratio in the financial market since the simpler models may be more robust to noise. These observations further demonstrate the hardness of this task.

Trading Simulation. In quantitative investment,"backtesting" refers to applying a trading strategy to historical data, simulating trading, and measuring the return of the strategy. For this task, we employ the top-k dropout strategy for each method, reporting the **annualized return**[4] (the geometric average of money earned by an investment strategy each year over a given time period), **max drawdown**[5] (maximum observed loss from a peak to a trough), and the **information ratio**[6] (ratio of returns above the returns of the CSI300 benchmark). The top-k dropout strategy is a straightforward quantitative investment approach: for each trading day, we hold k stocks, sell d stocks with the worst predicted 1-day return, and buy d unheld stocks with the best-predicted 1-day return. In our experiments, k is set to 50, and d is set to 5. The trading simulation task results on the test set of the China CSI300 market are displayed in Table 2. Our method surpasses all 16 baselines in annualized return and information ratio. To improve the stability of profitability, future research could explore modifications designed to reduce the max drawdown of our approach.

5 Conclusion

In this paper, we introduce a novel framework for stock trend prediction, suitable for quantitative analysis of financial markets and stock selection. To the best of our knowledge, our method is the first to consider (1) investors' subjective expectations, and (2) automatically mined dynamic topics that do not require additional knowledge. Through experiments on 16 baselines using the CSI 300 market, we demonstrate that our model achieves a stable annual return above 10%, outperforming all existing baselines and attaining the current state-of-the-art results for stock trend prediction and trading simulation tasks.

Future work could explore modifications to decrease the max drawdown of our method, resulting in more stable profitability. Additionally, since expectations are influenced by external factors such as financial reports or discussions on social media, future research could investigate incorporating this information into our model.

[4] https://www.investopedia.com/terms/a/annualized-rate.asp.
[5] https://www.investopedia.com/terms/m/maximum-drawdown-mdd.asp.
[6] https://www.investopedia.com/terms/i/informationratio.asp.

References

1. Arik, S.Ö., Pfister, T.: TabNet: attentive interpretable tabular learning. In: Proceedings of the AAAI Conference on Artificial Intelligence, vol. 35, pp. 6679–6687 (2021)
2. Bai, S., Kolter, J.Z., Koltun, V.: An empirical evaluation of generic convolutional and recurrent networks for sequence modeling. arXiv preprint arXiv:1803.01271 (2018)
3. Cho, K., et al.: Learning phrase representations using RNN encoder-decoder for statistical machine translation. arXiv preprint arXiv:1406.1078 (2014)
4. Du, Y., et al.: AdaRNN: adaptive learning and forecasting of time series. In: Proceedings of the 30th ACM International Conference on Information & Knowledge Management, pp. 402–411 (2021)
5. Fama, E.F.: The behavior of stock-market prices. J. Bus. **38**(1), 34–105 (1965)
6. Hochreiter, S., Schmidhuber, J.: Long short-term memory. Neural Comput. **9**(8), 1735–1780 (1997)
7. Jiang, J., Kim, J.B., Luo, Y., Zhang, K., Kim, S.: AdaMCT: adaptive mixture of CNN-transformer for sequential recommendation. arXiv preprint arXiv:2205.08776 (2022)
8. Ke, G., et al.: LightGBM: a highly efficient gradient boosting decision tree. In: Advances in Neural Information Processing Systems 30 (2017)
9. Kingma, D.P., Ba, J.: Adam: a method for stochastic optimization. arXiv preprint arXiv:1412.6980 (2014)
10. Kipf, T.N., Welling, M.: Semi-supervised classification with graph convolutional networks. arXiv preprint arXiv:1609.02907 (2016)
11. Poterba, J.M., Summers, L.H.: Mean reversion in stock prices: evidence and implications. J. Financ. Econ. **22**(1), 27–59 (1988)
12. Prokhorenkova, L., Gusev, G., Vorobev, A., Dorogush, A.V., Gulin, A.: CatBoost: unbiased boosting with categorical features. In: Advances in Neural Information Processing Systems 31 (2018)
13. Qin, Y., Song, D., Chen, H., Cheng, W., Jiang, G., Cottrell, G.: A dual-stage attention-based recurrent neural network for time series prediction. arXiv preprint arXiv:1704.02971 (2017)
14. Sawhney, R., Agarwal, S., Wadhwa, A., Derr, T., Shah, R.R.: Stock selection via spatiotemporal hypergraph attention network: a learning to rank approach. In: Proceedings of the AAAI Conference on Artificial Intelligence, vol. 35, pp. 497–504 (2021)
15. Sawhney, R., Agarwal, S., Wadhwa, A., Shah, R.: Deep attentive learning for stock movement prediction from social media text and company correlations. In: Proceedings of the 2020 Conference on Empirical Methods in Natural Language Processing (EMNLP), pp. 8415–8426 (2020)
16. Tang, H., Wu, L., Liu, W., Bian, J.: ADD: augmented disentanglement distillation framework for improving stock trend forecasting. arXiv preprint arXiv:2012.06289 (2020)
17. Tanimoto, T.T.: Elementary mathematical theory of classification and prediction (1958)
18. Vaswani, A., et al.: Attention is all you need. In: Advances in Neural Information Processing Systems 30 (2017)
19. Veličković, P., Cucurull, G., Casanova, A., Romero, A., Lio, P., Bengio, Y.: Graph attention networks. arXiv preprint arXiv:1710.10903 (2017)

20. Xu, W., Liu, W., Bian, J., Yin, J., Liu, T.Y.: Instance-wise graph-based framework for multivariate time series forecasting. arXiv preprint arXiv:2109.06489 (2021)
21. Xu, W., et al.: HIST: a graph-based framework for stock trend forecasting via mining concept-oriented shared information. arXiv preprint arXiv:2110.13716 (2021)
22. Yang, X., Liu, W., Zhou, D., Bian, J., Liu, T.Y.: Qlib: an AI-oriented quantitative investment platform. arXiv preprint arXiv:2009.11189 (2020)
23. Zhang, C., Li, Y., Chen, X., Jin, Y., Tang, P., Li, J.: DoubleEnsemble: a new ensemble method based on sample reweighting and feature selection for financial data analysis. In: 2020 IEEE International Conference on Data Mining (ICDM), pp. 781–790. IEEE (2020)

Let the Model Make Financial Senses: A Text2Text Generative Approach for Financial Complaint Identification

Sarmistha Das[1](✉), Apoorva Singh[1], Raghav Jain[1], Sriparna Saha[1], and Alka Maurya[2]

[1] Indian Institute of Technology Patna, Patna, India
sarmistha1515@gmail.com, {apoorva_1921cs19,sriparna}@iitp.ac.in
[2] Crisil Pvt. Ltd., Mumbai, India

Abstract. A complaint frequently expresses the complainer's dissatisfaction or objectionable notion to support a belief or claim against a party or parties. Financial loss, material inconvenience, and distress are sufficient examples to intensify the need for an automated complaint analysis tool in the financial domain, particularly on social media with diverse information-related affairs. Recently, advanced approaches like complaint detection with machine learning have escalated the research interest in the area of natural language processing. Earlier, the only research focus was on how complaints are identified linguistically. Substantial modern complaint analytical models attempt to bridge the gap between the interpretability and explainability of financial complaint detection tasks. To address this, we extend an existing complaint dataset X-FINCORP, with the rationale or cause annotations for the complaint/non-compliant labels. Each instance in the dataset is now associated with five labels: complaint, emotion, polarity, severity, and rationales. Our proposed model addresses the multi-task problem as a text-to-text generation task by utilizing a generative framework. Additionally, we introduce commonsense as external information to draw more informative intuitions and enhance the overall performance of the proposed generative model. The empirical results validate the generality of our proposed model over several evaluation metrics compared to state-of-the-art models and other baselines (Resources available at https://github.com/appy1608/Financial-Complaint-Identification.).

Keywords: Complaint Detection · Cause Analysis · Financial Complaint Corpus · Deep learning · Emotion Recognition · Multi-task learning · Severity Level Classification · Sentiment Analysis

1 Introduction

A complaint is an assertion that intensifies firm disappointment with facts and the unacceptable emotions related to expectations vs actuality entangled with

H. Kashima et al. (Eds.): PAKDD 2023, LNAI 13937, pp. 58–69, 2023.
https://doi.org/10.1007/978-3-031-33380-4_5

it. Previous research studies on real-world financial complaint identification [1] introduced standard severity levels, i.e., (a) no explicit reproach, (b) disapproval, (c) accusation, and (d) blame. Modern complaint analysis studies from social media [2] are constructed with automatic binary complaint identification. In finance, Explainable AI can be used to resolve digital service grievances in government institutions, investment managers, insurance companies, and retail banks by maintaining loans, credit cards, and other consumer financial products and service-related financial disputes. Recognizing the causal span of the expressed emotions is a fundamental research development in automatic reasoning about human emotions [3]. Using these associated studies as a guide, we intend to determine the cause or reason for the financial complaints posted on the social web. Human nature-wise, we tend to depend on commonsense knowledge and speculate the inferred conclusions. In Fig. 1, facts are demonstrated clearly. For instance, the complainer shared his information about how his day got ruined as his debit card stopped working during his payment process. Evidently from Fig. 1, we can say, though the complainer did not mention his emotions explicitly, we can identify the nature of the emotion as Anger(xReact), and he needs a feasible debit card (xNeed), whereas relying on commonsense we can also infer that the complainer is about to apply for a new debit card (xWant). Therefore we believe utilizing external knowledge we could better understand the complainer's feelings which leads to understanding more informative aspects. Moreover, several multi-tasking approaches have been introduced where auxiliary task improves the performance of the main primary task [4]. However, this approach initiates multitasking-related complications, especially in performance optimization [5] assign-weights and negative transfer during the training process. To deal with such multi-tasking issues, we introduce a text-to-text generation technique where complaint identification, cause extraction, and severity-level categorization are the primary tasks and emotion recognition and sentiment classification are the auxiliaries.

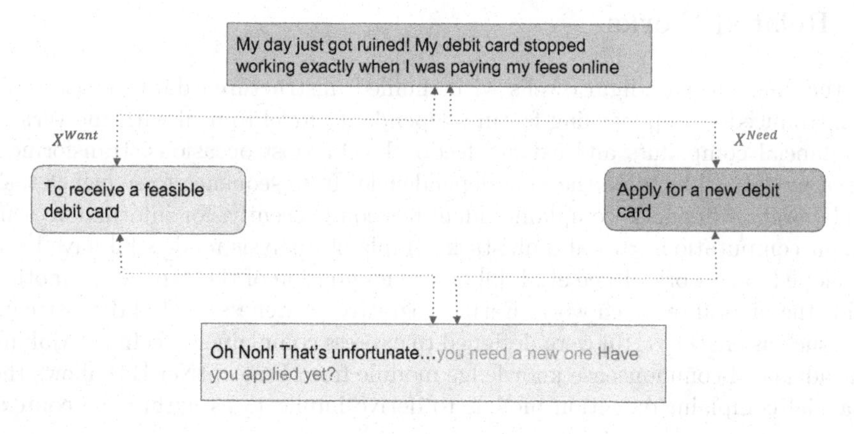

Fig. 1. Sample text from *X-FINCORP* dataset

Research Objectives: Following are the research objectives of the current study:

- At first, we intend to identify the rationale/causes which are responsible for categorizing financial data having complaints. Next, we try to understand the effect of using causal information on the financial complaint identification and severity level classification tasks in the proposed framework.
- We attempt to accomplish an explainable complaint identification task with additional commonsense knowledge through a text-to-text generation technique, especially in the financial domain.

Contributions: Our major contributions are as follows:

- Presumably, it is the first study on explainable complaint identification by focusing on the rationale/cause responsible for categorizing the financial complaints by investigating two aforementioned challenges i) explainable complaint identification and ii) Text-to-text generation as a multitasking problem.
- We presented *X-FINCORP* a publicly available dataset with manual annotation of causal spans for complaint/non-complaint labeled sentences and developed a benchmark approach for complaint cause identification, focusing on cause detection and extraction.
- We propose a generative modeling approach that jointly learns (a) binary complaint classification (CI), (b) emotion recognition (ER), and (c) sentiment categorization (SC) as the first sub-problem, and the second sub-problem involves (d) severity level classification (SLC), (e) cause extraction (CE) with a commonsense aware unified generative framework.

As briefly discussed in the result section, utilizing external commonsense information and generative modeling enhances the performance of the primary tasks.

2 Related Works

As the financial paradigm dwells with shuffled unstructured data, Singh et al. [6] introduced a deep learning feature-dependant model to deal with the variety of financial complaints and extract feedback. On most occasions, transformer-based models [7] handled the long dependencies from sequence to sequence task in elaborative financial complaints identification. Recently, for informative sentiment computation, several multi-task complaint analysis models [8] have been developed. The work of Lee et al. [9] on the foundation of the expressed emotion holds the pivotal research work. Earlier cognitive awareness-related downstream jobs such as chatbots [10] were designed to express complainers' feelings. Making this advanced commonsense knowledge module from ConceptNet [11] allows the financial complaint detection models to derive implications within the context

Table 1. Example instances of causal span annotation. Label: class labels for complaint, sentiment, and emotion tasks

Tweet text	Cause
$< USER >$ Authorization charge has not been refunded and it's more than a month now. Can someone help me ***Label:*** *complaint, sadness, negative*	Authorization charge has Have not been refunded
$< USER >$ Swap functionality is fantastic, great job. Can't wait for the credit card now ***Label:*** *non complaint, happiness, positive*	no cause

explicitly shared by the complainer. ATOMIC [12] is a commonsense knowledge base that contains everyday if-then events reasoning inferences. Besides several generative pre-trained models such as Decoder and Encoder Transformer, have been proposed to handle multi-tasking in text generation. BART [13], the decoder-encoder transformer can process the text bidirectionally and handle the noise simultaneously, as both are trained on bulky shuffled text data.

Even though there are multiple relatable fields in emotional complaint identification but there are no studies on cause detection and extraction through generating the texts. In order to address such a challenging task, we propose an explainable cause analysis localized on the dynamic attention mechanism at the word and sentence level by implementing a generative approach.

3 Corpus Extension

Numerous financial organizations and services maintain Twitter accounts for client support. As a result of such accounts and the micro-blogging nature of Twitter, customers are more comfortable in submitting their grievances and requesting assistance on Twitter, as opposed to the more time-consuming alternatives of sending emails or visiting in person to the businesses. This motivates us to utilize the *FINCORP* dataset[1], which is a Twitter-based financial complaint dataset and includes 3,133 typed samples of non-complaint and 3,149 complaint-typed samples in English. We selected this publicly available dataset because, unlike other datasets, it consists of complaints arising between financial organizations and their customers. Other than the complaint identification axis, it is also extended on three different associated axes: severity levels (*no explicit reproach, disapproval, accusation, blame, and non-complaints*), sentiment (*negative, neutral, positive*), and emotion (*anger, disgust, fear, happiness, sadness, surprise and other*), the 'other' emotion class depicts tweets that do not fall

[1] https://github.com/RohanBh23/FINCORP.

under the scope of Ekman's six basic emotions. Our work focuses on strengthening the available dataset by extending it with the manual annotation of causal spans for complaint/non-complaint labeled sentences to provide scope for multi-faceted research.

3.1 Cause Extraction Method

Task Definition. *Complaint Cause* is defined as a portion of the text that expresses why the user feels compelled to file a complaint. It is the speech act used by the individual to describe the circumstances in which their expectations have been violated.

Annotations. Three annotators (one doctoral and two undergraduate students in the computer science discipline) with adequate domain knowledge and expertise in developing supervised corpora were entrusted with annotating the causal span identification task for each sample in the dataset.

Annotators were directed to identify the causal span, X(I), that appropriately represented the basis of the complaint (C) for each instance (I) in the *X-FINCORP* dataset. For the non-complaint (NC) class, the annotators marked the sentence as 'no cause'[2]. Table 1 shows a few example instances of causal span annotation.

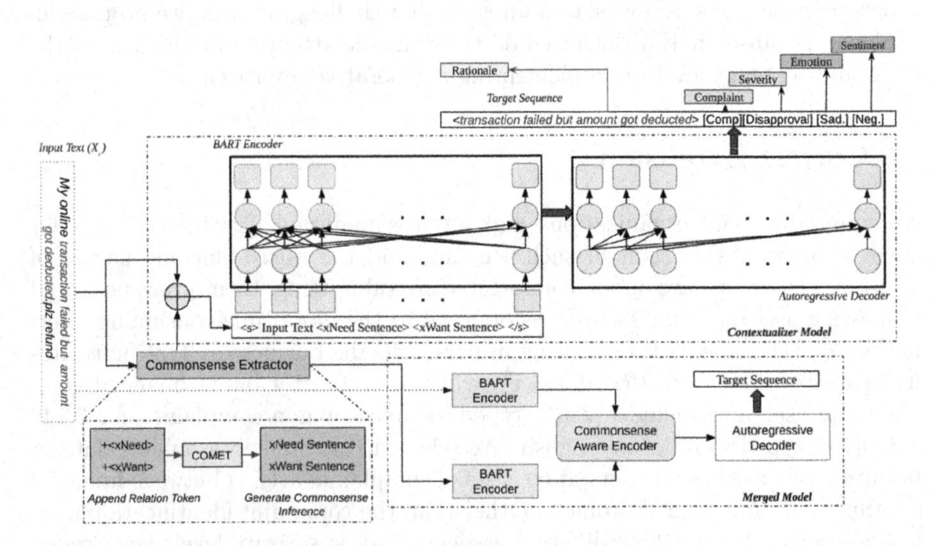

Fig. 2. The overall architecture of the proposed model (CS2I). The two variations of our proposed model, ConCS2I and MergeCS2I are depicted by the enclosed red and blue dotted boxes, respectively. (Color figure online)

[2] The authors of the work [8] mentioned the agreement ratings of complaint, severity level, emotion, and sentiment tasks as 0.83, 0.69, 0.68, and 0.82, respectively, suggesting good annotations.

4 Methodology

In this section, we illustrate our problem statement first then we explore further the proposed model. The architecture of the model is shown in Fig. 2

4.1 Problem Definition

An explainable financial complaint identification model should be robust enough to identify the nature of the financial complaint, i.e., whether it's a complaint or not, and if the sample belongs to complaint class then identifies its severity level, emotion category with the corresponding sentiment and most importantly the cause span responsible for making it a financial complaint. This is the procedural way of justifying the model's decision and making the results more interpretable. Considering customer complaint as input text $X_i = \{x_0, x_1, .., x_i, .., x_n\}$ where the length of the input complaint is n, we aim to learn five closely related tasks: (i) complaint identification (c), (ii) sentiment categorization (se), (iii) emotion recognition (e), (iv) severity level classification (s), (v) cause extraction (ce), where $c \in C$, C is the set of complaint classes, $se \in SE$, SE is the set of sentiment classes, $e \in E$, E is the set of emotion classes, $s \in S$, S is the set of severity levels and $ce(X_i) \in (X_i)$ that is relevant to the c, complaint label. Our proposed methodology is described in the following steps:

4.2 Construction of Text to Text Generation Task from Explainable Complaint Identification Task

To solve explainable complaint identification and other auxiliary tasks in a single unified manner, we introduce a text-to-text generation approach. In order to transform this problem into a text generation problem, we first construct Y_i, a natural language target sequence, with X_i, for the input sentence, we concatenate all task labels during the training phase. Hence the definition goes 1.

$$Y_i = \{< ce(X_i) > [c][s][e][se]\} \tag{1}$$

where c, s, e, se, and ce signify complaint identification, severity level prediction, emotion recognition, sentiment prediction, and cause extraction, respectively. After the predictions of each task, we added special characters, as shown in Eq. 1 so that during testing or inference we can extract task-specific predictions. Furthermore, we explicate the problem as: given an input sequence X_i, the task is to generate an output sequence, $Y_i^{'}$, that contains all the predictions defined in Eq. 1 using a generative model G; $Y_i^{'} = G(X_i)$.

4.3 Extraction of Casual Span

Initially, the CI task distinguishes between complaint and non-complaint tweets. Following that CE task performs the extraction based on the 0/1 label predicted by our model corresponding to each word token where $0 \in non - complaint$

and 1 ∈ *complaint*. After that *label* ∈ {0, 1}, will be mapped correspondingly to each word token in a given tweet text. This way the beginning and end of the casual span can be directly decoded from the token length in a sentence. For example: I/[0] have/[0] made/[0] a/[0], merchant/[0] transaction /[0] but/[0] the/[0] transaction/[1] was/[1] unsuccessful/[1], and/[1] amount/[1] has/[1] been/[1] debited/[1] from/[1] my/[1] account/[1].

4.4 Commonsense Aware Financial Complaint Identification

We propose CS2I (Commonsense aware Financial Complaint Identification), a cognitive commonsense unified framework to solve the task of explainable complaint identification. For better understanding, we divide our approach into two steps: 1) Commonsense aware extraction module, and 2) Commonsense aware transformer model.

Commonsense Aware Extraction Module. As complaints are often abbreviated or not thorough, the commonsense extraction module makes the insight information more sensible with contextual reasoning. Besides we make use of the ATOMIC dataset [12] as our knowledge base in the form (e, r, csr) where e denotes an event, r denotes a commonsense relation, and csr denotes the inferred commonsense reasoning.

Considering our financial complaint identification as an event, we want to understand the complainer's needs and what they want from the filled complaint. Hence we consider only two relations (xNeed and xWant). We applied a pre-trained BART-based language model, COMET [14] trained on the ATOMIC dataset to generate reasonable commonsense from the unseen complaints. Initially, for each complaint X_i, we append two commonsense relation tokens (xNeed and xWant). Later these concatenated inputs are fed to the pre-trained COMET model to generate two commonsense reasonings $csr^{r_{need}}$ and $csr^{r_{want}}$ for xNeed and xWant relation tokens, respectively, To obtain the final commonsense reasoning CSr for each review X_i, we concatenate the two generated commonsense reasonings, $CSr = csr^{r_{need}} \oplus csr^{r_{want}}$.

Commonsense Aware Transfer Module ConCS2I (Context Encoder CS2I): Following the previous work, ConCS2I generates the target sequence $Y_i^{'}$, with a given complaint input X_i, and corresponding commonsense reasoning, CSr; Along with the conditional probability for text generation model: $P_\theta(Y_i^{'}|X_i, cs)$, where θ is a set of model parameters. Initially, we concatenated complaint input tokens X_i with the commonsense reasonings CSr and then applied a unique token separator $< SEP >$. We get the final input sequence: $R_i = X_i \oplus csr$. Now, we have a pair of input sentence and target sequence (R_i, Y_i). First feed $R_i = \{x_0, x_1, .., x_n, < SEP >, csr_0.., csr_n\}$ to the encoder module to obtain the hidden input representations; $H_E = G_{Enco}(\{x_0, x_1, \ldots, x_n, < SEP >, csr_0. \ldots, csr_n\})$ here encoder computations are represented by $G_{Encoder}$.

Following that, we will feed H_E and all the output tokens to the decoder module till time step $t-1$ which is represented as $Y_{<t}$. For decoder module, at timestamp t, obtained hidden state will be: $H_D^t = G_{Deco}(H_E, Y_{<t})$ where G_{Deco} denotes the decoder computations.

Finally applying the softmax function over the hidden state, H_{DEC}^t, to calculate the output prediction for the given input and previous $t-1$ tokens at t^{th} time step with the conditional probability consideration;

$$P_\theta(Y_t^{'}|R, Y_{<t}) = F_{softmax}(\theta^T H_{DE}^t) \tag{2}$$

where softmax computation is represented by $F_{softmax}$ and θ denotes model weights.

MergeCS2I (Merged CS2I): We introduce another approach named Merge CS2I to merge commonsense knowledge in the model by applying a commonsense aware encoder module. Initially, we feed complaint input, X_i, along with commonsense reasoning, CSr, to a pre-trained BART encoder to obtain encoded representations, UB_x and UB_{csr}, respectively. We applied a commonsense aware encoder, an extension of the original transformer encoder to merge the information between UB_x and UB_{csr} representations. Unlike the traditional transformer encoder where input was projected as query, key, and value, we create UB_x and UB_{csr} corresponding to two triplets of queries, keys, and values matrices: (Q_x, K_x, V_x) and Q_{cs}, K_{cs}, V_{cs}). We also introduce a cross-attention layer that consists of one multi-head-cross attention and normalization layer that exchanges the key and value by taking (Q_x, K_{cs}, V_{cs}) and (Q_{cs}, K_x, V_x) as inputs to cross attention layer. The computed cross-merged vector representation is defined as follows: $Attention(Q, K, V) = softmax(\frac{QK^T}{\sqrt{d_k}})V$ where a set of query, key, and value is (Q, K, V) and the dimension of query and key is d_k.

The exchanged outputs of multi-head-cross attention contain information about each other ($U_{x->cs}$ and $U_{cs->x}$). Following this, we concatenate them and to obtain the output of the commonsense aware encoder, we pass concatenated output U_z through a self-attention layer, normalization layers, and fully connected layers with residual connections. After concatenation, Z becomes the final commonsense aware input representation vector. Further, we feed Z to an autoregressive decoder that follows the same computations defined in Eq. 2.

5 Experimental Results and Analysis

5.1 Baselines Setup

Multitask Systems: Inspired by recent work in multitask CI framework we aim to develop Baseline$_1$ [8] model as one of the multitask baselines. We implement the Baseline$_1$ model for the joint learning of CI, SC, and CE with SLC and ER as additional tasks and also keep the experimental setup the same as our current work. Furthermore, to understand the impact of different tasks, we evaluate them by keeping a few of them fixed and gauge their nature of impacted performance accordingly.

Table 2. Resultant difference of baselines with ConCS2I and MergeCS2I. For the CI and SC tasks, the results are in terms of macro-F1 score (F1) and Accuracy (A) values. F1, A metrics are given in %. JS: Jaccard Similarity, HD: Hamming distance, and ROS: Ratcliff-Obershelp Similarity. The maximum scores attained are represented by bold-faced values. The † denotes statistically significant findings.

	Complaint (CI)		Severity (SC)		Cause (CE)		
Model	F1	A	F1	A	JS	HD	ROS
SOTA [15]	86.6	87.6	59.4	55.5	–	–	–
ConCS2I$_{CI+SC+CE}$	89.7	90.2	73.3	73.4	83.7	75.0	88.7
ConCS2I$_{CI+SC+CE+ER}$	90.3	91.6	72.8	73.0	83.4	74.1	88.2
ConCS2I$_{CI+SC+CE+SLC}$	90.7	91.6	73.8	74.1	83.6	74.5	88.3
ConCS2I$_{All}$	**91.6**†	**92.9**†	**74.7**†	**74.9**†	**85.1**†	**77.2**†	**89.9**†
MergeCS2I$_{CI+SC+CE}$	89.9	90.1	72.6	73.2	77.4	71.1	85.8
MergeCS2I$_{CI+SC+CE+ER}$	90.1	90.4	73.1	73.4	77.4	70.7	84.2
MergeCS2I$_{CI+SC+CE+SLC}$	90.6	91.1	72.9	73.0	78.2	72.1	85.1
MergeCS2I$_{All}$	91.2	91.5	73.1	**73.5**	78.1	71.3	85.3
Baseline$_1$ [8]	81.4	82.8	60.3	62.8	76.1	68.2	81.8
BART	88.9	88.9	62.4	62.6	77.1	69.7	84.2
T5	86.7	86.6	68.7	69.3	84.1	74.1	89.1
SpanBERT	–	–	–	–	74.8	70.6	83.5

Cause Extraction Task Baselines: Since the cause span extraction is an advanced task in the complaint analysis area, we got our inspiration from the works of [3] in the emotion recognition domain. Exclusively we select the SpanBERT base model for the CE task though SpanBERT is fine-tuned on the SQuAD 2.0 dataset [16].

Text to Text Generation Model: We employ BART [13] and T5 [17] as the baseline text-to-text generation models. We fine-tune both these models on *X-FINCORP* dataset with complaint text as the input sequence and concatenated outputs (defined in Eq. 1) as the target sequence.

We conduct the ablation study to gauge the performance s of ConCS2I and MergedCS2I models and compare their essential components with the aforementioned baselines.

5.2 Experimental Setup

We have performed all the experiments on the Tyrone machine with Intel's Xeon W-2155 Processor having 196 Gb DDR4 RAM and 11 Gb Nvidia 1080Ti GPU. Our training data contains 80% of data samples with a nested 10-fold cross-validation approach. Adam optimizer is used to train the model with an adam epsilon value of 0.000001. For the CI and SC tasks, accuracy and macro-F1 metrics are used to evaluate predictive performance. For the quantitative assessment of the CE task, we used the Jaccard Similarity (JS), Hamming Distance (HD), and Ratcliff-Obershelp Similarity (ROS) metrics.

Table 3. Qualitative study of the CI and SC predictions by the SOTA [15] and the proposed model (ConCS2I). 'Actual Label': true labels for CI and SC tasks, the yellow highlighted text indicates the causal span annotation of the sentence. (Best viewed in color.)

Tweet text	SOTA	Proposed	Actual Label
8 months now, I have cancelled < $USER$ > ticket and still did not get any response or refund. This was not expected!	complaint blame	complaint disapproval	complaint disapproval
Thank you for your response. I already registered with UPI, it was working fine suddenly it is showing error and transaction getting failed.	complaint blame	complaint no explicit reproach	complaint no explicit reproach

5.3 Results and Discussions

- **Ablation Study:** From Table 2, evidently the proposed model ConCS2I$_{All}$ outperforms all the other baselines for primary tasks by a noteworthy margin. ConCS2I can capture more commonsense reasoning information compared to the other baselines as *X-FINCORP* dataset is a Twitter-based dataset with fixed character constraints. Additionally, we fixed the average sentence length of the used dataset to 15. Sample sentences from the dataset, such as *'Thanks'*, *'I need help'* render a lack of contextual information. From Table 2 we can also observe that all the ConCS2I multitask variants consistently outperform all the MergeCS2I variants on all the subtasks. Therefore we can illustrate that the model can learn a better-combined representation of commonsense. Even for input complaints in a direct concatenation setting, the model works precisely better than their merged vector encodings. Although both ConCS2I and MergeCS2I models are able to outperform the SOTA model on CI and SC tasks ConCS2I$_{All}$ outperforms the SOTA by a significant margin on CI and SC tasks, respectively. The prior reasons for such improvements can be implied to the fact: 1) ConCS2I and MergeCS2I both are leveraging the knowledge of the pre-trained BART model, which already has been trained on a huge corpus of data and 2) Having extra context leads them to make better predictions. Based on the tables mentioned earlier, it is evident that including extra tasks like ER and SLC enhances the model's performance. It also indicates that during the decoding or generating process, the model can learn the mapping between different tasks.
- **Qualitative Analysis:** We noticed that due to the imbalance nature of the higher number of instances from *Non-Complaint* class, the accurate classification shows skewness. Whereas tweets with vital financial complaint signs, such as accusatory expressions or blame-related terms, are less misclassified. Table 3 shows the qualitative study of the financial complaint severity predictions obtained by the SOTA [15] and the proposed Contextualizer system on a few sample test instances. The CE task combined with CI and commonsense reasoning impressively led to improved predictions where the SOTA system lacks these elements. It can also be observed from Table 3 that for the given instances, both the models correctly predict them as a complaint, but the proposed model only makes the correct prediction for the severity level.

5.4 Error Analysis

We investigate the possible reasons for the proposed model's errors:

- Scattered Causes: The proposed model is not able to identify multiple causes spread across a tweet instance. For example, *I requested to update my address via net banking, was not able to do. I was assured it will be done in 2 working days.* The causal span predicted: *I requested to update my address via net banking, was not able to do.* In the current work, the causes are annotated based on the first encounter with a strong expression of complaint reason in the tweet. Causes scattered across the complete tweet cannot be identified by the proposed model.
- Additional Cause Speculation: Instances of small sentences lead to unnecessary causual span generation. For example, *Credit card transaction failed* The actual span of cause is *transaction failed* but the model generates additional speculation like *credit limit exceeds*, which is not accurate.

6 Conclusion

Through this paper, we attempt to handle the complaint detection problem with explainability consideration. Enhanced performance of real-time explainable AI systems increases confidence with loyalty and integrity for robust customer support in the financial market. Our framework constructs two contributions: (a) development of the first explainable financial complaint detection dataset, which consists of the causal span annotations applied in decision-making (b) introduction of a commonsense-aware unified generative framework to simultaneously perform five tasks (CD, SC, CE, SLC, and ER). This work demonstrates how a multitasking problem could be reformulated as a text-to-text generation task just by utilizing the knowledge of sizable pre-trained sequence-to-sequence models. All the baselines and the SOTA for the three main tasks based on extensive evaluation are outperformed by our proposed framework.

Acknowledgement. Dr. Sriparna Saha gratefully acknowledges Crisil Pvt Ltd for carrying out this research. All authors highly appreciate the dataset annotators, Rik Biswas and Aakansha Prasad, for their contributions.

References

1. Trosborg, A.: Interlanguage Pragmatics: Requests, Complaints, and Apologies, vol. 7. Walter de Gruyter, Berlin (2011)
2. Preotiuc-Pietro, D., Gaman, M., Aletras N.: Automatically identifying complaints in social media. arXiv preprint arXiv:1906.03890 (2019)
3. Poria, S., et al.: Recognizing emotion cause in conversations. Cogn. Comput. **13**(5), 1317–1332 (2021)
4. Singh, A., Saha, S.: Are you really complaining? A multi-task framework for complaint identification, emotion, and sentiment classification. In: Lladós, J., Lopresti, D., Uchida, S. (eds.) ICDAR 2021. LNCS, vol. 12822, pp. 715–731. Springer, Cham (2021). https://doi.org/10.1007/978-3-030-86331-9_46

5. Wu, S.: Automating Knowledge Distillation and Representation from Richly Formatted Data. Stanford University, Stanford (2020)
6. Bhatia, R., Singh, A., Saha, S.: Complaint and severity identification from online financial content (2022)
7. Toutanova, K., et al.: In: Proceedings of the 2021 Conference of the North American Chapter of the Association for Computational Linguistics: Human Language Technologies (2021)
8. Singh, A., Nazir, A., Saha, S.: Adversarial multi-task model for emotion, sentiment, and sarcasm aided complaint detection. In: Hagen, M., et al. (eds.) ECIR 2022. LNCS, vol. 13185, pp. 428–442. Springer, Cham (2022). https://doi.org/10.1007/978-3-030-99736-6_29
9. Lee, S.Y.M., Chen, Y., Huang, C.-R.: A text-driven rule-based system for emotion cause detection. In: Proceedings of the NAACL HLT 2010 Workshop On Computational Approaches to Analysis and Generation of Emotion in Text, pp. 45–53 (2010)
10. Sabour, S., Zheng, C., Huang, M.: CEM: commonsense-aware empathetic response generation. In: Proceedings of the AAAI Conference on Artificial Intelligence, vol. 36, no. 10, pp. 11 229–11 237 (2022)
11. Speer, R., Chin, J., Havasi, C.: ConceptNet 5.5: an open multilingual graph of general knowledge. In: Thirty-First AAAI Conference on Artificial Intelligence (2017)
12. Sap, M., et al.: ATOMIC: an atlas of machine commonsense for if-then reasoning. In: Proceedings AAAI Conference on Artificial Intelligence, vol. 33, no. 01, pp. 3027–3035 (2019)
13. Lewis, M.,et al.: BART: denoising sequence-to-sequence pre-training for natural language generation, translation, and comprehension. arXiv preprint arXiv:1910.13461 (2019)
14. Rajpurkar, P., Jia, R., Liang, P.:Know What You Don't Know: unanswerable questions for squad. arXiv preprint arXiv:1806.03822 (2018)
15. Hwang, J.D., et al.: (comet-) atomic 2020: on symbolic and neural commonsense knowledge graphs. In: Proceedings AAAI Conference Artificial Intelligence, vol. 35, no. 7, pp. 6384–6392 (2021)
16. Roberts, A., et al.: Exploring the limits of transfer learning with a unified text-to-text transformer (2019)
17. Jin, M., Aletras, N.: Modeling the severity of complaints in social media. arXiv preprint arXiv:2103.12428 (2021)

Information Retrieval and Search

Web-Scale Semantic Product Search with Large Language Models

Aashiq Muhamed[1(✉)], Sriram Srinivasan[1], Choon-Hui Teo[1], Qingjun Cui[1], Belinda Zeng[2], Trishul Chilimbi[2], and S. V. N. Vishwanathan[1]

[1] Amazon, Palo Alto, CA, USA
{muhaaash,srirs,choonhui,qingjunc,vishy}@amazon.com
[2] Amazon, Seattle, WA, USA
{zengb,trishulc}@amazon.com

Abstract. Dense embedding-based semantic matching is widely used in e-commerce product search to address the shortcomings of lexical matching such as sensitivity to spelling variants. The recent advances in BERT-like language model encoders, have however, not found their way to realtime search due to the strict inference latency requirement imposed on e-commerce websites. While bi-encoder BERT architectures enable fast approximate nearest neighbor search, training them effectively on query-product data remains a challenge due to training instabilities and the persistent generalization gap with cross-encoders. In this work, we propose a four-stage training procedure to leverage large BERT-like models for product search while preserving low inference latency. We introduce query-product interaction pre-finetuning to effectively pretrain BERT bi-encoders for matching and improve generalization. Through offline experiments on an e-commerce product dataset, we show that a distilled small BERT-based model (75M params) trained using our approach improves the search relevance metric by up to 23% over a baseline DSSM-based model with similar inference latency. The small model only suffers a 3% drop in relevance metric compared to the 20x larger teacher. We also show using online A/B tests at scale, that our approach improves over the production model in exact and substitute products retrieved.

Keywords: Matching · Retrieval · Search · Pretrained Language Models

1 Introduction

An e-commerce product search engine typically serves queries in two stages—matching and ranking, for efficiency and latency reasons. In the matching stage, a query is processed and matched against hundreds of millions of products to retrieve thousands of products that are relevant to the query. In the subsequent ranking stage, the retrieved products are scored against one or more objectives

A. Muhamed and S. Srinivasan—Equal contribution.

H. Kashima et al. (Eds.): PAKDD 2023, LNAI 13937, pp. 73–85, 2023.
https://doi.org/10.1007/978-3-031-33380-4_6

and then sorted to increase the likelihood of satisfying the customer query in the top positions. Matching is therefore a critical first step towards a delightful customer experience in terms of search latency and relevance, and the focus of this paper. Lexical matching using an inverted index [1] has been the industry standard approach for e-commerce retrieval applications. This type of matching retrieves products that have one or more query keywords appear in their textual attributes such as title and description. Lexical matching is favorable because of its simplicity, explainability, low latency, and ability to scale to catalogs with billions of products. Despite the advantages, lexical matching has several shortcomings such as sensitivity to spelling variants (e.g. "grey" vs "gray") or mistakes (e.g. "sheos" instead of "shoes"), proneness to *vocabulary mismatch* (e.g. hypernyms, synonyms), and lack of semantic understanding (e.g. "latex free examination gloves" does not match the intent of "latex examination gloves"). These issues are largely caused by the underlying term-based distributional representation for query and product that fails to capture the fine-grained relationship between terms. Researchers and practitioners typically resort to *query expansion* techniques to address these issues.

Dense embedding based semantic matching [2] has been shown to significantly alleviate the shortcomings of lexical matching due to its distributed representation that admits granular proximity between the terms of a query-product pair in low dimensional vector space [3]. To fulfill the low latency requirement, these semantic matching models are predominantly shallow and use a bi-encoder architecture. Bi-encoders have separate encoders for generating query and product embeddings and use cosine similarity to define the proximity of queries and products. Such an architecture allows product embeddings to be indexed offline for fast approximate nearest neighbor (ANN) search [4] with the query embedding generated in realtime. Recently, BERT-based models [5] have advanced the state-of-the-art in natural language processing but due to latency considerations, their use in online e-commerce information retrieval is largely limited to the bi-encoder architecture [6–8] which does not benefit from the early interaction between the query and product representations.

In this work, we propose a multi-stage training procedure to train a small BERT-based matching model for online inference that leverages a large pretrained BERT-based matching model. A large BERT encoder (750M parameters) is first pretrained with the masked language modeling (MLM) objective on the product catalog data (details in Sect. 2.1), we refer to the trained model as DS-BERT. Next, the DS-BERT model is pre-finetuned using our novel query-product interaction pre-finetuning (QPI) task (see Sect. 2.2), the trained model is referred to as QPI-BERT. We find that interaction pre-finetuning greatly improves training stability of bi-encoders downstream as well as significantly improves generalization. QPI-BERT is then cloned into a bi-encoder model architecture and finetuned with query-product purchase signal, we refer to this model as QPI-BERT-FT (see Sect. 2.3). Finally, a smaller QPI-BERT bi-encoder student model (75M parameters) is distilled from the QPI-BERT-FT teacher by matching the cosine similarity score on the query-product pairs used for finetuning (see Sect. 2.4), we refer to this model as SMALL-QPI-BERT-DIS.

Through our offline experiments on a large e-commerce dataset, we show that the SMALL-QPI-BERT-DIS model (75M) suffers only a 3% drop in search relevance metric, compared to the QPI-BERT model with 20x its number of parameters (1.5B). This SMALL-QPI-BERT-DIS model improves search relevance by 23%, over a baseline DSSM-based matching model [2] with similar number of parameters and inference latency. Using an online A/B test we also show that the SMALL-QPI-BERT-DIS model outperforms the production model with 2% lift in both relevance and sales metrics.

Our work is closely related to the literature on semantic matching with deep learning. Some of the initial pre-BERT works include the Deep Semantic Similarity Model (DSSM) [2] , that constructs vector representations for queries and documents using a feedforward network and uses cosine similarity as the scoring function. DSSM-based models are widely used for real-time matching at web-scale [9,10]. This was later specialized for online product matching [3]. Post-BERT techniques leverage Pretrained Language Models (PLMs), such as BERT [5] to construct bi-encoders for matching tasks [8,11,12]. These techniques have broadly been applied to question-answering where the question and answer are from similar domains and interaction pre-finetuning is less essential. A recent work [13], proposes a multi-stage semantic matching training pipeline for web retrieval. However, unlike our approach, their focus is on deploying an ERNIE model (220M), while we study how large bi-encoders (1.5B) can be compressed to much smaller bi-encoders (70M) at web-scale using interaction pre-finetuning. In summary, the key contributions of this work are:

- We propose a multi-stage training procedure to effectively train a small BERT-based matching model for online inference from a much larger model (750 million to 1.5 billion parameters).
- We introduce a novel pre-finetuning task where a span masking and field permutation equivariant objective is used on joint query-product input text to help align the query and product representations. This task helps stabilize training and improve generalization of bi-encoders.
- We show using offline and online experiments at scale on an e-commerce website, that the proposed approach helps the small BERT SMALL-QPI-BERT-DIS model significantly outperform both a DSSM-based model (by 23%) in offline experiments and a production model in an online A/B test.

2 Methodology

In this section we describe our proposed four-stage training paradigm that consists of 1) domain-specific pretraining, 2) query-product interaction prefinetuning, 3) finetuning for retrieval, and 4) knowledge distillation to a smaller model. In the first three stages, we train a large BERT model for product matching and in the final stage we distill this knowledge to a smaller model that can be deployed efficiently in production (See Fig. 1).

2.1 Domain-Specific Pretraining

In the first stage of training, we pretrain a large BERT model on a domain specific dataset for product matching. The language used to describe products (catalog fields) in the e-commerce domain significantly differs from the language used on the larger web. Product titles and descriptions use a subset of the entire vocabulary, are often structured to follow a specific pattern, and in general have a different distribution from the sources that publicly available language models are trained on. Therefore, using an off-the-shelf pretrained BERT-based model does not perform well when finetuned for the product matching task.

Instead of using an off-the-shelf pretrained BERT model, we construct a BPE vocabulary [14] from the catalog corpus comprising of billions of products in e-commerce domain. We then pretrain the model on text from the catalog, and use all of the catalog text fields such as title and description of products available by concatenating them along with their field names. Our pretraining objective is the standard masked-language-modeling (MLM) loss [5,15,16]. We refer to the model trained with this strategy as the DS-BERT model (See Fig. 1a).

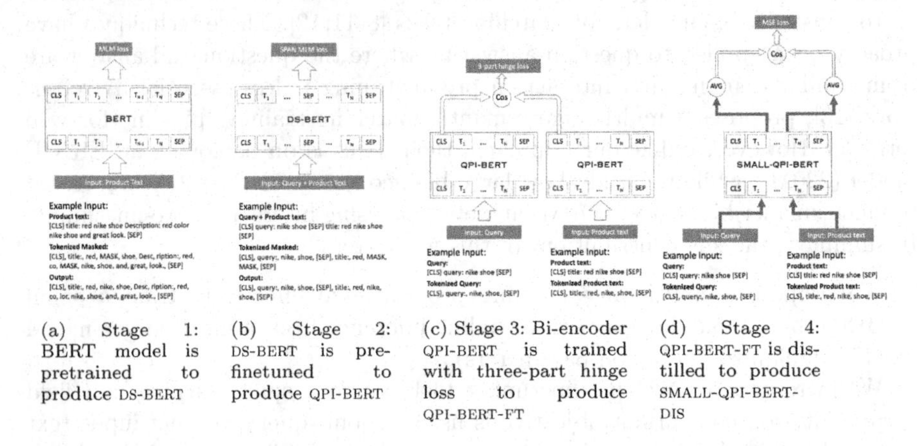

(a) Stage 1: BERT model is pretrained to produce DS-BERT

(b) Stage 2: DS-BERT is pre-finetuned to produce QPI-BERT

(c) Stage 3: Bi-encoder QPI-BERT is trained with three-part hinge loss to produce QPI-BERT-FT

(d) Stage 4: QPI-BERT-FT is distilled to produce SMALL-QPI-BERT-DIS

Fig. 1. This figure shows the four stages involved in training an effective deployable model for semantic matching.

2.2 Query-Product Interaction Pre-finetuning

Bi-encoders are preferred over cross-encoder models with full interaction for matching due to their efficiency and feasibility at runtime. Bi-encoders are however notoriously difficult to train on query-product pairs due to training instabilities arising from gradient variance between the two inputs. Losing the capability to explicitly model the interaction between queries and products also results in worse generalization than the cross-encoder.

In the second stage of training we propose a novel self-supervised approach to incorporate query-product interaction in the large encoder which is critical to improving the performance on the product matching task. We use query-product paired data to help the encoder learn the relationship between a query and a product using full cross-attention. To construct such a dataset, we first identify query-product pairs that share a relevant semantic relationship, for example, all products purchased for a given query can be considered relevant or query-product pairs can be manually labeled for relevance. In this paper, the dataset is constructed such that the query-product pairs are semantically relevant with a high probability $\alpha > 0.8$. The pre-finetuning dataset size (a few million examples) is much smaller than the pretraining dataset (a billion examples).

To perform pre-finetuning, we perform span MLM on the concatenated query and product text with a "[SEP]" token between them. At each iteration, we select spans from either the query text or product text (never both) to mask tokens. We sample span length (number of words) from a geometric distribution, till a predetermined percentage of tokens have been masked. The start of the span is uniformly sampled within the query or the product. During training we also observed that permuting the fields within the query and product, a form of field permutation equivariant training, also helped the model generalize better. We refer to the model trained with this strategy as QPI-BERT model (See Fig. 1b). Pre-finetuning with self-supervision on semantically relevant paired dataset boosts generalization for matching when a large noisy training set is available. This differs from previous works that use supervision on manually labeled data.

2.3 Finetuning for Matching

The third stage of training is to finetune the large teacher encoder QPI-BERT model in a bi-encoder setting for matching. We train a bi-encoder teacher as opposed to a cross-encoder teacher for retrieval as the extreme inefficiency in generating predictions for evaluation and slow training convergence rate makes it impractical to train cross-encoders for web-scale data and large models.

Let us denote the QPI-BERT model as M, query encoder as M_q, and product encoder as M_p, where the weights between query encoder and product encoder are shared. In our experiments sharing weights performed comparably to independently training them. For any query-product pair Q and P as inputs, we first generate the embedding Q_{emb} for query Q using M_q and embedding P_{emb} for product P using M_p using their "[CLS]" token representation. A cosine similarity score $s_{Q,P} = \cos(Q_{emb}, P_{emb})$ is used to compute relevance between them.

We train the bi-encoder using a three-part hinge loss. This loss requires the ground-truth data ($y_{Q,P}$) to be labeled with one of three possible values referred to as positive (1), hard negative (0) and random negative (−1). We use the purchased products for a given query as positive and any product uniformly sampled from the catalog as random negative. Identifying hard negatives is non-trivial [12,17], and in this work we choose a simple yet effective approach [3], where for a given query, all products that were shown to the user but did not receive any interaction is a hard negative. The loss takes the following form:

$$\text{loss}_{Q,P}(y_{Q,P}, s_{Q,P}) = \begin{cases} \max(\delta_{pos} - s_{Q,P}, 0), & \text{if } y_{Q,P} = 1. \\ \max(\delta_{hn}^{-} - s_{Q,P}, 0) & \text{if } y_{Q,P} = 0. \\ \quad + \max(s_{Q,P} - \delta_{hn}^{+}, 0), & \\ \max(s_{Q,P} - \delta_{rn}, 0), & \text{if } y_{Q,P} = -1. \end{cases} \tag{1}$$

where δ_{pos} and δ_{hn}^{-} are the lower thresholds for the positive and hard negative data scores respectively and δ_{hn}^{+} and δ_{rn} are the upper thresholds for the hard negative and random negative data scores respectively. We refer to the model trained with this strategy as the QPI-BERT-FT model (see Fig. 1c).

2.4 Distillation and Realtime Inference

The final stage of our framework is to distill the knowledge of teacher QPI-BERT-FT to a smaller student bi-encoder BERT model (75M to 150M parameters) that meets the online latency constraint. We first pretrain and prefinetune the small model similar to QPI-BERT to generate SMALL-QPI-BERT model M. Then we clone the encoder to create a query encoder $\tilde{M}_Q$ and a product encoder $\tilde{M}_P$. Unlike the large model case, for the small model we observe that sharing parameters between encoders helps improve performance significantly. The query embedding $\tilde{Q}_{emb}$ and product embedding $\tilde{P}_{emb}$ for the student model are computed by averaging all token embeddings in the query Q and product P respectively. The relevance score for a query-product pair is compute using cosine similarity i.e., $\tilde{s}_{Q,P} = cos(\tilde{Q}_{emb}, \tilde{P}_{emb})$. The model is trained by minimizing the distance between the scores generated by QPI-BERT-FT teacher and the model using the mean squared error (MSE) loss function.

$$\text{loss}_{Q,P}(s_{Q,P}, \tilde{s}_{Q,P}) = (s_{Q,P} - \tilde{s}_{Q,P})^2 \tag{2}$$

In practice we observed that simple score matching using MSE outperformed other approaches such as using L2 loss on the embeddings directly, Margin-MSE [18] with random negatives, or contrastive losses like SimCLR [19] with random negatives. We refer to the model distilled with this strategy as the SMALL-QPI-BERT-FT model (see Fig. 1d). At runtime, for every query entered by the customer, we compute the query embedding and then retrieve top K products using ANN search [4]. To serve traffic in realtime, we cache the product embeddings and compute only the query embedding online. The retrieved products are served directly to customers or mixed with other results and re-ranked before displaying to the customer.

3 Empirical Evaluation

3.1 Experimental Setup

Data. We use the following multilingual datasets for different stages of training:

Domain-Specific Pretraining Data: We use ∼1 billion product titles and descriptions from 14 different languages. This data is also used to construct a sentencepiece [20] tokenizer with 256K vocab size.

Interaction Pre-finetuning Data: We use ∼15M query-product pairs from 12 languages and use weak supervision in the form of rules to label them as relevant or irrelevant. ∼80% of the pairs are relevant query-product pairs.

Finetuning for Matching Data: We use ∼330M query-product pairs subsampled from a live e-commerce service to train the model for matching. We maintain a positive to hard negative to negative ratio of 1:10:11. The pairs are collected from multiple countries with at least 4 languages. We use a validation dataset to compute recall that contains 28K queries and 1M products from the subsampled catalog. Human evaluation (Sect. 3.1) uses a held-out set of 100 queries.

Models. We experiment with several model variants, both small and large summarized in Table 1. All large models we train are based on DS-BERT, which is a multilingual BERT model with 38 layers, 1024 output dimensions and 4096 hidden dimensions. When the parameters for the query and product encoder are not shared, the model has twice the parameters of the encoder. The small models we train are multilingual BERT models with 2 layers, 256 output dimensions, and 1024 hidden dimensions. In addition, we use DSSM and XLMROBERTA as baselines. • XLMROBERTA: Publicly available XLMRoberta [21] model which is finetuned for matching as described in Sect. 2.3. • DSSM: Bi-encoder model with a shared embedding layer (output dimension of 256) followed by batch norm and averaged token embedding to represent the query and product [3]. To ensure effective use of vocabulary for DSSM, we create a different sentencepiece model with 300k tokens using the matching training data.

Metrics. *R@100:* This is the average purchase recall computed on the validation data for the top 100 products retrieved.

Relevance Metrics: To understand the true improvement in the quality of matches retrieved by the model, we use Toloka (toloka.yandex.com) to label the results produced by our models. For every query we retrieve 100 results and ask the annotators to label them as exact match, substitute, or other. We report the average percentage of exact (E@100). substitute (S@100), and other (O@100). We use E@100 + S@100 (E+S) to measure semantic improvement in the model.

Table 1. Bi-encoder model variants. Differences are number of parameters (Params), embedding dimensionality (ED), embedding type (ET), domain-specific pretraining (DS PT), QPI prefinetuning (QPI PFT), whether encoders share parameters (Shared), whether model is distilled from QPI-BERT-FT (Dis).

Models	Params	ED	ET	DS PT	QPI PFT	Shared	Dist
Large Models							
XLMROBERTA	1.1B	1024	CLS	N	N	N	N
DS-BERT	1.5B	1024	CLS	Y	N	N	N
QPI-BERT-FT	1.5B	1024	CLS	Y	Y	N	N
QPI-BERT-FT*	1.5B	1024	CLS	Y	Y[a]	N	N
QPI-BERT-FT-SH	750M	1024	AVG	Y	Y	Y	N
Smaller Models							
SMALL-QPI-BERT-FT	150M	256	CLS	Y	Y	N	N
SMALL-QPI-BERT-FT-AVG	150M	256	AVG	Y	Y	N	N
SMALL-QPI-BERT-FT-SH	75M	256	CLS	Y	Y	Y	N
SMALL-QPI-BERT-FT-SH-AVG	75M	256	AVG	Y	Y	Y	N
SMALL-QPI-BERT-DIS	75M	256	AVG	Y	Y	Y	Y
DSSM	75M	256	AVG	N	N	Y	N

[a] Classification objective instead of span masking objective on pre-finetuning data.

Training. We use Deepspeed (deepspeed.ai) and PyTorch for training models on AWS P3DN instances. We used LANS optimizer [22] with learning rate between $1e^{-4}$ and $1e^{-6}$ based on the model and for all models we use a batch size of 8192. During pre-finetuning, we use validation MLM accuracy to perform early stopping and for finetuning we use validation recall for stopping. When using the three-part hinge-loss in Eq. 1, $\delta_{pos} = 0.9$, $\delta^{+}_{hn} = 0.55$ and $\delta_{rn} = \delta^{-}_{hn} = 0.2$.

3.2 Offline and Online Results

Does our Training Strategy Help Improve Semantic Matching Performance Offline? For large models, we compare QPI-BERT-FT with XLMROBERTA, DS-BERT, and QPI-BERT-FT*, and for small models, we compare SMALL-QPI-BERT-FT SMALL-QPI-BERT-FT-SH with DSSM (Table 2). a) QPI-BERT outperforms other approaches both in R@100 and E+S. Among large models, the performance of DS-BERT is better than XLMROBERTA and QPI-BERT-FT* is better than DS-BERT. This clearly indicates progressive improvement with the different stages in our approach. b) We observe is that DSSM outperforms XLMROBERTA in all metrics indicating a vocabulary and domain mismatch between the catalog data and web data. Domain-specific pretraining is essential to performance when training the large models. c) We see that QPI-BERT-FT significantly outperforms QPI-BERT-FT* in all metrics, validating the importance of interaction pre-finetuning over mere supervision alone for matching. d) For small models, we observe that the performance of SMALL-QPI-BERT-FT is very similar to DSSM, with SMALL-QPI-BERT-FT showing ~45% relative lift in S@100 but, ~8% relative drop in E@100, ~4% relative lift in E+S, and ~1% relative drop in R@100. When sharing parameters between the query and product encoder, and averaging embeddings, SMALL-QPI-BERT-FT-SH-AVG outperforms DSSM by ~38% relative lift in S@100, ~2% relative lift in E@100, ~10% relative lift in E+S, and ~2% relative lift in R@100. The results indicate that our strategy helps improve the performance overall and the improvements are higher for larger models (~23% relative lift in E+S over DSSM). This reinforces our proposed approach: train a large model and distill the knowledge to a smaller model, instead of directly training a smaller model.

Can Distillation Preserve Large Model Performance? Given the large improvement in matching metrics for large models, we would ideally like to retain this improvement in smaller models using distillation. We compare SMALL-QPI-BERT-DIS with QPI-BERT-FT (Table 2) and observe a ~3% relative drop in E+S and R@100. This shows that while there is small gap, it is possible to transfer most of the information from a 1.5B parameter large QPI-BERT-FT model to a 20x smaller SMALL-QPI-BERT-DIS model (75M parameter) using our approach.

Does Sharing Parameters in the Bi-encoder have an Impact on Retrieval Task Performance? To understand the effect of sharing parameters between query and product encoders in the bi-encoder setting, we compare QPI-BERT-FT-SH with QPI-BERT-FT among the large models and SMALL-QPI-BERT-FT-SHwith SMALL-QPI-BERT-FT, and SMALL-QPI-BERT-FT-SH-AVG, with SMALL-QPI-BERT-FT-AVGamong the small models (Table 2). We observe that sharing encoders has almost no impact on the performance of large models and the

Table 2. Offline metrics of models on a multi-lingual e-commerce dataset

Models	R@100	E@100	S@100	O@100	E+S
Large Models					
XLMROBERTA	68.43	29.52	17.46	53.02	46.98
DS-BERT	73.98	43.17	17.55	39.28	60.72
QPI-BERT-FT	82.2	50.36	**20.5**	29.14	70.86
QPI-BERT-FT*	75.6	48.35	19.66	31.99	68.01
QPI-BERT-FT-SH	**83.35**	**51.08**	19.81	**29.11**	**70.89**
Smaller Models					
SMALL-QPI-BERT-FT	77.06	40.28	18.16	41.56	58.44
SMALL-QPI-BERT-FT-AVG	50.84	33.63	13	53.37	46.63
SMALL-QPI-BERT-FT-SH	79.3	43.98	18.52	37.5	62.5
SMALL-QPI-BERT-FT-SH-AVG	80.17	44.45	17.33	38.22	61.78
SMALL-QPI-BERT-DIS	**80**	**48.04**	**20.78**	**31.18**	**68.82**
DSSM	78.1	43.56	12.49	43.95	56.05

maximum relative drop in E+S and recall is ~1% with QPI-BERT-FT-SH winning marginally. However, in the smaller models we observe that sharing parameters gives a large boost in performance with a relative lift of upto ~32% in the E+S metric and ~60% in R@100. When the model size is large enough, it is capable of learning independent encoders for both inputs. But, when the model is small, the model benefits from sharing parameters.

How Does our Approach Improve over a non-BERT-Based Model? To visualize the difference in matching quality between our BERT-based model and DSSM, we look at results for two queries, with DSSM retrieving more relevant products on one query and vice-versa on the other (Fig. 2). We observe that for query *"sailor ink"* QPI-BERT-FT performs better as all results are relevant products. For this query, DSSM behaves like a lexical matcher and fetches results for both *"sailor"* and *"ink"*. For query "omron sale bp monitor machine", DSSM retrieves all relevant matches. QPI-BERT-FT however, retrieves an irrelevant product (a fitness watch). While irrelevant, it still falls into the product type of "personal health" implying an error in semantic generalization. The significantly higher increase in S@100 compared to E@100 indicates that QPI-BERT-FT is a better semantic model as the representations must incorporate high-level concepts to match substitutes, that token-level exact matches cannot achieve.

What is the Latency Improvement of the Smaller BERT Model Compared to the Large Model? We have seen earlier that the large model can be effectively compressed to a 20x smaller model that incurs much lower inference latency. We compare the inference latencies of our models while generating query embeddings which is representative of realtime latency as the product embeddings are generated offline and indexed for ANN. We ignore the ANN latency as modern ANN search can be computed effectively in realtime (~1 ms)

(a) Query: *"sailor ink"*; Method: DSSM.

(b) Query: *"sailor ink"*; Method: QPI-BERT-FT.

(c) Query: *"omron sale bp monitor machine"*; Method: DSSM.

(d) Query: *"omron sale bp monitor machine"*; Method: QPI-BERT-FT.

Fig. 2. Top 6 results obtained by DSSM and QPI-BERT-FT for queries *"sailor ink"* and *"omron sale bp monitor machine"*

[4]. Figure 3 shows the time it takes to compute query embedding (inference) for different query lengths (query length computed as number of tokens after tokenization) on an r5.4xlarge AWS instance. As expected, DSSM has the lowest inference time and QPI-BERT has the largest. Both SMALL-QPI-BERT and DSSM have embedding generation time of under 1ms upto 32 tokens making it feasible to serve realtime traffic. SMALL-QPI-BERT reduces the latency time by ~60× compared to QPI-BERT with a relevance metric performance drop of only ~3%.

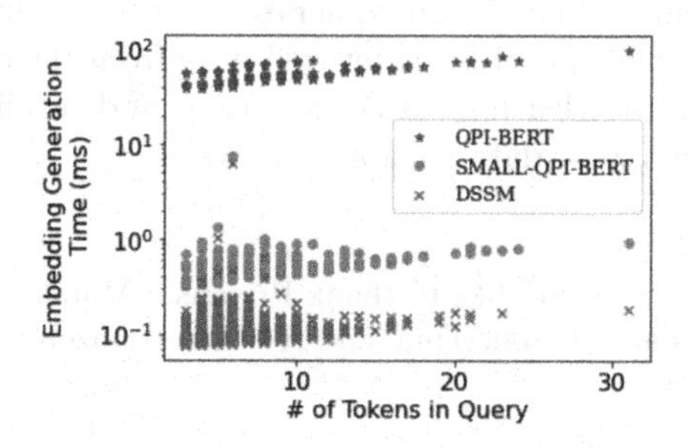

Fig. 3. Inference time for QPI-BERT, SMALL-QPI-BERT, and DSSM on r5.4xlarge

Table 3. A/B test results for SMALL-QPI-BERT-DIS rel. to production system.

PS	Units	E@16	S@16	E+S@16	SR	Latency P99
+2.07%	+1.47%	−1.19%	+3.37%	+2.18%	−16.9%	+4 ms

How Well Does the Approach Perform Online? To measure the impact of our approach online, we experiment with SMALL-QPI-BERT-DIS in a multilingual large e-commerce service. The service augments matching results from several sources like lexical matchers, semantic matchers, upstream machine learning models, and advertised products. We replace only the production semantic matcher with our SMALL-QPI-BERT-DIS and perform an A/B test. We measure both customer engagement metrics and relevance quality metrics. For customer engagement metrics, we look at the change in number of units purchased and the amount of product sales (PS). For quality metric, we look at the change in user evaluated E@16, S@16, E+S@16 and sparse results (SR) which is the percentage of queries with less than 16 products retrieved. We observe (Table 3) that our approach significantly improves over the production semantic matcher and lead to a significant drop in SR. The reduction in E@16 and increase in S@16 suggests that our approach is learning latent semantic meaning to increase substitutes displayed to customers. We also observe that our model does not have a significant impact on latency (~4 ms) and can be used at runtime.

4 Conclusion

In this work we develop a four-stage training paradigm to train an effective BERT model that can be deployed online to improve product matching. We introduce a new pre-finetuning task that incorporates the interaction between queries and products prior to training for retrieval which we show is critical to improving performance. Using a simple yet effective approach, we distill a large model to a smaller model and show through offline and online experiments that our approach can significantly improve customer experience. As future work, it would be interesting to incorporate other structured data from the e-commerce service to enhance representation learning, such as brand and product dimensions, as well as customer interaction data such as reviews.

Acknowledgement. We would like to thank Priyanka, Mutasem, Huajun, Jaspreet, Dhivya, Giovanni, Hemant, Anton, Tina, and RJ from Amazon Search.

References

1. Schütze, H., Manning, C.D., Raghavan, P.: Introduction to Information Retrieval. Cambridge University Press, Cambridge (2008)
2. Huang, P.-S., He, X., Gao, J., Deng, L., Acero, A., Heck, L.P.: Learning deep structured semantic models for web search using clickthrough data. In: CIKM (2013)
3. Nigam, P., et al.: Semantic product search. In: SIGKDD (2019)
4. Malkov Y.A., Yashunin, D.A.: Efficient and robust approximate nearest neighbor search using hierarchical navigable small world graphs. IEEE Trans. Pattern Anal. Mach. Intell. **42**(4), 824–836 (2018)
5. Devlin, J., Chang, M.W., Lee, K., Toutanova, K.: BERT: pre-training of deep bidirectional transformers for language understanding. In: NACCL (2019)
6. Reimers, N., Gurevych, I.: Sentence-BERT: sentence embeddings using Siamese BERT-networks. In: EMNLP-IJCNLP (2019)
7. Khattab, O., Zaharia, M.: Colbert: efficient and effective passage search via contextualized late interaction over BERT. In: SIGIR (2020)
8. Lu, W., Jiao, J., Zhang, R.: TwinBERT: distilling knowledge to twin-structured BERT models for efficient retrieval. ArXiv, vol. abs/2002.06275 (2020)
9. Huang, J.-T., et al.: Embedding-based retrieval in Facebook search. In: SIGKDD (2020)
10. Li, S., et al.: Embedding-based product retrieval in Taobao search. In: SIGKDD (2021)
11. Karpukhin, V., et al.: Dense passage retrieval for open-domain question answering. In: EMNLP (2020)
12. Xiong, L., et al.: Approximate nearest neighbor negative contrastive learning for dense text retrieval. In: ICLR (2021)
13. Liu, Y., et al.: Pre-trained language model for web-scale retrieval in baidu search. In: SIGKDD (2021)

14. Sennrich, R., Haddow, B., Birch, A.: Neural machine translation of rare words with subword units. In :Proceedings of the 54th Annual Meeting of the Association for Computational Linguistics (Volume 1: Long Papers), Berlin, Germany, pp. 1715–1725, Association for Computational Linguistics, Aug. 2016
15. Liu, Y., et al.: RoBERTa: a robustly optimized BERT pretraining approach, ArXiv, vol. abs/1907.11692 (2019)
16. Lan, Z., Chen, M., Goodman, S., Gimpel, K., Sharma, P., Soricut, R.: AlBERT: a lite BERT for self-supervised learning of language representations. In: ICLR (2020)
17. Ji, S., Vishwanathan, S.V.N., Satish, N., Anderson, M.J., Dubey, P.: BlackOut: speeding up recurrent neural network language models with very large vocabularies. In: ICLR (2016)
18. Hofstätter, S., Althammer, S., Schröder, M., Sertkan, M., Hanbury, A.: Improving efficient neural ranking models with cross-architecture knowledge distillation. ArXiv, vol. abs/2010.02666 (2020)
19. Chen, T., Kornblith, S., Norouzi, M., Hinton, G.: A simple framework for contrastive learning of visual representations. In: ICML (2020)
20. Kudo, T., Richardson, J.: SentencePiece: a simple and language independent subword tokenizer and detokenizer for neural text processing. In: EMNLP (2018)
21. Conneau, A.: Unsupervised cross-lingual representation learning at scale. In: ACL (2020)
22. Zheng, S., Lin, H., Zha, S., Li, M.: Accelerated large batch optimization of BERT pretraining in 54 minutes. ArXiv, vol. abs/2006.13484 (2020)

Multi-task Learning Based Keywords Weighted Siamese Model for Semantic Retrieval

Mengmeng Kuang[1], Zhenhong Chen[1], Weiyan Wang[1,2](✉), Lie Kang[1], Qiang Yan[1], Min Tang[1], and Penghui Hao[1]

[1] Tencent Holdings Ltd., Shenzhen, China
{mengkuang,hollischen,liekang,rolanyan,kevinmtang,terryhao}@tencent.com
[2] Hong Kong University of Science and Technology, Hong Kong, China
wwangbc@cse.ust.hk

Abstract. Embedding-based retrieval has drawn massive attention in online search engines because of its semantic solid feature expression ability. Deep Siamese models leverage the powerful dense embeddings from strong language models like BERT to better represent sentences (queries and documents). However, deep Siamese models can suffer from a sub-optimal relevance prediction since they can hardly identify keywords due to late interaction between the query and document. Although some studies tried to adjust weights in semantic vectors by inserting some global pre-computed prior knowledge, like TF-IDF or BM25 scores, they neglected the influence of contextual information on keywords in sentences. To retrieve better-matched documents, it is necessary to identify the keywords in queries and documents accurately. To achieve this goal, we introduce a keyword identification model to detect the keywords from queries and documents automatically. Furthermore, we propose a novel multi-task framework that jointly trains both the deep Siamese model and the keywords identification model to help improve each other's performance. We also conduct comprehensive experiments on both online A/B tests and two famous offline benchmarks to demonstrate the significant advantages of our method over other competitive baselines.

Keywords: Text matching · multi-task learning · Siamese model · semantic retrieval · keywords identification

1 Introduction

In the era of information explosion, it is more and more critical to quickly and accurately find query-related information from a large number of documents. Representation learning based retrieval has impressively improved the retrieval accuracy and reformed this critical field researched for decades [25]. Based on the deep matching models [4] and the state-of-the-art pre-trained frameworks, semantic retrieval has thrived as a typical application of representation learning.

H. Kashima et al. (Eds.): PAKDD 2023, LNAI 13937, pp. 86–98, 2023.
https://doi.org/10.1007/978-3-031-33380-4_7

An extensive collection of works, especially the deep Siamese models [4,6], have been proposed to tackle the semantic retrieval task [11]. DSSM [8], CLSM [23], ARC-I [5] explore adopting the traditional neural networks, while Sentence-BERT [18], ColBERT [10], TwinBERT [13] take a further step to employ the pre-trained language models like BERT [2]. All these works highlight the charm of deep Siamese structures. Especially, pre-trained BERT can effectively capture the contextual semantic meanings in the query or document with the self-attention [9], which significantly enhances the accuracy of the Siamese semantic retrieval model.

However, it is hard for deep Siamese models to directly infer the keywords in the query since there is no interaction between the query and the document. Keywords have been proven to play a unique and important role in information retrieval applications [19]. To realize the pre-computation for massive documents, the deep Siamese structure has the independent query encoder and document encoder, which have no interaction until the last layer computing the similarity. But unfortunately, the query encoder itself can have difficulties in adequately weighting different words in the query without any context information about documents. And the document encoder has a similar problem without any query information. Therefore, semantic representations without keyword identification will directly impact semantic similarity computing and thus affect the overall matching process.

In many previous studies, the global statics of context information is used to improve the query representations by introducing some pre-computed prior knowledge, like BM25 [19] or TF-IDF [20] scores. However, such statics can not take the contextual semantics into consideration to reflect the word weights precisely. A word is significant in one sentence, but may be not in another. Apparently, if we use the pre-computed statics as the prior knowledge, it can lead to a sub-optimal and even poor decision.

To remedy the limitations, we introduce a multi-task learning based keywords weighted Siamese model (MKSM) for semantic retrieval in this work. We propose a novel keywords identification model joined with the Siamese retrieval model to explicitly model the weights of the adaptable keywords and get better representations for the retrieval. Specifically, we model the keyword identification as a regression learning problem to consider contextual semantics instead of rule-based statistics. Furthermore, The keyword identification model shares the same neural network model with the Siamese model but has different training loss functions. Therefore, we train both the keyword identification model and the deep Siamese model jointly in the style of multi-task learning to improve each other's performance. The multi-task learning enables our solution to learn better keyword weights from retrieval signals and the regression target. Therefore, we can get a better representation containing the semantic meaning of keyword weights to conduct the matching process better.

To verify its effectiveness, we evaluate our proposed MKSM in the online production environment and on famous and public benchmark datasets. Specifically, MKSM has been deployed for the online service search scenario of a popular social application frequently used by over 100 million users. The online A/B test in real production shows that MKSM concretely improves the user experiences for the service search in terms of click-through rate (CTR) and retrieval rate (RR) for real production. Furthermore, the empirical results on public searching benchmarks have also demonstrated considerable improvements over baselines.

To summarize, our contributions are three-fold:

- We introduce an adaptable keywords identification model to learn better representations for queries and documents.
- We propose a novel semantic retrieval framework MKSM which joins the keywords identification method to a Siamese model for semantic retrieval in the form of multi-task learning.
- Extensive experiments on online A/B tests and two offline public benchmarks verify the effectiveness of our proposed model.

2 Related Works

A variety of deep matching models have been proposed for the information retrieval problems [15]. Siamese models applied to semantic retrieval started from the Deep Structured Semantic Model (DSSM) [8], which mapped both query and document to the same semantic space, and achieved the purpose of retrieval by maximizing the cosine similarity. Then, ARC-I [5] and CLSM [23] used Convolutional neural networks (CNNs) and max pooling to replace the fully connected networks to extract features, which could capture more contextual information for semantic vector representation. Further, LSTM-DSSM [17] proposed to use Long Short-term Memory (LSTM) networks to replace CNNs to obtain contextual information over longer sequences accurately. Sentence-BERT [18], a modification of the pre-trained BERT network that used Siamese and triplet network structures to derive semantically meaningful sentence embeddings that could be compared using cosine-similarity, which reduced the effort of finding the most similar pair from 65 h with BERT/RoBERTa to about 5 s, while maintaining the accuracy from BERT. Recently, TwinBERT [13] used twin-structured BERT-like encoders to encode the query and document, respectively, and a crossing layer to combine the two embeddings to produce a similarity score. Additionally, ColBERT [10] introduced a late interaction architecture that independently encoded the query and the document using BERT and then employed a cheap yet powerful interaction step that modeled their fine-grained similarity. Furthermore, [14] proposed a simple neural model that combined the efficiency of dual encoders with some of the expressiveness of more costly attentional architectures and explored sparse dense hybrids to capitalize on the precision of sparse retrieval. Accurately representing the text and its contextual information has always been a hot research direction, which is our concern in MKSM. Neither the models mentioned above nor the [7] (Facebook), MOBIUS [3] (Baidu), etc. that have been applied in the actual business has learned global contextual word weights.

3 Methods

In this section, we first provide the problem definition of the semantic retrieval task. Then we propose a comprehensive overview of our framework and further introduce the implementations of each component in the MKSM framework. Finally, we describe the multi-task training process of our structure.

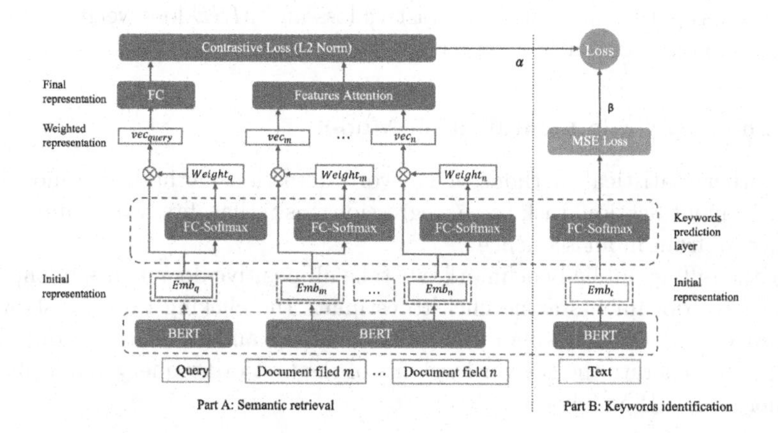

Fig. 1. The hierarchical framework of MKSM. *The whole framework can be divided into two parts. the left part, Part A, illustrates the semantic retrieval process, while the right part, part B, represents the keywords identification task.*

3.1 Problem Definition

The semantic retrieval task can be described as a matching problem M that gives a matching score for each query q and document d pair. Here, we use a single symbol d to stand the entire document which usually contains not only one field (e.g., name, description, etc.). Before calculating the matching score m_s, every string needs to be embedded as a semantic vector by some embedding methods E, like the BERT language model. In our framework, in addition to simply embedding the query and document, the keywords identification model can be regarded as an independent function K. Hence, the keywords weighted Siamese model for semantic retrieval can be represented as Eq. (1).

$$m_s = match(q, d) = M(K(E_q(q)), K(E_d(d))). \tag{1}$$

3.2 Framework Overview

From a horizontal view, MKSM comprises three parts, BERT semantic representation, Keyword weight correction, and Matching score calculation, as shown in Fig. 1. The framework, divided into semantic retrieval and keywords identification parts, starts with the query, document, and text represented by a shared BERT-pertained language model to get the corresponding embedding (Emb). A shared fully connected (FC) layer and softmax are appended to learn the word weights ($Weight$) in the keywords identification module. Moreover, the $Weight$ is utilized for weighing the embeddings of the query and various fields in the document to get better feature vectors (vec). Then, the contrastive loss (with L_2 normalization) is performed to measure the relevance of the query and document representations in the semantic retrieval part. The mean squared error (MSE) loss is applied as the objective of the keywords identification module. The total loss is the sum of the contrastive loss and MSE loss weighted by two hyperparameters, α, and β, respectively.

3.3 The Keywords Identification Model

Unlike other statistical methods or keyword detection methods, we model the keywords identification task as a regression task that fits word importance sequence s_k from input sequence s.

For the offline public benchmark, we take all positive documents as one click and negative documents as no click to estimate the clicking rate. As shown in Fig. 2, we consider the clicking rates a_i of documents and all related second-order queries s' to generate the keyword weights of the first-order query s. Specifically, we generate labels as follows:

1. We collect a large amount of high-relevance retrieval logs containing the query s and remove all stop words in the logs.
2. For the first-order query s, we dig out all clicked documents d and their corresponding click rates a.
3. For any clicked document d_j, we find all second-order queries that retrieve it. And we think the queries retrieving the same document have similar semantic meanings.
4. Then we count the word frequency $f_{w_i}^{d_j}$ in the second-order queries of the document d_j and then normalized all word frequencies as $f_{w_i}^{d_j} = f_{w_i}^{d_j} / \sum_i f_{w_i}^{d_j}$.
5. Finally, we combine all normalized word frequencies of all documents by weight averaging to generate the keyword weight as $f_{w_i} = \sum_j a_j f_{w_i}^{d_j} / \sum_i f_{w_i}^{d_j}$

Similarly, we generate the document keyword label by assuming that the clicked documents for the same query have similar semantic meanings. Specifically, we dig out all queries that retrieve the document and all other documents related to those queries. We count word frequencies in all related documents and then normalize the frequency according to click rates and the sum of frequencies.

As presented in the right part of Fig. 1, the keywords identification model includes three components, (1) representation component, (2) weight layer (L_k), and (3) loss calculation. We use the embedding of the "[CLS]" token in the BERT sentence embeddings as the initial representation. Then, a fully connected layer is supplemented with a hidden size equal to the padding size used to learn the word weights. Finally, the MSE loss function is computed for optimizing the parameters in the weight layer by Eq. (2).

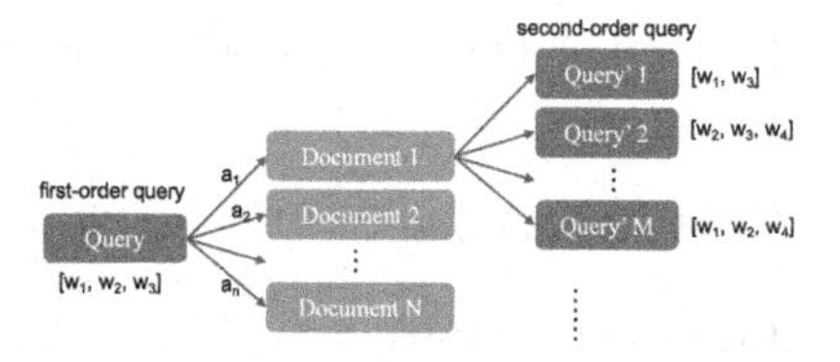

Fig. 2. Keyword weight label mining from second-order queries in history logs

$$loss_{keywords} = \frac{1}{P}\sum_{i=1}^{P}\left(l_i - L_k(BERT(s)_{[CLS]})_i\right)^2, \tag{2}$$

where l stands for the observed values of s.

3.4 The Siamese Retrieval Model

Like most deep matching models [3,7,13,18,24], the retrieval model also employs a twin-structured Siamese framework as shown in the left part of Fig. 1. The structure is a two-part design formed by the representation part and the matching part. In the representation part, there are three layers of representation, (1) initial representation, (2) weighted representation (L_k'), and (3) final representation (L_f). The initial representation is the average BERT embeddings of all the tokens. The weighted representation (L_k') is the initial representation associated with the weight layer (L_k) in the keywords identification model. The final representation is used as input of the matching part to calculate the matching scores minimized by a loss function, introduced in the rest of this section. We optimize the model to acquire a better matching score by minimizing the contrastive loss ($loss_{matching}$) as presented in Eq. (3).

$$loss_{matching} = \frac{1}{2N}\sum_{i=1}^{N} y(\mathrm{D}(r_q, r_d)_i)^2 + (1-y)\max(m - \mathrm{D}(r_q, r_d)_i, 0)^2, \tag{3}$$

where y is the relevance label with equals to 1 (relevant) or 0 (irrelevant), D represents the Euclidean distance, which can be expressed by Eq. (4), and m is a margin threshold.

$$\mathrm{D}(r_q, r_d) = \|r_q - r_d\|_2 = \left(\sum_{i=1}^{P}\left(r_q^i - r_d^i\right)^2\right)^{\frac{1}{2}} \tag{4}$$

3.5 The Multi-task Learning Strategy

As stated in Sects. 1, 3.3 and 3.4, to make the keywords identification model can learn adaptive keywords weights, we propose to train the keywords identification model and the Siamese retrieval model together. The combined loss function can be represented as Eq. (5).

$$loss = \alpha \times loss_{matching} + \beta \times loss_{keywords}. \tag{5}$$

where α and β are two hyper-parameters.

The training repeats the following back-propagation processes until the Siamese model can learn the representations of queries and documents well.

1. Firstly, back-propagating on the keyword identification model.
2. Secondly, back-propagating on the Siamese retrieval model with fixed parameters of shared weight layer.

4 Experiments

In this section, we first introduce the details of the experiment settings. Then we discuss the experiment results, including offline performance, online evaluation, and ablation study. Case study results and discussion on our work are in the supplementary materials.

Table 1. Dataset Statistics

Datasets	MS MARCO	Private
Training set	367,000 queries 3,200,000 documents	260,000 queries 1,300,000 documents
Validation set	519,300 pairs	80,000 pairs
Fields	title body	account name service name service description
Average length	1137	96

4.1 Experiments Setup

Evaluation Datasets. We validate the performance of MKSM on two datasets, where one is a public benchmark and the other is private. **MS MARCO** [16] is a famous benchmark from Microsoft, which is sampled from Bing's search query logs. To construct our **Private** dataset, we extract the daily service search logs from a popular instant messaging application and manually label the relevance. And we implement a label noise detection method based on confident learning [12] to purify this dataset. **MS MARCO** is an English dataset while **Private** is a Chinese benchmark. Table 1 summarizes the detailed statistics of such two datasets from three aspects.

Evaluation Metrics. Because our approach focuses on the matching stage in semantic retrieval, we choose Normalized Cumulative Gain (NCG) [22] as the evaluation metric. It is the best empirical metric for query-document matching, because it reflects the number of relevant documents returned without casing the specific ranking. NCG is computed as

$$\mathrm{NCG} = \frac{\mathrm{CG}}{i\mathrm{CG}}, \tag{6}$$

where CG(Cumulative Gain) is the sum of all the relevance scores in the recall set, and iCG is the ideal CG, which is the sum of relevance scores of the ideal document recall set. Specifically, CG is defined as

$$\mathrm{CG} = \sum_{i=1}^{T} relevance\_score_i, \tag{7}$$

Baselines. We compare our proposed MKSM framework with 6 representative retrieval baseline models[1]. Such methods can be categorized into different classes as follows (the detailed discussion of these methods is presented in Sect. 2),

- *Classical retrieval methods*: **TF-IDF** [20] and **BM25** [19].
- *Deep Siamese models*: **CLSM** [23] and **USE** [1].
- *Pre-trained language model*: **BERT** [2].
- *Keywords weighted model*: **BERT+TF-IDF** [21].

Implementation Details All the implementations mentioned in this paper are based on TensorFlow. We train MKSM with 4 NVIDIA Tesla V100 GPUs paralleled. The semantic vector representations for queries and documents are based on BERT pre-trained language model with padding size 128 or 1024 in the two datasets. The α and β are set as 0.6 and 0.4, respectively. AdamW, as an improved Adam optimizer, is used in the training processes in the MKSM framework. To make the inference phase more efficient, the embedding of documents is offline performed ahead. The cosine similarity of the query and document representations is utilized as the matching score.

4.2 Overall Performance

Table 2 illustrates the comparison results of NCG@T, where underlined numbers are the best results of baselines and bold numbers are the best results of all models. The difference between MKSM and $\mathrm{MKSM_{SEP}}$ is whether training the keywords identification model and the Siamese retrieval model separately.

From the results of Table 2, we can conclude the following observations,

[1] We don't compare with other baselines listed in the Sect. 2 since they are not open-sourced or fine-tuned for different retrieval scenarios.

– Our proposed MKSM framework obtains the best performance over other retrieval models in both benchmarks. Specifically, in MS MARCO and Private datasets, MKSM receives at least 0.9% and 0.7% promotion in terms of NCG, respectively. These results indicate the superiority of our proposed MKSM framework over baselines in both English and Chinese benchmarks, with different lengths of documents.

Table 2. Overall performance on MS MARCO and Private datasets

Models	MS MARCO				Private			
	NCG@10	NCG@20	NCG@50	NCG@100	NCG@5	NCG@10	NCG@20	NCG@30
TF-IDF	0.4154	0.5178	0.6258	0.7158	0.8398	0.8752	0.9190	0.9455
BM25	0.4360	0.5465	0.6736	0.7564	0.8332	0.8674	0.9158	0.9431
CLSM	0.4016	0.5245	0.6541	0.7155	0.7446	0.8146	0.8930	0.9330
USE	0.3746	0.4045	0.6045	0.6620	0.8376	0.8784	0.9253	<u>0.9521</u>
BERT	<u>0.4574</u>	0.5745	0.6920	0.7841	0.8332	0.8763	0.9186	0.9487
BERT+TF-IDF	0.4562	<u>0.5771</u>	<u>0.6938</u>	<u>0.7864</u>	<u>0.8432</u>	<u>0.8773</u>	<u>0.9268</u>	0.9507
MKSM_{SEP}	0.4619	0.5809	0.6992	0.7896	0.8452	0.8841	0.9302	0.9582
MKSM	**0.4630**	**0.5868**	**0.7041**	**0.7934**	**0.8521**	**0.8904**	**0.9336**	**0.9621**
Impr	1.2%	1.7%	1.5%	0.9%	1.1%	1.5%	0.7%	1.1%

"Impr." presents the improvement of MKSM over the best baseline.

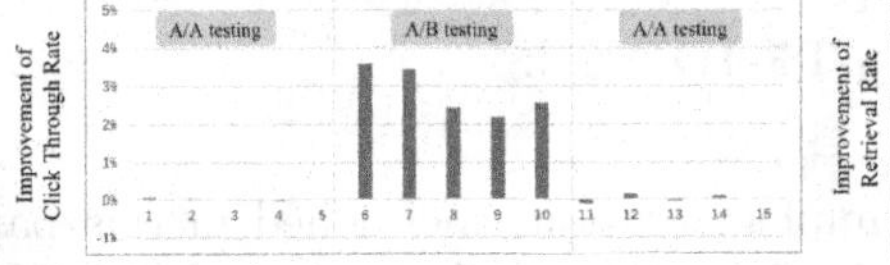

(a) Online experimental results of click-through rate.

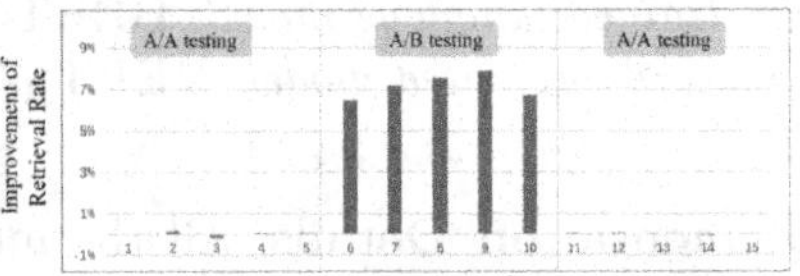

(b) Online experimental results of retrieval rate.

Fig. 3. Online evaluations.

– BERT+TF-IDF performs better than BERT in most empirical metrics, which indicates that the leverage of prior knowledge in query representations significantly improves the retrieval performance. Besides, our proposed MKSM and MKSM_{SEP} both perform better than BERT+TF-IDF. It demonstrates that keyword identification performs better than the traditional statistical information TF-IDF as the prior knowledge, no matter whether in separating training or multi-task training. It is because that our proposed keywords identification model can provide the keywords weight information, which is essential in the retrieval task.
– MKSM performs better than MKSM_{SEP} in terms of all metrics, which indicates that the training strategy of MKSM can influence the performance, and multi-task learning can introduce the prior knowledge to the Siamese model effectively.

4.3 Online Evaluation

We conduct A/B testing in the service retrieval scenario, comparing the proposed model MKSM with the current baseline, a distilled BERT.

The whole online experiment lasts 15 days. We monitor the results of A/A testing for the first five days, conduct A/B testing for the following five days, and conduct A/A testing again in the last five days. 15% of the users are randomly selected as the experimental group, and another 15% of the users are in the control group. During A/A testing, all the users are served by the BERT. During A/B testing, users in the control group are presented with retrieval results by the BERT, while users in the experimental group are presented with the MKSM semantic retrieval results. Note that the click experiment of MKSM shares the same exposure with the distilled BERT to verify whether the improvement is caused by the new semantic retrieval design.

Figures 3(a) and 3(b) show the improvement of the experimental group over the control group with respect to click-through rate (CTR) and retrieval rate (RR), which are defined as Eq. (8). We can see that the system is relatively stable in terms of CTR and RR during the A/A testing. As for the A/B testing, which starts from day 6, a significant improvement over the baseline BERT can be clearly observed. Specifically, the improvements concerning CTR and RR received by MKSM are at least 2% and 6%, respectively. In the final five days, we conduct A/A testing again, which replaces the MKSM framework with the distilled BERT. The improvement obtained by MKSM decays rapidly, which further proves the effectiveness of A/B testing.

$$\mathrm{CTR} = \frac{\#click}{\#exposure}, \mathrm{RR} = \frac{\Delta exposure}{\#exposure}, \tag{8}$$

where # means the number of *click* and *exposure*, and *Δexposure* means the increment of good results in *exposure*.

4.4 Ablation Study

In this subsection, to study the effectiveness of each component and certify that MKSM is the best combination, we conduct several models which are different from MKSM in terms of each component, such as $\mathrm{MKSM}_{\mathrm{SEP}}$, BERT, and $\mathrm{MKSM}_{[\mathrm{CLS}]}$. Specifically, $\mathrm{MKSM}_{\mathrm{SEP}}$ trains the keywords identification model and the Siamese model separately. BERT is the pure pre-trained language model without any prior knowledge. $\mathrm{MKSM}_{[\mathrm{CLS}]}$ uses the embedding of the [CLS] token in BERT as the initial representation of queries and documents. The performance comparison in **Private** benchmark is presented in Table 3.

From the results, we can confirm that:

- Compared with $\mathrm{MKSM}_{\mathrm{SEP}}$, MKSM shows a better performance, which indicates the multi-task learning manner can make the Siamese model and the keywords identification model interact more effectively.

– To demonstrate the superiority of the keywords identification model, we compare MKSM with BERT, a pure pre-trained language model for retrieval tasks without prior knowledge. The results reflect the effectiveness of our proposed keywords identification model. Besides, in contrast to fixed statistical information(referred to as "BERT+TF-IDF" in Table 2), our proposed MKSM with keywords identification model shows a better retrieval performance, which further indicates the superiority of our proposed MKSM framework.
– MKSM achieves better performance than $\text{MKSM}_{[\text{CLS}]}$. It means that using the average of all token embeddings as the initial representation of queries and documents in MKSM is slightly better than using only [CLS] token embedding. The reason is, compared with [CLS] token embedding, the average embedding of the BERT Encoder can provide much more useful information for keyword identification and retrieval.

Table 3. Ablation Study of MKSM

Methods	NCG@5	NCG@10	NCG@20	NCG@30
MKSM_{SEP}	0.8452	0.8841	0.9302	0.9582
BERT	0.8332	0.8763	0.9186	0.9487
$\text{MKSM}_{[\text{CLS}]}$	0.8447	0.8803	0.9277	0.9547
MKSM	0.8521	0.8904	0.9336	0.9621

5 Conclusion

In this paper, we propose a novel semantic retrieval model MKSM, which utilizes a keywords identification model and multi-task learning strategy to introduce practical prior knowledge in a Siamese model. MKSM can automatically learn the keywords in queries and documents by integrating the keyword weight layer and providing better final representations for calculating matching scores. We conduct extensive experiments and rigorous analysis in online A/B tests and offline public benchmarks to demonstrate that MKSM outperforms other modern deep matching models on semantic retrieval.

References

1. Cer, D., et al.: Universal sentence encoder. arXiv (2018)
2. Devlin, J., Chang, M.W., Lee, K., Toutanova, K.: Bert: pre-training of deep bidirectional transformers for language understanding. arXiv (2018)
3. Fan, M., Guo, J., Zhu, S., Miao, S., Sun, M., Li, P.: Mobius: towards the next generation of query-ad matching in baidu's sponsored search. In: Proceedings of the 25th ACM SIGKDD International Conference on Knowledge Discovery & Data Mining, pp. 2509–2517 (2019)

4. Guo, J., Fan, Y., Ai, Q., Croft, W.B.: A deep relevance matching model for ad-hoc retrieval. In: Proceedings of the 25th ACM International on Conference on Information and Knowledge Management, pp. 55–64 (2016)
5. Hu, B., Lu, Z., Li, H., Chen, Q.: Convolutional neural network architectures for matching natural language sentences. In: Advances in Neural Information Processing Systems, pp. 2042–2050 (2014)
6. Huang, C., Liu, Q., Chen, Y.Y., et al.: Local feature descriptor learning with adaptive siamese network. arXiv (2017)
7. Huang, J.T., et al.: Embedding-based retrieval in facebook search. In: Proceedings of the 26th ACM SIGKDD International Conference on Knowledge Discovery & Data Mining, pp. 2553–2561 (2020)
8. Huang, P.S., He, X., Gao, J., Deng, L., Acero, A., Heck, L.: Learning deep structured semantic models for web search using clickthrough data. In: Proceedings of the 22nd ACM International Conference on Information & Knowledge Management, pp. 2333–2338 (2013)
9. Jawahar, G., Sagot, B., Seddah, D.: What does bert learn about the structure of language? In: ACL 2019–57th Annual Meeting of the Association for Computational Linguistics (2019)
10. Khattab, O., Zaharia, M.: Colbert: efficient and effective passage search via contextualized late interaction over bert. In: Proceedings of the 43rd International ACM SIGIR Conference on Research and Development in Information Retrieval, pp. 39–48 (2020)
11. Klyuev, V., Oleshchuk, V.: Semantic retrieval: an approach to representing, searching and summarising text documents. Int. J. Inf. Technol. Commun. Convergence **1**(2), 221–234 (2011)
12. Kuang, M., Wang, W., Chen, Z., Kang, L., Yan, Q.: Efficient two-stage label noise reduction for retrieval-based tasks. In: Proceedings of the Fifteenth ACM International Conference on Web Search and Data Mining, pp. 526–534 (2022)
13. Lu, W., Jiao, J., Zhang, R.: Twinbert: distilling knowledge to twin-structured compressed bert models for large-scale retrieval. In: Proceedings of the 29th ACM International Conference on Information & Knowledge Management, CIKM 2020, pp. 2645–2652. Association for Computing Machinery, New York (2020)
14. Luan, Y., Eisenstein, J., Toutanova, K., Collins, M.: Sparse, dense, and attentional representations for text retrieval. Trans. Assoc. Comput. Linguist. **9**, 329–345 (2021)
15. Mitra, B., Craswell, N., et al.: An introduction to neural information retrieval. Now Foundations and Trends (2018)
16. Nguyen, T., et al.: Ms marco: a human generated machine reading comprehension dataset. In: CoCo@ NIPS (2016)
17. Palangi, H., et al.: Semantic modelling with long-short-term memory for information retrieval. arXiv (2014)
18. Reimers, N., Gurevych, I.: Sentence-bert: sentence embeddings using siamese bert-networks. arXiv (2019)
19. Robertson, S., Zaragoza, H.: The Probabilistic Relevance Framework: BM25 and Beyond. Now Publishers Inc., Norwell (2009)
20. Salton, G., Buckley, C.: Term-weighting approaches in automatic text retrieval. Inf. Process. Manag. **24**(5), 513–523 (1988)
21. Shan, X., et al.: Glow: global weighted self-attention network for web search. arXiv (2020)
22. Shan, X., et al.: Glow : global weighted self-attention network for web search. In: 2021 IEEE International Conference on Big Data (Big Data), pp. 519–528 (2021)

23. Shen, Y., He, X., Gao, J., Deng, L., Mesnil, G.: A latent semantic model with convolutional-pooling structure for information retrieval. In: Proceedings of the 23rd ACM International Conference on Conference on Information and Knowledge Management, pp. 101–110 (2014)
24. Sun, X., Tang, H., Zhang, F., Cui, Y., Jin, B., Wang, Z.: Table: a task-adaptive bert-based listwise ranking model for document retrieval. In: Proceedings of the 29th ACM International Conference on Information & Knowledge Management, pp. 2233–2236 (2020)
25. Xiong, L., et al.: Approximate nearest neighbor negative contrastive learning for dense text retrieval. arXiv (2020)

Relation-Aware Network with Attention-Based Loss for Few-Shot Knowledge Graph Completion

Qiao Qiao(✉), Yuepei Li, Kang Zhou, and Qi Li

Iowa State University, Ames, IA, USA
{qqiao1,liyp0095,kangzhou,qli}@iastate.edu

Abstract. Few-shot knowledge graph completion (FKGC) task aims to predict unseen facts of a relation with few-shot reference entity pairs. Current approaches randomly select one negative sample for each reference entity pair to minimize a margin-based ranking loss, which easily leads to a zero-loss problem if the negative sample is far away from the positive sample and then out of the margin. Moreover, the entity should have a different representation under a different context. To tackle these issues, we propose a novel Relation-Aware Network with Attention-Based Loss (RANA) framework. Specifically, to better utilize the plentiful negative samples and alleviate the zero-loss issue, we strategically select relevant negative samples and design an attention-based loss function to further differentiate the importance of each negative sample. The intuition is that negative samples more similar to positive samples will contribute more to the model. Further, we design a dynamic relation-aware entity encoder for learning a context-dependent entity representation. Experiments demonstrate that RANA outperforms the state-of-the-art models on two benchmark datasets.

Keywords: Few-shot learning · Knowledge graph completion

1 Introduction

Knowledge graphs (KGs) contain rich triples (facts), where each triple (h, r, t) illustrates a relation r between a head entity h and a tail entity t. KGs such as Wikidata [16] and NELL [3] have been applied to various downstream applications such as relation extraction [25], named entity recognition [24], and node classification [10].

Knowledge Graph Completion (KGC) is proposed to solve the issue of incompleteness caused by missing entities or relations in the KGs. KG embedding [2,15] has achieved considerable performance on KGC. These models perform well with enough training triples, but a large portion of relations in KGs follow a long-tail distribution. For example, around 10% of relations in Wikidata [4] have no more than 10 triples. Relations that do not have enough training triples are known as few-shot relations. It is crucial and challenging for the model to predict relations with limited training triples.

Few-shot knowledge graph completion (FKGC) methods have been proposed to address the few-shot relation issue. Given the relation r and few-shot reference entity pairs (h, t), the FKGC aims to rank candidate tail entities t for each query $(h, ?)$.

H. Kashima et al. (Eds.): PAKDD 2023, LNAI 13937, pp. 99–111, 2023.
https://doi.org/10.1007/978-3-031-33380-4_8

These few-shot reference entity pairs form a support set, and queries form a query set. One line of the existing methods focuses on designing metric learning algorithms to compute the similarity between entity pairs [18,23]. Another line leverages model agnostic meta-learning algorithm (MAML) [5] to learn the optimal parameters of the model [4,9,17].

To train the model, current FKGC methods apply a margin-based ranking loss function that aims to separate the score of the positive triple from the score of the negative triple by a margin. One negative triple is formed for each positive triple by replacing the true tail entity with a randomly selected candidate tail entity. This loss function does not effectively utilize the negative samples. Furthermore, an irrelevant negative sample is likely to be selected due to a large number of candidates. These irrelevant negative samples lead to zero loss because the negative triple is far away from the positive triple. Therefore these irrelevant negative samples would not contribute to the training and slow down the convergence rate [11]. For example, given a true triple *(Kobe Bryant, WorkIn, California)*, the model can select negative tail entities, such as *New York, Thailand, London*, etc. Because *Thailand* is irrelevant to the true tail entity *California*, the distance between *California* and *Thailand* is greater than a predefined margin, and the corresponding loss is zero. Thus *Thailand* may not contribute to the training.

To address the above limitations, we propose a framework called RANA (**R**elation-**A**ware **N**etwork with **A**ttention-Based Loss). To improve the quality of negative samples, we propose to filter irrelevant candidate tail entities first and then randomly sample multiple negative samples instead of one. Since the importance of negative samples is different and depends on their similarities to the positive sample, we apply an attention mechanism to assign a weight to each negative sample, where the weights of the most relevant negative samples are higher than the weights of the less relevant negative samples. The attention-based weighted loss function can enable the model to effectively avoid zero-loss issues and thus learn a better decision boundary.

Further, we propose a context-dependent dynamic relation-aware entity encoder to learn different representations of an entity in different relations. Specifically, given a target relation and its support set, the entity encoder uses the similarities between the target relation and neighboring relations to differentiate the impact of neighboring entities and dynamically encode the local connections of the entity. Finally, RANA employs meta-learning to enable the model to perform well on a new relation with a few training triples in a small number of gradient steps.

In summary, our main contributions are:

1. We propose a new negative sampling strategy and a novel attention-based loss function to solve the zero-loss and slower convergence issues.
2. We propose a dynamic entity encoder to learn a context-dependent entity representation and reduce the influence of unrelated neighboring entities.
3. Experiment results on benchmark datasets show that RANA consistently and significantly outperforms other baseline methods.

2 Related Work

2.1 Embedding Based Knowledge Graph Completion

Knowledge graph embedding aims to embed entities and relations into a low-dimensional continuous vector space while preserve their semantic meaning. Existing methods can be divided into the following categories: (1) Translation-based models calculate the Euclidean distance between entities and relations as the plausibility of a fact, such as TransE [2], RotatE [13], and TransMS [20]; (2) Semantic matching-based models calculate the semantic similarity between entities and relations as the plausibility of a fact, such as RESCAL [8], DistMult [19], and PUDA [14]; and (3) Neural network-based models take entities and relations into a deep neural network to fuse the graph network structure and content information of entities and relations, such as SME [1], CompLEx [15], and BertRL [22]. All above models require sufficient training triples and thus impair their performance on few-shot relations.

2.2 Few-Shot Knowledge Graph Completion

FKGC requires the model to predict new facts with a few training facts. Existing methods fall into two categories: (1) Metric-based models aim to learn the matching metrics by calculating the similarity between the query set and the support set. GMathching [18] focuses on one-shot KGC by considering both the learned embeddings and local graph structures. FSRL [23] and FAAN [12] extend GMatching to few-shot scenarios. (2) Optimization-based models aim to learn a set of good initial model parameters so that the learned model can be generalized to the new relation quickly. MetaR [4], GANA [9], and HiRe [17] focus on extracting relation-specific meta information from the embeddings of entities in the support set and transferring it to the query set.

However, all these methods use a margin-based ranking loss, which can not effectively avoid the low-quality negative sample, leading to a zero-loss issue and influencing the convergence rate. Negative sampling has been proven as important as positive sampling in determining the optimization objective [21]. Especially under the few-shot setting, given limited positive samples, how to select high-quality negative samples based on the corresponding positive sample is crucial.

3 Preliminary

3.1 Problem Definition

Knowledge Graph $\mathcal{G}$. A knowledge graph $\mathcal{G}$ is a set of triples $\mathcal{T} = \{(h, r, t) \subseteq \mathcal{E} \times \mathcal{R} \times \mathcal{E}\}$, where $\mathcal{E}$ and $\mathcal{R}$ represent the entity set and relation set, respectively. The relation set $\mathcal{R}$ contains few-shot relations and high-frequency relations. The background knowledge graph $\mathcal{G}_{background}$ is a set of triples associated with all high-frequency relations.

Knowledge Graph Completion. The KGC task is to either predict the tail entity t given the head entity h and the query relation r: $(h, r, ?)$ or predict unseen relation r between two existing entities: $(h, ?, t)$. In this work, we focus on tail entity prediction.

Few-shot Knowledge Graph Completion. Given a relation $r \in \mathcal{R}$ and its few-shot support set $\mathcal{S} = \{(h_i, t_i) \in \mathcal{T}\}$, the FKGC task aims to predict tail entity t for each query $\mathcal{Q} = \{(h_i, ?) \in \mathcal{T}\}$.

A Few-shot Relation's Neighborhood. Given a triple (h, r, t) of a few-shot relation r, the neighborhood of r is defined as $\{h, t, \mathcal{N}_h, \mathcal{N}_t\}$, where $\mathcal{N}_h$ and $\mathcal{N}_t$ are the sets of one-hop neighbors of h, t, respectively. All $\mathcal{N}_h$ and $\mathcal{N}_t$ are from the background knowledge graph $\mathcal{G}_{background}$. A neighbor in $\mathcal{N}_h$ or $\mathcal{N}_t$ is composed of a neighboring relation r_i and a neighboring entity c_i. We denote the neighbor of each entity (h or t) as $\mathcal{N}_e = \{(r_i, c_i) | (e, r_i, c_i) \in \mathcal{G}_{background}\}$.

3.2 Meta-learning Settings

Meta-learning aims to train a model on several related tasks so that the model can quickly learn a new task using a few training data. We leverage an optimization-based meta-learning algorithm called MAML [5], which aims to learn a task-specific parameter set $\mathbf{\Theta}_i$ by using well-initialized meta-model parameter set $\mathbf{\Theta}$. It can be divided into two stages, meta-training and meta-testing. During meta-training, given a task $\mathcal{T}_i$, a support set $\mathcal{S}_i$ and a query set $\mathcal{Q}_i$ are first sampled from $\mathcal{T}_i$. Then, the model learns a task-specific parameter set $\mathbf{\Theta}_i$ by one gradient descent update on the support set $\mathcal{S}_i$:

$$\mathbf{\Theta}_i = \mathbf{\Theta} - \eta * \nabla \mathcal{L}_{\mathcal{S}_i}(\mathbf{\Theta}). \tag{1}$$

Finally, meta-optimization across all query sets of tasks is performed to learn the meta-model parameter set $\mathbf{\Theta}$ by using task-specific parameter set $\mathbf{\Theta}_i$. During meta-testing, the model can quickly adapt to a new task using only a support set $\mathcal{S}$.

In FKGC, each task is defined as predicting new triples for a specific few-shot relation. All the relations in the meta-training form a meta-training set $\mathcal{R}_{meta-training}$. Since the goal is to predict facts of unseen relations, the relations in meta-validation $\mathcal{R}_{meta-validation}$, meta-testing $\mathcal{R}_{meta-testing}$, and $\mathcal{R}_{meta-training}$ are distinct.

4 Methodology

In this section, we first introduce triple representation, which aims to learn a context-dependent entity representation and a good initialization few-shot relation representation. Then we introduce a novel negative sampling strategy, which aims to filter irrelevant candidate tail entities and use an attention mechanism to differentiate the importance of each negative sample. Finally, we introduce meta-learning, which aims to learn well-generalized parameters so that the model can quickly adapt to a new task using few reference triples. Figure 1 shows the framework of RANA for a few-shot relation *WorkIn*.

4.1 Triple Representation

Dynamic Relation-Aware Entity Encoder. The entity representation should be context-dependent. For example, *(Kobe Bryant, California)* can involve in two different relations, such as *WorkIn* and *DieIn*, so *Kobe Bryant* should have different embeddings in these two different relation contexts.

Besides, given a few-shot relation, different neighbors should have a different impact on the entity itself. For example, in Fig. 1, given the few-shot relation *WorkIn* and the head entity *Kobe Bryant*, its neighbor *(AthleteOf, Lakers)* should get more attention since it reveals work information about *Kobe Bryant*, but the neighbor *(HasSpouse, Vanessa)* should get less attention since it reveals family information of *Kobe Bryant* which is irrelevant to the few-shot relation *WorkIn*.

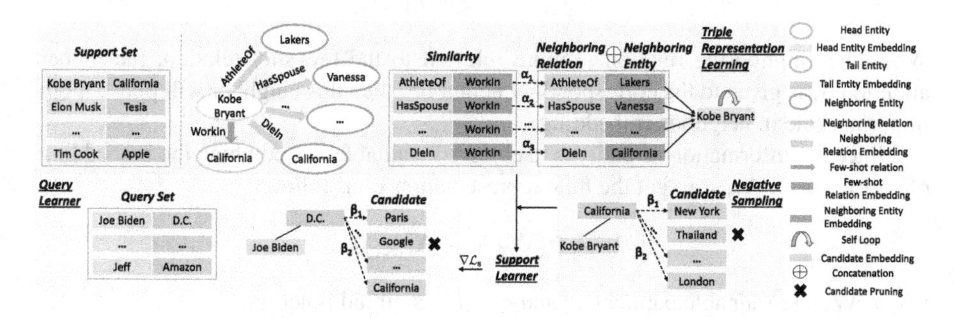

Fig. 1. The framework of RANA for a few-shot relation *WorkIn*

To address these issues, we design a dynamic relation-aware entity encoder, which incorporates neighboring relations to learn different embeddings of an entity in different relations and differentiates the importance of each neighbor by an attention mechanism.

Given an entity pair (h, t) from a support set $\mathcal{S}$, the embedding of few-shot relation r is defined as:

$$\mathbf{r} = \mathbf{t} - \mathbf{h}, \tag{2}$$

where $\mathbf{h}$ and $\mathbf{t}$ are the pretrained embeddings by TransE [2].

Here, we use the head entity h as an example to illustrate the entity encoding procedure, and this procedure also holds for the tail entity t.

To differentiate the impact of each neighbor, we use a Multilayer Perceptron (MLP) network to calculate the relevance score between the few-shot relation r and each neighboring relation r_i.

The relevance score is defined as follows:

$$m(r, r_i) = \mathbf{W_2}[\mathbf{tanh}(\mathbf{W_1}[\mathbf{r} \oplus \mathbf{r_i}])], \tag{3}$$

where $\oplus$ denotes the concatenation operation, $\mathbf{r_i}$ denotes the embedding of neighboring relations, and $\mathbf{W_1}$ and $\mathbf{W_2}$ are trainable parameters. A higher relevance score between the neighboring relation and the few-shot relation means that this neighboring relation is more important to the few-shot relation.

To learn the different representations of an entity in different relations, we design a dynamic neighbor embedding $\mathbf{A}_{\mathbf{r_i},\mathbf{c_i}}$ of the head entity h as follows:

$$\mathbf{A}_{\mathbf{r_i},\mathbf{c_i}} = \sum_{(r_i,c_i)\in\mathcal{N}_h} \alpha_i \mathbf{W_3}[\mathbf{r_i} \oplus \mathbf{c_i}], \tag{4}$$

where $\mathbf{W_3}$ are trainable parameters, and α_i is the attention score of each neighbor:

$$\alpha_i = \frac{exp(m(r, r_i))}{\sum_{r_i\in\mathcal{N}_h} exp(m(r, r_i))}. \tag{5}$$

When the neighboring relation is more relevant to the few-shot relation, the higher attention α_i is given to the corresponding neighbor. Then this neighbor will play a more important role in neighbor embedding.

Since the information of entity h itself is still valuable, we combine the embedding of entity h with $\mathbf{A}_{\mathbf{r_i},\mathbf{c_i}}$ to get the final representation $\mathbf{h}'$ as follows:

$$\mathbf{h}' = \sigma(\mathbf{W_4}(\mathbf{h} + \mathbf{A}_{\mathbf{r_i},\mathbf{c_i}})), \tag{6}$$

where $\mathbf{W_4}$ are trainable parameters and $\sigma(\cdot)$ is a sigmoid function.

Few-Shot Relation Representation. The same entity pair may involve in different relations, so the learning of relation representation is necessary, and it can further help triple representation learning.

The relation representation from a specific entity pair in the support set S is:

$$\mathbf{R}_{(\mathbf{h_i},\mathbf{t_i})} = FC^{\sigma}_{\mathbf{W_5}}[\mathbf{h}'_\mathbf{i} \oplus \mathbf{t}'_\mathbf{i}], \tag{7}$$

where the fully connected layer $FC^{\sigma}_{\mathbf{W_5}}$ is parameterized by $\mathbf{W_5}$ and activated by a LeakyReLU function $\sigma(\cdot)$.

The relation representation from the support set $\mathbf{R^s}$ is then the average of all representations from entity pairs in S,

$$\mathbf{R^s} = \frac{\sum_{i=1}^{I} \mathbf{R}_{(\mathbf{h_i},\mathbf{t_i})}}{I}, \tag{8}$$

where I is the number of entity pairs in the support set S.

4.2 Negative Sampling

Since the positive sample is limited under the few-shot setting, how to take advantage of negative samples is more critical. Previous FKGC methods use a margin-based ranking loss and randomly select one negative sample for each positive sample [4,9,12,18,23]. But the random selection is likely to select an irrelevant negative sample and lead to a zero-loss issue. Further, regardless of their relevance to the positive samples, all negative samples will have the same impact on the model training. To address these issues, RANA filters irrelevant negative samples and uses an attention mechanism to distinguish the importance of each negative sample.

Candidate Pruning. The candidate set of negative samples constructed by [18] limits the candidate entities to those have the same types as the true tail entities in the support set, but this broad candidate set includes many irrelevant candidates as negative samples. For example, given a fact *(Kobe Bryant, WorkIn, California)*, the previous candidate set is limited to location and company types of entities because the types of tail entities in the support set are company or location. However, a candidate such as *Thailand* is irrelevant to *California* and thus is not helpful in the model training.

To reduce the number of irrelevant candidates and enable the model to select high-quality negative samples during the training stage, RANA filters irrelevant candidates by the similarity of the true tail entity t and a candidate tail entity t^-. The similarity is calculated by:

$$f(\mathbf{t},\mathbf{t}^-) = \mathbf{t}^{-\mathbf{T}}\mathbf{t}, \tag{9}$$

where $\mathbf{t}$ is the embedding of a true tail entity and $\mathbf{t}^-$ is the embedding of a candidate tail entity. If $f(\mathbf{t},\mathbf{t}^-) < \tau$, where τ is a threshold, then t^- should be filtered.

Attention of Negative Samples. To fully utilize the negative samples, RANA selects multiple negative samples instead of one and differentiates each negative triple's contribution by an attention mechanism.

Intuitively, if a negative sample is more relevant to the positive sample, this negative sample should play a more important role in model training. Therefore, higher attention should be given to this negative sample. As shown in Fig. 1, given a positive sample *(Kobe Bryant, California)*, the negative sample *(Kobe Bryant, New York)* is more relevant to the positive sample than the negative sample *(Kobe Bryant, London)*, and thus the model should pay more attention to the former.

We define a scaled-dot product function $f(\mathbf{p_i},\mathbf{n_{ij}})$ to calculate the similarity between the positive sample (h_i,t_i) and each of its negative sample (h_i,t_{ij}^-):

$$\mathbf{p_i} = \mathbf{h_i} \oplus \mathbf{t_i}, \qquad \mathbf{n_{ij}} = \mathbf{h_i} \oplus \mathbf{t_{ij}^-}, \qquad f(\mathbf{p_i},\mathbf{n_{ij}}) = \frac{\mathbf{n_{ij}^T}\mathbf{p_i}}{\sqrt{|p|}}, \tag{10}$$

where $|p|$ is the dimension of $\mathbf{p_i}$. The attention of each negative triple is defined by:

$$\beta_{ij} = \frac{exp\, f(\mathbf{p_i},\mathbf{n_{ij}})}{\sum_{j=1}^{J} exp\, f(\mathbf{p_i},\mathbf{n_{ij}})}, \tag{11}$$

where J is the number of negative samples.

The Loss of RANA. Negative sampling is as valuable as positive sampling in determining the optimization object, but it has been overlooked in the margin-based ranking loss [21]. To alleviate zero-loss and slower convergence issue, we sample multiple negative triples instead of one to increase the probability of generating a relevant negative triple.

Motivated by TransE [2], we first calculate the distance of each entity pair (h_i,t_i) as follows:

$$d_{(h_i,t_i)} = ||\mathbf{h_i} + \mathbf{R} - \mathbf{t_i}||_{L2}, \tag{12}$$

Because the smaller distance indicates the triple is more likely to be true, the triple should lead to a higher score. The score function of each triple is designed as:

$$s_{(h_i,t_i)} = \gamma - d_{(h_i,t_i)}, \tag{13}$$

where γ is a hyperparameter.

Our log-based loss function is:

$$\mathcal{L} = -\sum_{i=1}^{I} log\,\sigma(s_{(h_i,t_i)}) - \sum_{i=1}^{I}\sum_{j=1}^{J} \beta_{ij} log\,\sigma(-s_{(h_i,t_{ij}^-)}), \tag{14}$$

where $\sigma(\cdot)$ is a sigmoid function, and β_{ij} is the attention of each negative triple calculated by Eq. (11). Since a more relevant negative triple has higher attention (β_{ij}), this loss function will make those relevant negative triples impact more in model training.

4.3 Meta Learning

To learn a new relation quickly with a support set, RANA employs MAML [5] to optimize the model parameters that can be adapted for few-shot relations.

Algorithm 1. Training framework

Input: Training tasks $\mathcal{R}_{meta-training}$, initial model parameter $\boldsymbol{\Theta}$
Pre-trained KG embedding (excluding relation in $\mathcal{R}_{meta-training}$)

1: **while** not done **do**
2: Sample a task $\mathcal{T}_i = \{\mathcal{S}_i, \mathcal{Q}_i\}$ from $\mathcal{R}_{meta-training}$
3: Get $\mathbf{R^s}$ from $\mathcal{S}_i$ by Eqs. (2)–(8)
4: Get negative sample of $\mathcal{S}_i$ by Eqs. (9)–(11)
5: Calculate the loss of $\mathcal{S}_i$ by Eqs. (12)–(14)
6: Update the embedding of the task-specific relation $\mathbf{R^q}$ with gradient descent by Eq. (15)
7: Get negative samples of $\mathcal{Q}_i$ by Eqs. (9)–(11)
8: Calculate the loss of $\mathcal{Q}_i$ by Eqs. (12)–(14)
9: Update whole model parameters $\boldsymbol{\Theta} \leftarrow \boldsymbol{\Theta} - \mu\nabla\mathcal{L}$
10: **end while**

Support Learner. Support learner aims to learn a representation $\mathbf{R^s}$ of the few-shot relation and $\mathbf{R^s}$ can be calculated by Eqs. (2)–(8).

Query Learner. Following the MetaR [4] assumption, the relation is the key common information between support and query set. So we aim to transfer the support relation R^s to the query relation R^q by minimizing a loss function via gradient descending.

In RANA, the relation embedding $\mathbf{R^q}$ can be updated by the gradient descent,

$$\mathbf{R^q} = \mathbf{R^s} - \eta * \nabla\mathcal{L}_\mathbf{s}, \tag{15}$$

where the hyperparameter η refers to the step size and $\mathcal{L}_s$ refers to the loss of the corresponding support set, which is calculated by Eqs. (12)–(14).

To update all parameters of RANA, we use the updated relation embedding $\mathbf{R}^{\mathrm{q}}$ to calculate the loss of the corresponding query set $\mathcal{L}_q$ by Eqs. (12)–(14) as well.

Objective and Training Process. During the meta training-stage, the objective of RANA is to minimize the sum of query loss for all tasks, and the overall loss is:

$$\mathcal{L} = arg \min_{\mathbf{\Theta}} \sum \mathcal{L}_q, \tag{16}$$

where $\mathbf{\Theta}$ represents all trainable parameters.

4.4 Algorithm of RANA

We summarize the overall training procedure in Algorithm 1.

4.5 Difference from RotatE

RotatE [13] is an embedding-based KGC method that uses a self-adversarial negative sampling technique to effectively optimize the model. Our approach differs from RotatE in a major way: We consider the similarity between the positive triple and negative triple as the weight of each negative triple, but RotatE considers the distribution of negative triples and treats the probability as the weight of each negative triple. Therefore, the weights of the negative samples in RotatE are independent of the positive samples. As we will show in the experiments (Sect. 5.5), RANA can achieve a better performance than RotatE's self-adversarial negative sampling under the few-shot setting.

Table 1. Statistics of the Datasets. Columns 2-7 represent the number of entities, relations, triples, relations in $\mathcal{R}_{meta-training}$, relations in $\mathcal{R}_{meta-validation}$, and relations in $\mathcal{R}_{meta-testing}$, respectively.

Dataset	#Ent	#Rel	#Triples	#Train Rel	#Valid Rel	#Test Rel
NELL-One	68, 545	358	181, 109	51	5	11
Wiki-One	4, 838, 244	822	5, 859, 240	133	16	34

5 Experiments

5.1 Datasets and Evaluation Metrics

We conduct experiments on NELL-One and Wiki-One, constructed by [18]. In both datasets, relations with more than 50 but less than 500 triples are selected as few-shot relations, and the remaining relations are treated as background relations. We use 51/5/11 and 133/16/34 few-shot relations for training/validation/testing in NELL-One and Wiki-One, respectively. The statistics of both datasets are shown in Table 1.

To evaluate the performance of RANA and all baselines, we utilize two metrics: mean reciprocal rank (MRR) and Hits@K. MRR is the mean reciprocal rank of correct entities, and Hits@K is the proportion of correct entities ranked in the top k.

5.2 Baseline

Traditional embedding-based methods aim to learn entity and relation embeddings by modeling relational structure in KG. We consider the following widely used methods as baselines: TransE [2], DistMult [19], ComplEx [15], SimplE [6], and RotatE [13]. All these methods require sufficient training triples for each relation and do not use local graph structure to update entity embeddings.

FKGC methods aim to learn long-tail and unseen relations by utilizing deep neural networks to explore the connection between the support set and the query set. We consider the following models as baselines: GMatching [18], MetaR [4], FSRL [23], FAAN [12], GANA [9], and HiRe [17]. We run RANA 5 times and report the average results.

5.3 RANA Setups

The pre-trained entity and relation embeddings are obtained from TransE. The embedding dimension is set to 50 and 100 for NELL-One and Wiki-One, respectively. We use Adam [7] with the initial learning rate of 0.01 to update parameters. The number of negative samples is 5, the margin γ is 12.0, the step size η is 1, and the number of neighbors is 25 on both datasets. The model with the highest MRR on the validation set is applied as the final model. The optimal hyperparameters are tuned on the validation set by grid search. We conduct RANA on a server with a Tesla V100 GPU (32G).

5.4 Overall Evaluation Results and Analysis

The performances of all models on NELL-One and Wiki-One are reported in Table 2. Compared to the traditional embedding-based methods, incorporating graph neighbors is effective for learning entity embedding. RANA outperforms the other FKGC models on both datasets. Compared with the runner-up results, the improvements obtained by RANA in terms of MRR, Hits@10, Hits@5, and Hits@1 are 4.9%, 10.2%, 8.2%, 2.8% on NELL-One, and 2.2%, 2.3%, 4.3%, 3.1% on Wiki-One, respectively.

Table 2. Results of 5-shot KGC. **Bold** numbers represent the best results and underline numbers denote the runner-up results. † cites the result from [12], $*$ cites the result from their original papers.

	NELL-One				Wiki-One			
Model	MRR	Hits@10	Hits@5	Hits@1	MRR	Hits@10	Hits@5	Hits@1
TransE†	0.174	0.313	0.231	0.101	0.133	0.187	0.157	0.100
DistMult†	0.200	0.311	0.251	0.137	0.071	0.151	0.099	0.024
ComplEx†	0.184	0.297	0.229	0.118	0.080	0.181	0.122	0.032
SimplE†	0.158	0.285	0.226	0.097	0.093	0.180	0.128	0.043
RotatE†	0.176	0.329	0.247	0.101	0.049	0.090	0.064	0.026
GMatching†	0.176	0.294	0.233	0.113	0.263	0.387	0.337	0.197
MetaR†	0.209	0.355	0.280	0.141	0.323	0.418	0.385	0.270
FSRL†	0.153	0.319	0.212	0.073	0.158	0.287	0.206	0.097
FAAN†	0.279	0.428	0.364	0.200	0.341	0.463	0.395	0.281
GANA*	<u>0.344</u>	0.517	0.437	<u>0.246</u>	0.351	0.446	0.407	0.299
HiRe*	0.306	<u>0.520</u>	<u>0.439</u>	0.207	<u>0.371</u>	<u>0.469</u>	<u>0.419</u>	<u>0.319</u>
RANA	**0.361±0.011**	**0.573±0.009**	**0.475±0.010**	**0.253±0.013**	**0.379±0.008**	**0.480±0.012**	**0.437±0.008**	**0.329±0.011**

5.5 Ablation Study

RANA is composed of two modules, including a dynamic relation-aware entity encoder and negative sampling. To investigate the contributions of each component, we conduct the 5-shot KGC with different settings. The results are summarized in Table 3.

Entity Encoder Variants: We analyze the impact of the neighboring relation in Eqs. (4) and (5) by removing r_i from Eq. (4) or adding c_i in Eq. (5). Besides, we remove the attention mechanism in Eq. (4). The results show that neighboring relation and attention mechanism can benefit model performance. It illustrates semantic information of relations can improve the entity representation, and different relations should have different impacts on the entity itself. Since the effect of the attention mechanism depends on neighbors, Wiki-One has much sparser neighbors than NELL-One [9], so the attention mechanism plays a small role in Wiki-One.

Negative Sampling Variants: To inspect the effectiveness of the negative sampling and attention-based loss functions, we conduct five different experiments. (A) We use only one negative sample in Eq. (14). (B) We remove the negative attention mechanism in Eq. (14). (C) We remove the candidate pruning stage. (D) We remove the candidate pruning stage and negative attention mechanism. (E) We replace Eq. (14) with RotatE [13] self-adversarial negative sampling loss. Experimental results show that the negative sampling strategy plays a key role in the success of RANA.

5.6 Influence of Size of Few-shot Support Set and Negative Sample

To analyze the impact of support set size, we compare RANA with GANA on NELL-one. Figure 2a shows the performances with support set size from 1 to 8. RANA outperforms GANA under different sizes of support sets, showing the effectiveness of RANA. After the 5-shot, the improvement of RANA is not significant. We randomly select 20 facts from the relation *teamcoach* to analyze the errors in the 5-shot setting. RANA predicts 12 out of 20 true tail entities in top 10. Among the other 8 facts, 4 of them have incorrect ground truth tail entities, and 3 of them have neighbors fewer than 10. For these cases, increasing the size of the support set is unlikely to change the results.

Table 3. Ablation Study

Model	NELL-One				Wiki-One			
	MRR	Hits@10	Hits@5	Hits@1	MRR	Hits@10	Hits@5	Hits@1
whole model	**0.372**	**0.580**	**0.477**	**0.257**	**0.387**	**0.486**	**0.443**	**0.339**
Eq. (4) w/o r_i	0.339	0.535	0.427	0.222	0.362	0.468	0.410	0.299
Eq. (5) with c_i	0.358	0.573	0.471	0.256	0.367	0.477	0.424	0.302
Eq. (4) w/o α_i	0.326	0.526	0.407	0.235	0.377	0.483	0.433	0.315
Eq. (14) with one negative sample	0.294	0.520	0.428	0.210	0.349	0.451	0.417	0.311
Eq. (14) w/o negative attention	0.293	0.494	0.416	0.213	0.298	0.387	0.371	0.257
w/o candidate pruning	0.298	0.507	0.425	0.217	0.311	0.445	0.360	0.243
w/o candidate pruning and negative attention	0.257	0.447	0.396	0.192	0.286	0.363	0.321	0.242
RotatE self-adversarial negative sampling	0.268	0.479	0.365	0.165	0.310	0.389	0.401	0.255

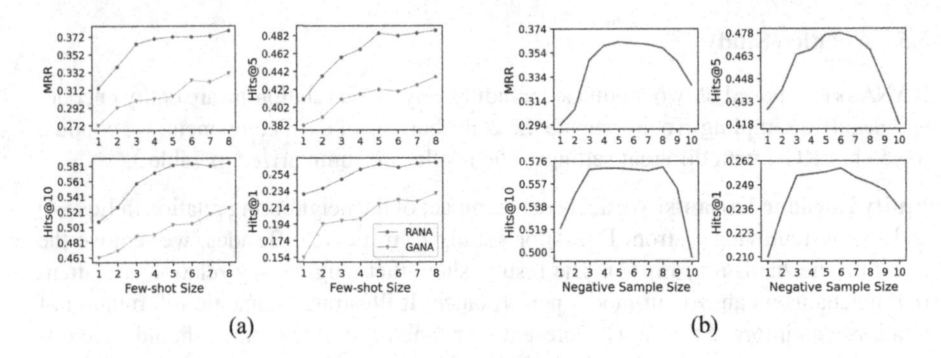

Fig. 2. (a) Influence of Few-shot Support Set Size,(b) Influence of Negative Sample Size

We conduct an experiment to analyze the influence of the negative sample size. Figure 2b shows the performance of RANA on NELL-One with the negative sample size from 1 to 10. The performance improves initially when increasing the negative sample size. After size 6, the performance begins to drop due to the class imbalance issue. Empirically, we recommend a negative sample size of 3 to 5.

6 Conclusion

In this paper, we propose a relation-aware network with attention-based loss for FKGC tasks. We strategically select multiple negative samples instead of one and propose an attention-based loss to differentiate the importance of each negative sample. A dynamic relation-aware entity encoder is designed to learn a context-dependent entity representation. The experimental results demonstrate that RANA outperforms other SOTA baselines on two benchmark datasets.

Acknowledgement. The work is supported in part by NSF IIS-2007941.

References

1. Bordes, A., Glorot, X., Weston, J., Bengio, Y.: A semantic matching energy function for learning with multi-relational data. Mach. Learn. **94**(2), 233–259 (2014)
2. Bordes, A., Usunier, N., Garcia-Duran, A., Weston, J., Yakhnenko, O.: Translating embeddings for modeling multi-relational data. In: NeurIPS vol. 26, pp. 2787–2795 (2013)
3. Carlson, A., Betteridge, J., Kisiel, B., Settles, B., Hruschka, E., Mitchell, T.: Toward an architecture for never-ending language learning. In: AAAI, vol. 24, pp. 1306–1313 (2010)
4. Chen, M., Zhang, W., Zhang, W., Chen, Q., Chen, H.: Meta relational learning for few-shot link prediction in knowledge graphs. In: EMNLP-IJCNLP, pp. 4217–4226 (2019)
5. Finn, C., Abbeel, P., Levine, S.: Model-agnostic meta-learning for fast adaptation of deep networks. In: ICML, pp. 1126–1135. PMLR (2017)
6. Kazemi, S.M., Poole, D.: Simple embedding for link prediction in knowledge graphs. In: NeurIPS 31 (2018)
7. Kingma, D.P., Ba, J.: Adam: a method for stochastic optimization. In: ICLR (2015)

8. Nickel, M., Tresp, V., Kriegel, H.P.: A three-way model for collective learning on multi-relational data. In: ICML (2011)
9. Niu, G., et al.: Relational learning with gated and attentive neighbor aggregator for few-shot knowledge graph completion. In: SIGIR, pp. 213–222 (2021)
10. Rong, Y., Huang, W., Xu, T., Huang, J.: DropEdge: towards deep graph convolutional networks on node classification. In: ICLR (2019)
11. Schroff, F., Kalenichenko, D., Philbin, J.: Facenet: A unified embedding for face recognition and clustering. In: CVPR, pp. 815–823 (2015)
12. Sheng, J., Guo, S., Chen, Z., Yue, J., Wang, L., Liu, T., Xu, H.: Adaptive attentional network for few-shot knowledge graph completion. In: EMNLP, pp. 1681–1691 (2020)
13. Sun, Z., Deng, Z.H., Nie, J.Y., Tang, J.: Rotate: Knowledge graph embedding by relational rotation in complex space. In: ICLR (2018)
14. Tang, Z., et al.: Positive-unlabeled learning with adversarial data augmentation for knowledge graph completion. In: IJCAI, pp. 1935–1942 (2022)
15. Trouillon, T., Welbl, J., Riedel, S., Gaussier, É., Bouchard, G.: Complex embeddings for simple link prediction. In: ICML, pp. 2071–2080. PMLR (2016)
16. Vrandečić, D., Krötzsch, M.: Wikidata: a free collaborative knowledgebase. In: CACM (2014)
17. Wu, H., Yin, J., Rajaratnam, B., Guo, J.: Hierarchical relational learning for few-shot knowledge graph completion. arXiv preprint arXiv:2209.01205 (2022)
18. Xiong, W., Yu, M., Chang, S., Guo, X., Wang, W.Y.: One-shot relational learning for knowledge graphs. In: EMNLP, pp. 1980–1990 (2018)
19. Yang, B., Yih, S.W.T., He, X., Gao, J., Deng, L.: Embedding entities and relations for learning and inference in knowledge bases. In: ICLR (2015)
20. Yang, S., Tian, J., Zhang, H., Yan, J., He, H., Jin, Y.: TransMS: knowledge graph embedding for complex relations by multidirectional semantics. In: IJCAI, pp. 1935–1942 (2019)
21. Yang, Z., Ding, M., Zhou, C., Yang, H., Zhou, J., Tang, J.: Understanding negative sampling in graph representation learning. In: SIGKDD, pp. 1666–1676 (2020)
22. Zha, H., Chen, Z., Yan, X.: Inductive relation prediction by BERT. In: AAAI (2022)
23. Zhang, C., Yao, H., Huang, C., Jiang, M., Li, Z., Chawla, N.V.: Few-shot knowledge graph completion. In: AAAI **34**, 3041–3048 (2020)
24. Zhou, K., Li, Y., Li, Q.: Distantly supervised named entity recognition via confidence-based multi-class positive and unlabeled learning. In: ACL, pp. 7198–7211 (2022)
25. Zhou, K., Qiao, Q., Li, Y., Li, Q.: Improving distantly supervised relation extraction by natural language inference. arXiv preprint arXiv:2208.00346 (2022)

MFBE: Leveraging Multi-field Information of FAQs for Efficient Dense Retrieval

Debopriyo Banerjee(✉), Mausam Jain, and Ashish Kulkarni

Rakuten Institute of Technology, Rakuten India Enterprise Pvt. Ltd., Bengaluru, India
{debopriyo.banerjee,mausam.jain,ashish.kulkarni}@rakuten.com

Abstract. In the domain of question-answering in NLP, the retrieval of Frequently Asked Questions (FAQ) is an important sub-area which is well researched and has been worked upon for many languages. Here, in response to a user query, a retrieval system typically returns the relevant FAQs from a knowledge-base. The efficacy of such a system depends on its ability to establish semantic match between the query and the FAQs in real-time. The task becomes challenging due to the inherent lexical gap between queries and FAQs, lack of sufficient context in FAQ titles, scarcity of labeled data and high retrieval latency. In this work, we propose a bi-encoder-based query-FAQ matching model that leverages multiple combinations of FAQ fields (like, question, answer, and category) both during model training and inference. Our proposed Multi-Field Bi-Encoder (MFBE) model benefits from the additional context resulting from multiple FAQ fields and performs well even with minimal labeled data. We empirically support this claim through experiments on proprietary as well as open-source public datasets in both unsupervised and supervised settings. Our model achieves around 27% and 23% better top-1 accuracy for the FAQ retrieval task on internal and open datasets, respectively over the best performing baseline.

Keywords: Information Retrieval · FAQ Retrieval · Question-Answering · Multi-field · BERT · Bi-encoder

1 Introduction

Customer support (CS) is critical to any business and plays an important role in customer retention, new customer acquisition, branding, and in driving a better experience. In a typical online customer support setting, customers reach out with their queries and are attended to by human agents. This requires businesses to hire and maintain a team of CS agents that scales as a function of the query volume and the productivity of agents that, in turn, translates to operational cost for the business. Customer support automation [19] can help save on this operational cost by providing automated responses to queries and by

D. Banerjee and M. Jain—These authors contributed equally to this work.

H. Kashima et al. (Eds.): PAKDD 2023, LNAI 13937, pp. 112–124, 2023.
https://doi.org/10.1007/978-3-031-33380-4_9

improving support agent productivity. One of the ways businesses typically try to achieve this is by automatically responding to customer queries from a repository of frequently asked questions (FAQs), thereby, insulating human agents from high query volumes. The success of such a system, measured as the fraction of customer queries that it automatically responds to, then depends on the effectiveness of the user query to FAQ matching.

Given a collection of FAQs where, each FAQ is a multi-field tuple $\langle Q, A, C \rangle$ of question Q, answer A, and question category C, the problem of FAQ retrieval [6,9,24,25] is to retrieve the top-k FAQs in response to a user query q. Similar to a typical retrieval problem, FAQ retrieval too suffers from the problem of lexical gap between a user query expressed in natural language and the corresponding matching FAQs. This is typically addressed by learning a relevance function between user queries and FAQs using labeled query-FAQ pairs for supervision. Unfortunately, such labeled data is often unavailable or scarce, especially in low-resource settings like Japanese query-FAQ retrieval, which is the domain of our interest. Curating large amounts of such labeled data through manual labeling is often expensive and requires domain knowledge for labeling.

The main contributions of this paper are summarized as follows:

- We propose a bi-encoder based retrieval model - MFBE that leverages multi-field information (question, answer and/or categories) in FAQs.
- We use different combinations of user query and FAQ fields to create an extended set of pseudo-positive pairs for training.
- We employ multiple FAQ representations during inference for query-FAQ scoring.

2 Related Work

Question answering [10,16] task has been the area of interest in NLP community for a long time and shares the concepts of Information Retrieval (IR) where relevant information from a corpus of documents is retrieved in response to a search query. FAQ retrieval is an example of IR which is the focus of this work. Traditional retrieval methods [12,23] mainly depend on lexical features for the retrieval task which limits them to capture the semantics of the query. In order to address the challenge of lexical gap between user queries and answers, there is a body of work [3,17,29] that trains a semantic retrieval model from labeled data in the form of user queries and matching responses. In recent years, there has been an increasing research [8,14] on unsupervised learning techniques for text-encoder training eliminating the need for annotated data. They propose augmentation techniques based on paraphrases of input sentences to generate positive and negative samples for an anchor that are then used to train the retrieval model using a contrastive learning strategy [4].

The performance of FAQ retrieval task depends upon the (i) choice and design of model architecture and (ii) retrieval and re-ranking algorithms used. A combination of bi-encoder and cross-encoder is seen in [13,20], where the authors start with an unsupervised setting with zero labelled data. Then they

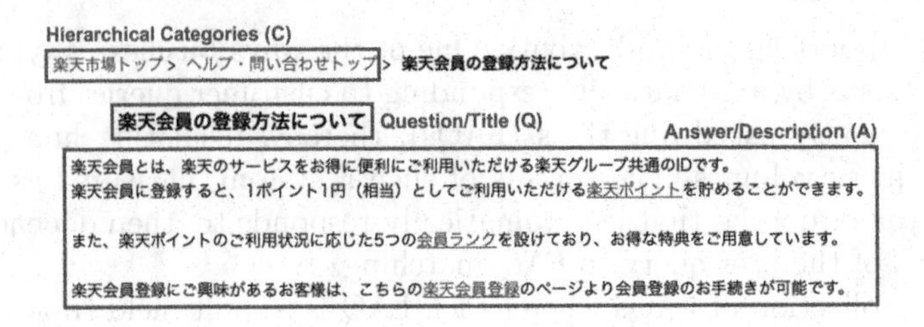

Fig. 1. Example of an FAQ in Japanese Language (JA).

iterate between bi-encoder and cross-encoder models generating more labelled samples in each iteration. This gives a powerful text-encoder model along with annotated dataset. The retriever and the re-ranker can also be jointly trained with the goal of achieving mutual improvement [22].

Previous works [6,24,25] on FAQ retrieval problem focused on query-question (q-Q) similarity using BM25 [23] and query-answer (q-A) similarity using BERT [5], where the BERT model parameters are fine-tuned on FAQ question-answer (Q-A) pairs. Sakata *et al.* [24] (close to our work) employ a two-stage method where they first retrieve a set of FAQs based on q-Q similarity and then re-rank these based on q-A similarity to obtain the final list of top-k FAQs as response. Here, question (Q) and answer (A) are typically referred to as fields. Fields can vary depending upon the dataset. For example, Wikipedia page title, content, abstract, etc. have been considered as fields in [16].

Dutta *et al.* [6] propose a seq-2seq model for extracting keywords in user queries to identify the intent of a user query for better retrieval of relevant FAQs. Another close work by Assem *et al.* [2] uses two separate deep learning architectures. They first learn latent lexical relationships between FAQ questions and their paraphrases to generate top-k most relevant similar questions from the collection. These top-k candidates are then fed to an LSTM-based architecture that captures fine-grained differences in semantic context between FAQ questions and their paraphrases thereby improving the accuracy@1. Liu *et al.* [15] discuss the difficulty in determining the relevance of query-answer pairs due to their heterogeneity in terms of syntax and semantics and propose to use synthetic data for increasing the positive training examples. Tseng *et al.* [26] cites that existing methods fail to attend to the global information specifically about an FAQ task and propose a graph convolution network-based method to cater to all relations of question and words to generate richer embeddings. They also explore domain specific knowledge graphs for improving question and query representations.

Some unsupervised sentence embedding methods that closely aligns with our work are [14] and [8], which are based on contrastive learning using augmentation techniques. Alternate representations of input sentences provide strong positives to the model. [11] is a self-supervised fine-tuning of BERT, which redesigned the contrastive learning objective to account for different views of the input sentence.

3 FAQ Retrieval for User Queries

FAQs are a pre-defined list of *question-answer* pairs available on web portals (*e.g.*, banking, e-commerce, telecom, *etc.*) that help in addressing user queries without human intervention. In some cases, FAQs are also associated with some hierarchical categories or tags. Figure 1 shows a sample FAQ that consists of question (or title), answer (or description) and hierarchical categories. In this paper, we mainly focus on improving the retrieval of FAQs conditioned on user queries using a neural text encoder [7]. In general, neural text encoders can be categorized into two types: bi-encoders [18] and cross-encoders [28]. **Bi-Encoder (BE):** It consists of two encoder branches (with optional weight sharing), where two sentences S_a and S_b are independently passed through each branch, resulting into two sentence embeddings f_a and f_b respectively. Their similarity can then be computed using a distance metric like cosine or dot-product of f_a and f_b.

Cross-Encoder (CE): The two sentences are first concatenated and then passed through an encoder. The resulting embedding vector is input to a classification head that is typically implemented as a shallow feed forward network. Here, computation of similarity between the two input sentences is modelled as a binary classification task with 1 (0) indicating as similar (not similar).

Generally, cross-encoders outperform bi-encoders in performance by leveraging the mutual attention among all the words in the concatenated sentence, but they also suffer from high inference latency. The bi-encoder architecture is inherently suited for the FAQ retrieval task as it allows for pre-computation and indexing of sentence embeddings of the FAQs before-hand, which is not possible with cross-encoders. We introduce Multi-Field Bi-Encoders (MFBE) with the aim of improving the performance of bi-encoders for the FAQ retrieval task by leveraging additional context from FAQ titles, description and categories. In this section, we first explain the notations, followed by our proposed approach.

Notations - Let $\mathcal{F} = \{F_i\}_{i=1}^{N}$ be the set of FAQs, where each FAQ F_i consists of a 3-tuple $\langle Q, A, C\rangle$ of question (or title) Q, answer (or description) A and categories (or tags) C. We define $M = \{Q, A, QA, QC, CA, QCA\}$ as the set of fields, where, each field $m \in M$ is obtained by concatenating one or more of Q, A or/and C. Let $M_s = \{Q, QC\} \subset M$ be a subset of fields present in M and $\mathcal{Q}$ be the set of all user queries. For a query $q \in \mathcal{Q}$, let F and $\tilde{F}$ denote the matching and non-matching FAQs, respectively, such that (q, F) forms a matching query FAQ pair and $(q, \tilde{F})$ corresponds to a non-matching pair. We denote the field m (or m_s) of an FAQ as F^m (or F^{m_s}). For example, the field QA of an FAQ is denoted by F^{QA}. In this case, for a query q, (q, F^{QA}) and $(q, \tilde{F}^{QA})$ are the matching and non-matching query FAQ pairs. We denote the text encoder as $E(\cdot)$, which maps any text to a d-dimensional real-valued vector. For every FAQ field F_i^m, we compute $E(F_i^m) = f_i^m$ and stack them in a matrix T_m, where row i in T_m corresponds to vector f_i^m. For a query q the corresponding embedding is computed as $E(q) = f_q$.

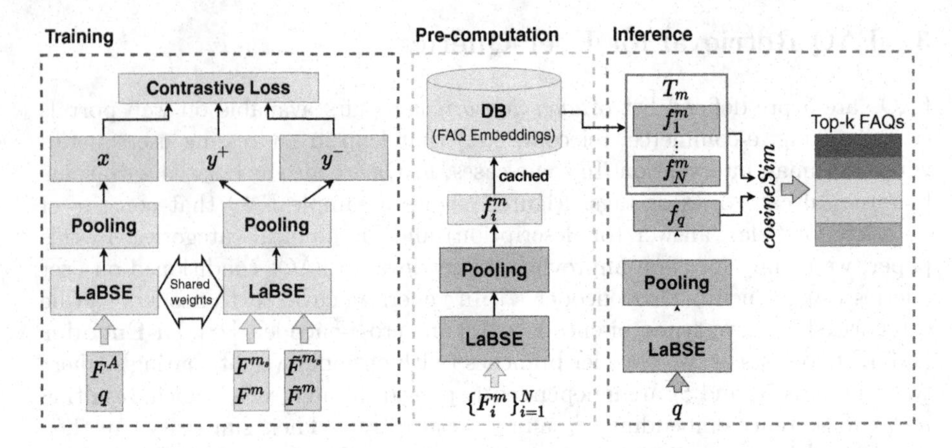

Fig. 2. Illustration of the working of MFBE model across different stages - training, pre-computation and inference.

Proposed Approach - We propose to learn the relevance function $rel(x, y; \theta)$ (where $x \in \{q, F^{m_s}\}$; when $x = q$, then $y \in \cup_{m \in M} F_m$, else $y \in \cup_{m \in M \setminus M_s} F^m$) using a pre-trained neural language model [7] as the text encoder with θ as the model parameters.

As shown in Fig. 2, MFBE consists of Language-agnostic BERT Sentence Embedding (LaBSE) [7] model as text encoder in two branches with shared weights.

We compute the similarity between a query q and an FAQ field F^m using the cosine-similarity, i.e., $sim(q, F^m) = cosineSim(f_q, f^m)$. As mentioned in [9], the similarity function should be decomposable and facilitate the pre-computation of representations of the FAQs. L2, inner product and cosine-similarity are some of the widely used similarity functions that are decomposable in nature. We choose cosine-similarity function, which is equivalent to inner product for normalized vectors.

Training Stage - The goal is to learn a latent space following metric learning [27], where matching query and FAQ pairs shall have smaller distance compared to the non-matching pairs. Let $\mathcal{D}_{train} = \{(x_i, y_i^+, y_{i,1}^-, \ldots, y_{i,n}^-)\}_{i=1}^N$ be the training data that consists of N instances, where each instance contains one query or an FAQ field $x_i \in \{q_i, F_i^{m_s}\}$, a matching (positive) FAQ field $y_i^+ \in \cup_{m \in M} F_i^m$, when $x_i = q_i$ (or $y_i^+ \in \cup_{m \in M \setminus M_s} F_i^m$, when $x_i = F_i^{m_s}$), along with n non-matching (negative) FAQ fields $y_{i,j}^-$.

We employ the contrastive loss function (Eq. 1) for fine-tuning the parameters of the MFBE.

$$L(x_i, y_i^+, y_{i,1}^-, \ldots, y_{i,n}^-) = -log \frac{e^{sim(x_i, y_i^+)}}{e^{sim(x_i, y_i^+)} + \sum_j e^{sim(x_i, y_{i,j}^-)}} \tag{1}$$

We train three variants of unsupervised and supervised MFBE models - MFBE$_{unsup}$, MFBE$_{sup}$, and MFBE$_{sup^*}$. Triplets used for each variant are as follows:

$$triplets_{unsup} = \bigcup_{m_s \in M_s} \{(F^A, F^{m_s}, \tilde{F}^{m_s}), (F^{m_s}, F^A, \tilde{F}^{m_s})\} \quad (2)$$

$$triplets_{sup} = \bigcup_{m \in M} \{(q, F^m, \tilde{F}^m), (F^m, q, \tilde{F}^m)\} \quad (3)$$

$$triplets_{sup^*} = triplets_{unsup} \cup triplet_{sup} \quad (4)$$

Positive and Negative FAQs - In FAQ datasets, positive examples are explicitly present from the manual annotation of query-FAQ, i.e., (q, F^m) pairs. For boosting the number of training samples, we consider both the pairs (F^Q, F^A) and (F^{QC}, F^A) from an FAQ as proxy for (q, F^m). But, negative examples are not explicitly present. However, the choice of negative samples play a decisive role in learning an effective text encoder. So, we consider *Gold* negatives [9], i.e., positive FAQs paired with other non-matching queries that appear in the training set. In order to make the training computation more efficient, we make use of *Gold* FAQs from the same mini-batch as negatives, termed as in-batch negatives [9]. In addition, we consider one negative FAQ sample for each matching query-FAQ pair.

Pre-computation of FAQ Embeddings - After training the text-encoder $E(\cdot)$, we use it to pre-compute and store T_m (Eq. 5) corresponding to the field m. We use the best performing field $m^{best} \in M$ (Eq. 6), in terms of *Acc* (accuracy@1, Eq. 7), evaluated on the test set $\mathcal{D}_{test}$ for the query-FAQ matching task.

$$T_m = \left[f_1^m f_2^m \dots f_N^m\right]^T \quad (5)$$

where $f_i^m = E(F_i^m)$, for $i = 1, 2, \dots, N$.

$$m^{best} = \arg\max_{m \in M} Acc(\mathcal{D}_{test}) \quad (6)$$

In Eq. 6, test set $\mathcal{D}_{test} = \{(q, F) | F \in \mathcal{F} \text{ is a matching FAQ for query } q \in \mathcal{Q}_{test}\}$, where $\mathcal{Q}_{test} \subset \mathcal{Q}$. As each FAQ consists of different FAQ fields F^m $(\forall m \in M)$, we define $Acc(\cdot)$ as below.

$$Acc(\mathcal{D}_{test}) = \frac{1}{|\mathcal{Q}_{test}|} \sum_{q \in \mathcal{Q}_{test}} \mathbb{1}[\{\arg\max_i cosineSim(f_q, f_i^m)\} \cap gt(q) \neq \Phi] \quad (7)$$

Here, $gt(q)$ denotes the set of FAQ indexes present in the ground truth labels corresponding to query q.

Inference - During inference stage, for a given query q, we first compute the input query embedding f_q. Then, we calculate cosine similarity scores of f_q and row vectors of $T_{m^{best}}$ (which are pre-computed embeddings of FAQs) and return the top-k candidates, sorted in descending order of their scores.

Table 1. Datasets used in this paper. For open datasets queries are separated into five folds. (JA = Japanese, EN = English)

Type	Name	Language	#FAQs	#Queries		#Avg. sentence length		
				Train	Test	Query	FAQ-Q	FAQ-A
Internal	IDS1	JA	795	1782	825	28.5	24.8	166.6
	IDS2	JA	510	481	139	28.6	31.3	514.6
	IDS3	JA	1129	528	152	14.9	33.1	473.8
	IDS4	JA	661	505	145	17.6	33.4	471.8
Open	LocalGov	JA	1786	749		26.1	31.1	357.2
	Stack-FAQ	EN	125	1249		73.3	55.7	513.8
	COUGH	EN	7115	1201		74.4	76.5	711.7

4 Experiments, Implementation and Results

Datasets - We conduct experiments on both internal proprietary and open datasets. Table 1 shows the list of datasets and their details used in this work.

Internal Datasets - IDS1, IDS2, IDS3 and IDS4 are Japanese language company internal datasets related to e-commerce, leisure, communication and payment domains respectively. These are carefully prepared by majority voting by five native Japanese speakers across query-FAQ annotations.

Open Datasets - *LocalGov*, introduced in [24], is a Japanese language dataset which is constructed from Japanese administrative municipality domain. *Stack-FAQ* is described in [1] as an English language dataset prepared from threads in StackExchange website concerning web apps domain. *COUGH* is another English dataset [30] constructed by scraping data from 55 websites (like, CDC and WHO) containing user queries and FAQs about Covid-19.

Implementation Details - We use single NVIDIA A100 GPU with 40G VRAM for all experiments. In all our experiments, we set maximum sequence length as 256 (for training and testing) and batch size as 32 which are constrained by the choice of GPU. For all datasets, the number of training epochs is set as 15. The choice of optimizer, learning rate and embedding dimension follows from [21] for all the experiments except for multi-domain fine-tuning experiments (Table 4 last 4 rows) where learning rate is set as $2e-7$. We set weight decay $\omega = 1e-5$ by employing grid search between 1e−1 and 1e−10 (reducing by 0.1×). For each matching query-FAQ pair, we consider 10 negative FAQs (i.e., n = 10 after experimenting with other values, such as 5, 10, and 20). All our baseline experiments follow same settings as the proposed models and are initialized with LaBSE checkpoint[1]. Code and relevant material of this work can be found here[2].

Results and Discussion - We report accuracy@1 (Acc), mean reciprocal rank @5 (MRR) and normalized discounted cumulative gain @5 (NDCG) and compare our proposed models with multiple baselines, such as, BM25 (lexical

[1] https://huggingface.co/sentence-transformers/LaBSE/tree/main.
[2] https://github.com/mausamsion/MFBE.

Table 2. Results on internal datasets.

Model	IDS1			IDS2			IDS3			IDS4		
	Acc	MRR	NDCG	Acc	MRR	NDCG	Acc	MRR	NDCG	Acc	MRR	NDCG
Baselines												
BM25 [23]	23.5	29.8	30.6	36.0	43.2	30.3	38.2	47.3	31.7	35.9	46.0	31.8
LaBSE [7]	29.9	38.6	40.1	36.0	48.4	35.8	49.3	59.2	41.7	51.0	59.0	40.4
DPR [9]	41.5	51.7	20.5	47.5	55.9	34.6	46.7	57.9	37.0	46.9	57.9	35.3
Proposed												
MFBE$_{unsup}$	33.2	43.9	46.2	51.1	60.3	44.7	51.3	59.7	43.8	51.0	61.7	43.3
MFBE$_{sup}$	51.0	60.6	62.4	60.4	**71.3**	**58.1**	57.9	67.8	52.8	64.1	72.2	54.9
MFBE$_{sup^*}$	**59.2**	**66.4**	**66.8**	**61.2**	69.8	55.1	**57.9**	**68.9**	**53.9**	**65.5**	**74.9**	**57.8**

Table 3. Results on open datasets.

Model	LocalGov			Stack-FAQ			COUGH		
	Acc	MRR	NDCG	Acc	MRR	NDCG	Acc	MRR	NDCG
Baselines									
BM25 [23]	26.4	33.4	26.5	40.0	49.1	52.9	39.1	48.0	26.1
LaBSE [7]	26.3	35.9	28.8	43.6	55.6	60.6	23.0	31.7	17.7
DPR [9]	46.5	55.2	44.7	88.8	91.9	93.0	42.8	51.5	29.3
Proposed									
MFBE$_{unsup}$	53.8	63.5	54.9	85.9	91.0	92.8	46.0	57.7	35.5
MFBE$_{sup}$	61.3	72.0	64.8	**98.0**	**98.8**	**99.0**	**53.1**	**65.2**	**40.3**
MFBE$_{sup^*}$	**62.9**	**72.3**	**65.7**	96.5	98.0	98.5	50.1	61.0	37.7

feature-based IR model), LaBSE (heavy-weight dense multilingual text encoder), and DPR (dense bi-encoder with independently learned encoders). Across all baselines, we keep the train settings related to field combinations same as our MFBE$_{sup^*}$ model. From Table 2, BM25 is the worst performing model, because of dependence on lexical features, hence fails to capture semantics among queries and FAQs. MFBE$_{sup^*}$ outperforms all baselines across different datasets. In Table 3, we report 5-fold cross validation results. MFBE$_{sup}$ and MFBE$_{sup^*}$ shows the best performance across all the baselines. MFBE$_{unsup}$, which has zero query knowledge, performs better than DPR, which is the best performing baseline of all, in two datasets out of three. Table 5 illustrates sample input query and output top-1 prediction of DPR and MFBE$_{unsup}$ models where we see that DPR fails to capture the semantic meaning of the input thus returning irrelevant response.

Cross-Domain - Table 4 shows the cross-domain results of MFBE$_{sup^*}$ where the model is trained on one dataset and evaluated on completely unknown ones (zero-shot setting). For example, the first row corresponds to the case where MFBE$_{sup^*}$ is trained on IDS1 and evaluated on IDS2, IDS3, and IDS4. We observe that the zero-shot performance of MFBE$_{sup^*}$ is the best, when trained

Table 4. Results of cross- and multi-domain experiments using $MFBE_{sup^*}$ model.

Model	IDS1			IDS2			IDS3			IDS4		
	Acc	MRR	NDCG	Acc	MRR	NDCG	Acc	MRR	NDCG	Acc	MRR	NDCG
Cross-domain												
IDS_1	–	–	–	47.5	62.2	55.2	55.3	66.0	53.0	53.1	65.7	53.6
IDS_2	37.3	48.3	50.3	–	–	–	47.4	56.9	39.2	48.3	58.3	40.5
IDS_3	42.4	53.1	55.5	44.6	57.0	44.2	–	–	–	57.2	63.6	45.8
IDS_4	40.5	51.9	54.3	44.6	58.5	44.8	50.0	61.4	46.1	–	–	–
Multi-domain												
IDS^*	55.5	64.2	65.5	62.6	71.2	54.8	55.3	66.2	50.8	59.3	67.9	50.8
Cross and Multi-domain												
IDS_1^*	55.0	64.0	65.2	62.6	71.3	54.9	55.3	66.2	50.7	59.3	68.0	50.7
IDS_2^*	54.4	63.3	64.7	63.3	71.9	55.4	54.6	65.6	50.6	59.3	68.2	51.6
IDS_3^*	54.8	63.8	65.1	61.9	70.5	54.5	55.9	67.1	51.4	57.9	67.2	51.0
IDS_4^*	54.2	63.0	64.2	61.2	70.3	53.4	56.6	67.0	50.6	61.4	69.8	52.9

Table 5. Example input and outputs from COUGH (EN) and LocalGov (JA) datasets of two baseline models and our $MFBE_{sup^*}$ model. In all the examples, $MFBE_{sup^*}$ returns the most relevant response in top-1.

Input	Top FAQ (question) prediction	
	DPR	$MFBE_{sup^*}$
Is personal protective equipment sufficient to protect others?	Are there exemptions to who has to wear a face covering?	Are cloth face coverings the same as personal protective equipment (PPE)?
How should i adjust my feeling during pendemic period?	I traveled and have been sick ever since I got back. What should I do?	During this time, it is important to be S.M.A.R.T. about staying active
国民年金の納付書を誤って捨ててしまいました。どうしたらいいでしょうか?	【児童手当現況届】間違って記入した場合は、どうしたらいいですか。	国民年金保険料を支払いたいのですが納付書をなくしてしまいました。
納税証明書が必要なのですが、どこで入手できますか?	介護保険料の納付書を紛失してしまった。再交付してほしいのですが?	納税証明書（法人市民税、事業所税を除く）を取得したい。

on IDS1, compared to others. This is because IDS1 consists of a large number of labelled user queries (nearly 1.7k) in the train split.

Multi-domain - Here $MFBE_{sup^*}$ model is trained on all the internal datasets IDS[1–4] denoted as IDS*. The average drop in Acc is only 4%, compared to the last row of Table 2. The model trained on IDS* is more robust across multiple domains with better performance in some cases (e.g., IDS2), making it useful for leveraging cross-domain knowledge.

Cross and Multi-domain - In this case, the model is first trained on IDS*, then fine-tuned on one dataset, and finally evaluated on all datasets. For example, the sixth row corresponds to the case, where $MFBE_{sup^*}$ is first trained on

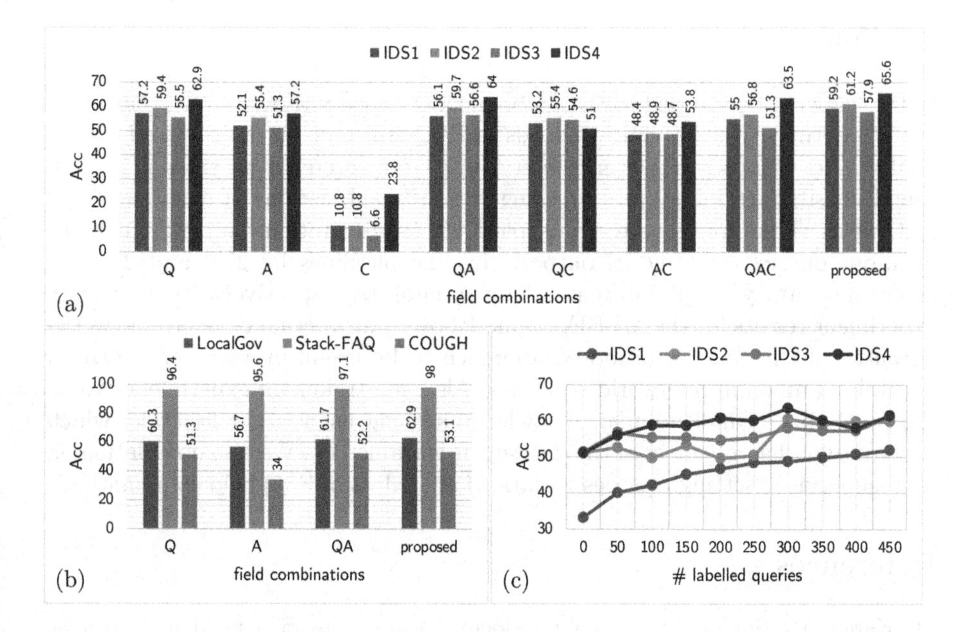

Fig. 3. Ablation experiments with MFBE_{sup^*} model. (a) Variation across multi-field combinations on internal datasets (b) Variation across multi-field combinations on open datasets (c) Variation in the number of training query-FAQ pairs

IDS*, then fine-tuned on IDS1 (denoted as IDS_1^*), and finally evaluated on all the datasets, i.e., IDS[1–4] (same for the last three rows). There is no significant change in performance when compared to results of fifth row because of prior exposure to the corresponding datasets.

Ablation Experiments - We train MFBE_{sup^*}, $\forall m \in M$ (taking one at a time) and varying the number of labelled queries. In Fig. 3-(a) and (b) we show the performance of our model as FAQ field combinations are changed (which consistent at both training and testing). It is observed that using category information adds noise and degrades performance which can be due to the inefficient usage of this field. The category field has keywords and hierarchy which needs to be leveraged but in this work, for simplicity, we concatenated these keywords to other input fields making it as a part of input string. The 'proposed' numbers are the best numbers across all of our proposed models (as discussed in Table 2 and 3). From Fig. 3-(c) it is observed that our MFBE_{sup^*} model is a good candidate for the scenarios where there is less annotated data with the accuracy flattening after around 300 query-FAQ pairs. This makes it suitable for bootstrapping to new domains where there are FAQ documents and no or less query-FAQ pairs.

5 Conclusion

In this paper, we proposed MFBE, a bi-encoder based retrieval model that make use of information from multiple fields in FAQs to improve the text embedding quality and thus better sentence matching. We also create an extended set of pseudo-positive training pairs by using various combinations of user-query and FAQ fields. Then we use these multiple FAQ representations to make inference on input queries. Our model outperforms the baselines by 27% and 23% (in terms of accuracy@1) on internal and open-datasets, respectively. Cross-domain experiment results for the MFBE_{sup^*} model over our internal datasets shows the potential of this kind of proposed approach to be useful in cold-start settings, which is common in real-world scenarios. Also, multi-domain experiment proves the possibility of multi-domain knowledge sharing using a single model which performs good across most of the datasets it is trained on. We also do ablation on semi-supervised setting (queries variation) and effect of FAQ field combinations.

References

1. Karan, M., Snajder, J.: Paraphrase-focused learning to rank for domain-specific frequently asked questions retrieval. Expert Syst. Appl. **91**, 418–433 (2018)
2. Assem, H., Dutta, S., Burgin, E.: DTAFA: decoupled training architecture for efficient FAQ retrieval. In: Proceedings of the 2021 Annual Meeting of the Special Interest Group on Discourse and Dialogue, pp. 423–430 (2021)
3. Bian, N., Han, X., Chen, B., Sun, L.: Benchmarking knowledge-enhanced commonsense question answering via knowledge-to-text transformation. In: Proceedings of the 2021 AAAI Conference on Artificial Intelligence, vol. 35, pp. 12574–12582 (2021)
4. Chen, T., Kornblith, S., Norouzi, M., Hinton, G.: A simple framework for contrastive learning of visual representations. In: Proceedings of the 2020 International Conference on Machine Learning, pp. 1597–1607 (2020)
5. Devlin, J., Chang, M.W., Lee, K., Toutanova, K.: BERT: pre-training of deep bidirectional transformers for language understanding. In: Proceedings of the 2019 Conference of the NAACL: HLT, vol. 1 (Long and Short Papers) (2019)
6. Dutta, S., Assem, H., Burgin, E.: Sequence-to-sequence learning on keywords for efficient FAQ retrieval. arXiv preprint arXiv:2108.10019 (2021)
7. Feng, F., Yang, Y., Cer, D., Arivazhagan, N., Wang, W.: Language-agnostic BERT sentence embedding. In: Proceedings of the 2022 Annual Meeting of the Association for Computational Linguistics, pp. 878–891
8. Gao, T., Yao, X., Chen, D.: SimCSE: simple contrastive learning of sentence embeddings. In: Proceedings of the 2021 Conference on Empirical Methods in Natural Language Processing, pp. 6894–6910
9. Karpukhin, V., et al.: Dense passage retrieval for open-domain question answering. In: Proceedings of the 2020 Conference on Empirical Methods in NLP, pp. 6769–6781 (2020)
10. Khattab, O., Potts, C., Zaharia, M.: Relevance-guided supervision for OpenQA with ColBERT. Trans. Assoc. Comput. Linguist. **9**, 929–944 (2021)
11. Kim, T., Yoo, K.M., Lee, S.G.: Self-guided contrastive learning for BERT sentence representations. arXiv preprint arXiv:2106.07345 (2021)

12. Kuzi, S., Zhang, M., Li, C., Bendersky, M., Najork, M.: Leveraging semantic and lexical matching to improve the recall of document retrieval systems: a hybrid approach. ArXiv abs/2010.01195 (2020)
13. Liu, F., Jiao, Y., Massiah, J., Yilmaz, E., Havrylov, S.: Trans-encoder: unsupervised sentence-pair modelling through self-and mutual-distillations. arXiv preprint arXiv:2109.13059 (2021)
14. Liu, F., Vulić, I., Korhonen, A., Collier, N.: Fast, effective, and self-supervised: transforming masked language models into universal lexical and sentence encoders. In: Proceedings of the 2021 Conference on Empirical Methods in NLP, pp. 1442–1459
15. Liu, L., Wu, Q., Chen, G.: Improving dense FAQ retrieval with synthetic training. In: Proceedings of the 7th IEEE International Conference on Network Intelligence and Digital Content, pp. 304–308 (2021)
16. Liu, Y., Hashimoto, K., Zhou, Y., Yavuz, S., Xiong, C., Yu, P.: Dense hierarchical retrieval for open-domain question answering. In: Findings of the Association for Computational Linguistics: EMNLP 2021, pp. 188–200 (2021)
17. Manzoor, A., Jannach, D.: Towards retrieval-based conversational recommendation. CoRR abs/2109.02311 (2021)
18. Mazaré, P.E., Humeau, S., Raison, M., Bordes, A.: Training millions of personalized dialogue agents. arXiv preprint arXiv:1809.01984 (2018)
19. Mesquita, T., Martins, B., Almeida, M.: Dense template retrieval for customer support. In: Proceedings of the 2022 International Conference on Computational Linguistics, pp. 1106–1115 (2022)
20. Qu, Y., et al.: RocketQA: an optimized training approach to dense passage retrieval for open-domain question answering. In: Proceedings of the 2021 Conference of the NAACL: HLT, pp. 5835–5847 (2020)
21. Reimers, N., Gurevych, I.: Sentence-BERT: sentence embeddings using Siamese BERT-networks. In: Proceedings of the 2019 Conference on Empirical Methods in NLP and the 9th International Joint Conference on NLP (EMNLP-IJCNLP), pp. 3982–3992
22. Ren, R., et al.: RocketQAv2: a joint training method for dense passage retrieval and passage re-ranking. In: Proceedings of the 2021 Conference on Empirical Methods in NLP, pp. 2825–2835
23. Robertson, S.E., Zaragoza, H.: The probabilistic relevance framework: BM25 and beyond. Found. Trends Inf. Retr. **3**(4), 333–389 (2009)
24. Sakata, W., Shibata, T., Tanaka, R., Kurohashi, S.: FAQ retrieval using query-question similarity and BERT-based query-answer relevance. In: Proceedings of the 2019 International ACM SIGIR Conference on Research and Development in Information Retrieval, pp. 1113–1116 (2019)
25. Seo, J., et al.: Dense-to-question and sparse-to-answer: hybrid retriever system for industrial frequently asked questions. Mathematics **10**(8), 1335 (2022)
26. Tseng, W.T., Wu, C.Y., Hsu, Y.C., Chen, B.: FAQ retrieval using question-aware graph convolutional network and contextualized language model. In: Proceedings of the 2021 Asia-Pacific Signal and Information Processing Association Annual Summit and Conference, pp. 2006–2012 (2021)
27. Wohlwend, J., Elenberg, E.R., Altschul, S., Henry, S., Lei, T.: Metric learning for dynamic text classification. In: Proceedings of the 2019 Workshop on Deep Learning Approaches for Low-Resource NLP, pp. 143–152 (2019)
28. Wolf, T., Sanh, V., Chaumond, J., Delangue, C.: TransferTransfo: a transfer learning approach for neural network based conversational agents. arXiv preprint arXiv:1901.08149 (2019)

29. Yamada, I., Asai, A., Hajishirzi, H.: Efficient passage retrieval with hashing for open-domain question answering. CoRR abs/2106.00882 (2021)
30. Zhang, X.F., Sun, H., Yue, X., Lin, S., Sun, H.: COUGH: a challenge dataset and models for COVID-19 FAQ retrieval. In: Proceedings of the 2021 Conference on Empirical Methods in NLP, pp. 3759–3769 (2020)

Isotropic Representation Can Improve Dense Retrieval

Euna Jung[1], Jungwon Park[1], Jaekeol Choi[1,2], Sungyoon Kim[4], and Wonjong Rhee[1,3](✉)

[1] GSCST at Seoul National University, Seoul, South Korea
{xlpczv,quoded97,jaekeol.choi,wrhee}@snu.ac.kr
[2] Naver Corporation, Seongnam-si, South Korea
[3] IPAI, AIIS at Seoul National University, Seoul, South Korea
[4] Electrical & Computer Engineering at Seoul National University, Seoul, South Korea
clifter0122@snu.ac.kr

Abstract. The latest Dense Retrieval (DR) models typically encode queries and documents using BERT and subsequently apply a cosine similarity-based scoring to determine the relevance. BERT representations, however, are known to follow an anisotropic distribution of a narrow cone shape and such an anisotropic distribution can be undesirable for relevance estimation. In this work, we first show that BERT representations in DR also follow an anisotropic distribution. We adopt unsupervised post-processing methods of Normalizing Flow and whitening to cope with the problem, and develop a token-wise method in addition to the sequence-wise method. We show that the proposed methods can effectively enhance the isotropy of representations, thereby improving the performance of DR models such as ColBERT and RepBERT. To examine the potential of isotropic representation for improving the robustness of DR models, we investigate out-of-distribution tasks where the test dataset differs from the training dataset. The results show that isotropic representation can certainly achieve a generally improved performance (The code is available at https://github.com/SNU-DRL/IsotropicIR.git).

Keywords: Dense Retrieval · Isotropic Representation · Normalizing Flow · Whitening · Robustness

1 Introduction

Recently, many Dense Retrieval (DR) models encode representations of queries and documents using BERT and estimate the relevance scores based on the simple similarity function such as cosine similarity or dot product. The representations of language models such as BERT, however, are known to follow an anisotropic distribution [5,15,17,27,29,32]. Anisotropic distribution refers to a

---

Supplementary Information The online version contains supplementary material available at https://doi.org/10.1007/978-3-031-33380-4_10.

H. Kashima et al. (Eds.): PAKDD 2023, LNAI 13937, pp. 125–137, 2023.
https://doi.org/10.1007/978-3-031-33380-4_10

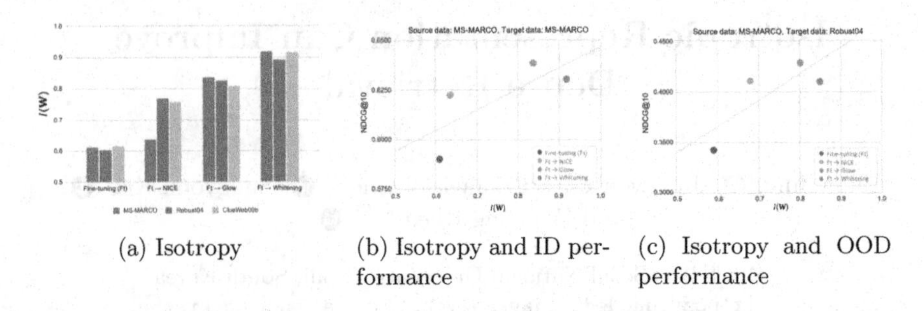

(a) Isotropy

(b) Isotropy and ID performance

(c) Isotropy and OOD performance

Fig. 1. Isotropy and re-ranking performance of ColBERT. For increasing isotropy, three post-processing methods are studied. In (a), a metric of isotropy ($I(\mathbf{W})$) is shown for BERT representations of three dense retrieval datasets. (b) and (c) show isotropy and re-ranking performance of ColBERT on MS-MARCO for In-Distribution (ID) setting and Out-Of-Distribution (OOD) setting, respectively.

directionally non-uniform distribution, such as a narrow cone [16]. If representations are anisotropically distributed, relevance estimation of DR models can be misleading. Because DR employs simple similarity functions for efficient computation, it is important to alleviate this anisotropy problem. In this study, we aim to show that BERT representations of DR also follow an anisotropic distribution and to improve the performance of BERT-based DR by enforcing isotropy to the representations.

To enhance the isotropy of DR representations, we adopt post-processing methods used in the field of sentence embedding such as Normalizing Flow [16] and whitening [23] and apply them on two representative DR models, ColBERT [12] and RepBERT [33]. Since the post-processing methods used for sentence embedding transform the representation of each sequence (i.e. sentence), they cannot be directly applied to multi-vector DR models, for example ColBERT, that compute cosine similarity among the token representations. Therefore, we consider a token-wise transformation of representations in addition to the sequence-wise transformation. As we will show later, the token-wise transformation turns out to be useful even for RepBERT that is a single-vector DR model. We empirically show the effectiveness of post-processing methods and compare the token-wise and the sequence-wise transformations when applicable.

By enforcing isotropy to the BERT representations, we show that we can significantly improve the re-ranking performance of both ColBERT and RepBERT. Adopting Normalizing Flow or whitening increases the performance of ColBERT by 5.2%–8.1% and the performance of RepBERT by 8.5%–23.3% on NDCG at 10 (NDCG@10) across three datasets. In the experiment of RepBERT, we have found that either token-wise method or sequence-wise method can perform better depending on the characteristics of the dataset.

To examine the potential of isotropic representation beyond the basic re-ranking task in the In-Distribution (ID) setting where the source data used for training and the target data used for the test are the same, we additionally investigate Out-Of-Distribution (OOD) setting where the source data is different from the target data. We evaluate the robustness of DR models for OOD tasks following Wu et al. [30]. With our experiments, we have found that enforcing isotropy on BERT representations can improve the robustness of DR by 5.0%–25.0% for NDCG@10. OOD performance of ColBERT trained on MS-MARCO can even surpass the ID performance of Robust04 and ClueWeb09b when the post-processing methods are applied.

We summarize our contributions in Fig. 1. In this paper, we focus on improving both ID and OOD performances of DR models by enforcing representation isotropy.

2 Related Works

2.1 Dense Retrieval and Similarity Function

Depending on whether DR uses a single vector or multiple vectors for encoding each of queries and documents, DR models are divided into *single-vector* and *multi-vector* models. RepBERT [33], ANCE [31], and RocketQA [22] are examples of single-vector DR models. ColBERT [12] and COIL [6] are examples of multi-vector DR models, and they are generally known to perform better. Both types of DR models estimate the relevance score using a similarity function between the representation vectors of query and document. For efficient computation, the similarity function needs to be decomposable such that the representations of the documents can be pre-computed and stored [11], and an efficient Approximate Nearest Neighbor (ANN) retrieval [9] can be performed. Most of the decomposable similarity functions are based on cosine similarity [20].

In this paper, we focus on two representative DR models, RepBERT and ColBERT[1]. RepBERT encodes *sequence representation* for each query and document by summing up token representations and estimates the relevance score using the dot product between the *sequence representations*. On the other hand, ColBERT estimates the relevance score by summing up the cosine similarity values between *token representations*.

2.2 Anisotropic Distribution of BERT Representations

Representations of large-scale language models are known to follow anisotropic distributions [5,15,17,27,29,32]. For example, some studies [15,17,27] show the existence of outlier dimensions having extreme values in representation vectors and attribute the anisotropic distribution to the outlier dimensions. As this anisotropic distribution can negatively affect the performance of sentence embeddings, Li et al. [16] and Su et al. [23] applied Normalizing Flow and whitening respectively for the sentence embedding, where the cosine similarity between two sentences' BERT representations is used for the relevance estimation. DR is different from sentence embedding in that query and document have distinct characteristics, and some multi-vector DR models utilize token-wise similarity. In

[1] To clearly compare RepBERT and ColBERT, we simplify ColBERT to use the same encoders for both queries and documents and skipped the final projection layer that reduces the representation dimension. Also, we use cosine similarity instead of the dot product for RepBERT.

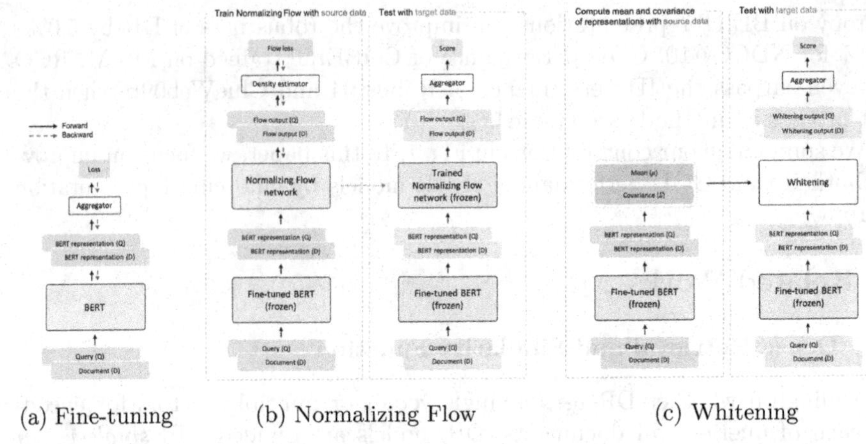

(a) Fine-tuning (b) Normalizing Flow (c) Whitening

Fig. 2. Train and test procedures of Normalizing Flow and whitening. To apply a post-processing method to DR, we first fine-tune the BERT for DR as shown in (a). In (b), we train a Normalizing Flow network while keeping the fine-tuned BERT frozen. For whitening in (c), we pre-compute the mean and covariance using the fine-tuned BERT and apply the whitening filter during the test time. "ggregator" in the test figures refers to the computation of the relevance score based on cosine similarity. For (b) and (c), the source and target data are the same for the ID re-ranking, and they are different for the OOD re-ranking.

this DR study, we show that the representations of BERT-based DR models also suffer from the anisotropic distribution and consider token-wise transformation as well as sequence-wise transformation.

2.3 Robustness of Ranking Models

The robustness of ranking models refers to the ability of models to operate properly in abnormal situations, which is an essential factor for ranking models in the real world [30]. While the robustness of the ranking models can be defined with multi-dimensional factors [7], OOD generalizability can be regarded as one of the most important factors for NRMs, which tend to show relatively poor performance for OOD tasks compared to the traditional ranking models [30]. Some of the existing works have focused on improving OOD generalizability of ranking models [26,30]. However, it was difficult to show a case where the OOD performance surpasses the ID performance. In this study, we show that the OOD performance of NRMs can be sufficiently improved by enforcing representation isotropy such that OOD outperforms the baseline ID.

3 Methodology

3.1 Enforcing Isotropy

The baseline method of training a BERT-based DR model is shown in Fig. 2(a). In our work, we enhance the isotropy of BERT representations with Normalizing Flow or whitening as shown in Fig. 2(b) and 2(c). The two post-processing processes follow entirely unsupervised frameworks.

Normalizing Flow. Normalizing Flow transforms a simple and tractable distribution into the target data distribution by applying a series of invertible and (almost everywhere) differentiable mappings [14,24,25]. By doing this, flow-based models can easily infer the target density as in Eq. (1) and perform sampling from it as in Eq. (2).

$$p_X(x) = p_Z(f(x)) \left|\det \mathbf{D}f(x)\right| \tag{1}$$

$$x = g(z) \quad \text{where} \quad z \sim p_Z(z) \tag{2}$$

where Z and X are random variables of the simple and target distributions respectively, and p_X and p_Z are the densities of X and Z, respectively. f is a function that maps X to Z, g is the inverse of f, and $\mathbf{D}f(x) = \frac{\partial f}{\partial x}$ is the Jacobian matrix of f.

As illustrated in Fig. 2(b)-left, we aim to transform BERT representations of query and document into the standard Gaussian distribution using the Normalizing Flow's density estimation. To do so, in Eq. (1), we set p_Z as the density of the standard Gaussian distribution. Then, we maximize the likelihood of representations, that is, p_X in Eq. (1). Consequently, the flow function f is trained to transform the representations $\mathbf{X}$ to follow the standard Gaussian distribution. Among the various flow-based models, we use NICE [2] and Glow [13].

For DR, we propose two different implementations of Normalizing Flow. In *token-wise* implementation, Normalizing Flow is applied to each *token representation* of BERT. In *sequence-wise* implementation, Normalizing Flow is applied to *sequence representations* made by aggregating token representations. Both implementations are applicable to single-vector models, such as RepBERT, but only the token-wise method is applicable to multi-vector models, such as ColBERT, because they do not utilize the sequence representation.

Whitening. Whitening is a linear transformation that renders the data distribution spherical. That is, it eliminates the structures of location, scale, and correlations in the distribution [4]. Let $\{x_i\}_{i=1}^N$ be a set of BERT representations where each x_i is a D-dimensional row vector. To apply whitening, a mean vector μ and an unbiased covariance matrix Σ of $\{x_i\}_{i=1}^N$ are computed as $\mu = \frac{1}{N}\sum_{i=1}^N x_i$ and $\Sigma = \frac{1}{N-1}\sum_{i=1}^N (x_i - \mu)^T (x_i - \mu)$. By SVD and with some following calculations, the whitened representation vector z_i can be presented as

$$z_i = (x_i - \mu)\, U\sqrt{\Lambda^{-1}}, \tag{3}$$

where $\Lambda \in \mathbb{R}^{D \times D}$ is a diagonal matrix with positive diagonals, $U \in \mathbb{R}^{D \times D}$ is an orthogonal matrix, and each z_i follows a distribution of zero mean and identity covariance [23].

For DR, we first pre-compute the mean and covariance using a train set of the source data as shown in Fig. 2(c)-left. For the inference shown in Fig. 2(c)-right, we obtain the whitened output z_i in Eq. (3) using the pre-computed mean and covariance. The following aggregator then computes the cosine similarities between those z_i's from queries and documents. As in the Normalizing Flow, we consider both token-wise and sequence-wise implementations of whitening.

Metrics. For a set of given representations, we measure the degree of isotropy by utilizing two metrics. Following Yu et al. [32] and Wang et al. [29], we employ $I(\mathbf{W})$ suggested by Mu et al. [19] for measuring the isotropy of representation vectors. The value of $I(\mathbf{W})$ ranges from 0 to 1, and representations having higher $I(\mathbf{W})$ tend to follow nearly isotropic Gaussian distribution.

Additionally, the average cosine similarity between representations, *avgcos*, is adopted for measuring isotropy following Ethayarajh et al. [3]. Representations vectors that are isotropically spread around the origin would have *avgcos* values close to zero.

3.2 Robustness for Out-of-Distribution Data

Under the hypothesis that NRMs having representations that follow isotropic distribution are less likely to overfit to the characteristics of the training dataset, we compare OOD performances of DR models with and without post-processing methods. In each case, we train the DR model on the source data and evaluate the re-ranking performance on the different target data. As we consider three different datasets in this work, we investigate six different combinations of OOD experiments where the source data differs from the target data. This concept of evaluating robustness is commonly addressed as OOD generalizability on an unforeseen corpus [30], and we will interchangeably use the terms OOD robustness and OOD generalizability in this work.

4 Experiments

4.1 Experimental Settings

Models, Datasets, and Ranking Metrics. We investigate two DR models, ColBERT [12] and RepBERT [33]. As for the datasets, we examined three popular document re-ranking datasets, Robust04 [28], WebTrack 2009 (ClueWeb09b) [1], and MS-MARCO [21] as in [18]. For Robust04, we used the document collections from TREC Disks 4 and 5[2] For ClueWeb09b, the document collections from ClueWeb09b[3] of WebTrack 2009 were used. We also used the

[2] 520K documents, 7.5K triplet data samples, https://trec.nist.gov/data-disks.html.
[3] 50M web pages, 4.5K triplet data samples, https://lemurproject.org/clueweb09/.

large document collections[4] from MS-MARCO. To evaluate the ranking models, we used NDCG@10 and MRR@10 as the performance metrics.

Training and Optimization. Following Huston et al. [8], we divided each of Robust04 and ClueWeb09b into five folds: three folds for training, one for validation, and the remaining one for the test. For MS-MARCO, we divided the dataset into training, validation, and test data. Using the three performance values for three different random seeds, we conducted one-tailed *t*-test under the assumption of homoscedasticity.

For fine-tuning BERT, we used a learning rate of 1e−4, batch size of 16, and an Adam optimizer. Following the advice from [34], we set the maximum epoch to be 30 for Robust04 and ClueWeb09b and 10 for MS-MARCO. We then selected the model with the highest validation performance among the checkpoints recorded after each epoch during training.

To train Normalizing Flow, we stack a Normalizing Flow network at the top of the fine-tuned BERT and train it in an unsupervised manner for either ten epochs (Robust04 and ClueWeb09b) or three epochs (MS-MARCO) with a learning rate of 1e−4, while keeping the fine-tuned BERT weights frozen. For the NICE network, we used a five-layer network with 1000 units in each layer. For the Glow network, we used two levels and depth of three following [16]. For both NICE and Glow, we used a learning rate of 1e−4.

For whitening, we pre-compute mean and covariance. As different documents are selected in each epoch, we collect training data for ten epochs (Robust04 and ClueWeb09b) or three epochs (MS-MARCO) and use the collected representation vectors for the pre-computation of mean and covariance. For all the experiments, we tuned the hyper-parameters only lightly.

Our experiments are implemented with Python 3.8, Torch 1.9.0, Transformers 4.12.5, and Huggingface-hub 0.2.1. We fine-tuned the pre-trained BERT-base-uncased model provided by the huggingface transformers. We used RTX3090 GPUs, each of which has 25.6G memory.

4.2 Experimental Results

Isotropic Representations Improve Re-Ranking Performance. In Table 1, we compare the re-ranking performance of fine-tuned ColBERT before and after applying Normalizing Flow or whitening to the BERT representations. Because ColBERT computes the cosine similarity between a query and a document's multiple token representations, we perform the token-wise post-processing. The results demonstrate that the degree of isotropy is enhanced and the re-ranking performance is improved after the post-processing. Across all three datasets, all of whitening, NICE, and Glow methods improve the re-ranking performance of the fine-tuned ColBERT by from 2.4% to 8.1% on NDCG@10. It can be confirmed that transforming BERT token representations to follow an isotropic distribution consistently improves the performance of ColBERT.

[4] 22G documents, 372K triplet data samples, https://microsoft.github.io/msmarco/TREC-Deep-Learning-2019.

Table 1. Re-ranking performance of ColBERT. We compare the performance of a fine-tuned model before and after post-processing BERT representations. NDCG at 10 (NDCG@10) and MRR at 10 (MRR@10) are used as re-ranking metrics, and $I(\mathbf{W})$ and $avgcos(\mathbf{W})$ are used to measure the isotropy of representations. Note: $^{*}p \leq 0.05$, $^{**}p \leq 0.01$ (1-tailed).

Dataset	Method	NDCG@10	MRR@10	$I(\mathbf{W})$	$avgcos(\mathbf{W})$
MS-MARCO	Fine-tuning (Ft)	0.590	0.866	0.611	0.223
	Ft → Whitening	0.630 (6.8%)*	0.901 (4.0%)*	0.918	0.009
	Ft → NICE	0.622 (5.4%)*	0.890 (2.8%)	0.636	0.191
	Ft → Glow	**0.638 (8.1%)***	**0.914 (5.5%)***	0.837	0.027
Robust04	Fine-tuning (Ft)	0.402	0.622	0.602	0.225
	Ft → Whitening	**0.423 (5.2%)****	**0.654 (5.2%)***	0.891	0.016
	Ft → NICE	0.412 (2.4%)*	0.643 (3.4%)**	0.769	0.064
	Ft → Glow	0.412 (2.4%)*	0.637 (2.4%)*	0.824	0.031
ClueWeb09b	Fine-tuning (Ft)	0.288	0.514	0.614	0.219
	Ft → Whitening	0.301 (4.7%)**	0.521 (1.4%)	0.915	0.009
	Ft → NICE	**0.305 (6.1%)****	**0.529 (3.0%)**	0.757	0.074
	Ft → Glow	0.300 (4.2%)**	0.522 (1.5%)	0.857	0.020

Table 2. Re-ranking performance of RepBERT. We compare the performance of a fine-tuned model before and after enforcing isotropy on BERT representations. For RepBERT, we compare the performances of token-wise and sequence-wise representation transformation methods. Note: $^{*}p \leq 0.05$, $^{**}p \leq 0.01$ (1-tailed).

Dataset	Method	Token-wise Method		Sequence-wise Method	
		NDCG@10	MRR@10	NDCG@10	MRR@10
MS-MARCO	Fine-tuning (Ft)	0.330	0.675	0.330	0.675
	Ft → Whitening	0.388 (17.3%)**	0.724 (7.3%)*	**0.406 (22.8%)****	0.724 (7.3%)*
	Ft → NICE	0.339 (2.6%)	0.695 (2.9%)	0.337 (2.1%)	0.687 (1.9%)
	Ft → Glow	0.365 (10.4%)*	0.718 (6.4%)	0.398 (20.4%)**	**0.728 (7.9%)***
Robust04	Fine-tuning (Ft)	0.344	0.543	0.344	0.543
	Ft → Whitening	**0.373 (8.5%)***	**0.601 (10.6%)****	0.347 (0.8%)	0.575 (5.8%)*
	Ft → NICE	0.352 (2.4%)	0.572 (5.2%)*	0.333 (-3.1%)	0.535 (-1.5%)
	Ft → Glow	0.371 (7.8%)*	0.589 (8.3%)*	0.354 (2.9%)	0.566 (4.2%)
ClueWeb09b	Fine-tuning (Ft)	0.193	0.373	0.193	0.373
	Ft → Whitening	**0.237 (23.3%)****	**0.451 (20.9%)****	0.219 (13.9%)*	0.429 (14.9%)**
	Ft → NICE	0.208 (8.0%)	0.403 (8.1%)	0.183 (-5.2%)	0.374 (0.1%)
	Ft → Glow	0.233 (21.0%)**	0.440 (17.9%)**	0.228 (18.2%)*	0.434 (16.4%)**

Table 2 presents the results for RepBERT. We compare the performance improvement of both token-wise and sequence-wise post-processing methods. The results in Table 2 show that the performance is improved for all the cases except for when NICE is applied as a sequence-wise processing method for Robust04 or ClueWeb09b. Between token-wise and sequence-wise methods, it can be observed that token-wise performs better for Robust04 and ClueWeb09b and sequence-wise method performs better for MS-MARCO. We provide a discussion on this issue in Sect. 5.2. Among the three post-processing methods, whitening almost always achieves the best performance for each dataset as long as a token-wise method is used for Robust04 or ClueWeb09b and a sequence-wise method is used for MS-MARCO. Overall, whitening improves the performance of the fine-tuned model by 8.5%–23.3% on NDCG@10 over the three datasets.

Table 3. Out-of-distribution generalizability of ColBERT. Performance and isotropy are evaluated on the target data using the models trained with the source data. Note: $^{*}p \leq 0.05$, $^{**}p \leq 0.01$ (1-tailed).

Source data	Target data	Method	NDCG@10	MRR@10	$I(\mathbf{W})$	$avgcos(\mathbf{W})$
MS-MARCO	Robust04	Fine-tuning (Ft)	0.343	0.561	0.588	0.248
		Ft → Whitening	0.412 (19.9%)**	0.644 (14.7%)**	0.849	0.022
		Ft → NICE	0.412 (20.1%)**	0.632 (12.7%)**	0.677	0.138
		Ft → Glow	**0.429 (25.0%)****	**0.658 (17.2%)****	0.799	0.041
MS-MARCO	ClueWeb09b	Fine-tuning (Ft)	0.296	0.535	0.546	0.323
		Ft → Whitening	0.310 (4.6%)	0.546 (2.1%)	0.894	0.012
		Ft → NICE	**0.311 (5.0%)**	**0.551 (3.0%)****	0.656	0.165
		Ft → Glow	0.307 (3.7%)	0.551 (2.9%)*	0.835	0.028
Robust04	MS-MARCO	Fine-tuning (Ft)	0.383	0.666	0.627	0.206
		Ft → Whitening	0.417 (8.9%)**	0.708 (6.2%)*	0.809	0.037
		Ft → NICE	0.419 (9.2%)**	0.710 (6.6%)*	0.761	0.065
		Ft → Glow	**0.419 (9.4%)****	**0.716 (7.6%)****	0.786	0.048
Robust04	ClueWeb09b	Fine-tuning (Ft)	0.219	0.408	0.618	0.219
		Ft → Whitening	0.242 (10.2%)**	0.445 (9.0%)**	0.787	0.047
		Ft → NICE	**0.245 (11.4%)****	**0.453 (11.0%)****	0.746	0.074
		Ft → Glow	0.243 (10.6%)**	0.447 (9.5%)**	0.769	0.057
ClueWeb09b	MS-MARCO	Fine-tuning (Ft)	0.451	0.795	0.611	0.232
		Ft → Whitening	0.469 (4.1%)*	0.804 (1.5%)	0.901	0.011
		Ft → NICE	**0.476 (5.7%)****	**0.820 (3.2%)***	0.765	0.069
		Ft → Glow	0.470 (4.3%)*	0.814 (2.4%)*	0.852	0.021
ClueWeb09b	Robust04	Fine-tuning (Ft)	0.359	0.571	0.606	0.237
		Ft → Whitening	0.402 (12.1%)**	**0.623 (9.1%)****	0.842	0.024
		Ft → NICE	0.398 (10.8%)**	0.610 (6.7%)*	0.743	0.080
		Ft → Glow	**0.405 (12.8%)****	0.620 (8.4%)**	0.800	0.041

Isotropic Representations Make DR Models Robust to OOD Data. We show the effectiveness of isotropic representations for OOD generalizability in Table 3. In this result for ColBERT, it can be observed that the baseline OOD generalizability of a fine-tuned model is always improved by applying a post-processing method that enforces isotropy to the BERT representations. Performance of NDCG@10 is improved by 5.0%–25.0%, and MRR@10 by 3.0%–17.2%, across the three datasets. The overall result indicates that isotropically distributed BERT representations are robust in the sense that they can perform better for an OOD dataset that has been unseen during the training.

In particular, when ColBERT is trained with MS-MARCO as the source data and tested with the target data of either Robust04 or ClueWeb09b, the OOD performance with isotropy enhancement even surpasses the ID performance. For example, an ID performance on NDCG@10 of ColBERT fine-tuned on Robust04 is 0.402, and the OOD performance when the source data is MS-MARCO is 0.343. However, when we post-process the model using Glow, the performance is improved to 0.429, which is even higher than the baseline ID performance. This result can be interpreted to be surprising because the OOD models trained *without* the target data outperformed the ID models that were trained *with* the target data. Although we did not include the results for RepBERT, similar results can be obtained for RepBERT as well. We present the OOD generalizability result for RepBERT in the supplementary material.

5 Discussion

5.1 Handling of Outlier Dimensions

The existence of outlier dimensions with extreme values has been known for BERT and other language models' representations [15,17,27]. For example, Luo et al. [17] pointed out that the outlier dimensions can have a negative impact on task performance and showed that an improvement was possible by simply clipping the outliers. In this study, we show that the outlier dimensions can also be observed in BERT representations fine-tuned for DR as shown in Fig. 3(b). Even though we didn't explicitly aim to handle such outlier dimensions, it can be seen in Fig. 3(c) and 3(d) that the outlier dimensions are tempered and become hardly observable after enforcing isotropy.

5.2 Token-wise Vs. Sequence-wise Transformation

As explained in Sect. 3.1, isotropy can be enforced with either token-wise or sequence-wise transformation. For RepBERT, which is a single-vector DR model, both methods can be applied. The results for RepBERT in Table 2 show that the token-wise method outperforms the sequence-wise method on Robust04 and ClueWeb09b, and vice versa on MS-MARCO. A possible explanation for the results is the length of the queries. While MS-MARCO tends to have long queries of complete sentences, Robust04 and ClueWeb09b tend to have keyword-based short queries [10]. Because the representations of the queries play an important role in NRMs, it might be natural for token-wise transformation to perform well on Robust04 and ClueWeb09b having short queries.

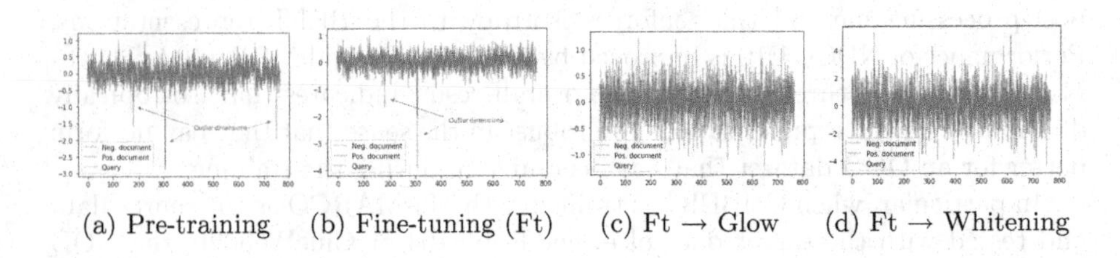

(a) Pre-training (b) Fine-tuning (Ft) (c) Ft → Glow (d) Ft → Whitening

Fig. 3. Visualization of sequence representation vectors of RepBERT: representations of a randomly sampled triplet of a query, positive document, and negative document from Robust04 are shown. The outlier dimensions with spiky values shown in representations of pre-trained BERT (a) and fine-tuned BERT (b) are tempered by enforcing isotropy as shown in (c) and (d).

5.3 Robustness and OOD Generalizability

As explained in Sect. 2.3, robustness is essential for DR, and OOD generalizability can be considered as one of the most important factors of robustness. From Table 3, it can be confirmed that a significant improvement in OOD generalizability can be attained by enforcing isotropy. When the source data is MS-MARCO, we were able to make the OOD performance with isotropy enhancement even surpass the baseline ID performance. While we have analyzed only the most fundamental scenarios, the results imply that it might be possible to further improve the robustness of DR models by enforcing isotropy and concurrently considering more complex schemes such as the use of multiple source datasets, advancement of the methods for enforcing isotropy, etc.

6 Conclusion

In this work, we have confirmed that the representations of BERT-based DR models are anisotropically distributed. Such an anisotropy can negatively affect the performance of DR models. We applied post-processing methods such as Normalizing Flow and whitening and have shown how to apply the methods in token-wise and sequence-wise manners in DR. With the proposed methods, we were able to improve the re-ranking performance of ColBERT and RepBERT for all the cases that we have studied. In addition to the commonly studied re-ranking with an in-distribution dataset, we have shown that isotropy of representations can be an essential factor for enhancing robustness of DR models. For the out-of-distribution tasks, we were able to achieve large improvements in many cases. Based on our results, isotropy can be deemed to be a crucial element for studying and improving the representations of DR models.

Acknowledgements. This work was supported by Naver corporation (Development of an Improved Neural Ranking Model.), National Research Foundation of Korea (NRF) grant funded by the Korea government (MSIT) (No. NRF-2020R1A2C2007139), and IITP grant funded by the Korea government (No. 2021-0-01343, Artificial Intelligence Graduate School Program (Seoul National University)).

References

1. Callan, J., Hoy, M., Yoo, C., Zhao, L.: Clueweb09 data set (2009)
2. Dinh, L., Krueger, D., Bengio, Y.: Nice: non-linear independent components estimation. arXiv preprint arXiv:1410.8516 (2014)
3. Ethayarajh, K.: How contextual are contextualized word representations? Comparing the geometry of BERT, ELMO, and GPT-2 embeddings. arXiv preprint arXiv:1909.00512 (2019)
4. Friedman, J.H.: Exploratory projection pursuit. J. Am. Stat. Assoc. **82**(397), 249–266 (1987)
5. Gao, J., He, D., Tan, X., Qin, T., Wang, L., Liu, T.Y.: Representation degeneration problem in training natural language generation models. arXiv preprint arXiv:1907.12009 (2019)

6. Gao, L., Dai, Z., Callan, J.: Coil: revisit exact lexical match in information retrieval with contextualized inverted list. arXiv preprint arXiv:2104.07186 (2021)
7. Goren, G., Kurland, O., Tennenholtz, M., Raiber, F.: Ranking robustness under adversarial document manipulations. In: The 41st International ACM SIGIR Conference on Research & Development in Information Retrieval, pp. 395–404 (2018)
8. Huston, S., Croft, W.B.: Parameters learned in the comparison of retrieval models using term dependencies. University of Massachusetts, IR (2014)
9. Johnson, J., Douze, M., Jégou, H.: Billion-scale similarity search with GPUs. IEEE Trans. Big Data **7**(3), 535–547 (2019)
10. Jung, E., Choi, J., Rhee, W.: Semi-Siamese Bi-encoder neural ranking model using lightweight fine-tuning. In: Proceedings of the ACM Web Conference 2022, pp. 502–511 (2022)
11. Karpukhin, V., et al.: Dense passage retrieval for open-domain question answering. arXiv preprint arXiv:2004.04906 (2020)
12. Khattab, O., Zaharia, M.: Colbert: efficient and effective passage search via contextualized late interaction over BERT. In: Proceedings of the 43rd International ACM SIGIR Conference on Research and Development in Information Retrieval, pp. 39–48 (2020)
13. Kingma, D.P., Dhariwal, P.: Glow: generative flow with invertible 1×1 convolutions. In: Advances in Neural Information Processing Systems, vol. 31 (2018)
14. Kobyzev, I., Prince, S.J., Brubaker, M.A.: Normalizing flows: an introduction and review of current methods. IEEE Trans. Pattern Anal. Mach. Intell. **43**(11), 3964–3979 (2020)
15. Kovaleva, O., Kulshreshtha, S., Rogers, A., Rumshisky, A.: BERT busters: outlier dimensions that disrupt transformers. arXiv preprint arXiv:2105.06990 (2021)
16. Li, B., Zhou, H., He, J., Wang, M., Yang, Y., Li, L.: On the sentence embeddings from pre-trained language models. arXiv preprint arXiv:2011.05864 (2020)
17. Luo, Z., Kulmizev, A., Mao, X.: Positional artefacts propagate through masked language model embeddings. arXiv preprint arXiv:2011.04393 (2020)
18. MacAvaney, S., Yates, A., Cohan, A., Goharian, N.: CEDR: contextualized embeddings for document ranking. In: Proceedings of the 42nd International ACM SIGIR Conference on Research and Development in Information Retrieval, pp. 1101–1104 (2019)
19. Mu, J., Bhat, S., Viswanath, P.: All-but-the-top: simple and effective postprocessing for word representations. arXiv preprint arXiv:1702.01417 (2017)
20. Mussmann, S., Ermon, S.: Learning and inference via maximum inner product search. In: International Conference on Machine Learning, pp. 2587–2596. PMLR (2016)
21. Nguyen, T., et al.: MS MARCO: a human generated machine reading comprehension dataset. In: CoCo@ NIPS (2016)
22. Qu, Y., et al.: RocketQA: an optimized training approach to dense passage retrieval for open-domain question answering. arXiv preprint arXiv:2010.08191 (2020)
23. Su, J., Cao, J., Liu, W., Ou, Y.: Whitening sentence representations for better semantics and faster retrieval. arXiv preprint arXiv:2103.15316 (2021)
24. Tabak, E.G., Turner, C.V.: A family of nonparametric density estimation algorithms. Commun. Pure Appl. Math. **66**(2), 145–164 (2013)
25. Tabak, E.G., Vanden-Eijnden, E.: Density estimation by dual ascent of the log-likelihood. Commun. Math. Sci. **8**(1), 217–233 (2010)
26. Thakur, N., Reimers, N., Lin, J.: Domain adaptation for memory-efficient dense retrieval. arXiv preprint arXiv:2205.11498 (2022)

27. Timkey, W., van Schijndel, M.: All bark and no bite: rogue dimensions in transformer language models obscure representational quality. arXiv preprint arXiv:2109.04404 (2021)
28. Voorhees, E.M., et al.: Overview of TREC 2004. In: TREC (2004)
29. Wang, L., Huang, J., Huang, K., Hu, Z., Wang, G., Gu, Q.: Improving neural language generation with spectrum control. In: International Conference on Learning Representations (2019)
30. Wu, C., Zhang, R., Guo, J., Fan, Y., Cheng, X.: Are neural ranking models robust? arXiv preprint arXiv:2108.05018 (2021)
31. Xiong, L., et al.: Approximate nearest neighbor negative contrastive learning for dense text retrieval. arXiv preprint arXiv:2007.00808 (2020)
32. Yu, S., Song, J., Kim, H., Lee, S., Ryu, W.J., Yoon, S.: Rare tokens degenerate all tokens: improving neural text generation via adaptive gradient gating for rare token embeddings. In: Proceedings of the 60th Annual Meeting of the Association for Computational Linguistics, vol. 1 (Long Papers), pp. 29–45 (2022)
33. Zhan, J., Mao, J., Liu, Y., Zhang, M., Ma, S.: RepBERT: contextualized text embeddings for first-stage retrieval. arXiv preprint arXiv:2006.15498 (2020)
34. Zhang, T., Wu, F., Katiyar, A., Weinberger, K.Q., Artzi, Y.: Revisiting few-sample BERT fine-tuning. arXiv preprint arXiv:2006.05987 (2020)

Knowledge-Enhanced Prototypical Network with Structural Semantics for Few-Shot Relation Classification

Yanhu Li[1], Taolin Zhang[2], Dongyang Li[1], and Xiaofeng He[1,3(✉)]

[1] School of Computer Science and Technology, East China Normal University, Shanghai, China
51215901017@stu.ecnu.edu.cn
[2] School of Software Engineering, East China Normal University, Shanghai, China
[3] NPPA Key Laboratory of Publishing Integration Development, ECNUP, Shanghai, China
hexf@cs.ecnu.edu.cn

Abstract. Few-shot relation classification (RC) aims to determine the labeled relation between two entities in a given sentence using only a few training instances. Previous studies integrate models with explicit triple knowledge, using the inherent concepts of entities to improve the instance representation. However, these studies neglect the implicit structural knowledge present in the knowledge graph (KG). In this paper, we present SKProto, a knowledge-enhanced prototypical network that leverages deep structured semantic knowledge from the multi-hop neighbors of entity-linked concepts. Specifically, we propose a concept-guided hybrid attention mechanism to learn implicit structural semantic knowledge for enhancing the context-aware instance representation. To further distinguish subtle semantic differences among the concepts, the multi-granularity semantic distinction approach is proposed to construct the negative samples with various difficulties (i.e. hard, medium, and easy) based on the conceptual hierarchical structure. Experimental results on the FewRel 2.0 benchmark show that SKProto outperforms state-of-the-art models. We also demonstrate that SKProto has better robustness than other competitive models in low-shot scenarios.

Keywords: Relation Classification · Few-shot Learning · Knowledge Graph · Contrastive Learning

1 Introduction

As the cost of data annotation grows and the long-tail distribution problem becomes more pronounced [8], few-shot relation classification (RC) [15] has been proposed as a solution for low-resource scenarios. This approach allows for the classification of relations using only a small amount of annotated data. Unlike data-driven methods [18,24,30], few-shot RC relies on discovering common implicit structures across a collection of similar tasks and then using learned internal knowledge to generalize to new scenarios [26].

H. Kashima et al. (Eds.): PAKDD 2023, LNAI 13937, pp. 138–149, 2023.
https://doi.org/10.1007/978-3-031-33380-4_11

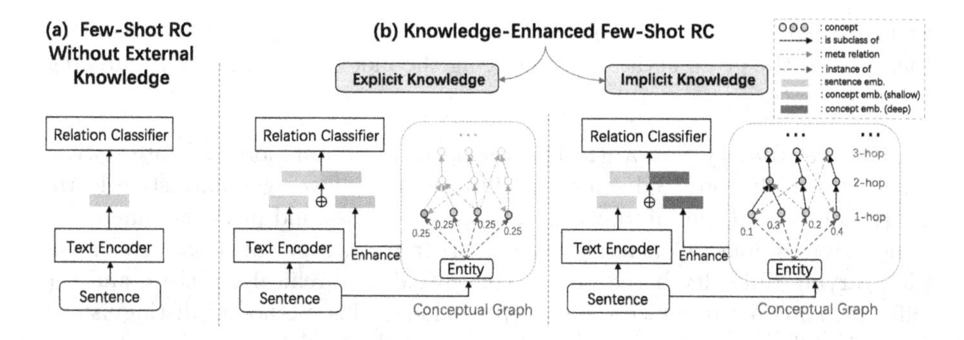

Fig. 1. Examples of the few-shot RC without external knowledge and the knowledge-enhanced few-shot RC methods. Existing knowledge-enhanced methods only utilize the explicit triple concepts, neglecting the implicit structural knowledge generated by structural multi-hop relation-based concepts. (Best viewed in color)

In the literature, few-shot RC methods can be divided into two categories: methods without external knowledge and knowledge-enhanced methods. The former relies on raw sentences, while the latter incorporates prior knowledge to improve performance. (1) Few-shot RC methods without external knowledge make inferences based solely on raw text, as there is insufficient labeled data to use. While these methods can work well in some cases [2,12,15], the limited information in raw text can reduce their performance. (2) Knowledge-enhanced few-shot RC methods incorporate external data as prior knowledge, such as textual descriptions [3,29] and structural knowledge graph (KG). In particular, conceptual graphs play a crucial role in semantic search, serving as a unique type of KG. It contains more stable implicit relations, providing richer semantic information to the text. Therefore, recent studies [28,29] integrate models with conceptual graphs, using entity-linked concepts to refine the instance representation. However, existing knowledge-enhanced few-shot RC methods only incorporate explicit triple knowledge, neglecting the multi-hop relation-based implicit structural knowledge present in the conceptual graphs. As shown in Fig. 1, incorporating 2-hop, 3-hop, or even higher-level concepts can provide more implicit structural semantic information than using only 1-hop concept information. Moreover, the subtle differences between similar concepts make it challenging for the model to inject knowledge accurately [5].

To alleviate the lack of semantic information and improve the accuracy of knowledge injection, we present a knowledge-enhanced prototypical network (SKProto) that incorporates implicit structural knowledge into the model. The two main modules are presented as follows:

Concept-Guided Hybrid Attention. We propose a hybrid attention mechanism to capture the implicit structural knowledge inherent in the conceptual graphs. The concept-level attention incorporates structural-semantic and textual-semantic information through multi-hop fusion for entity-linked concepts. The entity-level attention calculates the importance scores of entity-linked concepts for the given entities, encouraging the model to focus on the most relevant concepts.

Multi-Granularity Semantic Distinction. To obtain more accurate distinctions between fine-grained semantic differences, we leverage contrastive learning [14,22] to pull the information of positive samples and push the differences of negative samples. Concretely, we construct three types of negative samples with varying difficulty based on the conceptual hierarchical structure and the different domains[1] to which the concepts belong. This method of distinguishing semantic differences improves the robustness of the model.

We evaluate the effectiveness of our method on the Fewrel 2.0 benchmark, and experiment results show the superiority over other methods (by 2.21% on 1-shot and 0.99% on 5-shot). We further show that SKProto has better robustness than other competitive models in the low-shot scenarios.

2 Related Work

Few-Shot RC Without External Knowledge. Existing basic few-shot RC methods can be classified into two categories: metric-based methods and gradient-based methods.

- **Metric-Based Methods** focus on representing instances in a metric space to enable classification based on the distance between instances. Typically, Prototypical Networks [23] learn a metric space and take the average of the instance representation for each class of relation as its prototype representation. GMatching [27] uses a matching network to seek similar instances for the given instance. RSN [7] propose a siamese convolutional network to build the relation metric model.
- **Gradient-Based Methods** generalize the relation classifier and rapidly adapt the model to a new task via a few gradient update steps. MAML [11] adopts a model-agnostic gradient update strategy to produce class-agnostic initialization that can fast adapt to new relations. As an improvement of MAML, Reptile [19] reduces computational complexity by using first-order derivatives, which allows for faster and more efficient training of the model. Based on the above technologies, MLRC [21] and MLLRE [20] apply MAML and Reptile, respectively, to train a relation classifier.

These two types of few-shot RC methods do not completely resolve the problem of inadequate labeled training data. The raw sentences only provide limited semantic information, and further improvements of performance are mostly based on the following three methods, including knowledge-enhanced methods, pre-trained language model (PLM) based methods [1,24], and semi-supervised learning methods [6].

[1] Following KEFDA, we adopt a conceptual graph containing a general domain part and a specific domain part.

Knowledge-Enhanced Few-Shot RC. Based on the basic methods mentioned above, many methods introduce external knowledge to further improve the text understanding of the model. The external knowledge includes the structural conceptual graph, and textual descriptions of entities, relations, or concepts.

- **Concept-Enhanced Methods** introduce external conceptual graphs as prior knowledge for the model and incorporate them into the instance representation, allowing the model to understand the text precisely. Concept-FERE [28] selects the most suitable concepts for the entities by calculating the semantic similarity between instances and concepts. Meanwhile, a self-attention-based fusion module is proposed to bridge the gap between concept and sentence representation. KEFDA [29] adopt a two-view KG (includes the entity-level part and the concept-level part) embedding model proposed by JOIE [3], mainly utilizing the conceptual graph to capture semantic information.
- **Description-Enhanced Methods** regard textual descriptions as a feature to jointly optimize the text encoder and the knowledge-enhance module. KEFDA [29] fuses the entity description features with the sentence representation and uses concept descriptions to learn a relation-meta. CP+Proto [31] leverage relation descriptions to build prototype representation of the relations, jointly training both a prototype encoder and a sentence encoder.

In this paper, we propose a knowledge-enhanced few-shot RC method, to learn deep structured and fine-grained semantic knowledge from the conceptual graph to enhance the knowledge injection.

3 Method

3.1 Model Overview and Notations

The main architecture of the SKProto is shown in Figure 2. The model framework mainly includes two components: (1) Concept-guided hybrid attention, which enriches the implicit structural knowledge of concepts into the instance representation. (2) Multi-granularity semantic distinction, which distinguishes fine-grained semantic differences.

Definition 3.1 (Conceptual Graph). A conceptual graph can be represented as $\mathcal{G} = \{(e_\mathrm{h}, e_\mathrm{t}, r) \in \mathcal{E} \times \mathcal{E} \times \mathcal{R}_\mathcal{G}\}$, where $\mathcal{E}$ is the set of concepts, $\mathcal{R}_\mathcal{G}$ is the set of relations.

Definition 3.2 (Few-Shot RC). Following the N-way K-shot setting, each task $\mathcal{T} = \{\mathcal{R}, \mathcal{S}, q\}$ contains a support set $\mathcal{S}$, a query instance q and a relation set $\mathcal{R}$. The goal of few-shot RC predicts the relation $r \in \mathcal{R}$ for query instance q based on the $\mathcal{S}$. In details, support set $\mathcal{S} = \{S_1, S_2, ..., S_N\}$ has N relations and each relation r has $S_r = \{s_r^1, s_r^2, ..., s_r^K\}$ containing K instances. Each instance $s = (x, e_\mathrm{h}, e_\mathrm{t})$ contains sentence x, head entity e_h and tail entity e_t.

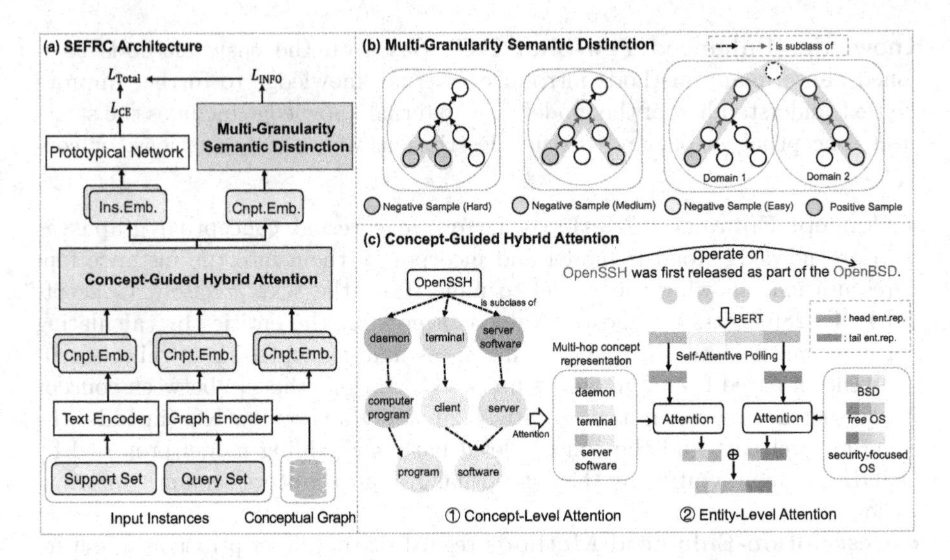

Fig. 2. Model overview of the SKProto. Part (a) is the model architecture. Part (b) and part (c) show the details of the multi-granularity semantic distinction module and the concept-guided hybrid attention module, respectively. (Best viewed in color)

3.2 Concept-Guided Hybrid Attention

We use the conceptual graph to enhance the instance representation. Concept-level attention captures implicit structural knowledge for each linked concept, fusing structural-semantic and textual-semantic information from the superclass concepts. Entity-level attention calculates the importance scores of the linked concepts for the given entities, prompting the model to pay more attention to the crucial concepts.

Concept-Level Attention. For the given concept c_i, we calculate the representation of its superclass concepts of τ-hop as $\mathbf{h}^{\tau} = \sum_{c_j \in \mathrm{S}^{\tau}(c_i)} \mathbf{h}_{c_j}$, where $\mathrm{S}^{\tau}(c_i)$ is the set of superclass concepts of c_i of τ-hop, and $\mathbf{h}_{c_i}$ is the concept representation encoded by graph encoder. Then, we fuse the structural and textual information to calculate the concept-level attention scores α'_{τ} using $\mathbf{h}^{\tau}$ and the transformed concept mention-span representation $\mathbf{h}_{c_i}^{\mathrm{span}}$ as follows:

$$\mathbf{h}_{c_i}^{\mathrm{span}} = \mathcal{LN}\left(\sigma\left(f_{\mathrm{sp}}\left(\mathbf{h}_1, \mathbf{h}_2, \ldots, \mathbf{h}_n\right)\right)\right),$$
$$\alpha'_{\tau} = \tanh\left(\mathbf{h}^{\tau}\mathbf{W}_t + \mathbf{h}_{c_i}^{\mathrm{span}}\mathbf{W}_{t'}\right)\mathbf{W}_a,$$

where $(\mathbf{h}_1, \mathbf{h}_2, \ldots, \mathbf{h}_n)$ is the hidden representation of concept mention-span encoded by the BERT model [17], $\mathcal{LN}$ is the LayerNorm function [10], f_{sp} is the self-attentive pooling [4] to generate the mention-span representation, σ is the non-linear activation function GELU [16]. $\mathbf{W}_t \in \mathbb{R}^{d_2 \times d_2}$, $\mathbf{W}_{t'} \in \mathbb{R}^{d_1 \times d_2}$ and $\mathbf{W}_a \in \mathbb{R}^{d_2 \times 1}$ are the weight matrix.

Finally, the concept-level attention weights $\boldsymbol{\alpha}$ for the linked concept c_i are obtained by normalizing the attention score α'_τ. The weighted concept representation of concept c_i is calculated as follows:

$$\widehat{\mathbf{h}}_{c_i} = \sum_{\tau=1}^{d} \sum_{c_j \in \mathrm{S}^\tau(c_i)} \alpha_\tau \mathbf{h}_{c_j}$$

where d is a hyperparameter that represents the maximum number of hops we consider, $\mathrm{S}^\tau(c_i)$ is the superclass concepts set of c_i of τ-hop.

Entity-Level Attention. For the given entity e, we calculate the semantic attention $\boldsymbol{\beta}$ for the linked concepts as:

$$\beta'_{c_i} = \frac{\left(\widehat{\mathbf{h}}_{c_i} \mathbf{W}_q\right)\left(\mathbf{h}_{\mathrm{e}}^{\mathrm{span}} \mathbf{W}_k\right)^{\mathrm{T}}}{\sqrt{d_2}}$$
$$\beta_{c_i} = \frac{\exp\left(\beta'_{c_i}\right)}{\sum_{c_j \in \mathcal{C}(e)} \exp\left(\beta'_{c_j}\right)}$$

where $\mathbf{h}_{\mathrm{e}}^{\mathrm{span}} = \mathcal{LN}\left(\sigma\left(f_{\mathrm{sp}}\left(\mathbf{h}_1, \mathbf{h}_2, \ldots, \mathbf{h}_n\right)\right)\right)$ is the transformed entity mention-span representation, $\widehat{\mathbf{h}}_{c_i}$ is the multi-hop relation-based concept representation calculated by the concept-level attention. $\mathcal{C}(e)$ is the set of linked concepts of entity e. $\mathbf{W}_q \in \mathbb{R}^{d_2 \times d_2}$ and $\mathbf{W}_k \in \mathbb{R}^{d_2 \times d_2}$ are the weight matrix. The representation of all linked concepts of entity e is aggregated in $\mathbf{h}_{\mathcal{C}(e)} = \sum_{c_j \in \mathcal{C}(e)} \beta_{c_j} \mathbf{h}_{c_j}$.

Instance Representation. For the instance $s = (x, e_{\mathrm{h}}, e_{\mathrm{t}})$, given its contextual sentence representation $\mathbf{h}^{\mathrm{sent}}$ extracted from the BERT encoder, we calculate the concept representation as:

$$\mathbf{h}^{\mathrm{cnpt}} = \sigma\left(\left(\mathbf{h}_{\mathcal{C}(e_{\mathrm{h}})} \oplus \mathbf{h}_{\mathcal{C}(e_{\mathrm{t}})}\right) \mathbf{W}_c + \mathbf{b}_c\right)$$

where $\mathbf{W}_c \in \mathbb{R}^{d_2 \times d_2}$ is the weight matrix, $\mathbf{b}_c \in \mathbb{R}^{d_2}$ is the bias vector. Then, we concatenate the sentence representation $\mathbf{h}^{\mathrm{sent}}$ and the aggregated concept representation $\mathbf{h}^{\mathrm{cnpt}}$, fusing them into the feedforward neural network (FFN) as:

$$\mathbf{h}' = \mathcal{LN}\left(\mathrm{FFN}\left(\mathbf{h}^{\mathrm{sent}} \oplus \mathbf{h}^{\mathrm{cnpt}}\right)\right)$$

3.3 Multi-granularity Semantic Distinction

In order to make knowledge injection more accurate, we propose multi-granularity semantic distinction to enhance fine-grained semantics discrimination of the model. Specifically, based on the conceptual hierarchical structure between concepts and the domain to which the concept belongs, we construct three types of negative samples with various difficulties. Meanwhile, for the concept $c_i \in \mathcal{C}(e)$, we take the other concept $c_j \in \mathcal{C}(e)$ as the positive sample.

Constructing Negative Samples. Among the three types of negative samples, the hard samples are the most similar to the positive samples, followed by the medium samples, and the easy samples are the least similar.

3.3.1 ***Hard Sample*** is the concept that has at least one same direct superclass with the positive sample. Formally, the hard sample c^{Hard} satisfy:

$$\mathrm{S}(c^{\text{Hard}}) \cap \mathrm{S}(c^{\text{Positive}}) \neq \varnothing$$

where $\mathrm{S}(c)$ is the set of direct superclass concepts directly related to c.

3.3.2 ***Medium Sample*** is the concept that exists within the same domain as the positive sample but does not have a common direct superclass concept. Formally, c^{Medium} satisfy:

$$\mathrm{S}(c^{\text{Medium}}) \cap \mathrm{S}(c^{\text{Positive}}) = \varnothing$$
$$\mathrm{S}^*(c^{\text{Medium}}) \cap \mathrm{S}^*(c^{\text{Positive}}) \neq \varnothing$$

where $\mathrm{S}^*(c)$ denotes the set of superclass concepts of c without a limit on the number of hops. As shown in Fig. 2, the two concepts have no common superclass concept indicating that they belong to different domains.

3.3.3 ***Easy Sample*** is constructed as the concept c^{Easy} that is not in the same domain as the positive sample. Formally, c^{Easy} satisfy:

$$\mathrm{S}^*(c^{\text{Easy}}) \cap \mathrm{S}^*(c^{\text{Positive}}) = \varnothing$$

As shown in Fig. 2, two concepts have a common superclass concept indicating that they belong to the same domain.

InfoNCE Loss. We employ the InfoNCE [25] loss function, which maximizes the similarity and minimizes the differences between positive and negative samples.

$$\mathcal{L}_{\text{Info}} = -\sum_i \log \frac{\exp\left(\cos\left(\widehat{\mathbf{h}}_{c^i}, \widehat{\mathbf{h}}_{c^{\text{Positive}}}\right)/t\right)}{\sum_{c^n \in \mathcal{N}} \exp\left(\cos\left(\widehat{\mathbf{h}}_{c^n}, \widehat{\mathbf{h}}_{c^{\text{Positive}}}\right)/t\right)}$$

where $\mathcal{N} = \{c^{\text{Hard}}, c^{\text{Medium}}, c^{\text{Easy}}\}$ is the set of negative samples. $\widehat{\mathbf{h}}_{c^{\text{Positive}}}$ and $\widehat{\mathbf{h}}_{c^n}$ are the multi-hop relation-based concept representation of the positive sample and negative samples. $\cos(\cdot)$ is the cosine similarity function and t is the temperature coefficient.

3.4 Prototypical Network

We calculate the prototype for each relation that exists in the support set by averaging the representation of all instances of relation r in the support $\mathcal{S}_r$ as $\mathbf{h}'_r = \frac{1}{|\mathcal{S}_r|}\sum_{s \in \mathcal{S}_r} \mathbf{h}'_s$. Then, the probability that the query instance q belongs to relation r can be calculated based on the distance between query instance representation $\mathbf{h}'_q$ and the relation prototype $\mathbf{h}'_r$,

$$p_\phi(y = r \mid q) = \frac{\exp\left(-\text{dis}\left(\mathbf{h}'_q, \mathbf{h}'_r\right)\right)}{\sum_{r' \in \mathcal{R}} \exp\left(-\text{dis}\left(\mathbf{h}'_q, \mathbf{h}'_{r'}\right)\right)}$$

where dis $(\cdot)$ is squared Euclidean distance function. We adopt the negation of distance as the scoring function and calculate the cross-entropy loss $\mathcal{L}_{\text{CE}}$ as

$$\mathcal{L}_{\text{CE}} = - \sum_{(\mathcal{S},\mathcal{R},q) \in \mathcal{T}_{\text{train}}} \sum_{r \in \mathcal{R}} I_r \log p_\phi(y = r \mid q)$$

where I_r is an indicator function. If the relation r is the ground-truth result, $I_r = 1$; otherwise, $I_r = 0$. Hence, the total loss function of SKProto can be denoted as $\mathcal{L}_{\text{Total}} = \lambda \mathcal{L}_{\text{CE}} + (1 - \lambda)\mathcal{L}_{\text{Info}}$, where λ is a hyperparameter.

4 Experiments

4.1 Data Source

Conceptual Graph. Similar to KEFDA, we use the general and domain-specific conceptual graph constructed on the Wikidata[2] and UMLS[3] knowledge bases. The embeddings of concepts in our model are generated by the open source toolkit OpenKE[4]. Details of the conceptual graph can be found in Table 1.

Table 1. Statistics of the conceptual graph. "Concept", "Relation", and "Triple" represent the number of concepts, relations, and triples, respectively.

Conceptual Graph	Domain	Concept	Relation	Triple
WikiData	General	6,409	39	8,057
UMLS	Specific	127	54	5,890

Benchmark Dataset. We use the FewRel 2.0 benchmark[5] to evaluate our model. Training tasks, validation tasks, and test tasks have no relation intersection. Table 2 shows the details of this benchmark.

4.2 Baselines

In this work, we compare our model with competitive baseline models. (1) "Proto" denotes the **Prototypical Networks** [23], which simply uses the raw sentence context semantic information to predict relation classes. "(CNN)"

[2] https://www.wikidata.org.
[3] https://www.nlm.nih.gov/research/umls/index.html.
[4] https://github.com/thunlp/OpenKE.
[5] https://github.com/thunlp/FewRel.

Table 2. Statistics of FewRel 2.0 benchmark. "Relation", "Instance", and "Entity" represent the number of relation classes, instances, and entities. "Concept.Link." represent the proportions of entities with concept links.

Task	Relation	Instance	Entity	Concept.Link.
Training	64	44,800	89,600	99.54%
Validation	10	1,000	2,000	98.70%
Testing	15	1,500	3,000	99.01%

and "(BERT)" denote using CNN and BERT as encoder [15]. The model with "-ADV" use adversarial training [13]. (2) **Bert-Pair** [2] utilizes the BERT sequence-pair model to measure the similarity between two instances. The model first concatenates the query instance with all support instances as a sequence of pairs, then identifies the relation class by the similarity of each pair. (3) **KEFDA** [29] incorporates general and domain-specific conceptual graphs, using a knowledge-enhanced prototypical network to conduct instance matching and a relation-meta learning network for implicit relation matching. (4) **CP+Proto** [31] leverage relation descriptions to build prototype representation of the relations, jointly training both a prototype encoder from the descriptions and a sentence encoder.

4.3 Experiment Settings

We experimented on the four few-shot settings: 5-way 1-shot, 5-way 5-shot, 10-way 1-shot, and 10-way 5-shot. We use classification accuracy to evaluate the model performance. To sufficiently learn the information in the training tasks, we set the number of training iterations to 30,000. For every 1000 training iterations, we validate the model 1000 iterations using the validation tasks and simultaneously save the model that performs best on the validation set. We use $\text{BERT}_{\text{BASE}}$[6] as the sentence encoder, which has a text dimension of 768 and a maximum sequence length of 128, the dimensions $d_1 = 768$, $d_2 = 256$. The temperature coefficient $t = 0.5$, the hyperparameter λ is 0.8. Also, we use DistMult [9] as the graph encoder, with node embedding of dimension 256.

4.4 Overall Results

Experiment results are shown in Table 3. From the results, we observe that: (1) On the Fewrel 2.0 benchmark, the performance of our SKProto model outperforms all strong baseline models, achieving a new state-of-the-art result. (2) The performance of SKProto is improved compared with the strongest baseline KEFDA on the four few-shot settings (i.e. +2.25%, +2.17%, +0.83%, +1.15%). Additionally, when switching to the low-shot scenarios, the accuracy of SKProto only decreases by 6.77% in comparison to the 7.99% decrease of KEFDA. The above improvement proves the effectiveness of capturing deep-structured semantics for the model to infer relation classes.

[6] https://github.com/huggingface/transformers.

Table 3. Accuracy (%) of models on FewRel 2.0 testing tasks under N-way K-shot settings. "Cnpt". and "Desc." denotes conceptual graphs and descriptions, respectively. "✓" indicates the use of the above external data.

Model	External.Data.		1-Shot			5-Shot		
	Cnpt.	Desc.	5-Way	10-Way	Avg.	5-Way	10-Way	Avg.
Proto (CNN)	–	–	35.09	22.98	29.04	49.37	35.22	42.30
Proto (BERT)	–	–	40.12	26.45	33.29	51.50	36.93	44.22
Proto-ADV (BERT)	–	–	41.90	27.36	34.63	54.74	37.40	47.40
Proto-ADV (CNN)	–	–	42.21	28.91	34.63	58.71	44.35	51.53
BERT-PAIR	–	–	67.41	54.89	61.20	78.57	66.85	72.71
CP+Proto	–	✓	83.11	73.02	78.07	90.80	83.08	86.94
KEFDA	✓	✓	87.81	81.84	84.83	95.00	90.63	92.82
SKProto	✓	–	**90.06**	**84.01**	**87.04**	**95.83**	**91.78**	**93.81**

4.5 Detail Analysis

Table 4. Accuracy (%) of models on FewRel 2.0 validation tasks under N-way K-shot settings, and the best results are in bold. "w/o Distinction.", "w/o Concept." denote the absence of multi-granularity semantic distinction and the concept-guided hybrid attention, respectively.

Model	1-Shot			5-Shot		
	5-Way	10-Way	Avg.	5-Way	10-Way	Avg.
SKProto	**88.48**	**82.94**	**85.71**	**95.69**	**92.57**	**94.13**
w/o Distinction	88.05	81.67	84.86	94.85	91.78	93.32
w/o Concept	85.84	80.03	82.94	93.85	90.34	92.10

Ablation Study. To verify the effectiveness of our framework, we conduct ablation experiments on the Fewrel 2.0 validation tasks. Specifically, we remove the following contributions to evaluate their impact on performance: the concept-guided hybrid attention and the multi-granularity semantic distinction. The final results are shown in Table 4. From the results, we observe that: (1) The impact of multi-granularity semantic distinction on model performance is 1.07% on average (0.85% for 1-shot and 1.28% for 5-shot). This proves that the performance of the model can be improved by learning the fine-grained semantics differences. (2) The impact of concept-guided hybrid attention on model performance is 2.1% on average (2.17% for 1-shot and 2.03% for 5-shot). This proves that concept-guided hybrid attention learns the implicit structural knowledge in the conceptual graph, providing clues for the model to infer the relation class.

Impact of the Maximum Number of Hops. As the number of hops in the conceptual graph increases, the semantics of the concepts become more abstract,

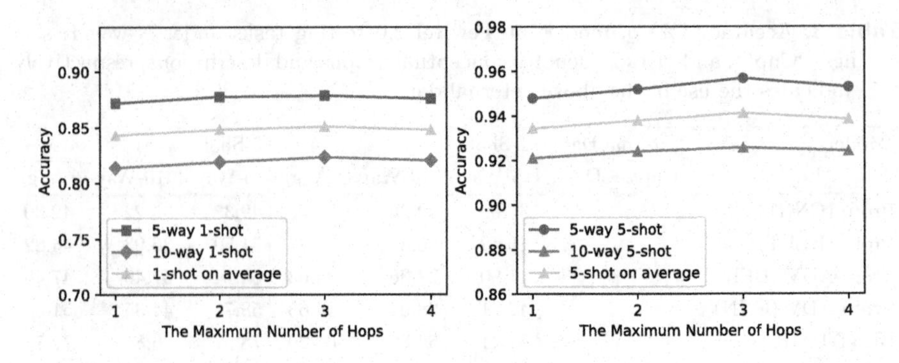

Fig. 3. Performance comparison of the maximum concept level under four settings.

and the subclass concepts it can cover also increase exponentially. When the maximum number of hops is set too large, excessive concepts may cause the model to become confused. Conversely, if the maximum number of hops is too small, the conceptual graph may not be able to provide enough semantic information. To investigate the effect of the number of hops on model prediction accuracy, we designed an experiment. As shown in Fig. 3, the model achieved its highest prediction accuracy when the maximum number of hops is set to 3.

5 Conclusion

In this paper, we propose a knowledge-enhanced prototypical network for few-shot RC, incorporating implicit structural knowledge from the conceptual graph. Specifically, we propose a concept-guided hybrid attention mechanism to learn implicit structural semantic knowledge. Meanwhile, we propose multi-granularity semantic distinction to make the model distinguish subtle semantic differences. As for future work, we attempt to use reinforcement learning methods for the dynamical knowledge injection of entities.

References

1. Bansal, T., Gunasekaran, K., Wang, T., Munkhdalai, T., McCallum, A.: Diverse distributions of self-supervised tasks for meta-learning in NLP. In: EMNLP, pp. 5812–5824 (2021)
2. Gao, T., et al.: FewRel 2.0: towards more challenging few-shot relation classification. In: EMNLP-IJCNLP, pp. 6250–6255 (2019)
3. Hao, J., Chen, M., Yu, W., Sun, Y., Wang, W.: Universal representation learning of knowledge bases by jointly embedding instances and ontological concepts. In: SIGKDD, pp. 1709–1719 (2019)
4. Lin, Z., et al.: A structured self-attentive sentence embedding. arXiv preprint arXiv:1703.03130 (2017)
5. Mahony, N.O., Campbell, S., Krpalkova, L., Carvalho, A., Walsh, J., Riordan, D.: Representation learning for fine-grained change detection. Sensors **21**, 4486 (2021)

6. Zhang, N., et al.: Long-tail relation extraction via knowledge graph embeddings and graph convolution networks. In: NAACL (2019)
7. Wu, R., et al.: Open relation extraction: relational knowledge transfer from supervised data to unsupervised data. In: EMNLP (2019)
8. Wang, H., Qin, K., Zakari, R.Y., Lu, G., Yin, J.: Deep neural network-based relation extraction: an overview. Neural. Comput. Appl. **34**, 1–21 (2022)
9. Yang, B., Yih, W.T., He, X., Gao, J., Deng, L.: Embedding entities and relations for learning and inference in knowledge bases. In: ICLR (2015)
10. Ba, J.L., Kiros, J.R., Hinton, G.E.: Layer normalization. In: arXiv preprint arXiv:1607.06450 (2016)
11. Finn, C., Abbeel, P., Levine, S.: Model-agnostic meta-learning for fast adaptation of deep networks. In: ICML (2017)
12. Gao, T., Han, X., Liu, Z., Sun, M.: Hybrid attention-based prototypical networks for noisy few-shot relation classification. In: AAAI, vol. 33, pp. 6407–6414 (2019)
13. Goodfellow, I.J., Shlens, J., Szegedy, C.: Explaining and harnessing adversarial examples. In: STAT (2015)
14. Hadsell, R., Chopra, S., LeCun, Y.: Dimensionality reduction by learning an invariant mapping. In: CVPR, vol. 2, pp. 1735–1742 (2006)
15. Han, X.: Fewrel: a large-scale supervised few-shot relation classification dataset with state-of-the-art evaluation. In: EMNLP, pp. 4803–4809 (2018)
16. Hendrycks, D., Gimpel, K.: Gaussian error linear units (GELUs). arXiv preprint arXiv:1606.08415 (2016)
17. Kenton, J.D.M.W.C., Toutanova, L.K.: BERT: pre-training of deep bidirectional transformers for language understanding. In: NAACL-HLT, pp. 4171–4186 (2019)
18. Miwa, M., Bansal, M.: End-to-end relation extraction using LSTMs on sequences and tree structures. In: ACL, pp. 1105–1116 (2016)
19. Nichol, A., Achiam, J., Schulman, J.: On first-order meta-learning algorithms. arXiv: Learning (2018)
20. Obamuyide, A., Vlachos, A.: Meta-learning improves lifelong relation extraction. In: ACL (2019)
21. Obamuyide, A., Vlachos, A.: Model-agnostic meta-learning for relation classification with limited supervision. In: ACL (2019)
22. Oord, A., Li, Y., Vinyals, O.: Representation learning with contrastive predictive coding. arXiv preprint arXiv:1807.03748 (2018)
23. Snell, J., Swersky, K., Zemel, R.S.: Prototypical networks for few-shot learning. In: NeurIPS (2017)
24. Soares, L., FitzGerald, N., Ling, J., Kwiatkowski, T.: Matching the blanks: distributional similarity for relation learning. In: ACL (2019)
25. Wan, C., Zhang, T., Xiong, Z., Ye, H.: Representation learning for fault diagnosis with contrastive predictive coding. In: SAFEPROCESS, pp. 1–5 (2021)
26. Wang, Q., Van Hoof, H.: Model-based meta reinforcement learning using graph structured surrogate models and amortized policy search. In: ICML (2022)
27. Xiong, W., Yu, M., Chang, S., Guo, X., Wang, W.Y.: One-shot relational learning for knowledge graphs. In: EMNLP, pp. 1980–1990 (2018)
28. Yang, S., Zhang, Y., Niu, G., Zhao, Q., Pu, S.: Entity concept-enhanced few-shot relation extraction. In: ACL-IJCNLP, pp. 987–991 (2021)
29. Zhang, J., Zhu, J., Yang, Y., Shi, W., Zhang, C., Wang, H.: Knowledge-enhanced domain adaptation in few-shot relation classification. In: SIGKDD (2021)
30. Zhang, Y., Qi, P., Manning, C.D.: Graph convolution over pruned dependency trees improves relation extraction. In: EMNLP, pp. 2205–2215 (2018)
31. Zhenzhen, L., Zhang, Y., Nie, J.Y., Li, D.: Improving few-shot relation classification by prototypical representation learning with definition text. In: NAACL (2022)

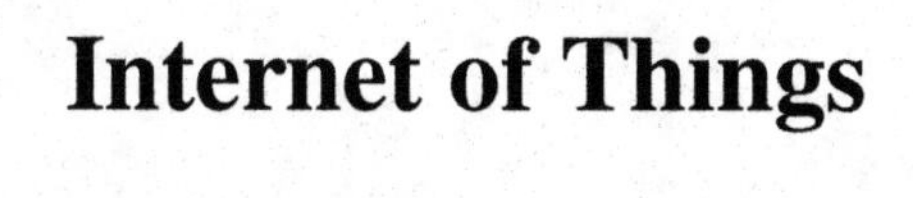

Internet of Things

MIDFA: Memory-Based Instance Division and Feature Aggregation Network for Video Object Detection

Qiaochuan Chen[1,2], Min Zhou[1], and Hang Yu[1(✉)]

[1] School of Computer Engineering and Science, Shanghai University, Shanghai, China
{qcchen,zhoumin,yuhang}@shu.edu.cn

[2] Key Laboratory of Silicate Cultural Relics Conservation (Shanghai University), Ministry of Education, Shanghai, China

Abstract. Previous video object detection methods focus on aggregating the features of other frames into the current frame to alleviate the image degradation, but they rarely focus on multi-class scenes. Aggregating features of different classes will generate confusing information and affect network performance. This problem can be solved by using a classifier to divide the features. However, classifier method has three problems: (a) Heterogeneous high-similarity objects and homogeneous low-similarity objects affect the accuracy of the classifier. (b) Two objects whose positions overlap also affect the classifier. (c) Previous classifier method did not exploit sufficient global information. Therefore, we propose a new method that divides the features of different instances to deal with the problems of (a) and (b). Then we designed two new memories (one is Init Memory and the other is MDR) to solve problem (c). These three parts constitute the MIDFA network. Experiments show that our method achieves 83.76% mAP on the ImageNet VID dataset based on ResNet-101, and 84.6% mAP on ResNeXt-101. In addition, we also conduct experiments on a custom-designed multi-class VID dataset, and adding Instance Division and MDR can increase the mAP of the network by 0.6% compared to using only Init Memory.

Keywords: Video Object Detection · Instance Division · Memory · Feature Aggregation

1 Introduction

The core task of the object detection is to identify and locate one or more objects in an image or a video. However, many studies have shown that object detection in images is very different from that in videos. Compared with image, video has an additional dimension of time. Therefore, video object detection network can gather temporal context information to enhance the feature representation of the current frame. As mentioned by FGFA [23], when encountering unfavorable factors such as motion blur, rare poses, defocusing, occlusions, etc., the current

H. Kashima et al. (Eds.): PAKDD 2023, LNAI 13937, pp. 153–164, 2023.
https://doi.org/10.1007/978-3-031-33380-4_12

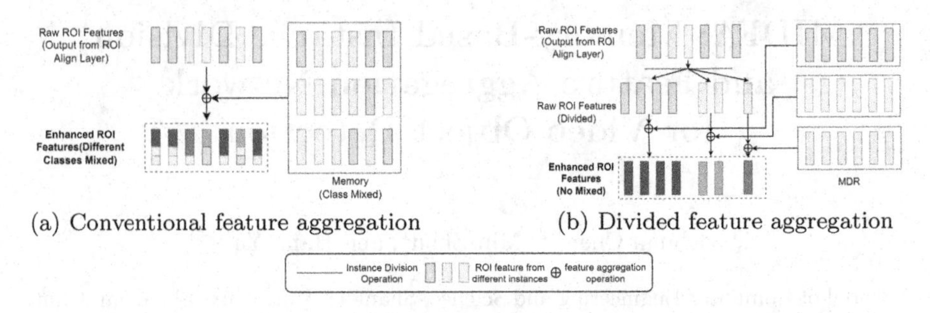

(a) Conventional feature aggregation (b) Divided feature aggregation

Fig. 1. Comparison of two memory-based feature aggregation methods. (a) This will produce confusing ROI features. (b) This will achieve pure feature aggregation.

frame will be extremely degraded, and the image object detection network cannot effectively handle it.

In recent years, A mechanism utilizing memory is proposed to solve the video object detection problem. The mechanism uses a memory structure to store the features of previous frames. The features stored in memory can be provided to the current frame to achieve feature enhancement. For example, Chen et al. [3] proposes a memory structure called LRM to support the current frame capture longer-distance temporal semantic information. Considering the existence of redundant information in memory, Sun et al. [19] proposed a memory bank method to store features in a more fine-grained manner. However, none of these memory-mechanism methods take into account the multi-class video scenarios. As shown in Fig. 1(a), if features of different classes (different colors represent features of different classes) are mixed in memory, the enhanced features obtained after feature aggregation will contain confusing information. This will affect the accuracy of the network in multi-class videos.

For this problem, Han et al. [8] uses a classifier to divide the features. It can make feature aggregation more pure and no longer generate mixed features. But when encountering features with heterogeneous high similarity, the classification accuracy of the classifier will drop. Assuming that there is a perfect classifier, the features of the same class can always be put together. However, if the semantic information they contain is quite different, the purpose of pure feature aggregation will also be defeated. In addition, if there are two objects of different classes and their positions overlap, the classifier will also not be able to accurately give the right category information of these two objects. From another point of view, object tracking can find all bounding boxes of the same instance in a video. Similarly, if we can find all ROI features of the same instance, putting them into the same memory can achieve pure feature aggregation. Therefore, we propose an Instance Division method, and store the features after feature division in the Memory Division Repository (MDR), as shown in Fig. 1(b). In addition, there is no need to store its features in memory after detecting the current frame. We can capture and store all the required features in memory before detecting a video. The contributions of this paper are as follows:

1. We propose an Instance Division method to enable the video object detection network to better handle multi-class videos. Specifically, we divide all the ROI feature according to the positional proximity and the appearance similarity.
2. We propose an MDR structure to store the ROI features of different instances to support the Instance Division method. Through the memory mechanism, the network can perceive more global temporal semantic information.
3. We design a new feature storage method to construct a new structure called Init Memory. Init Memory can allow the network to obtain rich global semantic information while occupying less GPU memory.
4. We have conducted extensive experiments on the ImageNet VID dataset and multi-class datasets. Results show that our method achieves good performance in terms of accuracy and speed.

2 Related Work

2.1 Video Object Detection

So far, video object detection tasks can be roughly divided into three categories: box-level, frame-level, and feature-level.

Box-Level. Connect bounding boxes across frames. Seq-NMS [10] re-scores all boxes in the video to obtain the best bounding box links. The BLR (Box Linking with Relations) method in RDN [5] is similar to seq-nms. Although the Box-Level method can effectively improve the detection frame, it will increase the calculation time and cannot achieve online detection due to the need to use future frames.

Frame-Level. The detection results are concatenated by trajectories across frames to provide the final prediction. STL [2] detects sparse keyframes and propagates predicted bounding boxes to non-keyframes through motion and scaling. DorT [15] combines detection results and tracking results to achieve efficient detection.

Feature-Level. Aggregate feature maps from neighboring frames. Due to the image difference between frames, the feature maps of adjacent frames need to be aligned before aggregation. FGFA [23] use FlowNet [6] to generate optical flow fields for feature map alignment, so as to adjust the adjacent frame feature maps to align to the current frame. Specifically, there are many methods [3,5,24] proposing to use attention mechanism to find suitable other frame features for feature aggregation for current frame features. These methods make a breakthrough in video object detection.

3 Method

3.1 Framework Overview

The proposed network framework is based on the traditional image object detection network Faster R-CNN. The right half of Fig. 2 can be divided into three

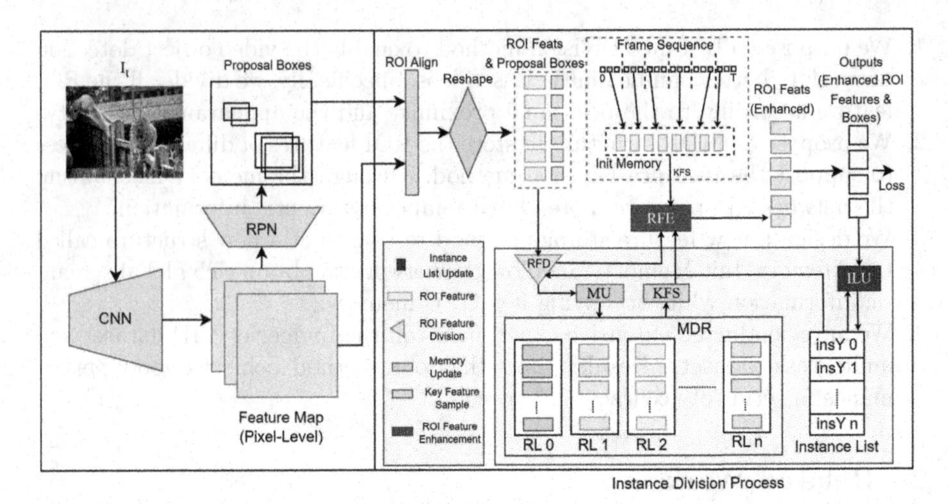

Fig. 2. Framework overview. The left half of the figure is the half process of Faster R-CNN. The right half is our design. The two parts together constitute our MIDFA network. The loss function of MIDFA is the same as Faster R-CNN. (Color figure online)

parts: Init Memory(marked by red dashed box), Instance Division Process, and ROI Feature Enhancement(RFE) module. Init Memory provides time-global semantic information. The purpose of the Instance Division process is to aggregate the features of different instances separately. The RFE module aggregates the ROI features provided by the Init Memory and Instance Division processes into the ROI features of the current frame.

3.2 Init Memory Sampling Strategy

The Memory Bank [19] method provides a more fine-grained feature storage method. But it needs to store the features of the current frame after each detection. In reality, it is very likely to have videos that are several hours or even a dozen hours long. Therefore, the original method is likely to have significant limitations. We propose Init Memory to solve this problem.

As shown in red dashed box of Fig. 2, the video has a total of T frames, and the $i * T$ (default $i = 5\%$ in the experiment) frame are used to construct the Init Memory. The video is divided into $i*T$ regions. Each region has $1/i$ frames, and one frame is randomly sampled from one region at a time. Since $i = 5\%$ in default, then $1/i = 20$, that is, random sampling is performed every 20 frames. 20 frames is about one second in the video. We don't need to collect too many repeated frames in one second, one frame is enough. This strategy can quickly collect diverse features from the global scope and there is no need to increase the memory size subsequently.

3.3 Memory-Based Instance Division and Feature Aggregation

ROI Feature Enhancement Module. Previous methods [3,5,20,24] cannot divide features into different groups and then aggregate them. Therefore, we introduce the ROI Feature Enhancement(RFE) module. Regarding the specific implementation, given the ROI feature set $Q = \{q_0, q_1, ..., q_N\}$ of a certain class at time t and the ROI feature set S_c from any memory, calculate the M relational features between $q_i \in Q$ and S_c, obtained by calculating the semantic similarity between q_i and S_c, the m-th relational feature formula of q_i is defined as follows:

$$f_r{}^m(q_i, S_c) = \sum_j \omega_{ij} \cdot (W_V{}^m \cdot S_{c_j}), m = 1, ..., M \tag{1}$$

$$\omega_{\text{ij}} = \frac{\exp(Sim(q_i, S_{c_j}))}{\sum_{j=1}^{n} \exp(Sim(q_i, S_{c_j}))}, Sim(q_i, S_{c_j}) = \frac{(W_K^m S_{c_j})^T (W_Q^m q_i)}{\left\|W_K^m S_{c_j}\right\| \left\|W_Q^m q_i\right\|} \tag{2}$$

where W_V^m, W_K^m, W_Q^m are the linear transformation matrix, ω_{ij} is the weight of the relationship between q_i and S_{cj}. $Sim(\cdot)$ calculates the degree of similarity between two vectors. Finally, by connecting M relational features and adding them to the original ROI feature q_i, the enhanced ROI feature at time t can be obtained:

$$RFE(q_i, S_c) = q_i + concat[\{f_r{}^m(q_i, S_c)\}_{m=1}^{M}] \tag{3}$$

where the $RFE(\cdot)$ is the enhanced feature. Therefore, whether such memory is an Init Memory or a sub-repository in the MDR, the RFE operation can be directly applied.

Instance Division Method. We propose a novel feature division method that makes up for the inadequacies of previous classifier methods. It contains four modules, which are RoI Feature Division(RFD), Memory Update(MU), Key Feature Sample(KFS) and Instance List Update(ILU). And includes two structures, namely MDR and Instance List.

ILU Module. We propose to separate the ROI features of different instances. The bounding boxes and ROI features of different instances at the previous moment are stored separately in the Instance List, which are represented as $IL = \{insY_0, ..., insY_i, ..., insY_n\}$, $insY_i = (box_i, r_i)$. Each $insY$ represents a tuple of box and ROI feature. At the frame t, we define the Detected Instance List of current frame from the Outputs in Fig. 2 as $DIL = \{insX_0, ..., insX_j, ..., insX_m\}$, $insX_j = (box_j, r_j)$. We use DIL^t to update IL^t as follows:

$$\text{IL}^t = \{insY_0^t, ..., insY_i^t, ..., insY_n^t\} \tag{4}$$

$$insY_i^t = Max(IoU(insY_i^{t-1}, Sim(insY_i^{t-1}, DIL^t))) \tag{5}$$

where the update of IL^t consists of the above two steps. Find the $insX_i^t$ in the DIL^t that has a high feature similarity (Sim($\cdot$)) and highest box IoU value (Max(IoU($\cdot$))) with each $insY_i^{t-1}$ of IL^{t-1}. And then the $insX_i^t$ becomes the

$insY_i^t$ instead of $insY_i^{t-1}$. All $insY$ of the previous moment in IL are replaced by $insX$ of this moment. An instance may disappear from the video over time. Then an $insY$ in IL cannot find a matching $insX$. Thus this $insY$ should be removed from the IL until the vanished instance comes back or a new one appears.

RFD Module. The process of RFD and ILU is the same, though the difference is the source of their input and the destination of the output. RFD matches the ROI features of the intermediate process with each $insY$ in IL through the same instance division step as ILU. RFD will label these ROI features with instance numbers. RFD stores these features in MDR according to their numbers. RFD also feeds these features into RFE so that they can be aggregated with features of the same instance number in MDR.

MU Module. The MU process is very simple, just store the grouped ROI features passed by RFD into the corresponding sub-repository of MDR.

KFS Module. KFS is shared between MDR and Init Memory. A small number of ROI features are randomly sampled from the memory structure, which is consistent with the Memory Bank [19] method.

4 Experiments

4.1 Dataset and Evaluation Setup

We evaluate the proposed method on the ImageNet VID and the subset of DET dataset [17] with the same categories as VID. The ImageNet VID dataset is dedicated to the task of video object detection and has a total of 30 categories. The ImageNet VID dataset has a total of 3,862 video clips for training, 555 video clips for validation. Following the standard of previous methods, we evaluate the model on the validation set of the VID dataset using mean precision (mAP) with an IoU threshold of 0.5.

4.2 Implementation Details

Detection Network. ResNet-101 [12] and ResNeXt-101 [21] are used as feature extractors. Notably, the input to the RPN head is the output of the penultimate layer of the feature extraction network. The anchors of RPN have 3 aspect ratios {1:2, 1:1, 2:1} and 4 scales {642, 1,282, 2,562, 5,122}, a total of 12 different anchors for each spatial position. During training and inference, 300 proposal boxes per frame are generated with an NMS threshold of 0.7 IoU. After generating the boxes, we apply the ROI Align layer and the 1024-dimensional fully connected layer after the last layer of the feature extraction network to extract the ROI features of each proposal box.

Experiment Setting. Following the previous experimental configuration, we train the proposed model weights on the training sets of the VID dataset and DET dataset. During training and inference, the input images are resized to 600 pixels on the short side and 1,000 pixels on the long side. The network is trained on 4 RTX 3060 GPUs, the optimizer is SGD, each GPU has a mini-batch, and each mini-batch contains two images (the current frame and a randomly sampled frame). The training process goes through 120k iterations, and the learning rates for the first 80k and the last 40k are 10^{-3} and 10^{-4}, respectively. Duplicate boxes were suppressed using NMS with a 0.5 IoU threshold.

4.3 Main Results

Table 1. End-to-end method comparison on the ImageNet VID validation set. "symmetric" represents the symmetric model of LRTR. † denotes using data augmentations.

Methods	Backbone	mAP(%)
Faster R-CNN [16]	ResNet-101	75.4
FGFA [23]	ResNet-101	76.3
THP [22]	ResNet-101+DCN	78.6
STSN [1]	ResNet-101+DCN	78.9
OGEMN [4]	ResNet-101+DCN	80.0
SELSA [20]	ResNet-101	80.3
$LRTR_{symmetric}$ [18]	ResNet-101	81.0
TCENet [11]	ResNet-101	81.0
RDN [5]	ResNet-101	81.8
SELSA [20]+TROI [7]	ResNet-101	82.0
LSTS [13]	ResNet-101	82.1
$CFANet_{ins}$ [8]	ResNet-101	82.6
MEGA [3]	ResNet-101	82.9
CSMN [9]	ResNet-101	83.1
$DSFNet_{ins}$ [14]	ResNet-101	83.3
$MAMBA_{ins}$ [19]	ResNet-101	83.7
RDN [5]	ResNeXt-101	83.2
$LRTR_{symmetric}$ [18]	ResNeXt-101	84.1
MEGA [3]	ResNeXt-101	84.1
CSMN [9]	ResNeXt-101	84.3
SELSA [20]+TROI [7]	ResNeXt-101	84.3
$MAMBA_{ins}$	ResNeXt-101	84.36
MIDFA(ours)	ResNet-101	83.76
MIDFA (ours)	ResNeXt-101	84.6
$MIDFA^{\dagger}$ (ours)	ResNeXt-101	**85.22**

We compare our method with previous "SOTA" methods in Table 1. In order to maintain the fairness of the result comparison, we use the same backbone. When the backbone is ResNet, our method is higher than the second Memory Bank method by 0.06% mAP. When the backbone is changed to ResNeXt, MIDFA outperforms the second place by 0.24% mAP. Since only the ROI features are stored in our memory, we compare it with all the "ins" level networks ("ins" means that memory only stores ROI features). Compared with the $MAMBA_{ins}$ model, Init Memory carries fewer features and can also capture enough global information. When compared with CFANet, which also aims to solve the multi-class situation, MIDFA can perceive more global information and have a higher accuracy due to the memory mechanism. After adding random crop and random scale data augmentation during MIDFA training, its mAP can reach 85.22% based on ResNeXt101.

4.4 Ablation Study

Table 2. Ablation study of MIDFA Network based on a single-frame baseline network.

Strategy	a	b	c	d
Init Memory		√	√	√
Instance Division			√	√
MDR				√
mAP(%)	75.40	$83.72_{\uparrow 8.32}$	$83.73_{\uparrow 8.33}$	$83.76_{\uparrow 8.36}$
Runtime(ms)	55.56	86.2	103.1	104.17

In Table 2, the effects of Init Memory, Instance Division method and MDR are investigated by gradually adding them. ResNet101 is used as the backbone.

(a) Baseline: When no components are added, the mAP reaches 75.4%. This shows the lack of image object detection network in VID task. The inference speed of the Baseline network reached the fastest, that is, 55.56ms per frame.

(b) Effectiveness of Init Memory: When only Init Memory is added, the network improves mAP by 8.35% compared to the baseline. It can be seen that with the help of a small amount of time global features, the overall accuracy of the network can be improved a lot.

(c) Effectiveness of Instance Division: After adding Instance Division processing, the network performance improved slightly, but the processing speed dropped a lot. MDR and Instance Division can be separated, because $insY$ in the IL also stores the ROI features. Without MDR, the ROI features of the current frame are aggregated with the features in the IL.

(d) Effectiveness of MDR: After continuing to add MDR, the mAP value improves about 0.04% compared to strategy b. With the help of Instance Division and MDR, the accuracy of object detection can be further improved. However, since the proportion of multi-class videos in the VID dataset is not very high (23 of the 555 videos have multi-class objects, accounting for 4.1%), the improvement effect of MDR in the complete VID dataset is not obvious.

Table 3. Multi-class video object detection results. "ALL" indicates test the MIDFA in the complete multi-class dataset.

Video Set	Memory Bank	Init Memory	Init Memory+MDR
ILSVRC2015_val_00004000	44.81	44.88	46.90
ILSVRC2015_val_00022000	93.98	94.01	94.24
ILSVRC2015_val_00025000- ILSVRC2015_val_00025001	80.38	80.36	80.67
ILSVRC2015_val_00026000- ILSVRC2015_val_00026003	76.61	76.55	76.67
ILSVRC2015_val_00037000- ILSVRC2015_val_00037008	99.57	99.66	99.67
ILSVRC2015_val_00123000- ILSVRC2015_val_00123004	57.90	57.94	58.02
ILSVRC2015_val_00130000	93.41	93.26	94.09
ALL	70.48	70.56	71.16

4.5 Experiment on Multi-class Dataset

We design a new multi-class dataset to test MIDFA, which comes from the full VID dataset and only contains multi-class videos. It includes a total of 23 videos, 10,258 pictures. Each video is grouped according to the scene, and the objects appearing in the same group are of the same class and background. For example, the videos in the middle of ILSVRC2015_val_00025000-ILSVRC2015_val_00025001 are all in the same group. In Table 3, after adding MDR, the accuracy for all videos is improved by 0.6% mAP compared to Memory Bank [19] and Init Memory. Init Memory can provide global information, while MDR can make different classes of ROI features more distinguishable from each other.

4.6 Comparison of GPU Memory Consumption of Different Memory

Table 4. Comparison of GPU memory consumption of different memory.

Memory	LRM			MB			IM		
Size(MB)	s	m	l	s	m	l	s	m	l
	20	20	20	18	86	94	4	20	44
mAP(%)	82.9			83.7			83.72		

In this part, we compared the GPU memory consumption of different memories. These memories are LRM [3] (Long Range Memory of MEGA), MB [19] (instance Memory Bank of MAMBA) and IM (Init Memory). As shown in Table 4, we tested these three memories in short, medium and long videos. Firstly, the LRM is not affected by the video duration, and the GPU usage has been kept at 20 MB. Though the mAP of the MEGA is lower than the latter two methods. Although the GPU usage of MB is smaller than LRM in short videos, it far exceeds LRM in medium and long videos. Compared with the first two, IM not only surpasses

LRM in terms of accuracy, but also occupies much less GPU memory than MB in all lengths of video. The duration of long videos in Table 4 is only 1 min and 35 s, which is relatively long to the VID dataset. The video length in the real scene is generally much longer than this.

4.7 Analysis of How RFE Handles Key Sets of Two Memories

Table 5. Results of different ways RFE handles Key Sets of Init Memory and MDR.

Different Ways	Init Memory first	MDR first	Together
mAP (complete dataset)	83.61	83.71	83.76
mAP (multi-class dataset)	70.98	71.16	71.09

The MIDFA network will send the Key Sets of two memories into the RFE module. Here, it is necessary to determine which key sets of two kinds of memory are aggregated with the original ROI feature by the RFE module first, or both are aggregated at the same time. In Table 5, the best results can be achieved by concatenating the two Memory Key Sets together in complete VID dataset. The MDR-first approach is optimal in multi-class datasets. Given that feature diversity has a higher priority in the full dataset, the more features captured in the RFE module, the higher the mAP value.

4.8 Comparison of Different Methods for Instance Division

Fig. 3. Comparison of different methods for Instance Division.

Instance Division method can effectively distinguish two overlapping objects. We visually compare the reality detection results of the two different feature division methods. In Fig. 3, each row represents a feature division method, and each column is a chronologically captured video frame containing detection boxes from a video. The labels of the boxes marked squirrel and cat in the first row are cat, so that the purpose of aggregating the features of multi-class objects separately cannot be achieved. The results in the second row show that the Instance Division method can more effectively distinguish different instances when they are overlapped. In the first frame in row 2, the squirrel is marked as object 3. And it is still marked as object 3 in subsequent frames.

5 Conclusion

We explore how to solve the multi-class problem in the video object detection task. In order to avoid several defects of the classifier method, we propose an Instance Division method to more effectively divide ROI features. Combining the two kinds of memory, our MIDFA network can perform better than previous methods on both the ImageNet VID dataset and the custom multi-class dataset.

The MIDFA network mainly relies on Init Memory to achieve good performance on the complete VID dataset. However, Instance Division and MDR can only play an obvious role in multi-class videos. In the maritime ship object detection scene, it is often encountered that various kinds of ships appear in the same shot, so our method will be very useful in such task.

Acknowledgement. This work is supported by Shanghai "Science and Technology Innovation Action Plan" Venus Project (Sailing Special Project) under Grant 23YF1412900.

References

1. Bertasius, G., Torresani, L., Shi, J.: Object detection in video with spatiotemporal sampling networks. In: Ferrari, V., Hebert, M., Sminchisescu, C., Weiss, Y. (eds.) ECCV 2018. LNCS, vol. 11216, pp. 342–357. Springer, Cham (2018). https://doi.org/10.1007/978-3-030-01258-8_21
2. Chen, K., et al.: Optimizing video object detection via a scale-time lattice. In: Proceedings of the IEEE Conference on Computer Vision and Pattern Recognition, pp. 7814–7823 (2018)
3. Chen, Y., Cao, Y., Hu, H., Wang, L.: Memory enhanced global-local aggregation for video object detection. In: Proceedings of the IEEE/CVF Conference on Computer Vision and Pattern Recognition, pp. 10337–10346 (2020)
4. Deng, H., et al.: Object guided external memory network for video object detection. In: Proceedings of the IEEE/CVF International Conference on Computer Vision, pp. 6678–6687 (2019)
5. Deng, J., Pan, Y., Yao, T., Zhou, W., Li, H., Mei, T.: Relation distillation networks for video object detection. In: Proceedings of the IEEE/CVF International Conference on Computer Vision, pp. 7023–7032 (2019)
6. Dosovitskiy, A., et al.: FlowNet: learning optical flow with convolutional networks. In: Proceedings of the IEEE International Conference on Computer Vision, pp. 2758–2766 (2015)
7. Gong, T., et al.: Temporal ROI align for video object recognition. In: Proceedings of the AAAI Conference on Artificial Intelligence, vol. 35, pp. 1442–1450 (2021)

8. Han, L., Wang, P., Yin, Z., Wang, F., Li, H.: Class-aware feature aggregation network for video object detection. IEEE Trans. Circuits Syst. Video Technol. **32**, 8165–8178 (2021)
9. Han, L., Wang, P., Yin, Z., Wang, F., Li, H.: Context and structure mining network for video object detection. Int. J. Comput. Vision **129**(10), 2927–2946 (2021)
10. Han, W., et al.: Seq-NMS for video object detection. arXiv preprint arXiv:1602.08465 (2016)
11. He, F., Gao, N., Li, Q., Du, S., Zhao, X., Huang, K.: Temporal context enhanced feature aggregation for video object detection. In: Proceedings of the AAAI Conference on Artificial Intelligence, vol. 34, pp. 10941–10948 (2020)
12. He, K., Zhang, X., Ren, S., Sun, J.: Deep residual learning for image recognition. In: Proceedings of the IEEE Conference on Computer Vision and Pattern Recognition, pp. 770–778 (2016)
13. Jiang, Z., et al.: Learning where to focus for efficient video object detection. In: Vedaldi, A., Bischof, H., Brox, T., Frahm, J.-M. (eds.) ECCV 2020. LNCS, vol. 12361, pp. 18–34. Springer, Cham (2020). https://doi.org/10.1007/978-3-030-58517-4_2
14. Lin, L.: Dual semantic fusion network for video object detection. In: Proceedings of the 28th ACM International Conference on Multimedia, pp. 1855–1863 (2020)
15. Luo, H., Xie, W., Wang, X., Zeng, W.: Detect or track: towards cost-effective video object detection/tracking. In: Proceedings of the AAAI Conference on Artificial Intelligence, vol. 33, pp. 8803–8810 (2019)
16. Ren, S., He, K., Girshick, R., Sun, J.: Faster R-CNN: towards real-time object detection with region proposal networks. In: Advances in Neural Information Processing Systems, vol. 28 (2015)
17. Russakovsky, O., et al.: ImageNet large scale visual recognition challenge. Int. J. Comput. Vision **115**(3), 211–252 (2015)
18. Shvets, M., Liu, W., Berg, A.C.: Leveraging long-range temporal relationships between proposals for video object detection. In: Proceedings of the IEEE/CVF International Conference on Computer Vision, pp. 9756–9764 (2019)
19. Sun, G., Hua, Y., Hu, G., Robertson, N.: MAMBA: multi-level aggregation via memory bank for video object detection. In: Proceedings of the AAAI Conference on Artificial Intelligence, vol. 35, pp. 2620–2627 (2021)
20. Wu, H., Chen, Y., Wang, N., Zhang, Z.: Sequence level semantics aggregation for video object detection. In: Proceedings of the IEEE/CVF International Conference on Computer Vision, pp. 9217–9225 (2019)
21. Xie, S., Girshick, R., Dollár, P., Tu, Z., He, K.: Aggregated residual transformations for deep neural networks. In: Proceedings of the IEEE Conference on Computer Vision and Pattern Recognition, pp. 1492–1500 (2017)
22. Zhu, X., Dai, J., Yuan, L., Wei, Y.: Towards high performance video object detection. In: Proceedings of the IEEE Conference on Computer Vision and Pattern Recognition, pp. 7210–7218 (2018)
23. Zhu, X., Wang, Y., Dai, J., Yuan, L., Wei, Y.: Flow-guided feature aggregation for video object detection. In: Proceedings of the IEEE International Conference on Computer Vision, pp. 408–417 (2017)
24. Han, M., Wang, Y., Chang, X., Qiao, Yu.: Mining inter-video proposal relations for video object detection. In: Vedaldi, A., Bischof, H., Brox, T., Frahm, J.-M. (eds.) ECCV 2020. LNCS, vol. 12366, pp. 431–446. Springer, Cham (2020). https://doi.org/10.1007/978-3-030-58589-1_26

Medical and Biological Data

Vision Transformers for Small Histological Datasets Learned Through Knowledge Distillation

Neel Kanwal[1(✉)], Trygve Eftestøl[1], Farbod Khoraminia[2], Tahlita C. M. Zuiverloon[2], and Kjersti Engan[1]

[1] Department of Electrical Engineering and Computer Science, University of Stavanger, Stavanger, Norway
{neel.kanwal,trygve.eftestol,kjersti.engan}@uis.no

[2] Department of Urology, University Medical Center Rotterdam, Erasmus MC Cancer Institute, Rotterdam, The Netherlands
{f.khoraminia,t.zuiverloon}@erasmusmc.nl

Abstract. Computational Pathology (CPATH) systems have the potential to automate diagnostic tasks. However, the artifacts on the digitized histological glass slides, known as Whole Slide Images (WSIs), may hamper the overall performance of CPATH systems. Deep Learning (DL) models such as Vision Transformers (ViTs) may detect and exclude artifacts before running the diagnostic algorithm. A simple way to develop robust and generalized ViTs is to train them on massive datasets. Unfortunately, acquiring large medical datasets is expensive and inconvenient, prompting the need for a generalized artifact detection method for WSIs. In this paper, we present a student-teacher recipe to improve the classification performance of ViT for the air bubbles detection task. ViT, trained under the student-teacher framework, boosts its performance by distilling existing knowledge from the high-capacity teacher model. Our best-performing ViT yields 0.961 and 0.911 F1-score and MCC, respectively, observing a 7% gain in MCC against stand-alone training. The proposed method presents a new perspective of leveraging knowledge distillation over transfer learning to encourage the use of customized transformers for efficient preprocessing pipelines in the CPATH systems.

Keywords: Artifact Detection · Computational Pathology · Deep Learning · Knowledge Distillation · Vision Transformer · Whole Slide Images

1 Introduction

Histological examination of tissue samples is conducted by studying thin slices from a tumor specimen mounted on a glass slide. During the laboratory procedures, the preparation of glass slides may introduce artifacts and variations causing loss of visual features [15,27]. Artifacts, such as air bubbles, occur when

H. Kashima et al. (Eds.): PAKDD 2023, LNAI 13937, pp. 167–179, 2023.
https://doi.org/10.1007/978-3-031-33380-4_13

air is trapped under the cover slip due to improper mounting procedure [16]. Eventually, the presence of air bubbles leaves an altered and fainted appearance [16,27]. During the manual assessment, pathologists usually ignore regions containing artifacts as they are irrelevant for diagnosis.

Computational Pathology (CPATH) systems are automated systems working with a digitized glass slide, called Whole Slide Image (WSI), as input. CPATH systems have the potential to automate diagnostic tasks and provide a second opinion or localize the Regions of Interest (ROIs) [14]. Different types of artifacts, like air bubbles, might be present on the WSI [16] and can deteriorate diagnostic CPATH results if included in the analysis. Therefore it has been proposed to detect and exclude artifacts as a first step before using more relevant tissue in a diagnostic or prognostic system [15,16]. The detection and exclusion of artifacts can be regarded as (a part of) a *preprocessing pipeline*, which also might include color normalization and patching [16]. A complete preprocessing pipeline should detect folded tissue, damaged tissue, blood, and blurred (out of focus) areas, as well as air bubbles [16]. This might be done by an ensemble of models, one for each artifact, or by a multiclass model. In this paper, we consider detecting *air bubbles* artifact, which is not given much attention in the literature.

Deep Learning (DL) methods have shown promising results in various medical image analysis tasks [4,28], and can be used for detecting artifacts in a preprocessing pipeline. Supervised learning for generalized DL models requires a significant amount of data and labels. In CPATH literature, little effort has been made to annotate artifacts; thus, publicly available datasets for histological artifacts are unavailable. Transfer Learning (TL) has been widely used for medical images to deal with the lack of labeled training data [6,21]. TL methods use the existing knowledge, such as ImageNet [2] weights, and fine-tune the model for a different task. Although TL on ImageNet weights is useful to cope with a lack of data, ImageNet weights are mostly available for complicated Deep Convolutional Neural Networks (DCNN) architectures and carry a strong texture bias [5]. However, such DCNNs are typically computationally complex, whereas a preprocessing pipeline, being a first step prior to diagnostic or prognostic models, should have generalized and efficient DL models with high throughput. This is especially true with an ensemble of DCNN models for the different artifacts.

After the success in natural language processing tasks, *transformers* have been given attention for vision tasks [3,17]. Vision Transformers (ViTs), using a convolution-free approach, have surpassed DCNNs in accuracy and efficiency on image classification benchmarks [1,3]. Unlike the convolution layer in DCNNs, which applies the same filter weights to all inputs, the multi-head attention [30] in ViTs attends to image-wide structural information [20]. Interestingly, ViTs are also shown to be more robust and generalized than DCNNs [1,20]; Unfortunately, the robustness and generalizability come from training on extremely large datasets [1,3,29], which contrasts with the biomedical scenario. These limitations bring us to the question: *how can we train generalized ViTs on a small histopathological dataset?*.

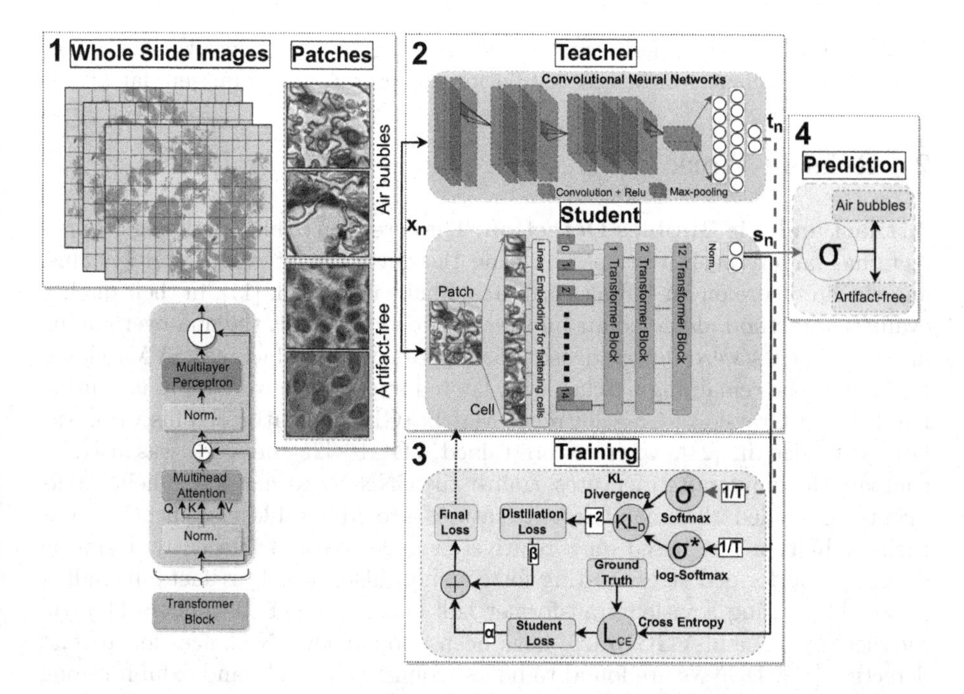

Fig. 1. An overview of our proposed air bubbles detection method by knowledge distillation: Predefined size patches for air bubbles and artifact-free classes are extracted from the WSI. A ViT student model is trained with the help of a DCNN teacher model by leveraging the transference of knowledge during the training process. The student-teacher recipe weights the teacher and student's outputs by the temperature (T). The overall training objective is to minimize the final loss, which is a linear combination of student loss and distillation loss. Finally, the student model is used to perform predictions for binary air bubbles detection task.

One possible answer lies in Knowledge Distillation (KD) [10], which transfers knowledge from a usually large teacher model to another, typically smaller, student model. Motivated by the KD idea, we present a student-teacher recipe, as shown in Fig. 1. We propose to use KD in combination with TL for detecting air bubbles on WSIs using a small training set. In short, we let the teacher model be a complex ImageNet pretrained DCNN and using KD, we train a small student model, which is a ViT. In the inference stage, we only need the small ViT, which is computationally efficient enough for a preprocessing pipeline implementation.

Our contributions in this paper can be summarized as follows:

- We train several state-of-the-art DCNNs and ViTs to compare their performance on a binary air bubbles detection task. Later, we choose suitable architectures to test our student-teacher framework.
- We conduct an in-depth comparison by initializing models with and without ImageNet weights and training ViT under a standalone vs. a student-teacher framework. We also assess the improvements in ViT's generalization capability over ImageNet transfer learning.

- We run extensive experiments to test the student ViT's performance under different teacher models and distillation configurations on unseen data.

2 Related Work

Artifact and Air Bubbles Detection: The detection of histopathological artifacts has largely been overlooked during the development of CPATH systems, and the literature on air bubbles is scarce. Shakhawat et al. [11], in their quality evaluation method, detected air bubbles in two steps. First, the non-overlapping affected patches were detected using a Support Vector Machine (SVM) classifier. Later, the remaining patches with fainted appearance were separated using handcrafted Gray-level Co-occurrence Matrix (GLCM) features. This work was later extended in [24], where a pretrained VGG16 [25] network was used to compare the handcrafted features against the CNN-based method. Their experiments concluded that handcrafted features provide stable classification, but their evaluation was based on a relatively smaller dataset. Recently, Raipuria et al. [22] performed stress testing for common histological artifacts, including air bubbles, using a vision transformer [29] and a ResNet [9] model. Though, MobiletNet [12] and VGG16 [25] have been popular DCNN choices for artifact detection [15]. DCNNs are found to be less robust than ViTs and exhibit strong texture bias [20,22].

Knowledge Distillation (KD): Originally proposed by Hinton et al. [10] for model compression, KD sought to extract knowledge from an ensemble of CNN experts to a smaller two-layer CNN generalist network to make it perform equally well. In short, KD aims to train a small student model under the guidance of a complicated teacher model, where the student model optimizes its learning by absorbing the hidden knowledge from the teacher. This transference of knowledge can be accomplished by minimizing output logits of student and teacher networks through some distillation methods, such as logit-based, feature-based, and relationship-based distillation methods [19].

KD helps make computationally friendly deployment algorithms, making it interesting for many biomedical imaging algorithms. Lingmei et al. [18] proposed a CNN model for glioma classification. They used the KD approach to compress the model and make it suitable for deployment on medical equipment. Salehi et al. [23] used a VGG16 [25] cloner network to calculate multi-level loss from a source network for detecting anomalies. Their method relied on distilling intermediate knowledge from the ImageNet pretrained source network. In a similar approach, He et al. [8] used the KD technique to boost the performance of CNN for ocular disease classification. They used fundus images and clinical information to train a ResNet [9] teacher first and used only the fundus images to train a similar student network later. Guan et al. [7] detected Alzheimer's disease by leveraging multi-modal data to train a teacher network. Their distillation scheme improved the prediction performance of the ResNet [9] student using a single imaging modality.

However, all these works focused on using only CNN as a student network and did not explore the effects of different configurations and teacher networks on the final classification outcome. In addition, the use of KD for histological artifacts has not been investigated yet.

3 Data Materials and Method

Figure 1 provides an overview of our air bubbles detection method using KD [10] in a student-teacher recipe. We exploit KD for data-efficient training by leveraging the transference of knowledge from the teacher model to the student model. Our proposed method uses a complex DCNN as the pre-trained teacher and a small ViT as the student when a small histological dataset is available. We are doing a logit-based distillation [19] since our teacher and student models are very different. The steps of our method are further described below.

3.1 Dataset

The air bubbles dataset was prepared from 55 bladder biopsy WSIs, provided by Erasmus Medical Center (EMC), Rotterdam, The Netherlands. The glass slides were stained with Hematoxylin and Eosin (H&E) dyes and scanned with Hamamatsu Nanozoomer at 40× magnification. WSIs are stored in *ndpi* format with a pixel size of 0.227 μm × 0.227 μm. These WSIs were manually annotated for air bubbles and artifact-free tissue by a non-pathologist who has received training for the task. To prevent data leakage, the dataset was later split into 35/10/10 training, validation, and test WSIs, respectively.

3.2 Foreground Segmentation and Patching

Let $I^{40x}_{\mathrm{WSI}(i)}$ correspond to a WSI at magnification level 40x (sometimes referred to as 400x). I^{40x}_{WSI} are very large gigapixel images, and it is not feasible to process the entire WSI at once. As such, all CPATH systems resort to patching or tiling of the image, or the ROI in the image, before further processing. Let $\mathcal{T} : I^{40x}_{\mathrm{WSI(i)}\in R} \rightarrow \{\mathbf{x}^i_j; j = 1 \cdots J\}$ represent the process of patching a ROI denoted by R of the image $I^{40x}_{\mathrm{WSI}(i)}$ into a set of J patches, where $\mathbf{x}^i_j \in \mathbb{R}^{W \times H \times C}$ and W, H, C present the width, height, and channels of the image, respectively. In the patching process, foreground-background segmentation was performed first by transforming (Red, Green, Blue) RGB images to (Hue, Saturation, Value) HSV color space. Later, Otsu thresholding was applied to the value channel to obtain the foreground with tissue. The extracted foreground was later divided over a non-overlapping square grid, and patches with at least 70% overlap to the annotation region (R) were extracted.

Let $\mathcal{D} = (\mathbf{X}, \mathbf{y}) = \{(\mathbf{x}_n, \mathbf{y}_n)\}^N_{n=1}$ denote our prepared dataset of N patches from a set of WSIs and $y_n \in \{0, 1\}$ is the binary ground truth for the n-th instance, where 1 indicates a patch within a region marked as air bubbles. Figure 1 (step 1) shows the patches $\mathbf{x}_n$ of 224 × 224 × 3 pixels with air bubbles and artifact-free classes obtained from a WSI at 40x magnification.

3.3 Selecting Student-Teacher Architectures

Let's symbolize the student model $\boldsymbol{\xi}$ with parameters θ providing the prediction output logits $\mathbf{s}_n = \boldsymbol{\xi}_\theta(\mathbf{x}_n)$, and correspondingly, the teacher model $\boldsymbol{\varphi}$ parameterized by ϕ providing the output logits $\mathbf{t}_n = \boldsymbol{\varphi}_\phi(\mathbf{x}_n)$.

Our student model is a ViT, similar to the pioneering work [3], which leverages multi-head self-attention mechanism [30] to capture content-dependant relations across the input patch. At the image pre-processing layer, the patches of 224×224 pixels are split into the non-overlapping cells of 16×16 pixels. Later, the linear embedding layer flattens these cells, and positional encodings are added before feeding the embeddings to the pile of transformer blocks, as illustrated in Fig. 1 (step 2). Since convolutional networks have shown their efficacy in image recognition tasks, transferring knowledge from a DCNN network can help the ViT absorb inductive biases. Therefore, we rely on popular state-of-the-art DCNNs for selecting teacher architecture. Nevertheless, WE systemically discover appropriate student and teacher candidates during the experiments later to demonstrate the approach's effectiveness over TL.

3.4 Training Student Under Knowledge Distillation

After selecting student and teacher architectures, we begin the process of training the student $\boldsymbol{\xi}$. The goal is to train $\boldsymbol{\xi}$ with the assistance of a $\boldsymbol{\varphi}$ to improve the $\boldsymbol{\xi}$'s generalization performance using additional knowledge beyond the labels. Our approach is similar to Hinton et al. [10] where model outputs $\mathbf{s}$, and $\mathbf{t}$ are normalized by a temperature T parameter before using the softmax function σ. The increasing value of T softens the impact of the fluctuations in the output probability distribution; therefore, more knowledge can be devolved with each input $\mathbf{x}_n$. Instead of using softmax on $\mathbf{s}_n$, we take advantage of the log-softmax function σ^*, which stabilizes the distillation process by penalizing for incorrect class. σ^* also adds efficiency by optimizing gradient calculations.

The output logits for input patch $\mathbf{x}_n$ can be written as:

$$\mathbf{s}_n = \boldsymbol{\xi}_\theta(\mathbf{x}_n) \quad \text{and} \quad \mathbf{t}_n = \boldsymbol{\varphi}_\phi(\mathbf{x}_n) \tag{1}$$

Let the log-softmax and softmax on logits, $\sigma^*(\mathbf{s}/T)$ and $\sigma(\mathbf{t}/T)$, for each element can be defined as (see Eq. (2)):

$$\sigma^*(s_i/T) = log\left(\frac{\exp{(s_i/T)}}{\sum_{j=1}^{c} \exp{(s_j/T)}}\right) \quad \text{and} \quad \sigma(t_i/T) = \frac{\exp{(t_i/T)}}{\sum_{j=1}^{c} \exp{(t_j/T)}} \tag{2}$$

where c is the total number of classes and T is the temperature. The class probabilities at the output of the $\boldsymbol{\xi}$ and $\boldsymbol{\varphi}$ model can thus be written as:

$$p_{\xi} = \sigma^*(\mathbf{s}/T) = \sigma^*(\boldsymbol{\xi}_\theta(\mathbf{x})) \quad \text{and} \quad p_{\varphi} = \sigma(\mathbf{t}/T) = \sigma(\boldsymbol{\varphi}_\phi(\mathbf{x})) \tag{3}$$

The student loss $L_{student}$ (Eq. (4)) provides hard targets and is obtained by applying cross entropy L_{CE} on ground truth y, and $\mathbf{s}$ when T is set to 1:

$$L_{student} = L_{CE}(y, \mathbf{s}) = -\sum_{i=1}^{c} y_i \cdot log(\sigma^*(s_i)) \tag{4}$$

Distillation loss $L_{distillation}$ provides the soft targets and is computed from the p_{ξ} and p_{φ} by applying Kullback-Leibler divergence KL_D. Since the outputs from ξ and φ were normalized by T, we multiply the loss with T^2 to maintain their relative contribution:

$$L_{distillation} = T^2 \times KL_D(p_{\xi} \| p_{\varphi}) = T^2 \cdot \sum_{i=1}^{c} p_{\xi_i} log \frac{p_{\xi_i}}{p_{\varphi_i}} \tag{5}$$

The final loss function, as shown in Eq. (6), is a weighted average of student and distillation losses where $\alpha \in [0, 1)$:

$$L_{Final} = \alpha \times L_{student} + \beta \times L_{distillation} \quad \because \beta = 1 - \alpha \tag{6}$$

High entropy in soft targets offers significantly more information per training patch than hard targets [10], allowing the student ViT to train with fewer data and a higher learning rate. Therefore, using a smaller alpha can be beneficial if the ξ is trained from scratch. Our standalone training setup for baseline comparison can be obtained by putting α and T equal to one and replacing log softmax with softmax function.

3.5 Prediction

Once the final loss is minimized based on the experimental setup (defined in Sect. 4), we find predictions from the student ξ by setting T equal to one. For an unseen test patch $\mathbf{x}_*$, output can be defined as (7):

$$\hat{y}_s = \arg\max(\sigma(\mathbf{s}_*)) = \arg\max(\sigma(\xi_{\theta}(\mathbf{x}_*))) \quad \in \{0, 1\} \tag{7}$$

4 Experimental Setup

Implementation Details: The patch extraction was accomplished using the HistoLab library. Extracted patches were normalized to ImageNet [2] mean and standard deviation. We augmented data at every training epoch using random geometric transformations, such as rotations, horizontal and vertical flips. ViTs were borrowed from Timm Library, and the experimental setup was built on the Pytorch. We used four variants of ViTs with different parametric depths from [3,29], where the classifier was replaced by a fully connected (FC) layer. We used four state-of-the-art DCNNs with varying parametric complexity. All DCNN backbones were initialized with ImageNet [2] weights, and classifiers were replaced with three-layer FC classifiers. All classifiers were initialized with random weights. After hyper-parameter exploration, the final parameters were set to a batch size of 64, SGD optimizer, ReduceLROnPlateau scheduler with a

Table 1. Results from Exp. 1: Four variants of Deep Convolutional Neural Networks (DCNNs) and Vision Transformers (ViTs), with increasing parametric complexity, are trained for the air bubbles detection task. The best outcomes in every section are bolded. ViT-tiny and MobileNet architectures provide the best results on the test set.

Architecture	Param. (#)	Validation Set			Test Set		
		Acc.(%)	F1	MCC(⇑)	Acc.(%)	F1	MCC(⇑)
Deep Convolutional Neural Networks (DCNNs)							
MobileNetv3 [12]	3.52M	98.28	0.983	0.965	**93.88**	**0.945**	**0.876**
EfficientNet [26]	20.89M	96.52	0.966	0.931	92.54	0.935	0.851
DenseNet161 [13]	27.66M	98.12	0.982	0.962	91.32	0.925	0.828
VGG16 [25]	136.42M	**98.34**	**0.984**	**0.966**	92.31	0.932	0.846
Vision Transformers (ViTs)							
ViT-tiny [29]	5.52M	**98.67**	**0.987**	**0.973**	**92.35**	**0.933**	**0.847**
ViT-small [29]	21.66M	97.01	0.971	0.941	91.16	0.922	0.822
ViT-large [3]	303.30M	98.12	0.982	0.962	92.08	0.928	0.839
ViT-huge [3]	630.76M	95.85	0.962	0.918	91.43	0.925	0.829
Results from Literature (Validation Accuracy (%))							
DeiT-S in [22] 91.5-92		ResNet-50 in [22] 88-89			VGG16 in [24] 87.33		

learning rate of 0.001, dropout of 0.2, cross-entropy loss, and early stopping with the patience of 20 epochs on validation loss to prevent over-fitting. For KD parameters, values of $T \in \{2, 5, 10, 20, 40\}$ and $\alpha \in \{0.3, 0.5, 0.7\}$ were explored. The best model weights are used to report the results. The NVIDIA GeForce A100 SXM 40GB GPU was utilized for training all models.

Evaluation Metrics: We evaluate the presented method using accuracy, F1-score, and Mathew Correlation Coefficient (MCC). Let TP, FN, FP, and TN describe true positive, false negative, false positive, and false negative predictions. The accuracy, termed as $(TP + TN)/(TP + FN + FP + TN)$, is the ratio of correct predictions by the model. F1 is the harmonic mean, defined as $2\cdot(\text{precision}\cdot\text{recall})/(\text{precision}+\text{recall})$ where Recall $= TP/(TP+FN)$ and Precision $= TP/(TP+FP)$. MCC is an informative measure in binary classification over imbalanced datasets and is defined as Eq. (8):

$$MCC = \frac{TP \cdot TN - FP \cdot FN}{\sqrt{(TP + FP) \cdot (TP + FN) \cdot (TN + FP) \cdot (TN + FN)}} \in [-1, 1] \quad (8)$$

5 Results and Discussion

5.1 *Exp. 1*: Baseline Experiments for Architecture Decision

In this experiment, we only apply TL to a set of architectures. We evaluate state-of-the-art DCNNs, namely MobileNetv3 [12], EfficientNet [26], DenseNet161 [13] and VGG16 [25] architectures and a family of four ViTs [3,29], with increasing

Table 2. Results from Exp. 2: Knowledge Distillation (KD) outcome for selected teacher and student candidates from Exp.1. The values of α, T are fixed at 0.5 and 10, respectively. The best results in every part are marked in bold, and the second best is underlined. ViT-tiny, with scratch and ImageNet initializations, is used for baseline comparisons. Two teachers (MobileNet and VGG16) with air bubbles knowledge are used. While MobileNet is also initialized with knowledge of other domains to evaluate the importance of teachers' knowledge.

Architecture (Initial.)	Validation Set			Test Set		
	Acc.(%)	F1	MCC(⇑)	Acc.(%)	F1	MCC(⇑)
Baseline (Initial.) - Standalone training						
ViT-tiny (Scratch)	96.13	0.963	0.922	91.51	0.925	0.829
ViT-tiny (ImageNet [2])	**98.67**	**0.987**	**0.973**	**92.35**	**0.933**	**0.847**
Teacher (Initial.) - Student [ViT-tiny (Scratch)]						
MobileNet (Scratch)	96.13	0.962	0.924	87.92	0.889	0.756
MobileNet (ImageNet [2])	95.58	0.957	0.914	92.31	0.927	0.848
MobileNet (Damaged [15])	76.8	0.785	0.533	49.23	0.608	-0.075
MobileNet (Air bubbles)	**98.01**	**0.981**	**0.960**	**95.25**	**0.957**	**0.904**
VGG16 (Air bubbles)	<u>97.18</u>	<u>0.973</u>	<u>0.944</u>	<u>93.42</u>	<u>0.940</u>	<u>0.867</u>
Teacher (Initial.) - Student [ViT-tiny (ImageNet)]						
MobileNet (Scratch)	**98.73**	0.983	0.971	93.38	0.941	0.866
MobileNet (ImageNet [2])	98.62	**0.987**	<u>0.972</u>	93.40	0.942	0.867
MobileNet (Damaged [15])	50.08	0.211	0.09	35.51	0.116	-0.294
MobileNet (Air bubbles)	98.61	**0.987**	**0.973**	**95.60**	**0.961**	**0.911**
VGG16 (Air bubbles)	<u>98.67</u>	<u>0.986</u>	<u>0.972</u>	<u>94.19</u>	<u>0.948</u>	<u>0.882</u>

architecture size. Exp 1 provides a baseline as well as helps to choose architectures for the KD setup in later experiments. Table 1 reports the results of the validation and test set. DCNNs largely exceed the performance of ViTs, where top-performing ViT lags the generalization performance of top-performing DCNNs by 3% in MCC. Moreover, architectures with sizeable parameters like VGG16 and ViT-tiny, and MobileNet, despite being architectures with fewer parameters, emerge as appropriate student and teacher candidates, respectively, based on the test results and outperform the results from the literature.

5.2 *Exp. 2*: How Important is Teacher's Knowledge?

This experiment evaluates the impact of existing teacher knowledge in the KD process to assess the real-life analogy where good teachers make good students. Therefore, we initialize MobileNet teachers with no knowledge (scratch), knowledge from a general domain (ImageNet), knowledge from another WSI artifact (damaged tissue [15]), and finally, domain-relevant knowledge (air bubbles) from the previous experiment. In addition, we also select VGG16 with air bubble knowledge as a teacher to assess the effect of highly parametric DCNN in the KD process. For this experiment, the α, T values are fixed at 0.5 and 10, respectively. The student is a ViT-tiny architecture initialized with random and ImageNet weights separately.

Table 2 exhibits that KD remarkably improves ViT's classification ability. Even without ImageNet knowledge, ViT-tiny, under the KD framework, surpasses all metrics under both MobileNet and VGG16 teachers. However, the best results are obtained using the MobileNet teacher, ascertaining that hidden knowledge can be easily distilled from a simpler architecture. Interestingly, teachers with knowledge other than the relevant domain (air bubbles) produce poorly performing student. Although the student with ImageNet knowledge does not indicate gain on the validation results relative to the baseline, it achieves 3% and 7% improvement in F1 and MCC scores on the test set, respectively.

Overall, the test results demonstrate that the KD is promising to train generalized ViT-tiny with little data, even without pretrained weights. ViT significantly enhances its generalization against the baseline when trained in a standalone setting. Especially when the teacher is enriched with the knowledge related to the task. KD, on top of ImageNet TL, provides a marginal gain in the performance of ViT-tiny, overcoming the reliance on pretrained weights.

5.3 *Exp. 3*: Influence of KD Parameters

Since the initialization of teachers with air bubbles knowledge has been shown to improve the learning process, it would be interesting to assess the influence of DCNN teachers under the different KD parameters (T and α). In this experiment, we chose $T \in \{2, 5, 10, 20, 40\}$ and $\alpha \in \{0.3, 0.5, 0.7\}$ to estimate the influence of teacher's output on ViT student, trained from scratch. The baseline experiment corresponds to $\alpha, T = 1$ and uses sigmoid on ViT outputs. Figure 2 (a) and (b) show MCC values as the effect of temperature on simple DCNN like MobileNet and complex DCNN like VGG16. Though the ViT-tiny student trained under the VGG16 teacher scores better on the validation set when T is high, the MobileNet teacher reveals better transference of hidden knowledge on all T values on the test set. Figure 2 (c) depicts the effect of α on ViT's generalization results. All α values give better results than the baseline, concluding that including distillation loss improves training compared to only student loss.

To sum up, the teacher's outcome strongly influences the student's generalizability in the KD process. Most of the T and α values deliver a noticeable gain over the standalone training in our case. However, *intermediate* T values and assigning *equal weight* to student and distillation loss is the most advantageous.

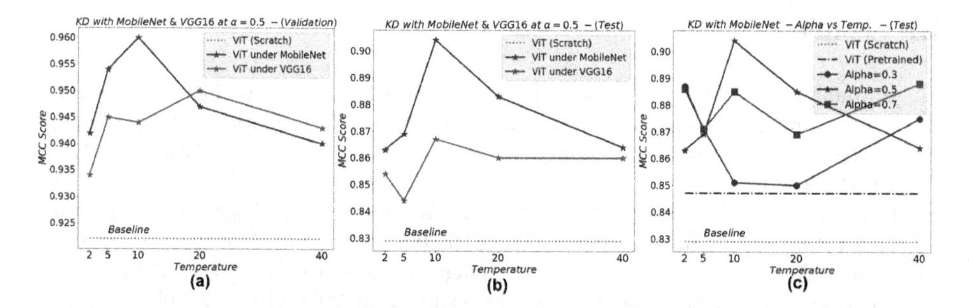

Fig. 2. Results from Exp. 3: Knowledge Distillation (KD) improves the performance of the Vision Transformer (ViT-tiny) under the supervision of both MobileNet and VGG16 teachers. (a) and (b) shows an improved performance from the baseline (standalone training from scratch), under all temperature (T) values, on validation and test set. (c) depicts the influence of giving higher/lower weightage to distillation loss from the teacher network (see Sect. 3). The MobileNet teacher, despite being simpler architecture, enriches ViT-tiny's generalization capability on all chosen α and T values.

6 Conclusion and Future Work

This paper presents the Knowledge Distillation (KD) to boost the generalization performance of small Vision Transformers (ViTs) on a small histopathological dataset. The main motivation is to create a well-performing and efficient preprocessing pipeline that requires a generalized and computationally-friendly model. We evaluated various pretrained DCNNs and ViTs for the air bubbles artifact detection task. ViTs, trained in a standalone setting, underperform DCNNs on unseen data. Our approach exploits the KD, in the absence of pretrained weights, to enhance the performance of ViT by training under the guidance of a DCNN teacher. Our analysis found that KD provides significant gain under most distillation settings when the teacher holds the knowledge of the same task. In conclusion, the ViT, when trained under KD, outperforms its state-of-the-art DCNN teacher and its counterpart in standalone training.

In future work, the method can be developed and tested on larger cohorts of histological data with stain variations and to detect multiple artifacts. Moreover, artifact detection by ViT trained under the student-teacher recipe can be combined as a preprocessing step with a diagnostic or prognostic algorithm in the computational pathology system.

Acknowledgment. This research is supported by the European Horizon 2020 program under Marie Skłodowska-Curie grant agreement No. 860627 (CLARIFY). The authors have no relevant financial or non-financial interests to disclose.

References

1. Bhojanapalli, S., Chakrabarti, A., Glasner, D., Li, D., Unterthiner, T., Veit, A.: Understanding robustness of transformers for image classification. In: Proceedings of the IEEE International Conference on Computer Vision (ICCV), pp. 10231–10241 (2021)
2. Deng, J., Dong, W., Socher, R., Li, L.J., Li, K., Fei-Fei, L.: ImageNet: a large-scale hierarchical image database. In: 2009 IEEE ICCV, pp. 248–255. IEEE (2009)
3. Dosovitskiy, A., Beyer, L., et al.: An image is worth 16×16 words: transformers for image recognition at scale. arXiv preprint arXiv:2010.11929 (2020)
4. Fuster, S., Khoraminia, F., et al.: Invasive cancerous area detection in non-muscle invasive bladder cancer whole slide images. In: IEEE 14th Image, Video, and Multidimensional Signal Processing Workshop (IVMSP), pp. 1–5. IEEE (2022)
5. Geirhos, R., Rubisch, P., Michaelis, C., Bethge, M., Wichmann, F.A., Brendel, W.: ImageNet-trained CNNs are biased towards texture; increasing shape bias improves accuracy and robustness. arXiv preprint arXiv:1811.12231 (2018)
6. Golatkar, A., Anand, D., Sethi, A.: Classification of breast cancer histology using deep learning. In: Campilho, A., Karray, F., ter Haar Romeny, B. (eds.) ICIAR 2018. LNCS, vol. 10882, pp. 837–844. Springer, Cham (2018). https://doi.org/10.1007/978-3-319-93000-8_95
7. Guan, H., Wang, C., Tao, D.: MRI-based Alzheimer's disease prediction via distilling the knowledge in multi-modal data. Neuroimage **244**, 118586 (2021)
8. He, J., Li, C., Ye, J., Qiao, Y., Gu, L.: Self-speculation of clinical features based on knowledge distillation for accurate ocular disease classification. Biomed. Signal Process. Control **67**, 102491 (2021)
9. He, K., Zhang, X., Ren, S., Sun, J.: Deep residual learning for image recognition. In: Proceedings of the IEEE CVPR, pp. 770–778 (2016)
10. Hinton, G., Vinyals, O., Dean, J., et al.: Distilling the knowledge in a neural network. arXiv preprint arXiv:1503.02531 2(7) (2015)
11. Hossain, M.S., Nakamura, T., Kimura, F., Yagi, Y., Yamaguchi, M.: Practical image quality evaluation for whole slide imaging scanner. In: Biomedical Imaging and Sensing Conference, vol. 10711, pp. 203–206. SPIE (2018)
12. Howard, A., et al.: Searching for MobileNetV3. In: Proceedings of the IEEE International Conference on Computer Vision, pp. 1314–1324 (2019)
13. Huang, G., Liu, Z., Van Der Maaten, L., Weinberger, K.Q.: Densely connected convolutional networks. In: Proceedings of the IEEE Conference on Computer Vision and Pattern Recognition, pp. 4700–4708 (2017)
14. Kanwal, N., Amundsen, R., Hardardottir, H., Janssen, E.A., Engan, K.: Detection and localization of melanoma skin cancer in histopathological whole slide images. arXiv preprint arXiv:2302.03014 (2023)
15. Kanwal, N., Fuster, S., et al.: Quantifying the effect of color processing on blood and damaged tissue detection in whole slide images. In: IEEE 14th Image, Video, and Multidimensional Signal Processing Workshop (IVMSP), pp. 1–5. IEEE (2022)
16. Kanwal, N., Pérez-Bueno, F., Schmidt, A., Engan, K., Molina, R.: The devil is in the details: Whole slide image acquisition and processing for artifacts detection, color variation, and data augmentation. IEEE Access **10**, 58821–58844 (2022)
17. Kanwal, N., Rizzo, G.: Attention-based clinical note summarization. In: Proceedings of the 37th ACM Symposium on Applied Computing, pp. 813–820 (2022)
18. Lingmei, A., et al.: Noninvasive grading of glioma by knowledge distillation base lightweight convolutional neural network. In: IEEE 2021 AEMCSE, pp. 1109–1112

19. Meng, H., Lin, Z.E.A.: Knowledge distillation in medical data mining: a survey. In: 5th International Conference on Crowd Science and Engineering, pp. 175–182 (2021)
20. Naseer, M., Ranasinghe, K., Khan, S., Hayat, M., Shahbaz Khan, F., Yang, M.H.: Intriguing properties of vision transformers. NeurIPS **34**, 23296–23308 (2021)
21. Noorbakhsh, J., Farahmand, S., et al.: Deep learning-based cross-classifications reveal conserved spatial behaviors within tumor histological images. Nat. Commun. **11**(1), 1–14 (2020)
22. Raipuria, G., Singhal, N.: Stress testing vision transformers using common histopathological artifacts. In: Medical Imaging with Deep Learning (2022)
23. Salehi, M., Sadjadi, N., Baselizadeh, S., Rohban, M.H., Rabiee, H.R.: Multiresolution knowledge distillation for anomaly detection. In: Proceedings of the IEEE Conference on Computer Vision and Pattern Recognition, pp. 14902–14912 (2021)
24. Shakhawat, H.M., Nakamura, T., Kimura, F., Yagi, Y., Yamaguchi, M.: Automatic quality evaluation of whole slide images for the practical use of whole slide imaging scanner. ITE Trans. Media Technol. Appl. **8**(4), 252–268 (2020)
25. Simonyan, K., Zisserman, A.: Very deep convolutional networks for large-scale image recognition. arXiv preprint arXiv:1409.1556 (2014)
26. Tan, M., Le, Q.: EfficientNet: rethinking model scaling for convolutional neural networks. In: International Conference on Machine Learning, pp. 6105–6114. PMLR (2019)
27. Taqi, S.A., Sami, S.A., Sami, L.B., Zaki, S.A.: A review of artifacts in histopathology. J. Oral Maxillofacial Pathol. JOMFP **22**(2), 279 (2018)
28. Tomasetti, L., Khanmohammadi, M., Engan, K., Høllesli, L.J., Kurz, K.D.: Multi-input segmentation of damaged brain in acute ischemic stroke patients using slow fusion with skip connection. arXiv preprint arXiv:2203.10039 (2022)
29. Touvron, H., et al.: Training data-efficient image transformers & distillation through attention. In: International Conference on Machine Learning, pp. 10347–10357 (2021)
30. Vaswani, A., Shazeer, N., et al.: Attention is all you need. In: Advances in Neural Information Processing Systems (2017)

Cascaded Latent Diffusion Models for High-Resolution Chest X-ray Synthesis

Tobias Weber[1,2,3(✉)], Michael Ingrisch[2,3], Bernd Bischl[1,3], and David Rügamer[1,3]

[1] Department of Statistics, LMU Munich, Munich, Germany
tobias.weber@stat.uni-muenchen.de
[2] Department of Radiology, University Hospital, LMU Munich, Munich, Germany
[3] Munich Center for Machine Learning (MCML), Munich, Germany

Abstract. While recent advances in large-scale foundational computer vision models show promising results, their application to the medical domain has not yet been explored in detail. In this paper, we progress into the realms of large-scale modeling in medical synthesis by proposing *Cheff* - a foundational cascaded latent diffusion model, which generates highly-realistic chest radiographs providing state-of-the-art quality on a 1-megapixel scale. We further propose *MaCheX*, which is a unified interface for public chest datasets and forms the largest open collection of chest X-rays up to date. With *Cheff* conditioned on radiological reports, we further guide the synthesis process over text prompts and unveil the research area of report-to-chest-X-ray generation.

Keywords: latent diffusion model · chest radiograph · image synthesis

1 Introduction

Chest X-ray examinations are one of the most common, if not the most common, procedures in everyday clinical practice. This not only enables the collection of large databases but paves the way for lots of great opportunities to advance medical AI assistance in clinical workflows. Automated lung disease diagnosis [21], for example, can help to accelerate an examination and assist radiologists by reducing human error in a high-stress environment. Despite the availability of public datasets, many challenges in practical usefulness remain, partly induced due to inherent class imbalances and noisy labels. One possible approach to tackle this problem is by synthesizing underrepresented classes via generative modeling [30]. Generative models provide various other opportunities to improve clinical routines by, e.g., modeling the characteristics of diseases in different degrees using the patient's original thoracic scan [34] or suppressing bones to enhance soft tissue within the lung [8].

The basis for these methods is a stable and high-quality synthesis process. While previous evaluations in clinical practice focus on generative adversarial networks (GANs; see, e.g., [27]), the trend of generative architectures moves away

H. Kashima et al. (Eds.): PAKDD 2023, LNAI 13937, pp. 180–191, 2023.
https://doi.org/10.1007/978-3-031-33380-4_14

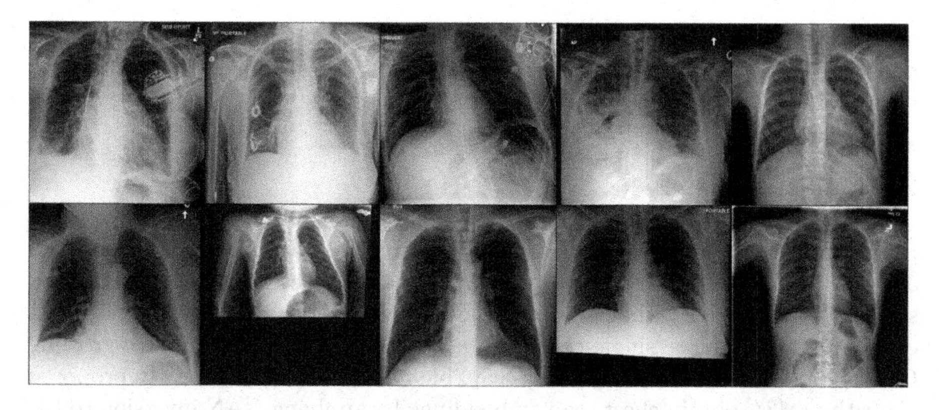

Fig. 1. High-resolution synthetic chest X-rays generated with our cascaded latent diffusion pipeline *Cheff*.

from an adversarial setting. Current state-of-the-art approaches tend to employ large-scale autoencoders with a focus on generating a prior in their latent space [5,24]. Moreover, diffusion models [10,29] have shown immense potential in image synthesization [4,17] resulting in a variety of proposed large-scale models including GLIDE [18], DALL·E 2 [22], Imagen [25] or Stable Diffusion [24].

Recently, GoogleAI released an API *CXR Foundation Tool* offering chest radiograph embeddings based on a contrastive neural network [28]. The underlying training dataset contains a significant proportion of non-disclosed proprietary data, though. The imminent risk of biases in foundational chest X-ray models has been recently discussed in [7] warning about the consequences in real-world applications. While these biases are inherent in the existing data, closed-source analyses make it difficult for the research community to counteract these problems.

Our Contributions. In this paper, we adapt large-scale learning concepts and models of general computer vision synthesis to chest radiography. To make this possible, our first contribution is to provide a unified interface to a massive collection of public chest X-ray datasets (*MaCheX*) with over 650,000 scans.[1] This dataset allows the training of *Cheff*, a cascaded **che**st X-ray latent **diff**usion pipeline, our second contribution.[2] The basis for *Cheff* are two foundational chest X-ray models: (i) an autoencoder for obtaining chest X-ray embeddings and (ii) a super-resolution diffusion model for refining low-resolution scans. A task-specific diffusion process in the latent space of the autoencoder leads to high-fidelity and diverse synthesis (see, e.g., Fig. 1). This further enables variable conditioning mechanisms while reducing computational training costs due to the usage of shared foundational models in the full pipeline. Finally, our model provides realistic report-to-chest-X-ray conversion by applying our model stack in a text-to-image setting by conditioning on radiologists' reports. Overall, our

[1] **MaChex:** https://github.com/saiboxx/machex.

[2] **Cheff:** https://github.com/saiboxx/chexray-diffusion.

approach pushes the limits in state-of-the-art synthesis of radiological images and fosters development of downstream clinical AI assistance tools by making both our model and dataset publicly accessible.

2 Related Work

In recent years, the rising popularity of generative models resulted in various approaches focusing on chest X-ray synthesis. [27] investigate the class-guided synthesis of targeted pathologies over classifier gradients. Their training is, however, only based on a single-center dataset while using a progressive-growing GAN ([13]; PGAN) known to be less capable than other more recently proposed algorithms. [34] obtain chest scan embeddings by applying GAN inversion to the generator of [27] and investigate characteristics of pathologies in this latent space. [9] utilize privacy-free synthetic chest X-rays for federated learning to ensure data protection across hospitals. [31] proposes a decomposition of diseases, ultimately producing anomaly saliency maps, by adding a second generative network exclusive for abnormal scans to produce a residual for a healthy version of this scan. Other use cases of generative methods include bone suppression [8] or improving classifier performance on underrepresented classes [30].

Apart from the classic adversarial literature, there is a surge in diffusion model research that also deals with chest X-rays. [35] propose a routine training on normal data and then utilizes a partial denoising process to detect anomalies. [3] fine-tune the foundational Stable Diffusion on a few thousand scans. While the synthesized scans lack a realistic style, it allows for a first glimpse into how report-to-scan generation can be used in medical imaging. [20] promote the usage of diffusion models over PGANs by fitting a latent diffusion model (LDM; [24]) on chest X-rays and observing a performance boost in a synthetically aided classifier. Additionally, the synthesizing process ensures anonymity with respect to the patient retrieval problem by employing a privacy-enhancing sampling algorithm [19]. A different strategy involves prompting DALL·E 2 to investigate zero-shot capabilities for medical imaging synthesis, however with limited success [1]. Generative modeling for automated radiological reports from chest X-rays is another highly active research area (cf. [14,15]), whereas the topic of creating images from reports is still an underrepresented, if not an unexplored, subject.

3 Methods

In the general framework of diffusion models [10,29], a Markov chain $(\boldsymbol{X}_t)_{t=1,\dots,T}$ with joint distribution defined by the density $q(\boldsymbol{x}_{1:T}) := \prod_{t=1}^{T} q(\boldsymbol{x}_t|\boldsymbol{x}_{t-1})$ is used to model a diffusion process of an uncorrupted image $\boldsymbol{x}_0 \in \mathcal{X}$. This is done by assuming $(\boldsymbol{X}_t \mid \boldsymbol{X}_{t-1} = \boldsymbol{x}_{t-1}) \sim \mathcal{N}(\sqrt{1-\beta_t}\boldsymbol{x}_{t-1}, \beta_t \boldsymbol{I})$ with time-varying variance $\beta_t \in (0,1)$, which is chosen by a pre-defined variance schedule. In other words, this autoregressive process incrementally applies Gaussian noise to an input $\boldsymbol{x}_0$ for a number of timesteps T. The learning task then is to revert this so-called *forward process*, i.e., inferring $\boldsymbol{x}_{t-1}$ from its corrupted version $\boldsymbol{x}_t$ for $t = 1, ..., T$. For large T the result $\boldsymbol{x}_T$ will approximate isotropic Gaussian noise and the learned *reverse process* can be utilized as a powerful iterative generative model to model structured information from noise.

A simplified training objective (c.f. [10,24]) can be formulated by minimizing

$$\mathbb{E}_{t \sim U(1,T), \boldsymbol{x}_0 \in \mathcal{X}, \boldsymbol{\epsilon}_t \sim \mathcal{N}(\mathbf{0},\mathbf{1})} \left[\|\boldsymbol{\epsilon}_t - \boldsymbol{\epsilon}_\theta(\boldsymbol{x}_t, t)\|_2^2 \right] , \tag{1}$$

where $\boldsymbol{\epsilon}_\theta$ usually is a time-conditional U-net [6] parameterized with θ to predict the noise $\boldsymbol{\epsilon}_t$ in $\boldsymbol{x}_t$ as a function of $\boldsymbol{x}_0$ for uniformly drawn t from $\{1,\ldots,\mathrm{T}\}$ using the fact that $\boldsymbol{x}_t = \sqrt{\bar{\alpha}_t}\boldsymbol{x}_0 + \sqrt{1-\bar{\alpha}_t}\boldsymbol{\epsilon}_t$ with $\bar{\alpha}_t = \prod_{s=1}^{t}(1-\beta_s)$.

Cheff. We now present our method *Cheff* to iteratively generate high-resolution images. Our multi-stage synthesis approach can be decomposed into three cascading phases: (i) Modeling a diffusion process in latent space, (ii) translating the latent variables into image space with a decoder, (iii) refinement and upscaling using a super-resolution diffusion process. Phase (i) and (ii) together define an LDM. Figure 2 summarizes this multi-stage approach again.

Every model can be trained in an encapsulated fashion as described in the subsequent subsections. In the following, we denote $\boldsymbol{x}_{\mathrm{HR}}$ as a high-resolution image and $\boldsymbol{x}_{\mathrm{LR}}$ as its lower-resolution variant (in our model stack these are 1024 and 256 pixels, respectively). We start by explaining the second (reconstruction) part of our model first.

Autoencoder Training. It has been shown that applying diffusion in a semantic latent space instead of a high-dimensional data space leads to a notable reduction in computational costs due to reduced input size with a neglectable loss in synthesis quality [24]. We construct a latent space by training an autoencoder with an encoder E and decoder D. The autoencoder's task is to reconstruct $\boldsymbol{x}_{\mathrm{LR}}$, where the reconstruction is $D(E(\boldsymbol{x}_{\mathrm{LR}}))$ and $\boldsymbol{z} = E(\boldsymbol{x}_{\mathrm{LR}}) \in \mathcal{Z}$ represents a

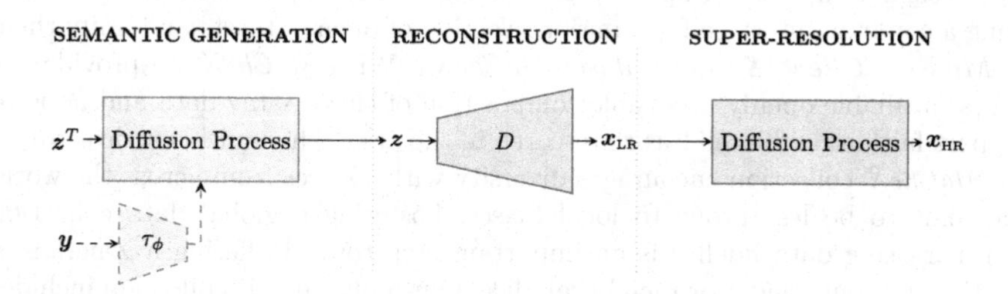

Fig. 2. Flow of our cascaded synthesis pipeline. A first diffusion model generates a latent sample $\boldsymbol{z} \in \mathcal{Z}$ from $\boldsymbol{z}^T \sim \mathcal{N}(\mathbf{0}, \boldsymbol{I})$ through its reverse process. Optionally, an embedded conditioning $\tau_\phi(\boldsymbol{y})$ can be added, where $\boldsymbol{y}$ in our application could, e.g., be pathology labels, radiological reports, or radiologist's annotations. The decoded image $\boldsymbol{x}_{\mathrm{LR}} = D(\boldsymbol{z})$ is subsequently refined by a second diffusion model to a high-resolution image $\boldsymbol{x}_{\mathrm{HR}}$. Foundational models are marked in red, whereas blue components are task-specific.

latent sample. We follow [5,24] for the autoencoder training and use a pixel-wise reconstruction loss next to a perceptual loss [36] and an adversarial objective.

Semantic Generation. The actual synthesis diffusion model is trained by first obtaining $\boldsymbol{z} = E(\boldsymbol{x}_{\text{LR}})$ and modifying the optimization objective for the semantic generation as

$$\mathbb{E}_{t\sim U(1,T),\boldsymbol{z}_0\sim\mathcal{Z},\boldsymbol{\epsilon}_t\sim\mathcal{N}(\boldsymbol{0},\boldsymbol{1})}\left[\|\boldsymbol{\epsilon}_t - \boldsymbol{\epsilon}_\theta(\boldsymbol{z}_t, \tau_\phi(\boldsymbol{y}), t)\|_2^2\right] . \tag{2}$$

where $\boldsymbol{\epsilon}_\theta$ again is a time-conditional U-net. The diffusion process in (2) is the same as described before, but operates only on samples in the latent space. Additionally, we allow the conditioning on some modality $\boldsymbol{y}$ to guide the synthesis process. $\boldsymbol{y}$ can represent information of various kinds, e.g., radiological reports, class conditioning, or annotations. This information is fed into the model using a conditioning embedding, which is obtained by $\tau_\phi(\boldsymbol{y})$, where τ_ϕ is a neural network tasked with creating an embedding for $\boldsymbol{y}$.

Super-Resolution. After synthesizing $\boldsymbol{z}$ and reconstructing $\boldsymbol{x}_{\text{LR}}$ from it, we apply an additional iterative refinement procedure to not only counteract blurriness, which is induced by the autoencoder but also to advance to realistic resolution domains in clinical practice. Inspired by SR3 [26], we condition a diffusion model with $\boldsymbol{x}_{\text{LR}}$ to infer an optimal high-resolution output $\boldsymbol{x}_{\text{HR}}$. Thus, the training objective is:

$$\mathbb{E}_{t\sim U(1,T),\boldsymbol{x}_{\text{HR}}\sim\mathcal{X}_{\text{HR}},\boldsymbol{\epsilon}_t\sim\mathcal{N}(\boldsymbol{0},\boldsymbol{1})}\left[\|\boldsymbol{\epsilon}_t - \boldsymbol{\epsilon}_\psi(\boldsymbol{x}_{\text{HR},t}, \boldsymbol{x}_{\text{LR}}, t)\|_2^2\right] , \tag{3}$$

where $\boldsymbol{\epsilon}_\psi$ is a denoising time-conditional U-net with conditioning on $\boldsymbol{x}_{\text{LR}}$.

4 MaCheX: Massive Chest X-ray Dataset

We present a large-scale, open-source, diverse collection of chest X-ray images using a common interface for a major selection of open datasets and unify them as ***Ma**ssive **Che**st **X**-Ray Dataset*(*MaCheX*). With *MaCheX* we provide the largest available openly accessible composition of chest X-ray data and hope to support fair and unbiased future research in the area. The multi-centric setup of the *MaCheX* collection encourages diversity with sources from across the world and aims to be less prone to local biases. Fostering a global data collection and increasing data fidelity is an important step toward offering a generalized solution without gender or racial prejudice. Our multi-centric collection includes various datasets described and summarized in Table 1.

In the current version of *MaCheX*, only frontal AP/PA scans are considered. By inclusion of the lateral position, another 250,000 chest X-rays could be added to the collection, totaling nearly a million samples. The present work, however, focuses on the analysis and synthesis of a frontal viewing position. All scans are rescaled so that the shortest edge meets a 1024px resolution and are then center-cropped to 1024 × 1024px. We do not apply histogram equalization or

Table 1. Overview of the components of *MaCheX*.

Dataset	No. patients	No. samples			%	Origin
		Train	Test	$\in$ *MaCheX*		
ChestX-ray14 [33]	23,152	86,523	25,595	86,523	13,28%	Bethesda, MD, USA
CheXpert [11]	65,240	223,414	235	191,027	29,32%	Palo Alto, CA, USA
MIMIC-CXR [12]	65,379	377,110	-	243,334	37,35%	Boston, MA, USA
PadChest [2]	67,625	160,868	-	96,278	14,78%	Alicante, Spain
BRAX [23]	18,529	40,967	-	19,309	2,96%	São Paulo, Brazil
VinDr-CXR [16]	-	15,000	3,000	15,000	2,30%	Hanoi, Vietnam

other standardization techniques. By combining the designated train subsets of the respective datasets, the full *MaCheX* dataset amounts to 651,471 samples. This includes over 440,000 labels, over 220,000 free-text radiological reports, and 15,000 coordinate bounding boxes for radiologist annotations.

We open-source our implementation of the pre-processing setup and provide an easy-to-use interface to access *MaCheX* in a deep learning setting with PyTorch. The structure of *MaCheX* is clear and simple, allowing to straightforwardly adapt the code to different frameworks and use cases.

5 Experiments

Our model stack effectively contains three models: The semantic diffusion model (SDM), the autoencoder (AE), and the super-resolution diffusion model (SR). AE and SR form a foundational basis that is trained on full *MaCheX* with a separate test set of size 25,000. A full description of our training routine and hyperparameter setup can be found in the supplementary material. For AE we use a KL-regularized continuous variational AE (VAE) with a downsampling factor 4 as in [24]. In SR, we found that conditioning on bicubic downsampled versions of $\boldsymbol{x}_{\mathrm{HR}}$ as $\boldsymbol{x}_{\mathrm{LR}}$ as in [26] does not perfectly reflect the structure of recovered latent samples $D(\boldsymbol{z})$ and results in artifacts and slight blurriness. To align SR with the synthesis pipeline, we fine-tune SR on $D(\boldsymbol{z})$.

SDM in the lower-dimensional (latent sample) domain is trained with a task-specific data subset of *MaCheX*. To guide the synthesis process for chosen conditionings, we induce the respective conditioning embedding $\tau_\phi(\boldsymbol{y})$ into the model over cross-attention [32] in the attention layers of the time-conditional U-net.

Reconstruction Quality. Using our AE and SR, a high-resolution image can be converted to the latent space and successfully reconstructed (cf. Figure 3). Despite a compression factor of 16, even small details are recovered in fine-grained quality and limited blurriness, some structures even gaining sharpness. Reconstruction of text annotations and small numbers, however, still proves to be difficult. Table 2 additionally provides a quantitative assessment, i.e., the efficacy of the various transitions in *Cheff* as well as the effect of fine-tuning

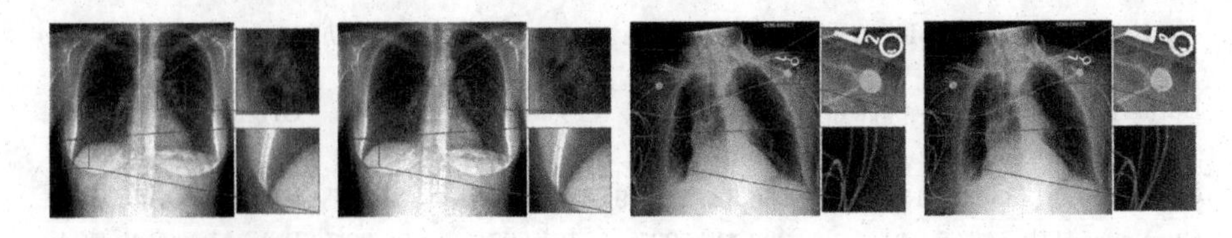

Fig. 3. Original images (the respective **left** image) compared to their reconstructions (the respective **right** image) from the latent space $\mathcal{Z}$.

Table 2. Test set performance for different reconstruction workflows. ① symbolizes the encoding of $\boldsymbol{x}_{\mathrm{HR}}$ and retrieval from $\boldsymbol{z}$ in the full *Cheff* pipeline, whereas ② and ③ analyze the reconstruction capacity of AE and SR separately.

	Reconstruction workflow	MSE ↓	PSNR ↑	SSIM ↑
①	$\boldsymbol{x}_{\mathrm{HR}} \xrightarrow{\text{bicubic}} \boldsymbol{x}_{\mathrm{LR}} \xrightarrow{E} \boldsymbol{z} \xrightarrow{D} \hat{\boldsymbol{x}}_{\mathrm{LR}} \xrightarrow{\mathrm{SR_{base}}} \hat{\boldsymbol{x}}_{\mathrm{HR}}$	0.0039	24.05	**0.9512**
	$\boldsymbol{x}_{\mathrm{HR}} \xrightarrow{\text{bicubic}} \boldsymbol{x}_{\mathrm{LR}} \xrightarrow{E} \boldsymbol{z} \xrightarrow{D} \hat{\boldsymbol{x}}_{\mathrm{LR}} \xrightarrow{\mathrm{SR_{fine}}} \hat{\boldsymbol{x}}_{\mathrm{HR}}$	**0.0026**	**25.77**	0.9510
②	$\boldsymbol{x}_{\mathrm{LR}} \xrightarrow{E} \boldsymbol{z} \xrightarrow{D} \hat{\boldsymbol{x}}_{\mathrm{LR}}$	0.0005	36.80	0.9926
③	$\boldsymbol{x}_{\mathrm{HR}} \xrightarrow{\text{bicubic}} \boldsymbol{x}_{\mathrm{LR}} \xrightarrow{\mathrm{SR_{base}}} \hat{\boldsymbol{x}}_{\mathrm{HR}}$	0.0035	24.59	0.9540
	$\boldsymbol{x}_{\mathrm{HR}} \xrightarrow{\text{bicubic}} \boldsymbol{x}_{\mathrm{LR}} \xrightarrow{\mathrm{SR_{fine}}} \hat{\boldsymbol{x}}_{\mathrm{HR}}$	**0.0025**	**26.09**	**0.9559**

SR, by analyzing the mean squared error (MSE), the structural similarity index measure (SSIM), and the peak signal-to-noise ratio (PSNR) on the test data set. The fine-tuned SR performs better than the base SR in almost every aspect, even in naive upscaling, while all transitions show high-quality reconstructions in general.

Unconditional Synthesis. Scaling up the data and model stack leads to a realistic synthesis with high fidelity and realism (Fig. 4e) in a 1-megapixel resolution setting including a variety of medical devices, e.g., chest tubes, pacemakers, ECG leads, etc. This is a major improvement over previously proposed approaches for chest X-ray synthesis (Fig. 4). The PGAN of [27] misses the capacity of medical devices (cf. [34]). While synthetic samples of [1] by prompting DALL·E 2 are comparatively realistic for a non-medical zero-shot setting, they involve color jittering. In contrast, Stable Diffusion is not able to generate clinically realistic chest X-rays, clearly showing its artistic style even when fine-tuned [3]. LDM training of [20] involves only a 256-pixel resolution of a single data source.

In comparison to our latent diffusion-based pipeline, we investigate a fully cascaded stack of diffusion models in the style of Imagen [25] and DALL·E 2 [22] with U-net configurations from [26] (see Fig. 4d). The cascade involves three models in an upscaling chain: 64px → 256px → 1024px, where the first model synthesizes samples from noise. This uncovers a potential downside of utilizing models with a progressive growing backend (same as in [27]). While the samples

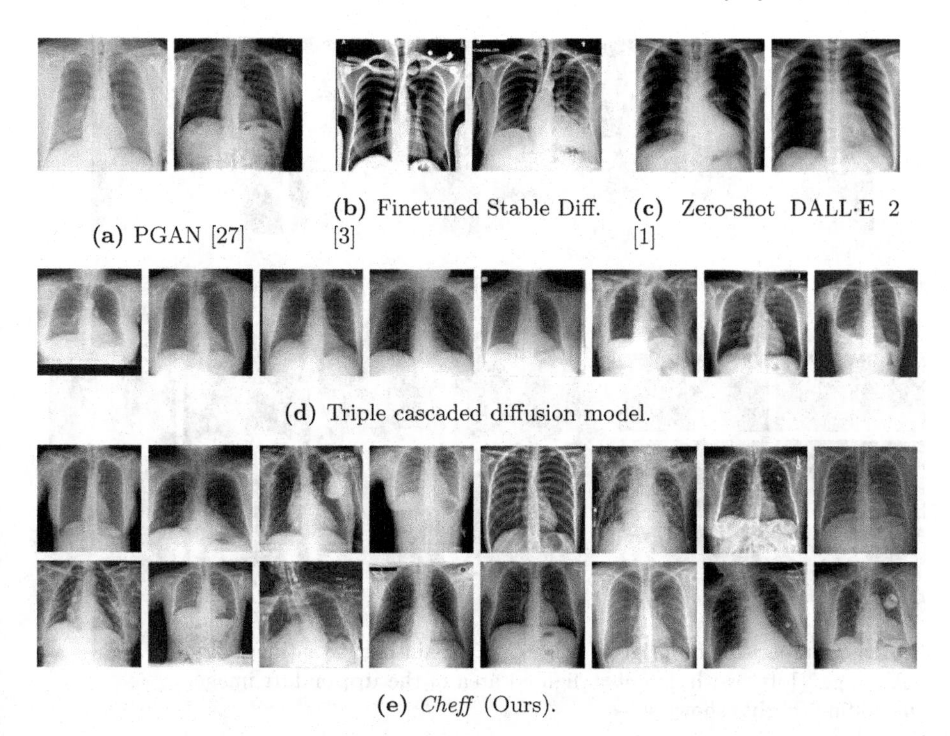

(a) PGAN [27] (b) Finetuned Stable Diff. [3] (c) Zero-shot DALL·E 2 [1]

(d) Triple cascaded diffusion model.

(e) *Cheff* (Ours).

Fig. 4. Synthesized chest X-rays using different methods.

maintain high quality, there is also an abundance of medical devices and minor foreign objects. This fact also reflects in the Fréchet Inception Distance (FID $\downarrow$) and Kernel Inception Distance (KID $\downarrow$), which is 46.54 and 0.0530 for the full cascade but reaches 11.58 and 0.0099, respectively, using *Cheff*. We hypothesize that details of this granularity are lost when synthesizing on the low-resolution levels of $\leq$ 64 pixels. The subsequent upsampling models cannot recover these already disregarded elements. Our approach works on semantic features instead of low-resolution images and thus is able to compress necessary information in a meaningful way, successfully circumventing this issue (Fig. 4e).

In- and Outpainting. The iterative nature of diffusion models allows for simple adaption of inpainting tasks, where the model is used to fill a designated marked area. This method can be utilized to, e.g., remove occluding and distracting elements like medical devices from a thoracic scan (Fig. 5). While *Cheff* has the ability to synthesize removed parts such as devices, object removal works well since the rest of the image is provided and acts as a prior. Thus *Cheff* fills the removed parts with the most likely content – a clear lung – and not the original devices. In *Cheff* or LDMs in general, this kind of masking is possible as the commonly used autoencoder implementation produces a spatially-aware convolutional embedding, which, despite living in a latent space, maintains structural

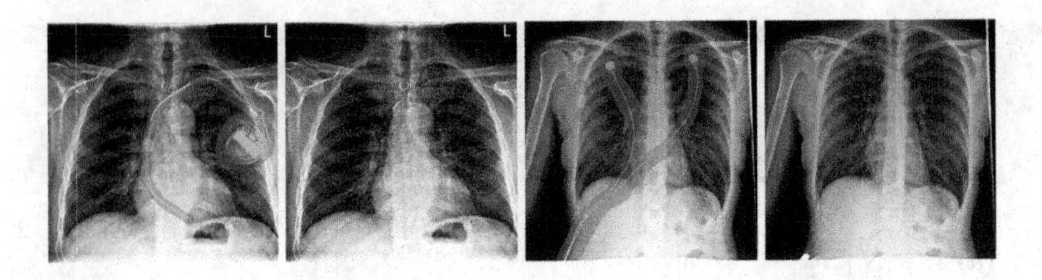

Fig. 5. Removal of support devices via image inpainting.

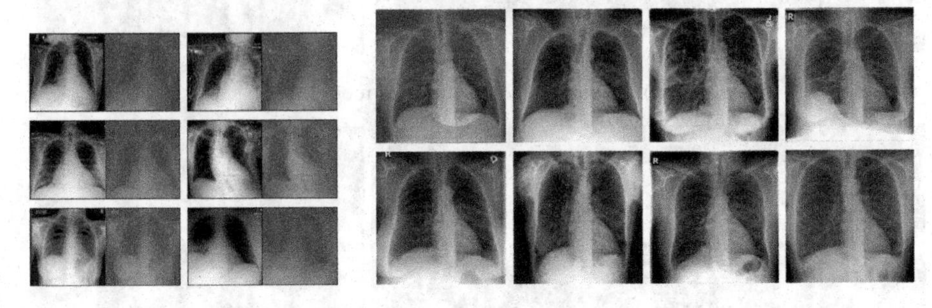

Fig. 6. Comparison of original image (**left**) with latent embedding (**right**) shows spatial coherency.

Fig. 7. Outpainting variations of the highlighted area in the **upper left** image.

information of the input data (Fig. 6). In another formulation, masking as outpainting serves to interpolate the remaining parts of an image. As seen in Fig. 7, a variety of chest X-rays can be synthesized sharing the same initial provided area. Notably, *Cheff* does not collapse to one solution but is able to explore various in-distribution options.

Radiological Report-to-Image Synthesis. MIMIC-CXR provides a range of radiological text reports for every study (over 220,000), which can be utilized as conditioning $\boldsymbol{y}$. The sections *Findings* and *Impressions* are extracted from the raw reports and concatenated before applying a tokenizer. Following [24], τ_ϕ is a trainable BERT-style encoder-only transformer. Examples in Fig. 8 show that the model has learned concepts of various pathologies and allows for a customized synthesis of a patient's condition. Conditioning on both *Findings* and *Impressions* allows controlling not only pathologies but also the creation of external devices like pacemakers. Furthermore, the freedom of text inputs allows for specifying the localization of the targeted disease or item.

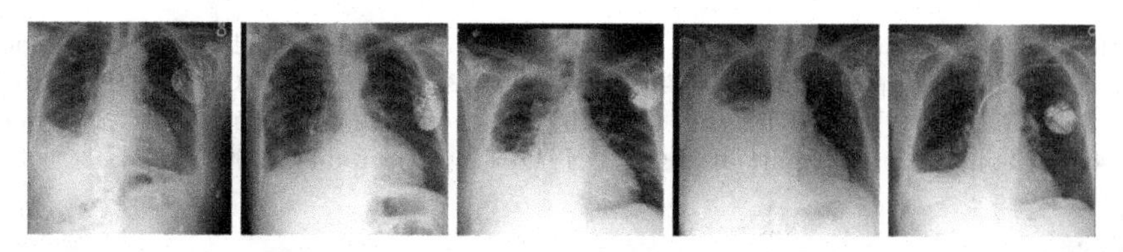

(a) *"Large pleural effusion is in the right lower lung. A pacemaker is in the left upper chest."*

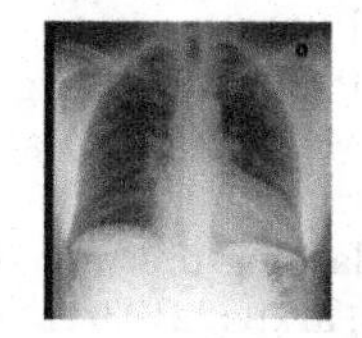

(b) *"Acute cardiomegaly."*

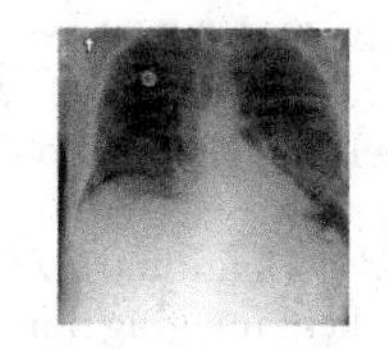

(c) *"Prominent left-sided atelectasis."*

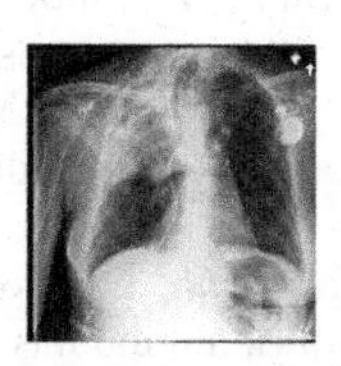

(d) *"Mass filling is in the upper zone of the right lung."*

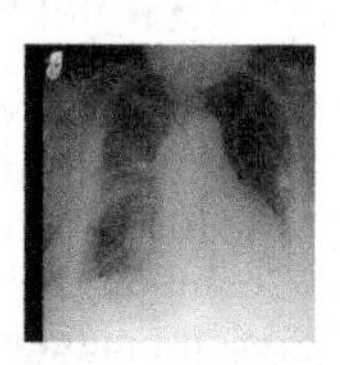

(e) *"Pneumonia is in the right lung."*

Fig. 8. Prompting a text-conditioned *Cheff* via radiological findings.

6 Conclusion

In this paper, we advance the state-of-the-art in Chest X-ray synthesis by proposing a multi-stage foundational cascaded latent diffusion model called *Cheff*. A success factor for the quality of *Cheff* is *MaCheX* - our large-scale multi-centric Chest X-ray collection from numerous, publicly available datasets with high diversity in phenotypes, medical conditions, diseases, and medical devices.

Approaching the terrain of indistinguishable synthetic patient samples, *Cheff* requires an increased awareness of potential harms and responsible utilization. Further, despite being able to generate realistic-looking 1-megapixel radiographs, modern X-ray scanners produce images with an up to 7-megapixel resolution and 14-bit depth, which is a requirement for radiological reading on dedicated monitors and is still out of reach for current chest X-ray synthesis. When developing downstream applications with *Cheff* for a clinical context, this should be kept in mind and potential biases should be addressed, e.g., through rigorous evaluation on real-world data.

By proposing *Cheff*, we offer a high-capacity generator for chest radiographs that forms the basis for a variety of use cases, e.g., image inpainting for removal of distracting medical devices or aiding existing efforts to increase reliability in classifiers by generating underrepresented classes. Our method does not only offer traditional synthesis but also enters the exciting area of radiological report-to-chest-X-ray generation, which allows fine-grained control over the diffusion process via text prompts.

Acknowledgments. The authors gratefully acknowledge LMU Klinikum for providing computing resources on their Clinical Open Research Engine (CORE). This work has been partially funded by the Deutsche Forschungsgemeinschaft (DFG, German Research Foundation) as part of BERD@NFDI - grant number 460037581.

References

1. Ali, H., Murad, S., Shah, Z.: Spot the fake lungs: generating synthetic medical images using neural diffusion models. arXiv:2211.00902 [cs, eess] (2022)
2. Bustos, A., Pertusa, A., Salinas, J.M., de la Iglesia-Vayá, M.: PadChest: a large chest X-ray image dataset with multi-label annotated reports. Med. Image Anal. **66**, 101797 (2020)
3. Chambon, P., Bluethgen, C., Langlotz, C.P., Chaudhari, A.: Adapting pretrained vision-language foundational models to medical imaging domains. arxiv:2210.04133 [cs] (2022)
4. Dhariwal, P., Nichol, A.: Diffusion models beat GANs on image synthesis. In: Advances in Neural Information Processing Systems, vol. 34 (2021)
5. Esser, P., Rombach, R., Ommer, B.: Taming transformers for high-resolution image synthesis. In: Proceedings of the IEEE/CVF Conference on Computer Vision and Pattern Recognition (2021)
6. Falk, T., et al.: U-net: deep learning for cell counting, detection, and morphometry. Nat. Meth. **16**, 67–70 (2019)
7. Glocker, B., Jones, C., Bernhardt, M., Winzeck, S.: Risk of bias in chest X-ray foundation models. arXiv:2209.02965 (2022)
8. Han, L., Lyu, Y., Peng, C., Zhou, S.K.: Gan-based disentanglement learning for chest x-ray rib suppression. Med. Image Anal. **77**, 102369 (2022)
9. Han, T., et al.: Breaking medical data sharing boundaries by using synthesized radiographs. Sci. Adv. **6**, eabb7973 (2020)
10. Ho, J., Jain, A., Abbeel, P.: Denoising diffusion probabilistic models. In: Advances in Neural Information Processing Systems, vol. 33 (2020)
11. Irvin, J., et al.: CheXpert: a large chest radiograph dataset with uncertainty labels and expert comparison. In: Proceedings of the AAAI Conference on Artificial Intelligence, vol. 33 (2019)
12. Johnson, A.E.W., et al.: MIMIC-CXR, a de-identified publicly available database of chest radiographs with free-text reports. Sci. Data **6**, 317 (2019)
13. Karras, T., Aila, T., Laine, S., Lehtinen, J.: Progressive growing of GANs for improved quality, stability, and variation. In: 6th International Conference on Learning Representations, ICLR (2018)
14. Liu, F., Wu, X., Ge, S., Fan, W., Zou, Y.: Exploring and distilling posterior and prior knowledge for radiology report generation. In: Proceedings of the IEEE/CVF Conference on Computer Vision and Pattern Recognition (2021)
15. Liu, G., et al.: Clinically accurate chest X-ray report generation. In: Machine Learning for Healthcare Conference (2019)
16. Nguyen, H.Q., et al.: VinDr-CXR: an open dataset of chest x-rays with radiologist's annotations. arXiv:2012.15029 [eess] (2022)
17. Nichol, A., Dhariwal, P.: Improved denoising diffusion probabilistic models. In: International Conference on Machine Learning (2021)
18. Nichol, A., et al.: GLIDE: towards photorealistic image generation and editing with text-guided diffusion models. In: International Conference on Machine Learning (2022)

19. Packhäuser, K., Gündel, S., Münster, N., Syben, C., Christlein, V., Maier, A.: Deep learning-based patient re-identification is able to exploit the biometric nature of medical chest X-ray data. Sci. Rep. **12**, 14851 (2022)
20. Packhäuser, K., Folle, L., Thamm, F., Maier, A.: Generation of anonymous chest radiographs using latent diffusion models for training thoracic abnormality classification systems. arxiv:2211.01323 [cs, eess] (2022)
21. Rajpurkar, P., et al.: CheXNet: radiologist-level pneumonia detection on chest X-rays with deep learning. arXiv:1711.05225 [cs, stat] (2017)
22. Ramesh, A., Dhariwal, P., Nichol, A., Chu, C., Chen, M.: Hierarchical text-conditional image generation with CLIP latents. arxiv:2204.06125 [cs] (2022)
23. Reis, E.P., et al.: BRAX, a Brazilian labeled chest X-ray dataset. Sci. Data **9**, 487 (2022)
24. Rombach, R., Blattmann, A., Lorenz, D., Esser, P., Ommer, B.: High-resolution image synthesis with latent diffusion models. In: Proceedings of the IEEE/CVF Conference on Computer Vision and Pattern Recognition (2022)
25. Saharia, C., et al.: Photorealistic text-to-image diffusion models with deep language understanding. arXiv:2205.11487 (2022)
26. Saharia, C., Ho, J., Chan, W., Salimans, T., Fleet, D.J., Norouzi, M.: Image super-resolution via iterative refinement. IEEE Trans. Pattern Anal. Mach. Intell. **45**, 4713–4726 (2022)
27. Segal, B., Rubin, D.M., Rubin, G., Pantanowitz, A.: Evaluating the clinical realism of synthetic chest X-rays generated using progressively growing GANs. SN Comput. Sci. **2**(4), 1–17 (2021). https://doi.org/10.1007/s42979-021-00720-7
28. Sellergren, A.B., et al.: Simplified transfer learning for chest radiography models using less data. Radiology **305**(2), 454–465 (2022)
29. Sohl-Dickstein, J., Weiss, E., Maheswaranathan, N., Ganguli, S.: Deep unsupervised learning using nonequilibrium thermodynamics. In: International Conference on Machine Learning (2015)
30. Sundaram, S., Hulkund, N.: GAN-based data augmentation for chest X-ray classification. arXiv:2107.02970 (2021)
31. Tang, Y., Tang, Y., Zhu, Y., Xiao, J., Summers, R.M.: A disentangled generative model for disease decomposition in chest X-rays via normal image synthesis. Med. Image Anal. **67**, 101839 (2021)
32. Vaswani, A., et al.: Attention is all you need. In: Advances in Neural Information Processing Systems, vol. 30 (2017)
33. Wang, X., Peng, Y., Lu, L., Lu, Z., Bagheri, M., Summers, R.M.: ChestX-ray8: Hospital-scale chest X-ray database and benchmarks on weakly-supervised classification and localization of common thorax diseases. In: Proceedings of the IEEE/CVF Conference on Computer Vision and Pattern Recognition (2017)
34. Weber, T., Ingrisch, M., Bischl, B., Rügamer, D.: Implicit embeddings via GAN inversion for high resolution chest radiographs. In: Medical Image Computing and Computer Assisted Intervention (MICCAI), Medical Applications with Disentanglement (2022)
35. Wolleb, J., Bieder, F., Sandkühler, R., Cattin, P.C.: Diffusion models for medical anomaly detection. In: Medical Image Computing and Computer Assisted Intervention (MICCAI), vol. 13438 (2022)
36. Zhang, R., Isola, P., Efros, A.A., Shechtman, E., Wang, O.: The unreasonable effectiveness of deep features as a perceptual metric. In: Proceedings of the IEEE/CVF Conference on Computer Vision and Pattern Recognition (2018)

DKFM: Dual Knowledge-Guided Fusion Model for Drug Recommendation

Yankai Tian, Yijia Zhang(✉), Xingwang Li, and Mingyu Lu(✉)

School of Information Science and Technology, Dalian Maritime University, Dalian 116024, Liaoning, China
{zhangyijia,lumingyu}@dlmu.edu.cn

Abstract. Drug recommendation is an AI healthcare task that has become increasingly important during the current COVID-19 pandemic. At present, the accumulation of a large number of electronic health records (EHR) provides strong data support for medical and scientific researchers. Most existing work uses splicing diagnosis and procedure to complete the recommendation task while ignoring the problem that the patient's diagnosis and procedure information are recorded in different levels of detail. Moreover, the splicing of different medical records cannot accurately characterize the patient's condition, thus reducing the accuracy of recommended drugs. In this paper, we propose a drug recommendation model, DKFM, which introduces a molecular knowledge-guided dual-level drug fusion mechanism. DKFM can integrate a patient's diagnosis and procedure important information and determine the impact on the current condition. Then the diagnosis and procedure information was respectively injected into the drug functional group encoder to obtain two candidate drug sets. Finally, select the drugs that are really suitable for the patient's condition at the sets level and the score matrix level as the final recommended drugs. We validate DKFM on the public MIMIC-III dataset, and experimental results show that the proposed model can outperform the state-of-the-art approaches.

Keywords: Drug recommendation · Electronic health record · Healthcare · Molecular knowledge

1 Introduction

In recent years, the accumulation of electronic health records (EHR) has provided researchers and doctors with a large number of reference materials and research data. The task of drug recommendation has also emerged as the times require [1, 2]. Especially in the context of the current COVID-19 pandemic, the drug recommendation task is becoming increasingly important because it is not convenient to go out for medical treatment due to the control in some areas. A good drug recommendation model can make it safer and more secure for patients to seek medical treatment at home, especially for patients with complex health conditions [2, 3]. The early drug recommendation model was based on examples and only on the current medical conditions of patients [4, 5], which was

H. Kashima et al. (Eds.): PAKDD 2023, LNAI 13937, pp. 192–203, 2023.
https://doi.org/10.1007/978-3-031-33380-4_15

unreasonable. The current status of the patient is important, but the drug recommended for the patient without regard to previous visit information may be inappropriate and may even have irreversible consequences. In order to remedy this deficiency, a number of longitudinal approaches [3, 6] are proposed. Those approaches integrated diagnosis and procedure information and relied on various clinical records of actual visits (e.g., diagnosis [2, 4, 7, 8], laboratory tests [11], and procedures [1, 3, 12–14]).

Through our investigation, it is found that doctors, in order to improve the efficiency of medical treatment, often write diagnosis records briefly. On the contrary, the specific symptoms and treatment methods in the records of the procedure are more detailed but not as comprehensive as the diagnosis records. We present an example of diagnosis and procedure information in MIMIC-III. As shown in Fig. 1, there is only one item in the patient's diagnosis record related to esophageal cancer, and it was very brief. In contrast, various procedure information about esophageal cancer is recorded in detail in the procedure record. Therefore, the two may interact with each other due to the different levels of detail. Eventually, this leads to reduces accuracy of the recommended drugs.

Diagnosis		Procedure	
esophageal	45.2	partial esophagotomy	42.41
URTI	06.01	partial colon separation	45.52
B group Salmonella enteritis	27.001	intracolonic replacement esophagostomy	42.55/42.65

Fig. 1. An example of an electronic health record.

In addition, the chemical properties of the drug are determined by its molecular substructure [15]. What really plays an important role in the treatment of patients' diseases is the functional group of the drug. However, many current studies [16, 17] focus on patient character-drug interactions to learn drug representation while ignoring the molecular substructure information of drugs. Different drugs may overlap in the molecular substructure so that patients can be given a wider variety of recommended drugs. When a patient has antagonistic action on a molecular substructure, drugs with the same substructure can be excluded from the recommended drugs for the patient [1]. Thus, the drugs recommended for the patient can be as safe as possible.

For the above mention problems, we propose a drug recommendation model that uses a molecular knowledge-guided dual-level drug fusion mechanism. This model is also equipped with Taylor binary cross entropy loss. Our contribution is as follows:

- We propose a drug recommendation model, DKFM, which introduces a *molecular knowledge-guided dual-level drug fusion* mechanism. DKFM can handle the problem of the unbalanced distribution of diagnosis and procedure information.
- We designed a dual-level drug fusion mechanism that first obtains the initial two drug candidate sets through diagnosis and procedure information. Then the final recommended drugs set is selected at the sets level and the score matrix level.
- We conduct comprehensive experiments on a public dataset MIMIC-III to demonstrate the effectiveness of the proposed DKFM.

2 Related Works

Due to the current environment of COVID-19, the implementation of the drug recommendations model has received more and more attention. According to the data used, the existing methods are roughly divided into rule-based methods, instance-based methods, and longitudinal methods.

2.1 Rule-Based Methods

Rule-based methods [18–21] rely on artificially designed recommendation protocols. For example, Chen et al. [19] used answer set programming (ASP) to code a complete set of clinical practice guidelines for chronic heart rhythm failure and generate treatment recommendations.

2.2 Instance-Based Methods

Instance-based methods [4, 8] only inject the information of the current visit as input into the model for the recommendation. For example, Gong et al. [8] constructed high-quality heterogeneous maps by bridging EMR and medical knowledge graphs (ICD-9 ontology and DrugBank) and decomposed recommendations into link prediction processes. However, they ignore the fact that the patient's historical visit information is inextricably linked to the patient's current visit status.

2.3 Longitudinal Methods

These methods [1, 3, 12, 13] use historical patient information and explore sequential dependence between visits. Most of these essential models use RNN to code longitudinal patient information (integrated diagnosis and procedures information). Le et al. [6] and Shang et al. [3] combined memory networks with RNNs to enhance memory. Yang et al. [1] further combined the molecular information of the drug and proposed a molecular graph encoder to represent the drug better. Yang et al. [12] explicitly modeled changes in patients' health status to strengthen the correlation between multiple visits through repeated residual learning. However, RNN approaches are challenging to deal with the tight relationship between multiple visits, and integrated diagnosis and procedure information also will reduce the accuracy of recommended drugs.

In this paper, inspired by Yang et al. [1], DKFM exploits the relationship between multiple visits and uses the molecular knowledge-guided dual-level drug fusion mechanism to obtain the recommended drug set.

3 The DKFM Model

As shown in Fig. 2, our DKFM consists of three components: (1) a patient representation encoder module, which captures important features of diagnosis and procedure information, respectively; (2) a molecular knowledge-guided drug candidate sets generation module, which respectively injects the diagnosis and procedure information into the drug functional group encoder to obtain the two initial candidate drug sets; (3) a dual-level drug fusion module, which does a rational fusion of the two at sets level and score matrix level.

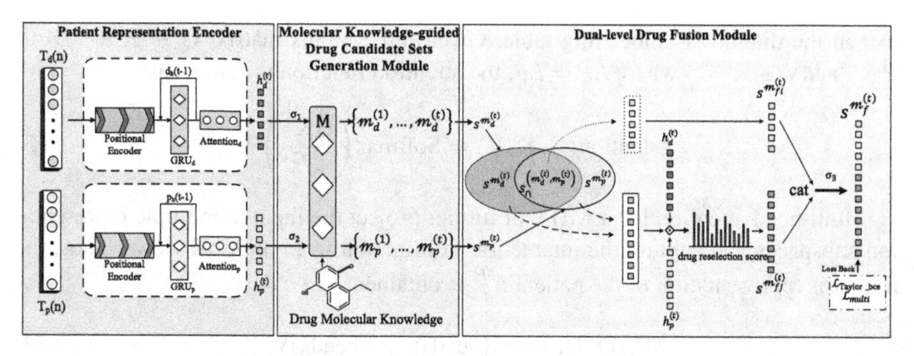

Fig. 2. An overview of our proposed DKFM.

3.1 Patient Representation Encoder

In order to capture the critical information in diagnosis and procedure and the strong correlation between multiple visits of patients, we used the GRU-Attention model, which can extract the important features of both to get embedding.

3.1.1 Diagnosis Embedding

Diagnosis information $d_{(t)}$ plays an indispensable role in the process of drug recommendation. A patient may have one or more medical records. For example, it is not appropriate to recommend the same drug for patients newly diagnosed with hyperglycemia and those diagnosed with hyperglycemia for as long as three years. Therefore, in order to recommend appropriate drugs to patients, historical diagnosis information must be integrated. We design a learnable diagnosis embedding table $T_d(n)$ for each patient, which can record and train all historical diagnosis information

$$T_d(n) = d_{(1)cat} d_{(2)cat} \ldots d_{(n)} \quad (1)$$

where cat is a connect operation.

Time series information is one of the most important attributes of diagnosis information. The whole diagnosis is meaningless if the time series information is wrong. Therefore, it is sent to the positional encoder layer after obtaining the diagnosis embedding table. After this, a representation of a unique location is assigned to all diagnosis information.

In this model, we use an independent GRU-Attention architecture to embed diagnosis information dynamically. The diagnosis embedding table $T_d(n)$ of the embedded position information will be injected into the GRU, and the corresponding output $d_{h1}, d_{h2}, \ldots d_{hn}$ will be obtained after the reset gate and update gate of the GRU,

$$d_{hn} = GRU_d(T_d(n), T_d(n-1)) = GRU_d(T_d(n), \ldots, T_d(1)) \quad (2)$$

Then, the Attention Layer is introduced into the next sub-layer to calculate the attention probability distribution value of each input and further extract the diagnosis features. And highlight the key historical information and mutual connections for this visit in the diagnosis embedding table. Given three inputs matrix, $Q \in \mathbb{R}^{L_Q \times s}$, $K \in \mathbb{R}^{L_K \times s}andV \in \mathbb{R}^{L_V \times s}$ where $L_K = L_V$, the attention function is defined as:

$$\text{Attention}(Q, K, V) = \text{Softmax}\left(\frac{QK^T}{\sqrt{s}}\right)V \tag{3}$$

Multi-head attention layer MH will further project the input to multiple representation subspaces and capture the interaction information from multiple views. Thus, the diagnosis representation of the patient $\boldsymbol{h}_d^{(t)}$ is obtained.

$$\text{MH}(Q, K, V) = [\text{head}_1; \ldots; \text{head}_h]W^O \tag{4}$$

$$\text{head}_i = \text{Attention}\left(QW_i^Q, KW_i^K, VW_i^V\right) \tag{5}$$

$W_i^Q, W_i^K, W_i^V \in \mathbb{R}^{s \times s/h}$ and $W^O \in \mathbb{R}^{s \times s}$ are the parameters to learn. h is the number of heads.

3.1.2 Procedure Embedding

The procedure information $p_{(t)}$ also plays a very important role in the process of recommending drugs to patients. The procedure information records the patient's operation records in the hospital, which can directly reflect the patient's symptoms. We also designed a procedure embedding table $T_p(n)$ for each patient, which records the important procedure information of the patient,

$$T_p(n) = p_{(1)cat}p_{(2)cat} \cdots p_{(n)} \tag{6}$$

Like the diagnosis information, we also add position coding to each program information so that it maintains a relative position during the training process. The corresponding output $p_{h1}, p_{h2}, \ldots\ldots p_{hn}$ Will be obtained after passing through the GRU model.

$$p_{hn} = \text{GRU}_p\left(T_p(n), T_p(n-1)\right) = \text{GRU}_p\left(T_p(n), \ldots, T_p(1)\right) \tag{7}$$

After the last layer of Attention, the attention probability distribution value of each input is calculated, and the procedure representation $\boldsymbol{h}_p^{(t)}$ is further extracted.

3.2 Molecular Knowledge-Guided Drug Candidate Sets Generation Module

3.2.1 Molecular Knowledge Encoder

Actually, some functional groups play a major role in drug molecules, and the rest of the molecular structure is its carrier or auxiliary. Therefore, functional groups can best reflect the function of drugs. Based on this, we designed a drug functional group encoder. Firstly, the BRICS molecular dismantling method was used to obtain the functional groups of drug molecules. Compared with other molecular resolution methods (RECAP, ECFP), the BRICS resolution method is different in that it is based on whether chemical bonds can be synthesized. So that it will not destroy the original molecular structure and get the wrong functional groups. This method returns a list after deduplication, with serial numbers on the atoms corresponding to specific reaction types ([1*]C(= O)C(C)C; [4*]C(= O)NN(C)C).

Then, we will generate a two-dimensional matrix $M \in \{0, 1\}^{|D| \times |G|}$ to represent the correspondence between drugs and functional groups, where D represents the number of drugs, G represents the number of crown groups,$M_{ij} = 1$ representing the i-th drug with the j-th functional group.

3.2.2 Encoding Patient Representation

The M is a better representation of the drug and its efficacy, and we hope it can help predict recommended drugs. Firstly, the diagnosis representation $\boldsymbol{h}_d^{(t)}$ and procedure representation $\boldsymbol{h}_p^{(t)}$ obtained above are converted into the same dimension as M through a Linear layer. The query representation q_d, q_p are obtained through the activation functions σ_1, σ_2,

$$q_d = \sigma_1\left(\text{Linear}\left(\boldsymbol{h}_d^{(t)}, M\right)\right) \tag{8}$$

$$q_p = \sigma_2\left(\text{Linear}\left(\boldsymbol{h}_p^{(t)}, M\right)\right) \tag{9}$$

q_d and q_p represent the current condition of the patient and the combination of drugs required by the patient. Therefore, the q_d is continuously pruned through the feedforward neural network $\text{FNN}_1(\cdot)$ to obtain the currently recommended drug set $s^{m_d^{(t)}}$ in the diagnosis background. Similarly, q_p gets the recommended drug set $s^{m_p^{(t)}}$ in the background of the procedure through $\text{FNN}_2(\cdot)$,

$$s^{m_d^{(t)}} = \text{FNN}_1(q_d; W_1 \odot M) \tag{10}$$

$$s^{m_p^{(t)}} = \text{FNN}_2\left(q_p; W_2 \odot M\right) \tag{11}$$

where W_1 and W_2 are trainable parameters. The M is used as a mask matrix in the neural network to dynamically encode the drug combination required by the patient.

3.3 Dual-Level Drug Fusion Module

In the molecular knowledge-guided drug candidate sets generation module, we obtain two sets of drug recommendations, $s^{m_d^{(t)}}$ and $s^{m_p^{(t)}}$, by injecting patient representations into a drug functional group encoder. Obviously, it is indispensable to make a reasonable selection of the two to recommend a drug set that really fits the patient's condition. In this section, we design a drug selection module to select the drugs in $s^{m_d^{(t)}}$ and $s^{m_p^{(t)}}$.

3.3.1 Selection at Sets Level

While a patient's condition is documented in varying degrees of detail in diagnosis and procedure, some severe illnesses are certainly documented in both. Therefore, there must be some same drugs in $s^{m_d^{(t)}}$ and $s^{m_p^{(t)}}$, and these drugs are likely to be the drugs that patients really need. We take the intersection of the two $s_{\cap}^{\left(m_d^{(t)},m_p^{(t)}\right)}$ as the fixed part $s^{m_{fi}^{(t)}}$ of the final recommended drug $s^{m_f^{(t)}}$, and the XOR set $s_{\oplus}^{\left(m_d^{(t)},m_p^{(t)}\right)}$ as the candidate drug sets.

$$s_{\cap}^{\left(m_d^{(t)},m_p^{(t)}\right)} = s^{m_d^{(t)}} \cap s^{m_p^{(t)}} \tag{12}$$

$$s_{\oplus}^{\left(m_d^{(t)},m_p^{(t)}\right)} = s^{m_d^{(t)}} \oplus s^{m_p^{(t)}} \tag{13}$$

where $\cap$ is an intersection operation,$\oplus$ is an XOR operation.

3.3.2 Selection at Score Matrix Level

In the XOR set of $s^{m_d^{(t)}}$ and $s^{m_p^{(t)}}$, there are also drugs that are essential to the patient's condition, so we design a score matrix to select drugs. $\boldsymbol{h}_{\boldsymbol{d}}^{(t)}$ and $\boldsymbol{h}_{\boldsymbol{p}}^{(t)}$ are the patient conditions, which also represent the drug combination required by the patient. Therefore, we use $\boldsymbol{h}_{\boldsymbol{d}}^{(t)}$ and $\boldsymbol{h}_{\boldsymbol{p}}^{(t)}$ as drug reselection criteria and operate it with the candidate drug set $s_{\oplus}^{\left(m_d^{(t)},m_p^{(t)}\right)}$ to get the drug reselection score matrix M_{score}.

$$\mathrm{M}_{score} = \mathrm{FNN}_3\left(\left(\boldsymbol{h}_{\boldsymbol{d}}^{(t)}, \boldsymbol{h}_{\boldsymbol{p}}^{(t)}\right); s_{\oplus}^{\left(m_d^{(t)},m_p^{(t)}\right)}\right) \tag{14}$$

We select the top j items with higher scores in the matrix M_{score} as supplementary drugs and add them to the final recommended drugs $s^{m_f^{(t)}}$,

$$s^{m_{fj}^{(t)}} = \max(\mathrm{j}, \mathrm{M}_{score}) \tag{15}$$

$$s^{m_f^{(t)}} = \sigma_3\left(\mathrm{cat}\left(s^{m_{fi}^{(t)}}; s^{m_{fj}^{(t)}}\right)\right) \tag{16}$$

where max is the operation that takes the first j of the set, and cat is the operation of the connection.

3.4 Training

We denote the recommendation task of this model as a multi-label classification task. Given the total number of drugs $|\mathcal{D}|$, we denote the real drug by $m_r^{(t)}$ and the drug predicted by this model by $m_f^{(t)}$. We treat each patient visit as a classification task and backpropagate based on the average loss of all visits for the current patient and use a Taylor binary cross-entropy loss (Taylor_bce), treating the loss function as a linear combination of polynomial functions. After a lot of experiments, we show that when the number of polynomials is **1**, the loss function is more suitable for the model.

$$\mathcal{L}_{bce} = -\sum_{i=1}^{|\mathcal{D}|} m_{ri}^{(t)} \log\left(m_{fi}^{(t)}\right) + \left(1 - m_{ri}^{(t)}\right) \log\left(1 - m_{fi}^{(t)}\right) \tag{17}$$

$$\mathcal{L}_{\text{Taylor\_bce}} = \mathcal{L}_{bce} + \mathcal{E} \cdot \left(1 - P^{(t)}\right) \tag{18}$$

$$P^{(t)} = sum\left(onehot\left(m_r^{(t)}, \text{size}\left(m_r^{(t)}\right)\right) \cdot \text{softmax}\left(m_f^{(t)}\right)\right) \tag{19}$$

where $\mathcal{E}$ is the hyperparameter, $\cdot$ is the product between the scalars, *onehot* is the conversion of the drug representation into a one-hot tensor, size is the operation of getting the dimensions of the drug representation, and $P^{(t)}$ is the model's prediction probability of the target ground-truth class.

In order to ensure the correctness of the results, we also adopt multi-label hinge loss to ensure that the truth labels have at least 1 margin larger than others,

$$\mathcal{L}_{multi} = \sum_{i,j:m_{ri}^{(t)}=1, m_{rj}^{(t)}=0} \frac{\max\left(0, 1 - m_{fi}^{(t)} + m_{fj}^{(t)}\right)}{|\mathcal{D}|} \tag{20}$$

a standard approach to training with multiple loss functions is by the weighted sum of the loss measuring terms [10],

$$\mathcal{L} = \alpha \mathcal{L}_{\text{Taylor\_bce}} + (1 - \alpha)\mathcal{L}_{multi} \tag{21}$$

where α is usually pre-defined hyperparameter.

4 Experiments

4.1 Dataset and Metrics

The experiments were conducted on the part of the dataset MIMIC-II [9], which contains 26 tables. We use three tables in it: DIAGNOSES_ICD (Diagnosis Information), PROCEDURES_ICD (Procedure Information) and PRESCRIPTIONS (Drug Information) to conduct research. After many experiments, we set the hyperparameter $\mathcal{E}$ in Formula 18 to 0.995 and the hyperparameter α in Formula 21 to 0.96 to achieve the best performance. We use three efficacy metrics: Jaccard similarity, F1 score, and PRAUC to evaluate the recommendation efficacy.

4.2 Baselines

We evaluate the performance of our approach by comparing DKFM to several different baseline models:

- LR, standard Logistic Regression;
- ECC [22], Ensemabled Classifier Chain, which uses the SVM classifier.
- LEAP [4], which prediction by instance-based LSTM model;
- DMNC [6], which proposes a new longitudinal memory enhancement network;
- GAMENet [3], which adds the graph neural networks to the model to predict drugs;
- SafeDrug [1], which adds molecular information to the prediction model by a bi molecular encoder.

4.3 Performance Comparison

Table 1 shows the results of some of the methods. Overall, the DKFM model outperformed all baseline models in terms of Jaccard, f1 scoring, and PRAUC. LR, ECC, and LEAP performed relatively poorly because they only considered the current patient's condition. DMNC, GAMENet and SafeDrug performed relatively well because they utilized longitudinal patient information in different ways. DKFM uses a molecular knowledge-guided dual-level drug fusion mechanism to make full use of the patient's historical visit information, so DKFM outperforms existing baseline models.

Table 1. Performance Comparison of Different Models

Model	Jaccard	F1-score	PRAUC
LR	0.4865	0.6434	0.7509
ECC	0.4996	0.6569	0.6844
LEAP	0.4521	0.6138	0.6549
DMNC	0.4864	0.6529	0.7580
GAMENet	0.5067	0.6626	0.7631
SafeDrug	0.5213	0.6768	0.7647
DKFM	**0.5327**	**0.6855**	**0.7723**

4.4 Ablation Study

To verify the effectiveness of each module of DKFM, we design the following ablation models:

- DKFMw/oPA: We remove the positional encoder and attention module from the model, which means that, like SafeDrug, only the GRU is retained to encode the patient's longitudinal medical records.
- DKFM$w/o\mathcal{D}$: We remove the diagnosis information for each patient, which means the drug selection module is useless.

- DKFM$w/o\mathcal{P}$: We remove the procedure information of each patient.
- DKFM$w/o\mathcal{DDF}$: We remove the dual-level drug fusion mechanism, which means that we merge diagnosis and procedure information before injecting it into the drug functional group encoder, as in the previous model.
- DKFMw/oTaylor_bce: We remove the Taylor binary cross-entropy loss, which means setting $\mathcal{E}$ in Formula 18 to 0.

Table 2. Ablation Study for Different Components of DKFM on MIMIC-III.

Model	Jaccard	F1-score	PRAUC
DKFMw/oPA	0.5215	0.6753	0.7578
DKFM$w/o\mathcal{D}$	0.4923	0.6529	0.7439
DKFM$w/o\mathcal{P}$	0.5117	0.6670	0.7505
DKFM$w/o\mathcal{DDF}$	0.5185	0.6743	0.7583
DKFMw/oTaylor_bce	0.5301	0.6812	0.7685
DKFM	**0.5327**	**0.6855**	**0.7723**

Table 2 shows the results for the different variants of DKFM. As expected, the results of DKFMw/oPA demonstrate that the Positional Encoder and Attention module is effective in integrating patients' longitudinal historical medical record information. The results of DKFM$w/o\mathcal{D}$ and DKFM$w/o\mathcal{P}$ are the worst among all ablation models, which indicates that both diagnosis and procedure information plays an important role in drug recommendation. The results of DKFM$w/o\mathcal{DDF}$ show that the dual-level drug fusion mechanism brings significant improvement to the base model. The results of DKFMw/oTaylor_bce show that our designed loss function makes the model converge better. Overall, the full DKFM outperforms all ablation models, implying that every component of our model is meaningful.

4.5 Case Study

We present an example patient in MIMIC-III and illustrate the effectiveness of the dual-level drug fusion mechanism in improving drug recommendations by comparing DKFM with several variants. We use Anatomical Therapeutic Chemical (ATC)5 classification system to represent medications. As shown in Table 3, the patient had two visits, in which the drugs marked in bold were the same as the actual drugs. According to the analysis, the drugs of the patient on the second visit overlapped with the first visit, and the quantity increased, indicating that the patient's previous disease was not cured and he also suffered from new diseases.

By contrast, we found that the correct drug ***A12B*** predicted by DKFM$w/o\mathcal{D}$ is not reflected in DKFM$w/o\mathcal{DDF}$, but it is successfully predicted inDKFM. Moreover, the prediction accuracy of DKFM is the highest among several variants. Therefore, the dual-level drug fusion mechanism has improved the drug recommendation task.

Table 3. Analysis of DKFM and its variants

Models	Visit1	Accuracy	Visit2	Accuracy
Actual Drugs	A01A,A02B,A06A,B05C,A12A A12C,C01C,A07A,C07A,C03C A12B,J01M,B01A,C01B	—	N02B,A01A,A02B,A06A,B05C A12A,A12C,C01C,A07A,N01A C03C,A12B,N07A,N02A,R03A R01A,N04B,S01E,M03B,H04A	—
DKFM$w/o\mathcal{P}$	**A01A,A02B,A06A,B05C**,J01D R03A,C03B,VO4C,**C07A,C03C** CO3X,M01C,P02B,**C01B**	50.0%	H04A,**A01A,A02B**,N05B,**B05C** **A12A**,R05C,**C01C**,D06A,**N01A** A03F,A12B,N07A,G04B,A07D **R01A,N04B**,P01A,**M03B**,H02A	55.0%
DKFM$w/o\mathcal{D}$	**A01A,A02B,A06A**,G03C,N06B **A12C**,LO1C,H03B,RO3D,C03C **A12B**,L01D,**B01A**,A11D	42.9%	H04A,**A01A,A02B**,L01X,**B05C** **A12A**,J05A,**C01C**,J02A,C03D **C03C**,P01A,J01F,**N02A**,D04A D06B,V03A,**S01E**,D01A,**H04A**	45.0%
DKFM$w/o\mathcal{RR}$	**A01A,A02B,A06A,B05C**,A16A **A12C**,A10B,N06D,**C07A,C03C** M03B,C08D,**B01A,C01B**	64.3%	H04A,**A01A,A02B**,N06D,**B05C** **A12A**,H05B,**C01C**,C08D,**N01A** **C03C,A12B,N07A**,M04A,**R03A** **R01A,N04B,S01E**,S01A,**H04A**	75.0%
DKFM	**A01A,A02B,A06A,B05C**,J01D **A12C,C01C**,N06D,**C07A,C03C** ***A12B***,C08D,**B01A,C01B**	**78.6%**	H04A,**A01A,A02B**,N06D,**B05C** **A12A,A12C,C01C**,C08D,**N01A** **C03C,A12B,N07A**,C09X,**R03A** **R01A,N04B,S01E,M03B,H04A**	**80.0%**

5 Conclusion

In this paper, we propose a new drug recommendation model, DKFM, to handle the problem of different levels of detail in recording diagnosis and procedure information. DKFM uses a molecular knowledge-guided dual-level drug fusion mechanism which first obtains the initial two drug candidate sets and then obtains the final recommended drugs set by selecting the two at the sets level and score matrix level. Experiment results on MIMIC-III show that DKFM is superior to existing drug recommendation models. Further ablation study and case study results also show that each module of DKFM is effective.

Acknowledgment. This work is supported by grant from the Natural Science Foundation of China (No. 62072070).

References

1. Yang, C., Xiao, C., Ma, F., Glass, L., Sun, J.: Safedrug: Dual molecular graph encoders for safe drug recommendations. arXiv preprint arXiv:2105.02711 (2021)
2. Shang, J., Ma, T., Xiao, C., Sun, J.: Pre-training of graph augmented transformers for medication recommendation. arXiv preprint arXiv:1906.00346 (2019)
3. Shang, J., Xiao, C., Ma, T., Li, H., Sun, J.: Gamenet: Graph augmented memory networks for recommending medication combination. In: Proceedings of the AAAI Conference on Artificial Intelligence, vol. 33, pp. 1126–1133 (2019)

4. Zhang, Y., Chen, R., Tang, J., Stewart, W.F., Sun, J.: Leap: learning to prescribe effective and safe treatment combinations for multimorbidity. In: Proceedings of the 23rd ACM SIGKDD International Conference on Knowledge Discovery and Data Mining, pp. 1315–1324 (2017)
5. Wang, M., Liu, M., Liu, J., Wang, S., Long, G., Qian, B.: Safe medicine recommendation via medical knowledge graph embedding. arXiv preprint arXiv:1710.05980 (2017)
6. Le, H., Tran, T., Venkatesh, S.: Dual memory neural computer for asynchronous two-view sequential learning. In: Proceedings of the 24th ACM SIGKDD International Conference on Knowledge Discovery & Data Mining, pp. 1637–1645 (2018)
7. Bajor, J.M., Lasko, T.A.: Predicting Medications from Diagnostic Codes with Recurrent Neural Networks (2016)
8. Gong, F., Wang, M., Wang, H., Wang, S., Liu, M.: SMR: medical knowledge graph embedding for safe medicine recommendation. Big Data Res. **23**, 100174 (2021)
9. Johnson, A.E., et al.: Mimic-iii a freely accessible critical care database. Sci. Data **3**(1), 1–9 (2016)
10. Dosovitskiy, A., Djolonga, J.: You only train once: Loss-conditional training of deep networks. In: International Conference on Learning Representations (2019)
11. Zheng, Z., et al.: Drug package recommendation via interaction-aware graph induction. In: Proceedings of the Web Conference 2021, pp. 1284–1295 (2021)
12. Yang, C., Xiao, C., Glass, L., Sun, J.: Change matters: Medication change prediction with recurrent residual networks. arXiv preprint arXiv:2105.01876 (2021)
13. Wang, S., Ren, P., Chen, Z., Ren, Z., Ma, J., de Rijke, M.: Order-free medicine combination prediction with graph convolutional reinforcement learning. In: Proceedings of the 28th ACM International Conference on Information and Knowledge Management, pp. 1623–1632 (2019)
14. Wang, Y., Chen, W., Pi, D., Yue, L., Wang, S., Xu, M.: Self-supervised adversarial distribution regularization for medication recommendation. In: IJCAI, pp. 3134–3140 (2021)
15. Carey, J.S., Laffan, D., Thomson, C., Williams, M.T.: Analysis of the reactions used for the preparation of drug candidate molecules. Org. Biomol. Chem. **4**(12), 2337–2347 (2006)
16. He, X., Folkman, L., Borgwardt, K.: Kernelized rank learning for personalized drug recommendation. Bioinformatics **34**(16), 2808–2816 (2018)
17. Zhang, Y., Zhang, D., Hassan, M.M., Alamri, A., Peng, L.: Cadre: Cloud-assisted drug recommendation service for online pharmacies. Mob. Netw. Appl. **20**(3), 348–355 (2015)
18. Almirall, D., Compton, S.N., Gunlicks-Stoessel, M., Duan, N., Murphy, S.A.: Designing a pilot sequential multiple assignment randomized trial for developing an adaptive treatment strategy. Stat. Med. **31**(17), 1887–1902 (2012)
19. Chen, Z., Marple, K., Salazar, E., Gupta, G., Tamil, L.: A physician advisory system for chronic heart failure management based on knowledge patterns. Theory Pract. Logic Program. **16**(5–6), 604–618 (2016)
20. Gunlicks-Stoessel, M., Mufson, L., Westervelt, A., Almirall, D., Murphy, S.: A pilot smart for developing an adaptive treatment strategy for adolescent depression. J. Clin. Child Adolesc. Psychol. **45**(4), 480–494 (2016)
21. Lakkaraju, H., Rudin, C.: Learning cost-effective and interpretable treatment regimes. In: Artificial Intelligence and Statistics, pp. 166–175. PMLR (2017)
22. Read, J., Pfahringer, B., Holmes, G., Frank, E.: Classifier chains for multi-label classification. Mach. Learn. **85**(3), 333–359 (2011). [14] Shang, J., Ma, T., Xiao, C., Sun, J.: Pre-training of graph augmented transformers for medication recommendation. arXiv preprint arXiv:1906. 00346 (2019)

Hierarchical Graph Neural Network for Patient Treatment Preference Prediction with External Knowledge

Quan Li[1], Lingwei Chen[2], Yong Cai[3,4](✉), and Dinghao Wu[1]

[1] Pennsylvania State University, University Park, PA, USA
{qbl5082,dwu12}@psu.edu
[2] Wright State University, Dayton, OH, USA
lingwei.chen@wright.edu
[3] California State University, Monterey Bay, CA, USA
[4] IQVIA Inc., Wayne, PA, USA
yong.cai@iqvia.com

Abstract. The healthcare industry has a wealth of data that can be used by researchers and medical professionals to infer a patient's condition and intention to receive treatment using machine learning models. However, this line of research generally suffers from some limitations: (1) struggling to leverage structural interactions among patients; (2) attending to learn patient representations from electronic medical records (EMRs) but rarely considering supplementary contexts; and (3) overlooking EMR data imbalance issue. To address these limitations, in this paper, we propose a hierarchical graph neural network for patient treatment preference prediction. Doctors' information and their viewing activities are first integrated as external knowledge with EMRs to construct the hierarchical graph, where a dual message passing paradigm is then devised to perform intra- and inter-subgraph aggregation to enrich patient representations and advance label propagation. To mitigate patient data imbalance issue, a community detection method is further designed to better prediction. Our experimental results demonstrate the state-of-the-art performance on patient treatment preference prediction.

Keywords: Hierarchical graph neural network · Oncology treatments · Preference prediction · Healthcare · Community detection

1 Introduction

In many oncology treatments, doctors and patients generally adopt watch-and-wait strategy [12]. After confirmed diagnoses, patients can wait for a long time to take aggressive treatment. For example, it takes 5 to 10 years on average for a Chronic Lymphocytic Leukemia (CLL) patient before taking treatment. But the treatment decision is highly dependent on patient condition and doctors'

H. Kashima et al. (Eds.): PAKDD 2023, LNAI 13937, pp. 204–215, 2023.
https://doi.org/10.1007/978-3-031-33380-4_16

scrutiny. Some patients may only take a short period of time on watch-and-wait. Estimating and predicting if a patient has been ready to take a treatment can serve as reminders and assist doctors and patients to make the right decisions. However, due to the high variation of treatment patterns in oncology area, predicting the likelihood for a patient to take treatment is a challenging problem.

With the rapid development in machine learning and deep learning [19], the healthcare industry has started to exploit these data-driven concepts and theories into practical products and applications to predict patient conditions and propensity for treatment and medication [10,22,29], which in turn facilitate doctors' analyses and decisions to plan treatments for patients. One of the widely used data for such tasks is electronic medical record (EMR), which maintains rich and important patient information, and keeps growing in its volume and diversity. This has thus attracted researchers in the healthcare industry to take EMRs as inputs to train machine learning models and make patient-specific predictions through them [31,32,35,37]. Though with the promising performance, these models trained on EMRs provide the successful principles to solve the high variation issues in patients, their inputs are inherently self-contained, and struggle to leverage structural interactions with other patients.

Graph neural networks (GNNs) have recently emerged as one of the most powerful techniques for graph mining [5,16,18]. These GNNs perform information aggregation to extract high-level features from the nodes and their neighborhoods [4], which have boosted the performances for various tasks over graphs. Therefore, a surge of effective research works build GNNs to learn structural semantics from EMRs and advance patient-specific models [3,8,9,24,26]. For example, GRAM [8] and KAME [26] constructed the knowledge graph over EMRs to depict the hierarchy of medical concepts in the form of a parent-child relationship and utilized GNNs to embed medical code to characterize each patient. Liu et al. [24] analyzed EMR using heterogeneous GNN to capture more diverse patient information (e.g., profile, symptoms, and visit history). However, these structured approaches still suffer from two limitations. (1) While attending to depict patients and learn higher-level patient representations from EMRs, this line of research rarely utilizes any supplementary contexts. As indicated by some surveys and case studies [7], the doctor-patient relationship may essentially impact on patients' treatment preferences; in other words, the external knowledge (e.g., doctor information) can be extracted to further assist in predicting if a patient would like to take treatment or not. (2) EMR data imbalance issue has been completely overlooked by current researches as well. Due to laborious process and delay effect on data annotation, the imbalance issue exists across common and rare diseases, and the downstream patient treatment distribution, which naturally enforces data-driven models to favor the majority class over the minority class and degrade their prediction performances. With this in mind, our goal here is to investigate how much patient treatment preference prediction can benefit from a structured imbalanced learning model with external knowledge.

To this end, in this paper, we propose a novel hierarchical graph neural network model with external knowledge for patient treatment preference prediction that can effectively mitigate the impact of data imbalance as well. More specifically, we introduce the doctors' information and their viewing activities (captured by website topics) as external knowledge to be integrated with EMRs to enrich patient representations and advance the structured learning model for better prediction performance. The hierarchical graph is first constructed to abstract the interactions of patients, doctors, and topics, where a dual message passing paradigm is devised to perform intra-subgraph and inter-subgraph neighborhood aggregation for node representation refinement and label propagation. To cope with imbalanced patient data, a community detection method is further designed to cluster the higher-level embeddings of negative and unlabeled patients to derive community-preserving patient graph, where the treatment preference predictions for patients are produced through communities.

2 Problem Statement

EMRs contain the medical and treatment history of the patients in different practices, which allow us to predict the propensity of patients to take treatments for different diseases. In this paper, we focus on the prediction of patient oncology treatments. Without loss of generality, we represent our data as $\mathcal{X} = \{(x_{pi}, y_i)\}_{i=1}^{l} \cup \{x_{di}\}_{i=1}^{m} \cup \{x_{ti}\}_{i=1}^{n}$ consisting $l+m+n$ samples, where l is the number of patient records, m is the number of doctor records, and n represents the number of topics retrieved from website data. Unlike existing works [3,11,25] that merely use EMRs to train the models for performing patient-specific tasks, we constructively consider doctor information out of EMRs to interact with patients and facilitate our prediction. Each patient record x_p is annotated with a ground truth $y \in \{0, 1\}$ for a specific treatment preference, where $y = 1$ indicates that the patient prefers to take the treatment and $y = 0$ denotes that the patient has no such intention. Note that, positives are much smaller than negatives in our data and also in the real-world scenario. We initially map $\mathcal{X}$ including patients, doctors, and topics into k-dimensional feature vectors and learn a patient representation function ϕ through hierarchical graph neural network to aggregate information from patients, doctors, and topics to obtain higher-level $\mathbf{X}_p = \phi(\mathcal{X}_p, \mathcal{X}_d, \mathcal{X}_t)$, $\mathbf{X}_p \subseteq \mathbb{R}^{l\times k}$. Resting on patient representations, we aim to learn a classification model $f : \mathbf{X}_p \rightarrow \mathbf{Y}$ to perform our prediction task. Thus, the treatment preference label for a given patient data x can be predicted as:

$$y^* = \underset{y\in\{0,1\}}{\text{argmax}}\ f_y(\mathbf{x}_p) \tag{1}$$

where $f_y(\mathbf{x}_p)$ is the confidence score of predicting patient $\mathbf{x}_p$ as treatment preference label y using the classification model f. From Eq. (1), we can see that the final label assigned to the input is the one with the highest confidence score.

3 Proposed Model

In this section, we present the technical details of our proposed model as follows, the overview of which is illustrated in Fig. 1.

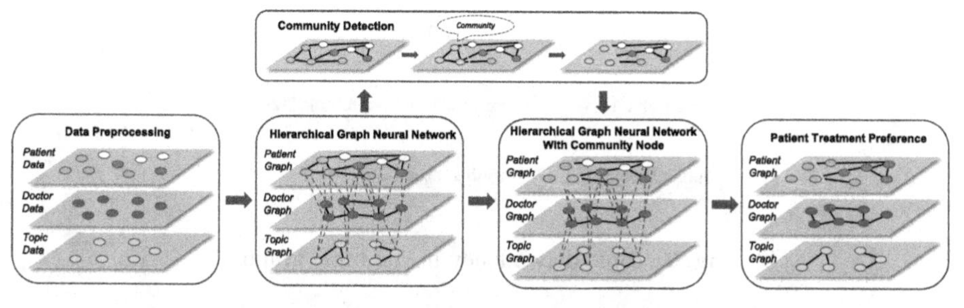

Fig. 1. The overview of our proposed model.

3.1 Hierarchical Graph Construction

To proceed with patient representation learning using GNNs, the first step is to construct the graph. As we introduce doctors' information and the website topics they have viewed as external knowledge, here we design a hierarchical graph to integrate patients, doctors, and topics.

Hierarchical Graph Notations. This hierarchical graph can be formalized as $G = (V, E, \mathbf{X})$, where V is the node set (i.e., patients, doctors, and topics), E is the edge set to connect the node pairs, and $\mathbf{X}$ is the initial feature matrix. More specifically, G can be further refined into three subgraphs: $G = \{G_p, G_d, G_t\}$. G_p is the patient graph with nodes V_p and edges E_p, G_d is the doctor graph with nodes V_d and edges E_d, and G_t is the topic graph with nodes V_t and edges E_t. In addition, E_{pd} connects patient graph and doctor graph when patients and doctors are associated with national patient identifiers (NPIs), and E_{dt} connects doctor graph and topic graph when doctors view the website topics.

Node Representations. The node feature matrix $\mathbf{X}$ is composed of three matrices $\mathbf{X}_p$, $\mathbf{X}_d$, and $\mathbf{X}_t$ such that $\mathbf{X} = \{\mathbf{X}_p, \mathbf{X}_d, \mathbf{X}_t\}$, where $\mathbf{X}_p$, $\mathbf{X}_d$, and $\mathbf{X}_t$ embed the feature spaces for patients, doctors, and topics respectively. Each patient feature vector $\mathbf{x}_p$ is initialized as $\mathbf{x}_p = \langle x_{p1}, x_{p2}, x_{p3}, \cdots, x_{pk} \rangle$, where $x_{pi} \in \{0, 1\}$ is a binary value indicting the absence or presence of a disease symptom i in patient $\mathbf{x}_p$. Each doctor $\mathbf{x}_d$ is represented as a set of profile attributes, where each attribute is directly converted into numerical feature values using one-hot encoding. Each topic $\mathbf{x}_t$ is represented as either a word or a phrase; in this regard, we leverage SBERT [28] to derive a fixed-size embedding for each topic. In order to keep the dimensionality of all nodes consistent for message passing yet the dimension of $\mathbf{x}_d$ and $\mathbf{x}_t$ is smaller than $\mathbf{x}_p$, we zero-pad $\mathbf{x}_d$ and $\mathbf{x}_t$ to be k-dimensional, and hence the node feature matrix $\mathbf{X} \subseteq \mathbb{R}^{(l+m+n)\times k}$.

Patient Graph. Given a set of patient records $\mathcal{X}_p$, we construct a fully-connected graph $G_p = (V_p, E_p, \mathbf{X}_p)$ to associate patients (both labeled and unlabeled). Manifold learning [23] is non-linear dimensionality reduction process which reveals the low-dimensional manifold embedded in the high-dimensional space, which can be feasibly exploited to build up the intrinsic neighborhood among patient representations. Thus, we formulate each edge $e_p \in E_p$ between

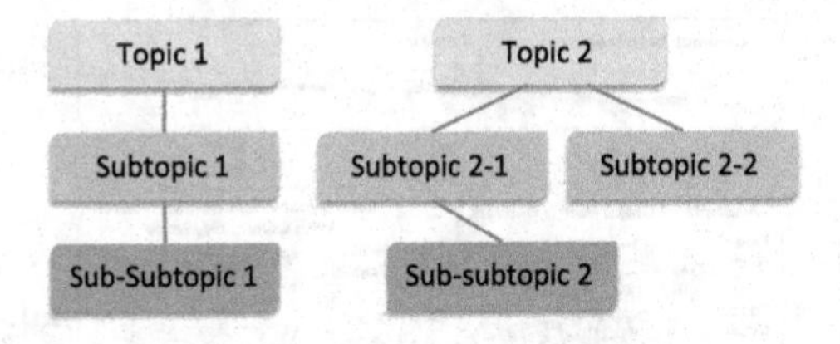

Fig. 2. Topic dependency in the topic graph.

v_{pi} and v_{pj} in G_p by a layerwise non-linear combination of distance between $\mathbf{x}_{pi}$ and $\mathbf{x}_{pj}$:

$$e_p = g_{\boldsymbol{\Theta}}(\mathbf{x}_{pi}, \mathbf{x}_{pj}) = \sigma(\cdots \sigma(|\mathbf{x}_{pi} - \mathbf{x}_{pj}|\boldsymbol{\Theta}^{(0)}) \cdots \boldsymbol{\Theta}^{(L-1)})\boldsymbol{\Theta}^{(L)} \tag{2}$$

where $\sigma(\cdot)$ is a non-linear activation function (e.g., ReLU), and $\boldsymbol{\Theta}$ is learnable weight matrix for each layer. As the constructed structure behaves differently regarding different patient representations, the learned edges do not specify a fixed patient graph, suggesting the graph can be refined when the embedding space across patient nodes is updated.

Doctor Graph. In addition to doctors' profile attributes, our collected doctor data $\mathcal{X}_d$ also record the numbers of patients shared with other doctors, which can be directly exploited to build the doctor graph. To be specific, if two doctors v_{di} and v_{dj} share greater than or equal to one patient in common, we create an edge $e_d \in E_d$ between v_{di} and v_{dj} in G_d, such that the doctor graph $G_d = (V_d, E_d, \mathbf{X}_d)$ can be easily derived with fixed structure. Afterwards, the doctor nodes V_d are further associated with the patient nodes V_p through $e_{pd} \in E_{dp}$ when the doctor v_d is the patient v_p's primary care doctor identified by NPI.

Topic Graph. To better characterize doctors, we integrate doctors' viewing activities into their profile attributes for doctor presentation learning, where these activities are captured by the website topics viewed by doctors. To this end, we build a topic graph $G_t = (V_t, E_t, \mathbf{X}_t)$ to model this data. As demonstrated in Fig. 2, all the topics are organized through layer-wise dependency; for example, a topic may contain another one or more subtopics, where some other topics may be listed under a subtopic. An edge $e_t \in E_t$ between v_{ti} and v_{tj} in G_t can be thus formulated when v_{tj} is v_{ti}'s subtopic. Naturally, the topic nodes V_t can be associated with the doctor nodes V_d through their viewing records.

3.2 Hierarchical Graph Neural Network with Dual Message Passing

Considering the constructed hierarchical graph with intra-subgraph and inter-subgraph neighborhood structures, we propose a hierarchical graph neural network to perform the dual message passing for node representation refinement and label propagation, including intra-message passing and inter-message passing.

Intra-message Passing. Intra-message passing is the propagation mechanism that aggregates the information from neighbors inside the patient graph, doctor graph, and topic graph, respectively, the data flow paths of which are specified as black lines in Fig. 1. A regular graph convolutional network (GCN) [16] is implemented for a single subgraph. Specifically, given a subgraph (i.e., patient graph, doctor graph, or topic graph), we build the adjacency matrix $\mathbf{A}^{(h)}$ using its edge information (edge matrix needs to be normalized first for patient graph). The message passing can be then formalized as multi-layer neighborhood information aggregation, which receives an input $\mathbf{X}^{(h)}$ and produces $\mathbf{X}^{(h+1)}$:

$$\mathbf{X}^{(h+1)} = \sigma(\widetilde{\mathbf{A}}^{(h)}\mathbf{X}^{(h)}\mathbf{W}_{intra}^{(h)}) \tag{3}$$

where at layer h, $\mathbf{W}_{intra}$ is weight matrix, $\widetilde{\mathbf{A}} = \mathbf{D}^{-\frac{1}{2}}\hat{\mathbf{A}}\mathbf{D}^{-\frac{1}{2}}$, $\hat{\mathbf{A}} = \mathbf{A} + \mathbf{I}$, and $\mathbf{D}$ is the diagonal degree matrix defined on $\hat{\mathbf{A}}$, i.e., $\mathbf{D}_{ii} = \sum_{j=1}^{n} \hat{\mathbf{A}}_{ij}$.

Inter-message Passing. Inter-message passing mechanism is used to propagate the information between two subgraphs, including patient-doctor and doctor-topic neighborhoods in our hierarchical graph, the data flow paths of which are specified as red lines in Fig. 1. Similarly, a GCN is implemented for a single inter-subgraph neighborhood, where an adjacency matrix $\mathbf{A}^{(h)}$ is first constructed based on the node set from both subgraphs and the edge set (i.e., E_{pd} or E_{dt}) connecting subgraphs, and then the message passing is performed:

$$\mathbf{X}^{(h+1)} = \sigma(\widetilde{\mathbf{A}}^{(h)}\mathbf{X}^{(h)}\mathbf{W}_{inter}^{(h)}) \tag{4}$$

where at layer h, $\mathbf{W}_{inter}$ is weight matrix for inter message passing. Different from intra-message passing, we do not add self-loops to the adjacency matrix in Eq. (4) to allow better aggregation of heterogeneous information.

Optimization. With dual message passing from topic graph to doctor graph, and then from doctor graph to patient graph, the output of the final GCN layer for intra-message passing over the patient graph can be defined as:

$$\mathbf{Z} = f_{\mathbf{W}}(\mathbf{A}, \mathbf{X}_p) = \mathrm{softmax}(\mathbf{X}_p^{(H)}) \tag{5}$$

where $\mathbf{W}$ refers to the complete trainable weights raised by intra- and inter-message passing. Therefore, the optimization of hierarchical GNN model can be formulated to minimize the training loss as follows:

$$\mathbf{W}^* == \underset{\mathbf{W}}{\mathrm{argmin}}\ \mathcal{L}(\mathbf{Z}, \mathbf{y}) + \lambda \|\mathbf{W}\|_2^2 \tag{6}$$

where $\mathcal{L}$ is the cross-entropy loss, and λ is the regularization parameter. This model can be applied under inductive and transductive settings. In this paper, we focus on transductive patient treatment preference prediction where all node connections and features are accessible during training.

3.3 Community Detection for Data Imbalance

As discussed in Sect. 1, another significant challenge for patient treatment preference prediction is the EMR data imbalance issue. This enforces GNN models to aggregate information from majority-class nodes and become less sensitive to under-represented positive samples, which leads to less-accurate prediction performance. Accordingly, different paradigms have been presented to address this issue, such as oversampling [1,14], undersampling [2,17], and cost-sensitive learning [13,21]. Due to the fact that sampling techniques tend to generate models with relatively low generalizability that either overfit on oversampled data or underperform for discarding potentially useful data, and cost-sensitive learning is easily impacted by weights, making it hard to select optimal cost values, these methods are still limited for our task.

In this paper, we explore a community detection method to cope with imbalanced patient data. The motivations behind this choice are that: (1) community detection [27,34] is one of the widely used approaches to analyze complex networks involving social interactions, which is perfectly applicable to the patient graph; (2) individuals are known by the community they keep, while patients in EMRs are natural individuals whose treatment behaviors and preferences can be represented by a group of others with very similar symptoms; and (3) community detection works as undersampling but can effectively mitigate the impact of information loss [20]. More specifically, the proposed community detection method to address data imbalance consists two steps:

- **Detecting Communities**. If we start community detection over the graph using the initial patient representations, we need to traverse the graph to reveal the community structure using algorithms such as infomap [30]. Instead, here we follow the strategy to first learn the higher-level patient representations using hierarchical GNN with dual message passing to embed semantics from patients and doctors, and abstract graph structure, such that we can then simply apply standard clustering algorithm such as k-means to cluster the embeddings of negative patients into K distinct communities, where K is equal to the number of positive patients, and cluster the embeddings of unlabeled patients into N communities, where N is dependent on test data size (we evaluate the impact of N on prediction performance in Sect. 4.4). Afterwards, all the edges ending with the patient nodes in a community are adjusted to be connected with this community as one node.
- **Training GNN using community-preserving patient graph**. With the new community-preserving patient graph, we continue performing dual message passing over hierarchical graph and train the hierarchical GNN by minimizing the cross-entropy loss in Eq. (6). During testing, the prediction label of a community node will be assigned to all patients in this community.

4 Experiments and Results

4.1 Experiment Setup

Datasets. Our experiments are tested on EMRs for CLL patients with doctor data and website topics provided by IQVIA. The patient data retain patients' records including their different symptom features and NPIs for their primary care doctors. The doctor data include doctors' profiles (e.g., age, gender, location,

Table 1. Statics of datasets

Dataset	#Distinct data	#Features	#Positives	#Negatives
Patients	93,474	2,016	773	92,701
Doctors	2,134	112	–	–
Website Topics	300	–	–	–

etc.) and the number of patients shared with other doctors. The topic data contain topics of websites viewed by doctors. The data statistics are shown in Table 1, illustrating that there are 93,474 patients (773 positives and 92,701 negatives), 2,134 valid doctors, and 300 website topics, respectively.

Baselines. To the best of our knowledge, we are the first to predict patient treatment preference; no previous work can thus be used as baselines. We select rare disease prediction models, traditional classification models, GNN models, and imbalanced learning models as our baselines. Note that, for GCN, GRAM, and RA-GCN designed for single graph, we only use the patient graph as input.

- **Support Vector Machine (SVM)**: This is one of the supervised learning methods which can be used to find a hyperplane for classification.
- **Random Forest (RF)**: This is an ensemble learning method for classification by constructing a number of decision trees.
- **Multi-Layer Perceptron (MLP)**: This is a fully connected class of artificial neural network, which is a traditional supervised classification model.
- **Graph Convolutional Network (GCN)** [16]: This is a semi-supervised learning model on graph-structured data with graph convolutional layers.
- **GRAM** [8]: GRAM is a graph-based attention model for healthcare representation learning, which leverages graph attention network [33] to get the information from neighbors with different importance. We use their attention mechanism to build the graph neural network and set it as our baseline.
- **HSGNN** [24]: Heterogeneous similarity GNN is designed for heterogeneous graphs with healthcare data. We reconstruct the graph with our patient and doctor data and set their model as our baseline.
- **Oversampling** [14]: This is an approach to deal with imbalanced data by increasing the minority class in the dataset. In our experiments, we simply add the minority class repeatedly for oversampling.
- **Undersampling** [17]: This is an approach to deal with imbalanced data by randomly removing the data from the majority class in the dataset.
- **XGBoost** [6]: It is one of the state-of-the-art and widely used machine learning models. It becomes the powerful machine learning model of many data scientists and can deal with irregularities of data, which has been justified as one of the most popular methods for dealing with imbalanced data.
- **Pseudo-labeling** [36]: This approach generates pseudo-labels from unlabeled data, which are injected into training data to address data imbalance.
- **RA-GCN** [15]: It sets different weights to different classes to address the data imbalance problem with GCN for disease prediction.

Table 2. Evaluation results with different baselinesb

Model	Precision (%)	Recall (%)	F1-score
Support Vector Machine	45.60	50.00	0.4769
Random Forest	45.73	50.10	0.4781
Multi-Layer Perceptron	46.30	50.24	0.4818
GCN	45.64	49.88	0.4744
GRAM	46.70	50.00	0.4952
HSGNN	46.58	48.20	0.4738
Oversampling	75.28	50.21	0.6024
Undersampling	23.40	50.03	0.3188
XGBoost	52.28	59.24	0.5554
Pseudo-labeling	53.41	51.86	0.5262
RA-GCN	47.22	42.46	0.4471
Our Model	**77.72**	**59.80**	**0.6759**

4.2 Data Preprocessing and Setting

We preprocess the data by filtering out those patient records whose primary doctors cannot be found in the doctor data and removing doctors whose profiles are missing, which leads to 19,176 patients (351 positives) and 2,134 doctors left, respectively. Due to resource limitation, we select all positives and 3,510 negatives as experimental data and randomly split it by 8:2 for training and testing. We use precision, recall, and F1-score as evaluation metrics, which are typically used for healthcare data. All the GCNs used for intra- and inter-message passing are set as a two-layer structure with 16 hidden units.

4.3 Comparisons with Baselines

In this section, we compare our model with the selected baselines. The competitive result is illustrated in Table 2. We can observe that among baselines, the traditional models (i.e., SVM, RF, and MLP) have the worst performance on the patient treatment preference prediction with single patient data input. GNN models perform slightly better than traditional ones with the precision increases by around 1%, which is limited for data imbalance issue. Most of the models dealing with imbalanced data achieve better performance than others, where oversampling delivers the best results, XGBoost also provides some promising performance boost, but undersampling significantly underperforms for information loss. Obviously, our model completely outperforms baselines with a large improvement margin of precision (2% – 30%), recall (0.6% – 11%), and F1-score (0.07 – 0.36). This confirms that (1) doctor information and viewing activities can serve as external knowledge to enrich patient representations; (2) community detection performed on negatives and unlabeled patients can effectively mitigate the data imbalance issue and better prediction performance.

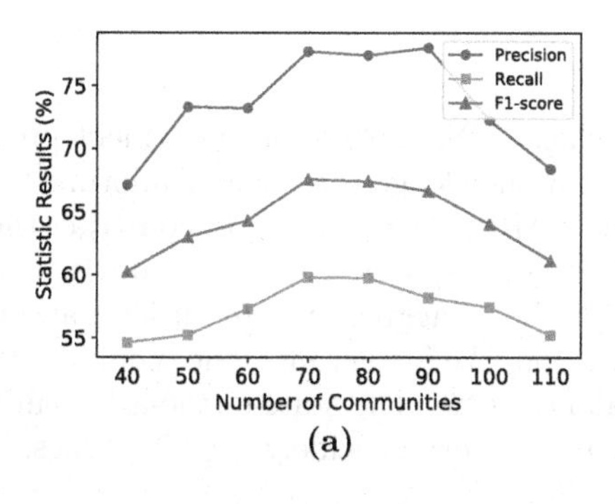

(a)

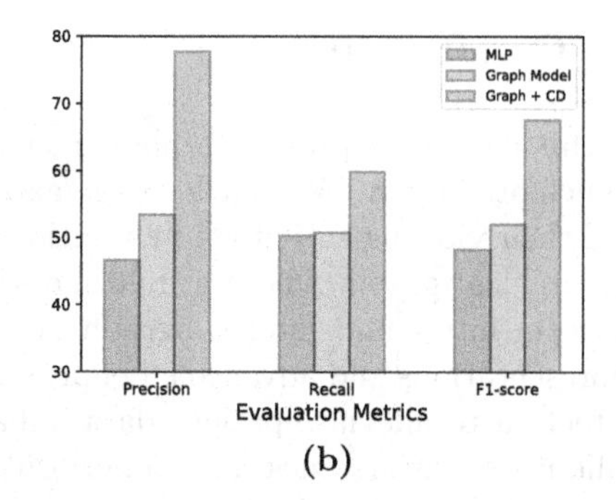

(b)

Fig. 3. Evaluation on model: (a)Number of Communities N (b) Different components

4.4 Impact of Community Number over Test Data

The community number K over negatives is decided by the number of positives, but the community number N over test data is adjustable. In this section, we analyze the impact of N on prediction performance. The results are shown in Fig. 3(a): the prediction results increase when N falls in the range of [40,70], then keep stable at [70,90], and decrease drastically when N rises to 110. The reason behind this could be that when $N \in [70, 90]$, the positive data and negative data are in a relative equilibrium, which alleviates the data imbalance impact and prevents the model from favoring any majority class; when N is too small, some unique patients tent to get misrepresented by communities; when N is too large, the imbalance issue may emerge to degrade the predictions.

4.5 Ablation Study

In this section, we conduct the ablation study to evaluate the performance contributed by different design parts. As our model proceeds with (1) hierarchical graph construction and (2) community detection, we construct three alternative models: (1) MLP: feeds patient features directly to MLP without any graph or community detection; (2) Graph Model: applies the hierarchical graph with dual message passing; (3) Graph + Community Detection (Graph + CD): leverages community detection to build community-preserving graph and trains the hierarchical GNN over that. The results are reported in Fig. 3(b).

As shown in Fig. 3(b), the performance becomes better with the components added to the model. The graph model increases the precision, recall, and F1-score from multi-layer perceptron by about 8%, 1%, and 4% respectively. Our model with hierarchical graph and community detection is able to further improve precision, recall, and F1-score to a higher level, which are around 77%, 60%, and 67% respectively. This reaffirms that the hierarchical graph enables doctor information and activities to be propagated to patients through dual message passing and increases the expressiveness of patient representations, while community detection successfully alleviates the effect of data imbalance and makes the model more effective on patient treatment preference prediction.

5 Conclusion

In this paper, we propose a novel hierarchical GNN for patient treatment preference prediction. We first leverage external knowledge (i.e., doctor information and their viewing activities) in addition to EMR patient records to construct the hierarchical graph, where a dual message passing paradigm is then devised to perform intra- and inter-subgraph neighborhood aggregation to enrich patient representations and advance label propagation. We further introduce community detection to alleviate patient data imbalance issue. The state-of-the-art results validate its effectiveness and superiority to the current widely used baselines.

References

1. Ando, S., Huang, C.Y.: Deep over-sampling framework for classifying imbalanced data. In: Ceci, M., Hollmén, J., Todorovski, L., Vens, C., Džeroski, S. (eds.) ECML PKDD 2017. LNCS (LNAI), vol. 10534, pp. 770–785. Springer, Cham (2017). https://doi.org/10.1007/978-3-319-71249-9_46
2. Buda, M., Maki, A., Mazurowski, M.A.: A systematic study of the class imbalance problem in convolutional neural networks. Neural Netw. **106**, 249–259 (2018)
3. Cai, D., Sun, C., Song, M., Zhang, B., Hong, S., Li, H.: Hypergraph contrastive learning for electronic health records. In: SDM, pp. 127–135. SIAM (2022)
4. Chen, J., Ma, T., Xiao, C.: Fastgcn: fast learning with graph convolutional networks via importance sampling. arXiv preprint arXiv:1801.10247 (2018)
5. Chen, L., Li, X., Wu, D.: Enhancing robustness of graph convolutional networks via dropping graph connections. In: Hutter, F., Kersting, K., Lijffijt, J., Valera, I. (eds.) ECML PKDD 2020. LNCS (LNAI), vol. 12459, pp. 412–428. Springer, Cham (2021). https://doi.org/10.1007/978-3-030-67664-3_25
6. Chen, T., Guestrin, C.: Xgboost: A scalable tree boosting system. In: SIGKDD, pp. 785–794 (2016)
7. Chipidza, F.E., Wallwork, R.S., Stern, T.A.: Impact of the doctor-patient relationship. The primary care companion for CNS disorders **17**(5), 27354 (2015)
8. Choi, E., Bahadori, M.T., Song, L., Stewart, W.F., Sun, J.: Gram: graph-based attention model for healthcare representation learning. In: SIGKDD (2017)
9. Choi, E., et al.: Learning the graphical structure of electronic health records with graph convolutional transformer. In: AAAI. vol. 34, pp. 606–613 (2020)
10. Chu, J., Dong, W., Wang, J., He, K., Huang, Z.: Treatment effect prediction with adversarial deep learning using electronic health records. In: BMC MIDM (2020)
11. Cui, L., Biswal, S., Glass, L.M., Lever, G., Sun, J., Xiao, C.: Conan: complementary pattern augmentation for rare disease detection. In: AAAI (2020)
12. Dossa, F., Chesney, T.R., Acuna, S.A., Baxter, N.N.: A watch-and-wait approach for locally advanced rectal cancer after a clinical complete response following neoadjuvant chemoradiation: a systematic review and meta-analysis. The lancet Gastroenterology & hepatology **2**(7), 501–513 (2017)
13. Elkan, C.: The foundations of cost-sensitive learning. In: IJCAI. vol. 17, pp. 973–978. Lawrence Erlbaum Associates Ltd (2001)
14. Fernández, A., Garcia, S., Herrera, F., Chawla, N.V.: Smote for learning from imbalanced data: progress and challenges, marking the 15-year anniversary. J. Artifi. Intell. Res. **61**, 863–905 (2018)

15. Ghorbani, M., Kazi, A., Baghshah, M.S., Rabiee, H.R., Navab, N.: Ra-gcn: graph convolutional network for disease prediction problems with imbalanced data. Med. Image Anal. **75**, 102272 (2022)
16. Kipf, T.N., Welling, M.: Semi-supervised classification with graph convolutional networks. arXiv preprint arXiv:1609.02907 (2016)
17. Lee, W., Seo, K.: Downsampling for binary classification with a highly imbalanced dataset using active learning. Big Data Res. **28**, 100314 (2022)
18. Li, Q., Li, X., Chen, L., Wu, D.: Distilling Knowledge on Text Graph for Social Media Attribute Inference. In: SIGIR, pp. 2024–2028 (2022)
19. Li, X., Chen, L., Wu, D.: Turning attacks into protection: Social media privacy protection using adversarial attacks. In: SDM, pp. 208–216. SIAM (2021)
20. Lin, W.C., Tsai, C.F., Hu, Y.H., Jhang, J.S.: Clustering-based undersampling in class-imbalanced data. Inf. Sci. **409**, 17–26 (2017)
21. Ling, C.X., Sheng, V.S.: Cost-sensitive learning and the class imbalance problem. Encyclopedia Mach. Learn. **2011**, 231–235 (2008)
22. Liu, R., Wei, L., Zhang, P.: A deep learning framework for drug repurposing via emulating clinical trials on real-world patient data. Nature Mach. Intell. **3**(1), 68–75 (2021)
23. Liu, Y., et al.: Learning to propagate labels: Transductive propagation network for few-shot learning. arXiv preprint arXiv:1805.10002 (2018)
24. Liu, Z., Li, X., Peng, H., He, L., Philip, S.Y.: Heterogeneous similarity graph neural network on electronic health records. In: IEEE Big Data (2020)
25. Ma, F., Wang, Y., Gao, J., Xiao, H., Zhou, J.: Rare disease prediction by generating quality-assured electronic health records. In: SDM, pp. 514–522. SIAM (2020)
26. Ma, F., You, Q., Xiao, H., Chitta, R., Zhou, J., Gao, J.: Kame: Knowledge-based attention model for diagnosis prediction in healthcare. In: CIKM (2018)
27. Papadopoulos, S., Kompatsiaris, Y., Vakali, A., Spyridonos, P.: Community detection in social media. Data Min. Knowl. Disc. **24**(3), 515–554 (2012)
28. Reimers, N., Gurevych, I.: Sentence-bert: Sentence embeddings using siamese bert-networks. arXiv preprint arXiv:1908.10084 (2019)
29. Ross, M.K., Yoon, J., van der Schaar, A., van der Schaar, M.: Discovering pediatric asthma phenotypes on the basis of response to controller medication using machine learning. Ann. Am. Thorac. Soc. **15**(1), 49–58 (2018)
30. Rosvall, M., Bergstrom, C.T.: Maps of random walks on complex networks reveal community structure. Proceedings of the national academy of sciences (2008)
31. Saqib, M., Sha, Y., Wang, M.D.: Early prediction of sepsis in emr records using traditional ml techniques and deep learning lstm networks. In: EMBC (2018)
32. Segura-Bedmar, I., Colón-Ruíz, C., Tejedor-Alonso, M.Á., Moro-Moro, M.: Predicting of anaphylaxis in big data emr by exploring machine learning approaches. J. Biomed. Inform. **87**, 50–59 (2018)
33. Thekumparampil, K.K., Wang, C., Oh, S., Li, L.J.: Attention-based graph neural network for semi-supervised learning. arXiv preprint arXiv:1803.03735 (2018)
34. Yang, J., McAuley, J., Leskovec, J.: Community detection in networks with node attributes. In: ICDM, pp. 1151–1156. IEEE (2013)
35. Yang, J., Liu, Y., Qian, M., Guan, C., Yuan, X.: Information extraction from electronic medical records using multitask recurrent neural network with contextual word embedding. Appl. Sci. **9**(18), 3658 (2019)
36. Yang, Y., Xu, Z.: Rethinking the value of labels for improving class-imbalanced learning. NeurIPS **33**, 19290–19301 (2020)
37. Zhao, J., Gu, S., McDermaid, A.: Predicting outcomes of chronic kidney disease from emr data based on random forest regression. In: Mathematical biosciences (2019)

16. [illegible] convolutional networks [illegible] with [illegible]. [illegible] (2015)
17. Kipf, T.N., Welling, M.: Semi-supervised classification with graph convolutional networks. arXiv preprint arXiv:1609.02907 (2016)
18. Lee, [illegible] binary classification [illegible] imbalanced [illegible] (2008)
19. Li, [illegible] Distributed [illegible] for [illegible] social [illegible] (2020)
20. [illegible] (2021)
21. [illegible] attacks [illegible] (2018)
22. [illegible] (2020)
23. [illegible] (2021)
24. [illegible] arXiv:1905.10 [illegible] (2019)
25. [illegible] graph [illegible] network [illegible] (2021)
26. [illegible] health records [illegible] (2019)
27. [illegible] (2019)
28. [illegible] graph [illegible] arXiv [illegible] (2019)
29. [illegible] (2019)
30. [illegible] quantum [illegible] (2016)
31. [illegible] (2018)
32. [illegible] (2018)
33. [illegible] arXiv [illegible] (2018)
34. [illegible] Community [illegible] networks with node attributes [illegible] (2013)
35. [illegible] (2019)
36. [illegible] random forest regression [illegible] (2019)

Multimedia and Multimodal Data

An Extended Variational Mode Decomposition Algorithm Developed Speech Emotion Recognition Performance

David Hason Rudd[1], Huan Huo[1(✉)], and Guandong Xu[1,2(✉)]

[1] The University of Technology Sydney, 15 Broadway, Ultimo, Australia
{david.hasonrudd,huan.huo,guandong.xu}@uts.edu.au
[2] Data Science Institute, 15 Broadway, Ultimo, Australia

Abstract. Emotion recognition (ER) from speech signals is a robust approach since it cannot be imitated like facial expression or text based sentiment analysis. Valuable information underlying the emotions are significant for human-computer interactions enabling intelligent machines to interact with sensitivity in the real world. Previous ER studies through speech signal processing have focused exclusively on associations between different signal mode decomposition methods and hidden informative features. However, improper decomposition parameter selections lead to informative signal component losses due to mode duplicating and mixing. In contrast, the current study proposes VGG-optiVMD, an empowered variational mode decomposition algorithm, to distinguish meaningful speech features and automatically select the number of decomposed modes and optimum balancing parameter for the data fidelity constraint by assessing their effects on the VGG16 flattening output layer. Various feature vectors were employed to train the VGG16 network on different databases and assess VGG-optiVMD reproducibility and reliability. One, two, and three-dimensional feature vectors were constructed by concatenating Mel-frequency cepstral coefficients, Chromagram, Mel spectrograms, Tonnetz diagrams, and spectral centroids. Results confirmed a synergistic relationship between the fine-tuning of the signal sample rate and decomposition parameters with classification accuracy, achieving state-of-the-art 96.09% accuracy in predicting seven emotions on the Berlin EMO-DB database.

Keywords: Speech emotion recognition (SER) · Variational mode decomposition (VMD) · Sound signal processing · Convolutional neural network (CNN) · Acoustic features

1 Introduction

Word meaning is often conveyed by the tone of voice, although human emotions are not solely conveyed through the words used, but also through by modifying facial expressions and vocal tone. Thus, changing voice characteristics is how

H. Kashima et al. (Eds.): PAKDD 2023, LNAI 13937, pp. 219–231, 2023.
https://doi.org/10.1007/978-3-031-33380-4_17

most humans express different emotions [25]. Consequently, considerable human-computer interaction research has analyzed speech signal emotion recognition (ER) where using other popular semantic analysis methods like wav2vec2.0 [5] are not trustworthy. Several applications employed variational mode decomposition (VMD) [11] in different fields such as medical science, structural engineering, and sound engineering [2,17,23]. Signal based ER employs various instantaneous signals, including electrodermal activity, blood volume pulse, galvanic skin response, electrocardiogram (ECG), Electroencephalography (EEG), and speech, are commonly categorized into several decomposed modes due to the complexity and nonstationary nature of them, which allows latent factors and patterns to be extracted more easily. Nonstationary signal properties and its components make mean short time Fourier transform (STFTs) are not always suitable, and previous studies have mostly considered these approaches in isolation [8]. VMD decomposes signals into modes with a narrowband around a center frequency; it can overcome STFT limitation and EMD mode mixing effects. Therefore, we were motivated to apply VMD for speech signal processing.

Acoustic feature selection is essential for SER to describe various voice signal aspects captured from different features [6]. Acoustic features include time-frequency, time, and frequency domain representations. Extracted features from time-frequency domains carry more informative data than the other domains, and better capture latent emotion content from speech signals [28]. Several previous studies used VMD method to analyze signals, extracting features from the decomposed signals. However, we proposed VGG-optiVMD, utilizing a VMD based feature augmentation method to enrich predictors and maximize emotion classification accuracy. Results from the proposed VGG-optiVMD approach on several common publicly available databases confirm significant ER improvement compared with previous approaches. The main contributions from this study can be summarized as follows.

- To our best knowledge, this study is the first to employ VMD as a dynamic acoustic feature augmentation method for SER performance.
- The proposed VGG-optiVMD algorithm automatically selects optimum decomposition parameters for VMD.
- A robust classification accuracy was achieved with a state-of-art result 96.09%.

2 Related Works

Dendukuri et al. [10] decomposed the speech signal into three components sampling 16000 Hz over 20 ms frames, then input various mode central frequency statistical parameters to a support vector machine (SVM) classifier. Lal et al. [20] empirically demonstrated VMD advantages to decompose speech signals in the correct central frequency and subsequently estimated epoch locations from noise degraded emotional speech signal. Zhang et al. [33] proposed multidimensional feature extraction for EEG signal emotion recognition combining wavelet packet decomposition (WPD) with VMD to break down an EEG signals and extract wavelet packet entropy, modified multiscale sample entropy, fractal dimension, and first difference of each emotional variational mode functions as feature components. They subsequently demonstrated robust results using a random forest (RF) classifier on the DEAP dataset [18]. Khare et al. [17] reduced reconstruction error using meta-heuristic techniques to condensing from 16 to 1 dimension using eigenvector centrality method channel selection on EEG signals. They subsequently improved Optimized variational mode decomposition (O-VMD) accuracy by 5% compared with traditional VMD on the dataset of four emotions that built by themselves.

Pandey [24] proposed subject-independent emotion recognition using VMD and deep neural networks (VMD-DNN) on the benchmark DEAP dataset. Two features, first difference and power-spectral-density used since were sufficient to recognize calm, happy, sad, and angry emotions. SVM and DNN classifier accuracy was improved by employing VMD based feature extraction.

Several previous studies considered STFT signal decomposition techniques for SER. Few previous studies employed VMD to analyze speech signals mainly processing EEG signals through VMD for ER. To the best of our knowledge, the current study is the first to employ VMD to enrich multidimensional feature vectors to enhance VGG-16 network learning.

3 Proposed Methodology

The main aim for decomposition-based speech signal processing via VMD method is to constrain noise and interference frequencies to enhance signal data decoding.

3.1 Variational Mode Decomposition

Variational mode decomposition is a popular technique for decomposing non-stationary signals into sub-signals or modes, where mode contains a specific meaningful property from the original signal in a narrow bandwidth around the center frequency. The VMD adaptive algorithm reduces the original signal complexity [11]. The VMD algorithm applies the Wiener filter, Hilbert transform, analytical signals, and frequency mixing. The two main VMD objects are to constrain the bandwidth for each IMF center frequency and reconstruct the original signal from the sum of all modes. First, the Hilbert transform filters frequencies on the negative side of the spectrum, and then shifts the obtained bandwidth to the modes central frequency. Second, the obtained spectrum is shifted to the baseband region via a modulator function to obtain bandwidth around central frequency ω. Finally, H1 Gaussian smoothness for the demodulation signal is used to estimate the bandwidth. Thus, constraining the L^2 norm squared gradient [11] defines the optimization problem (1),

$$\min_{\{g_k\},\{\omega_k\}} \left\{ \sum_{k=1}^{K} \left\| \frac{\partial}{\partial_t} \left[\left(\delta(t) + \frac{j}{\pi t}\right) * g_k(t) \right] e^{-j\omega_k t} \right\|_2^2 \right\}, \quad \text{subject to:} \quad \sum_{k=1}^{K} g_k(t) = g(t), \tag{1}$$

where the partial derivative $\frac{\partial}{\partial_t}[.]$ minimizes variation in the obtained bandwidth; $g(t)$ is the original speech signal frame; $g_k(t)$ is the kth mode for $g(t)$; K is the total number of modes; $\omega_k = \{w1, \ldots, wk\}$ is the mode center frequency, and a convenient way to reference the center frequencies for the set of K modes; $e^{-j\omega_k t}$ is a modulator function to shift the spectrum for each mode to the baseband.

The analytical signal generated by applying the Hilbert transform $\frac{j}{\pi t}$ and unit impulse function $\delta(t)$ as shown in equation (1). The $\delta(t)$ denotes to the Dirac delta distribution known as a unit impulse so that its value is zero everywhere and infinite at original signal. The original voice signal can be reproduced by solving the constraint optimization (1), which can be simplified using an augmented Lagrangian multiplier to transform it into an unconstrained problem (2),

$$\begin{aligned}\mathcal{L}\left(g_k, \omega_k, \lambda\right) := \alpha \textstyle\sum_{k=1}^{K} \left\| \frac{\partial}{\partial_t}\left[\left(\left(\delta(t) + \frac{j}{\pi t}\right) * g_k(t)\right) e^{-j\omega_k t}\right] \right\|^2 \\ + \left\| g(t) - \textstyle\sum_{k=1}^{K} g_k(t) \right\|_2^2 + \left\langle \lambda(t), g(t) - \textstyle\sum_{k=1}^{K} g_k(t) \right\rangle,\end{aligned} \tag{2}$$

where, λ is a time-dependent Lagrangian multiplier, and α is a bandwidth control parameter. The unconstrained Lagrangian problem (2) can be solved to obtain the frequency and the modes using the alternate direction method of multipliers (ADMM) [11,14,27] optimization in spectral domain. However, optimization outcomes are the same for the frequency and time domains. Hence, mode $g_k(\omega)$ can be updated in the spectral domain,

$$\hat{g}_k^{n+1}(\omega) \leftarrow \frac{\hat{g}(\omega) - \sum_{i<k} \hat{g}_i^{n+1}(\omega) - \sum_{i>k} \hat{g}_i^{n}(\omega) + \frac{\hat{\lambda}^n(\omega)}{2}}{1 + 2\alpha\left(\omega - \omega_k^n\right)^2}. \tag{3}$$

Updating is obtained using the Wiener filter for the current residual using the signal prior $1/(\omega - \omega_k)^2$ to restrain variation across the central frequency minimum, providing the updated mode center frequency ω_k as

$$\hat{\omega}_k^{n+1} = \frac{\int_0^\infty \omega \left|\hat{G}_k(\omega)\right|^2 \, d\omega}{\int_0^\infty \left|\hat{G}_k(\omega)\right|^2 \, d\omega} \tag{4}$$

where $\hat{G}_k(\omega)$ is the Fourier transformed for $g_k^{n+1}(t)$. A better decomposed signal can be obtained by reconstructing the original signal as the sum of modes and estimating bandwidth using the Wiener filter. Details of the VMD algorithm are provided in [11]. To leverage VMD effectiveness, we proposed the VGG-optiVMD algorithm for automatically selecting optimum α and K by analyzing different decomposition parameter effects on classification accuracy.

3.2 Proposed VGG-optiVMD

Reconstruction error for a decomposed signal can be reduced by selecting optimum K and α. Improper decomposition parameter selection will create duplicate modes, causing signal information losses consequently reduced classifier performance.

Algorithm 1. Proposed VGG-optiVMD algorithm

Input: $g(t)$ is a preprocessed speech signal converted to feature vectors.
Output: Decomposes of signal $g(t)$ and Optimum value of α and K
Initialization: The value of modes K and α;
the tolerance of convergence criterion τ; $\{\hat{g}_k^1\}, \{\hat{\omega}_k^1\}, \hat{\lambda}^1$; $n = 0$
Repeat:
1: $n = n + 1$,
2: **for** k=1 : K **do**
3: update $\hat{g}_k$ for all $\omega \geq 0$ by Eq. (3) and ω_k by Eq. (4)
4: **end for**
5: Upgrade the Lagrangian multiplier λ for the dual accent $\forall\omega 0$:

$$\lambda^n(\omega) = \lambda^n + \tau(g(\omega) - \sum_k g_k^{n+1}(\omega))$$

Until:
6: convergence: $\sum_{k=1}^{K} \|\hat{g}_k^{n+1} - \hat{g}_k^n\|_2^2 / \|\hat{g}_k^n\|_2^2 < \in$.
7: **return** $\{g_1(t), g_2(t), \ldots, g_K(t)\}$= IMFs; subtract of all sub signals
8: Set Parameters τ=0; DC=0; init=1; tol=1e-9; K=2; α=2000
9: Decompose signal $g(t)$
10: Record training set accuracy, and F1 score in VGG16 classifier.
11: **while** max(ACC) do
12: **if** ACC==max; $\alpha \leq 6000$; $K \leq 8$ **then**
13: The optimum value of K and α is obtained.
14: **else**
$K = K + 1$; α=α+1000 go to step 9
15: **end if**
16: **end while**
17: Identify optimum value of decomposition parameters α and K while tol=1e-9, DC=0, init=1, and τ=0

One drawback for VMD is that finding decomposition parameters K and α to provide optimum performance challenging. In contrast, in our method we automate optimum VMD decomposition parameter selection using a feedback loop from the VGG16 flattening output layer. Algorithm 1 shows the proposed optimized VMD algorithm (VGG-optiVMD). The key strength for VGG-optiVMD is generality and reproducibility across different databases for real-world multimedia applications, e.g., ER for customer satisfaction analysis.

3.3 Feature Extraction, Data Augmentation, and Classification

Essential and informative acoustic features in the time-frequency domain include the Mel spectrogram, chromograms, spectral contrasts, tonnetz, and Mel-frequency cepstral coefficients (MFCCs) [1,13] are extracted and subsequently employed in various combinations to generate multidimensional feature vectors. Figure 1 shows the proposed framework to train CNN-VGG16 [29] to extract enriched feature vectors and classify seven emotions: anger, boredom, happy, neutral, disgust, sadness, and fear on two databases EMODB [7] and RAVDESS [21].

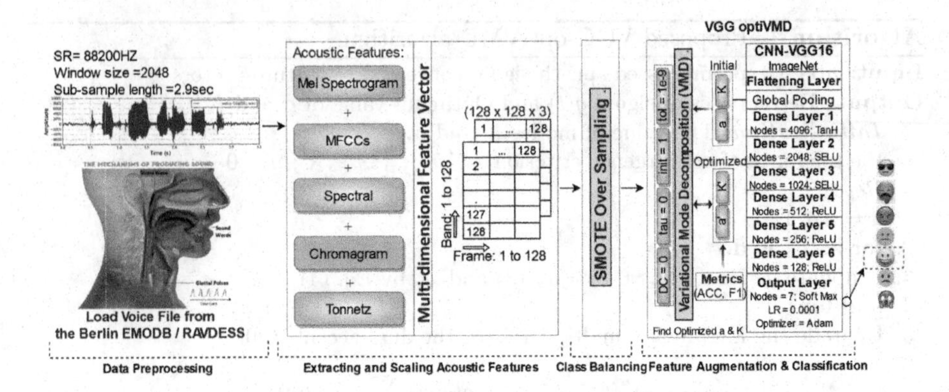

Fig. 1. Proposed model development workflow: extracted features are enriched using the VGG-optiVMD to automatically identify K and α.

Figure 1 shows the model development proceeds. First, the voice signal is sampled 88400 Hz and five well-known acoustic features extracted and reshaped into a single ($128 \times 128 \times 3$) feature vector and second the SMOTE [21] oversampling strategy is applied to compensate for minority classes and reduce model bias. Furthermore, the testing and training features are randomly partitioned into 20% and 80% sets, respectively. Subsequently, the proposed VGG-optiVMD algorithm is applied to decode frequency statistical properties at specific times that distinguish emotions within the feature vector. Finally, the VGG network is trained on the augmented feature vector to classify emotions into seven classes.

4 Experiment Setup

Several experiments were performed on nine different feature vectors to identify the proposed VGG-optiVMD algorithm effectiveness using. The details of network implementations are available in our GitHub repository[1].

4.1 Modelling

The aim of modeling was to enhance informative data within the feature vectors and avoid overfitting. Augmentation effects on classification accuracy were assessed using diverse K and α sets. Optimal K and α was assessed iteratively until robust classification accuracy was achieved or the break loop condition reached. K and α were set to a wide range of 3–8 and 1000–6000, respectively, based on empirical experiments since there was no significant improvement in prediction accuracy outside those ranges. The VGG16 architecture used the ADAM optimizer with learning rate = 0.0001; six fully connected hidden layers with ReLU, SELU, and TanH activation functions; epochs = 50, batch size = 4; and SoftMax function for the output layer.

[1] https://github.com/DavidHason/VGG-optiVMD.

5 Result and Discussion

To assess the effectiveness of our VMD-based feature augmentation method several evaluation metrics were employed including F1 score, training set accuracy, and confusion matrix. Analyzing the results of the baseline model, which is built with the same framework simply without VMD-based feature vector augmentation, helps us to justify the power of the VGG-optiVMD in SER. Therefore, we attempted to evaluate the model performance through variation of sample rate, window size, K and α without using VMD (baseline model) and with VMD (proposed model). As shown in Fig. 2, unlike the baseline model, the proposed model performed better with a larger sampling rate and window size. Moreover, the highest train set accuracy and F1 score were obtained via VGG-optiVMD, proving that our VMD-based feature augmentation method significantly improved the classification accuracy.

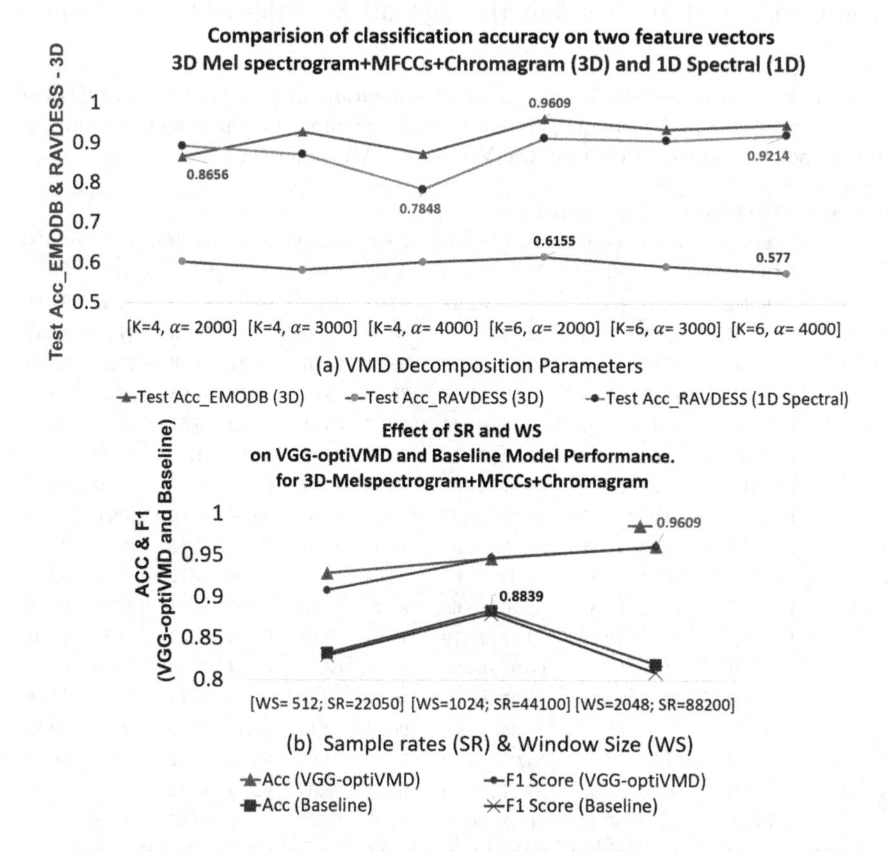

Fig. 2. The model performance is assessed by different signal sampling rates and VMD parameters K and α. Graph (a) The VGG-optiVMD identified the set of $K = 6$ and $\alpha = 2000$ as optimum value. Graph(b) represents the effect of various ranges of sample rate and window size on the proposed and baseline model in EMODB. The highest accuracy can be achieved by SR = 88200 and WS = 2048.

Based on the experiment results shown in Table 1, there is a correlation between the number of modes K, bandwidth control parameter α and classification accuracy. The different acoustic features are enriched with various sets of decomposition parameters. Results indicated that higher accuracy was obtained for K (4–6) and α (2000– 4000) in both datasets, although VGG-optiVMD is set to a limited range of α (1000–10000) and K (2–8) due to increasing a heavy computational load when K value is over 8 with sample rate 88400. This limitation can be considered a functional constraint of VGG-optiVMD. Nevertheless, a state-of-the-art result was achieved with the accuracy of 96.09% with K=6 and α=2000 as demonstrated in Table 1. The Fig. 3 shows the efficient functionality of VGG-optiVMD on the feature vector 3D-Mel Spectrogram+MFCCs+Chromagram. Figure (a) represents the feature before applying VMD based data augmentation, and figure (b) clearly shows that the informative frequencies are distinguished on the feature vector by acquiring higher distinction energies represented in time-frequency domain after applying the data augmentation method. Therefore, the implications of this finding can improve

Table 1. Empirical results (%) of emotion classification accuracy (ACC) and F1-score (F1) are demonstrated through different sets of decomposition parameters α and K, that are selected automatically by the VGG-optiVMD algorithm.

Features:	VMD Decomposition Parameters										
	Databases	K=4, α=2000		K=4, α=4000		K=6, α=2000		K=6, α=3000		K=6, α=4000	
	EMO/RAV	Acc	F1	Acc	F1	Acc	F1	Acc	F1	Acc	F1
CH	EMODB	68.54	68.37	81.63	81.47	94.05	94.88	94.90	91.10	95.41	95.11
	RAVDESS	70.23	70.55	82.73	82.96	85.21	85.92	79.81	79.79	47.49	46.53
MS	EMODB	91.84	91.86	93.15	93.07	95.19	95.07	95.34	94.98	95.92	94.89
	RAVDESS	64.21	64.69	71.36	71.55	75.28	75.95	84.19	84.68	87.25	88.11
MF	EMODB	48.1	46.92	65.16	64.42	64.87	65.18	56.12	56.57	67.64	66.9
	RAVDESS	42.64	41.77	53.29	52.14	55.61	56.80	51.81	51.44	41.86	40.46
SP	EMODB	94.27	93.11	93.01	92.95	93.88	93.07	93.44	93.37	94.02	93.87
	RAVDESS	89.25	90.11	78.48	79.21	91.28	92.88	90.70	90.10	**92.14**	93.55
TZ	EMODB	74.93	75.11	91.25	90.89	88.92	88.91	91.84	91.12	92.44	92.10
	RAVDESS	48.21	48.26	51.04	51.67	52.07	52.12	49.06	49.12	51.98	52.23
MS+SP	EMODB	89.62	90.85	88.76	89.08	88.2	88.13	95.92	96.11	95.41	95.12
	RAVDESS	78.33	78.12	74.37	74.79	78.52	78.78	81.38	81.42	81.84	81.91
MF+SP	EMODB	58.1	58.2	66.91	66.98	65.16	65.11	62.54	62.13	67.64	67.21
	RAVDESS	53.08	53.12	56.25	56.68	60.28	60.94	58.21	58.14	54.7	54.06
MF+CH	EMODB	85.21	85.2	84.35	84.36	90.14	90.13	87.41	87.52	90.82	90.82
	RAVDESS	51.29	51.35	54.25	54.89	53.65	54.66	55.13	55.12	56.08	56.84
M+M+C	EMODB	86.56	86.42	87.41	87.35	**96.09**	96.04	93.54	93.42	94.73	95.98
	RAVDESS	60.28	60.11	60.28	60.84	61.55	62.36	59.25	60.87	57.70	57.56

Features abbreviation: M+M+C: 3D-Mel Spectrogram+MFCCs+Chromagram; MS+SP: 2D-Mel Spectrogram+Spectral; CH: Chromagram; MF: MFCCs; TZ: 1D-Tonnetz;
The best results on both databases are indicated in bold font.

Table 2. Visualization of the model performance with confusion matrix (%) for the 3D-Mel Spectrogram+MFCCs+Chromagram with test accuracy = %96.09 on the Berlin EMO-DB dataset.

Emotion:	Anger	Boredom	Disgust	Fear	Happiness	Neutral	Sadness
Anger	**95.24**	0	0	0	4.76	0	0
Boredom	0	**95.24**	0	0	0	0	4.76
Disgust	0	0	**100.00**	0	8	0	0
Fear	0	0	0	**94.05**	0	0	0
Happiness	8.33	0	0	0	**91.67**	0	0
Neutral	0	2.38	0	0	1.19	**96.43**	0
Sadness	0	0	0	0	0	0	**100**

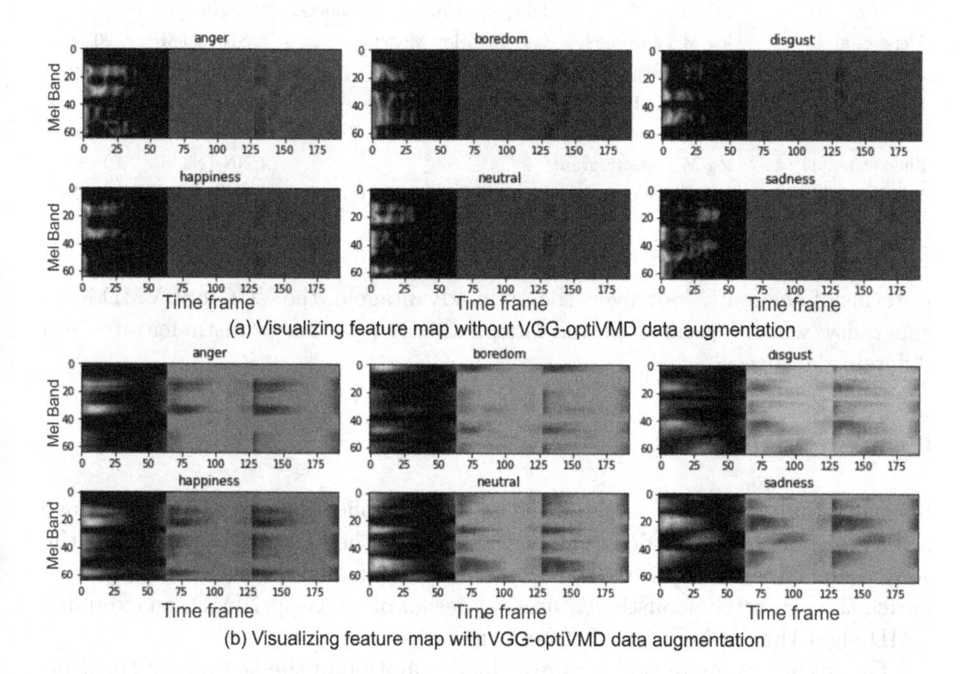

Fig. 3. The efficient functionality of VGG-optiVMD on the feature vector 3D-Mel Spectrogram+MFCCs+Chromagram clearly shows a higher distinction in the energy magnitude of frequencies in (b).

the learning process in VGG16 and result in better prediction accuracy. The confusion matrix in Table 2 demonstrates the high performance of the classification model with accuracy above 90% for all classes. Nevertheless, the model performs poorly when predicting happiness and anger emotions due to the similarity of signal attributes such as intensity, frequency and harmonic structure. The VGG-optiVMD method is compared with the most recent works, shown in Table 3, that our method outperforms previous models and achieves a state-of-the-art result

Table 3. Comparison of the proposed method with previous works on the EMODB and RAVDESS databases.

Method proposed by	Feature extraction strategy	Learning Net.	Acc(%)
Badshah et al. [3]	log Mel spectrogram	CNN	52
Dendukuri et al. [10]	45d- Mode statistical+MFCCs+Spectral	SVM-VMD	61.2
Zamil et al. [32]	13 MFCCs	Tree Model	70
Popova et al. [26]	Mel spectrograms	VGG16	71
Hajarol. et al. [12]	Mel spectrograms+MFCCs	CNN	72.21
Wang et al. [30]	Fourier Parameter+MFCCs	SVM	73.3
Kown et al. [19]	Spectrogram	Deep SCNN	79.50
Badsha et al. [4]	Spectrogram	CNN	80.79
Huang et al. [15]	Spectrogram	CNN	85.2
Issa et al. [16]	MFCCs+Chroma.+Mel spec.+Contrast+Tonnetz	VGG16	86.10
Meng et al. [22]	log Mel spec.+1st & 2nd delta(log Mel spec.)	CNN-LSTM	90.78
Wu et al. [31]	Modulation Spectral Features (MSFs)	SVM	91.60
Rudd et al. [28]	Harmonic-Percussive (HP)+log Mel spec	VGG16-MLP	92.79
Demircan et al. [9]	LPC+MFCCs	SVM	92.86
Zhao et al. [34]	log Mel spectrogram	CNN-LSTM	95.89
VGG-optiVMD	**3D-Mel spectrogram+MFCCs+Chromagram**	**VGG16-VMD**	**96.09**

in terms of accuracy. Moreover, the main advantage of the VGG-optiVMD is its generality, which can be employed independently for other acoustic features and different databases.

6 Conclusion

Speech signal processing is employed in some applications when we only have access to speech voice to detect emotions which is the first aim of this study, the second aim of this study is to introduce specific data augmentation techniques to enrich the extracted acoustic features by design of VGG-optiVMD, an extended VMD algorithm to improve SER performance.

The findings provide solid empirical confirmation of the key role of the sampling rate, the number of the decomposed mode, K and the balancing parameter of the data-fidelity constraint, α, in the performance of the emotion classifier. Taken together, these findings suggest that VMD decomposition parameters K (2–6) and α (2000–6000) are and EMODB databases. The proposed VGG-optiVMD algorithm improved the emotion classification to a state-of-the-art result with a test accuracy of 96.09% in the Berlin EMO-DB and 86.21% in the RAVDESS datasets. Further work needs to be done to establish whether extracting acoustic features only from informative decomposed modes can reduce computational load constraints. Therefore, the study should be repeated using the VMD algorithm before acoustic feature extraction process.

Acknowledgement. This work is partially supported by the Australian Research Council under grant number: DP22010371, LE220100078, DP200101374 and LP1701 00891

References

1. Aizawa, Kiyoharu, Nakamura, Yuichi, Satoh, Shin'ichi (eds.): PCM 2004. LNCS, vol. 3331. Springer, Heidelberg (2005). https://doi.org/10.1007/b104114
2. Alshamsi, H., Kepuska, V., Alshamsi, H., Meng, H.: Automated facial expression and speech emotion recognition app development on smart phones using cloud computing. In: 2018 IEEE 9th Annual Information Technology, Electronics and Mobile Communication Conference (IEMCON), pp. 730–738. IEEE (2018)
3. Badshah, A.M., Ahmad, J., Rahim, N., Baik, S.W.: Speech emotion recognition from spectrograms with deep convolutional neural network. In: 2017 International Conference on Platform Technology and Service (PlatCon), pp. 1–5 (2017)
4. Badshah, A.M., Rahim, N.: Ullah: Deep features-based speech emotion recognition for smart affective services. Multimedia Tools and Applications **78**(5), 5571–5589 (2019)
5. Baevski, A., Zhou, Y., Mohamed, A., Auli, M.: wav2vec 2.0: A framework for self-supervised learning of speech representations. In: Advances in Neural Information Processing Systems, vol. 33, pp. 12449–12460 (2020)
6. Basharirad, B., Moradhaseli, M.: Speech emotion recognition methods: A literature review. In: AIP Conference Proceedings, vol. 1891, p. 020105. AIP Publishing LLC (2017)
7. Burkhardt, F., Paeschke, A., Rolfes, M., Sendlmeier, W.F., Weiss, B., et al.: A database of german emotional speech. In: Interspeech. vol. 5, pp. 1517–1520 (2005)
8. Carvalho, V.R., Moraes, M.F., Braga, A.P., Mendes, E.M.: Evaluating five different adaptive decomposition methods for eeg signal seizure detection and classification. Biomed. Signal Process. Control **62**, 102073 (2020)
9. Demircan, S., Kahramanli, H.: Application of fuzzy c-means clustering algorithm to spectral features for emotion classification from speech. Neural Comput. Appl. **29**(8), 59–66 (2018)
10. Dendukuri, L.S., Hussain, S.J.: Emotional speech analysis and classification using variational mode decomposition. Int. J. Speech Technol, pp. 1–13 (2022)
11. Dragomiretskiy, K., Zosso, D.: Variational mode decomposition. IEEE Trans. Signal Process. **62**(3), 531–544 (2013)
12. Hajarolasvadi, N., Demirel, H.: 3d cnn-based speech emotion recognition using k-means clustering and spectrograms. Entropy **21**(5), 479–495 (2019)
13. Harte, C., Sandler, M., Gasser, M.: Detecting harmonic change in musical audio. In: Proceedings of the 1st ACM Workshop on Audio and Music Computing Multimedia, pp. 21–26 (2006)
14. Hestenes, M.R.: Multiplier and gradient methods. J. Optim. Theory Appl. **4**(5), 303–320 (1969)
15. Huang, Z., Dong, M., Mao, Q., Zhan, Y.: Speech emotion recognition using cnn. In: Proceedings of the 22nd ACM International Conference Media, pp. 801–804 (2014)
16. Issa, D., Demirci, M.F., Yazici, A.: Speech emotion recognition with deep convolutional neural networks. Biomed. Signal Process. Control **59**, 101894–101904 (2020)

17. Khare, S.K., Bajaj, V.: An evolutionary optimized variational mode decomposition for emotion recognition. IEEE Sens. J. **21**(2), 2035–2042 (2020)
18. Koelstra, S., Kolestra, S., et al.: Deap: a database for emotion analysis; using physiological signals. IEEE Trans. Affect. Comput. **3**(1), 18–31 (2011)
19. Kwon, S.: A cnn-assisted enhanced audio signal processing for speech emotion recognition. Sensors **20**(1), 183 (2019)
20. Lal, G.J., Gopalakrishnan, E., Govind, D.: Epoch estimation from emotional speech signals using variational mode decomposition. Circ. Syst. Signal Process. **37**(8), 3245–3274 (2018)
21. Livingstone, S.R., Russo, F.A.: The ryerson audio-visual database of emotional speech and song (ravdess): a dynamic, multimodal set of facial and vocal expressions in north american english. PLoS ONE **13**(5), e0196391 (2018)
22. Meng, H., Yan, T., Yuan, F., Wei, H.: Speech emotion recognition from 3d log-mel spectrograms with deep learning network. IEEE access **7**, 125868–125881 (2019)
23. Mousavi, M., Gandomi, A.H.: Structural health monitoring under environmental and operational variations using mcd prediction error. J. Sound Vib. **512**, 116370 (2021)
24. Pandey, P., Seeja, K.: Subject independent emotion recognition from eeg using vmd and deep learning. J. King Saud University-Comput. Inform. Sci. **34**(4), 1730–1738 (2019)
25. Pierre-Yves, O.: The production and recognition of emotions in speech: features and algorithms. Int. J. Hum Comput Stud. **59**(1–2), 157–183 (2003)
26. Popova, A.S., Rassadin, A.G., Ponomarenko, A.A.: Emotion recognition in sound. In: International Conference on Neuroinformatics, pp. 117–124 (2017)
27. Rockafellar, R.T.: A dual approach to solving nonlinear programming problems by unconstrained optimization. Math. Program. **5**(1), 354–373 (1973)
28. Rudd, D.H., Huo, H., Xu, G.: Leveraged mel spectrograms using harmonic and percussive components in speech emotion recognition. In: Pacific-Asia Conference on Knowledge Discovery and Data Mining, pp. 392–404. Springer (2022). https://doi.org/10.1007/978-3-031-05936-0_31
29. Russakovsky, O., Russakovsky, O., et al.: Imagenet large scale visual recognition challenge. Int. J. Comput. Vision **115**(3), 211–252 (2015)
30. Wang, K., An, N., Li, B.N., Zhang, Y., Li, L.: Speech emotion recognition using fourier parameters. IEEE Trans. Affect. Comput. **6**(1), 69–75 (2015)
31. Wu, S., Falk, T.H., Chan, W.Y.: Automatic speech emotion recognition using modulation spectral features. Speech Commun. **53**(5), 768–785 (2011)
32. Zamil, A.A.A., Hasan, S., Baki, S.M.J., Adam, J.M., Zaman, I.: Emotion detection from speech signals using voting mechanism on classified frames. In: 2019 International Conference on Robotics, Electrical and Signal Processing Techniques (ICREST), pp. 281–285. IEEE (2019)
33. Zhang, M., Hu, B., Zheng, X., Li, T.: A novel multidimensional feature extraction method based on vmd and wpd for emotion recognition. In: 2020 IEEE International Conference on Bioinformatics and Biomedicine (BIBM), pp. 1216–1220. IEEE (2020)
34. Zhao, J., Mao, X., Chen, L.: Speech emotion recognition using deep 1d & 2d cnn lstm networks. Biomed. Signal Process. Control **47**, 312–323 (2019)

Dynamically-Scaled Deep Canonical Correlation Analysis

Tomer Friedlander(✉) and Lior Wolf

Tel Aviv University, Tel Aviv, Israel
tomerf1@mail.tau.ac.il, wolf@cs.tau.ac.il

Abstract. Canonical Correlation Analysis (CCA) is a method for feature extraction of two views by finding maximally correlated linear projections of them. Several variants of CCA have been introduced in the literature, in particular, variants based on deep neural networks for learning highly correlated nonlinear transformations of two views. As these models are parameterized conventionally, their learnable parameters remain independent of the inputs after the training process, which limits their capacity for learning highly correlated representations. We introduce a novel dynamic scaling method for an input-dependent canonical correlation model. In our deep-CCA models, the parameters of the last layer are scaled by a second neural network that is conditioned on the model's input, resulting in a parameterization that is dependent on the input samples. We evaluate our model on multiple datasets and demonstrate that the learned representations are more correlated in comparison to the conventionally-parameterized CCA-based models and also obtain preferable retrieval results.

Keywords: Multimodal learning · Information retrieval · CCA

1 Introduction

Given two domains, the goal of CCA methods [2,20] is to recover highly correlated projections between them. The output of such methods is pairs of 1D projections, where each pair contains a single projection of each domain. Collectively, all projections from a single domain are uncorrelated, similarly to PCA.

All existing CCA models are parameterized by conventional static parameters, i.e. their learnable parameters remain independent of the inputs after being optimized in the training process. In this work we propose to apply dynamic scaling, i.e. scaling the parameterization of the feature extractors based on the specific inputs. Such dynamic scaling is able to increase the expressiveness of the model in a way that adjusts the learned representations specifically to the inputted paired views, resulting in more correlated representations. Importantly, to remain faithful to the CCA line of work, the representation of a vector in the first domain cannot be dependent on the input of the second domain. Each projection (linear or non-linear) has to be performed independently of the other

H. Kashima et al. (Eds.): PAKDD 2023, LNAI 13937, pp. 232–244, 2023.
https://doi.org/10.1007/978-3-031-33380-4_18

domain. Otherwise, the settings of the projection changes, and the complexity of performing, for example, retrieval, become quadratic in the number of samples.

The usage of dynamic scaling is shown to lead to favorable results in comparison to existing CCA methods, both classical [16,17,20] and modern deep CCA approaches [2,4,15,21,23]. Specifically, we observed an increase in total correlation scores across the standard benchmarks of the field. In addition, we were able to surpass the performance of the state-of-the-art CCA-based retrieval model, Ranking CCA [7], on the task of image to text retrieval.

1.1 Related Work

Many extensions of the CCA model [11] have been introduced in the literature [1,2,4,5,15–17,19–21,23]. Regularized CCA [20] employs ridge regression and Kernel CCA (KCCA) [1] applies non-linear transformations using kernel functions. Scalable KCCA versions such as FKCCA and NKCCA [16] employ random Fourier features and the Nyström approximation, respectively. Deep CCA (DCCA) by [2] models the transformation functions using deep neural networks. CorrNet is an encoder-decoder architecture for maximizing the correlation between the projections of two views [3], but it does not compute canonical components. Deep canonically-correlated Autoencoder (DCCAE) [23] was proposed in parallel as an encoder-decoder model that optimizes the CCA formulation together with reconstructing the input views. NCCA [17] is a non-parametric CCA model, which was demonstrated to match the performance of DCCA on some datasets without using neural networks. Soft-CCA [4] replaces the hard decorrelation constraint of the DCCA formulation with a softer constraint. Soft-HGR [21] is a neural framework for optimizing a softer formulation of the Hirschfeld-Gebelein-Reónyi (HGR) maximal correlation, which generalizes Pearson's correlation. CCA [11] can be regarded as optimizing the HGR objective in the restricted case of linear projections. ℓ_0-CCA [15] is a sparse variant of DCCA that multiplies the input views. The gates are static (independent of the inputs) and multiply the inputs and not the parameters of the model.

In the task of *cross-modality retrieval*, there is a source sample of one view and a set of target samples of another view. The goal of this task is to find the matching target sample of the source sample out of the set of all target samples. While retrieval can be successful even if it relies on capturing limited aspects of the data, CCA-based retrieval methods strive to construct a shared embedding space, which captures the maximal correlation of the two views. Deep CCA was used in [26]. Later, [7] proposed Ranking-CCA as an end-to-end CCA-based model that explicitly minimizes the pairwise ranking loss for retrieval and achieves improved retrieval results.

Dynamic Networks [10] are neural networks that adapt their parameters or architecture to the inputs. For example, Hypernetworks [9] use one neural network to generate the weights of another network.

2 Method

We first present the dynamically scaled layer and then discuss its use in Deep-CCA and Ranking-CCA models. For the purpose of parameterizing a layer with input-dependent parameters, we train another neural network, denoted by $\mathbf{h}$. The outputs of this neural network are used as scaling factors for dynamically adjusting the parameters of the conventional layer. We name such a layer a Dynamically-Scaled Layer (DSL). A schematic illutration of a DSL within a neural network is depicted in Fig. 1.

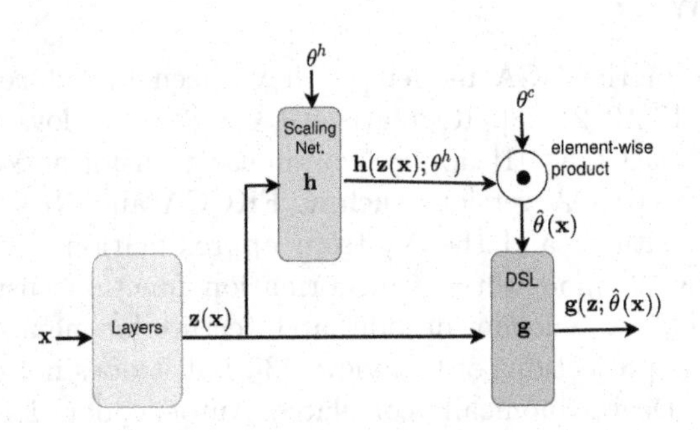

Fig. 1. Schematic illustration of a DSL following other neural network layers. Input vectors and the parameterization are represented by horizontal and vertical lines, respectively.

Suppose a given layer of a neural network is a DSL and it is modelled by the transformation function $\mathbf{g}$. Denote the parameters of this layer after applying the dynamic scaling by $\hat{\boldsymbol{\theta}}$. These parameters are obtained by scaling the conventional parameters of the layer, $\boldsymbol{\theta}^c$, with the outputs of the scaling network, $\mathbf{h}$.

The scaling network $\mathbf{h}$ is formally defined by $\mathbf{h}(\mathbf{z}(\mathbf{x}); \boldsymbol{\theta}^h)$, where $\boldsymbol{\theta}^h$ are the conventional parameters of the scaling network itself. $\mathbf{z}(\mathbf{x})$ is the vector of activations from the previous layer, which serves as the input of the DSL layer $\mathbf{g}$, similarly to the way information is passed between layers, and is the input of the scaling network $\mathbf{h}$. The notation of the original input of the network, denoted by $\mathbf{x}$, is explicitly used in order to emphasize that the vector $\mathbf{z}$ is a function of $\mathbf{x}$. The outputs of the scaling network, $\mathbf{h}$, are the same size as the conventional parameters, $\boldsymbol{\theta}^c$. The scaling network scales the conventional parameters according to the following equation:

$$\hat{\boldsymbol{\theta}}(\mathbf{x}) = \mathbf{h}(\mathbf{z}(\mathbf{x}); \boldsymbol{\theta}^h) \odot \boldsymbol{\theta}^c \tag{1}$$

where $\odot$ is the element-wise product operation. After applying the dynamic scaling, the parameters of $\mathbf{g}$ become input-dependent, i.e. dependent on $\mathbf{x}$. For a concrete example, let $\mathbf{g}$ be a fully-connected layer. Denote by $\mathbf{S_W} \in \mathbb{R}^{d_{in} \times d_{out}}$ and $\mathbf{S}_b \in \mathbb{R}^{d_{out}}$ the outputs of the scaling network for adjusting the weight matrix, $\mathbf{W}$, and the bias vector, $\mathbf{b}$, respectively. The parameters of the fully-connected DSL are computed by the following equation:

$$\hat{\theta}(\mathbf{x}) = (\hat{\mathbf{W}}, \hat{\mathbf{b}}) = (\mathbf{S_W} \odot \mathbf{W}, \mathbf{S_b} \odot \mathbf{b}) \tag{2}$$

The limitation of the DSL is the increased number of parameters and the training time due to the scaling network as a function of the dimension of $\mathbf{z}$ and the total number of parameters to be scaled.

Optimization. The optimization process consists of two phases. Firstly, the warm-up period is performed and the model is optimized similarly to a neural network that is conventionally-parameterized, i.e. the scaling network is not included in the model and $\hat{\boldsymbol{\theta}}(\mathbf{x}) = \boldsymbol{\theta}^c$ is used instead of Eq. 1. The warm-up period allows to obtain a better initialization of the conventional parameters ($\boldsymbol{\theta}^c$) prior to optimizing the scaling network. After a predefined number of T epochs, the scaling network is added to the model and its optimization starts as well. The optimization of the scaling network ($\boldsymbol{\theta}^h$) is done by gradient descent as for any other parameter since the scaling network is a differentiable neural network and the scaling operator is the product operator.

We next move from a single layer to introducing a novel Deep CCA model, whose transformation functions are parameterized by input-dependent parameters. Let $\mathbf{x}_1 \in \mathbb{R}^{n_1}$ and $\mathbf{x}_2 \in \mathbb{R}^{n_2}$ be random vectors of two views. The d-dimensional projections of the first and second views are computed by the transformation functions $\mathbf{f}_1^* : \mathbb{R}^{n_1} \rightarrow \mathbb{R}^d$ and $\mathbf{f}_2^* : \mathbb{R}^{n_2} \rightarrow \mathbb{R}^d$, respectively. Each transformation function $\mathbf{f}_j^*$ consists of a neural network feature extractor ($\mathbf{f}_j$) followed by a projection matrix ($\mathbf{A}_j$), as will be detailed next. In our method, the last layer of each feature extractor is modelled by a DSL, making them parameterized by input-dependent parameters.

Formally, denote the mapping function represented by the first stage of the feature extractor $\mathbf{f}_j$, i.e. without the final DSL, by $\tilde{\mathbf{f}}_j$ and its conventional static parameters by $\tilde{\boldsymbol{\theta}}_j^c$. The output of the first stage of the feature extractor of the jth view for the input $\mathbf{x}_j$ is denoted by $\mathbf{z}_j(\mathbf{x}_j) = \tilde{\mathbf{f}}_j(\mathbf{x}_j; \tilde{\boldsymbol{\theta}}_j^c)$.

The DSL scaling networks $\mathbf{h}_1$ and $\mathbf{h}_2$ with the conventional static parameters $\boldsymbol{\theta}_1^h$ and $\boldsymbol{\theta}_2^h$ are defined for the first and second views, respectively. In addition, the mapping function of the DSL of $\mathbf{f}_j$ and its conventional static parameters are denoted by $\mathbf{g}_j$ and $\boldsymbol{\theta}_j^c$, respectively. The scaling network $\mathbf{h}_j$ scales the corresponding conventional parameters $\boldsymbol{\theta}_j^c$ according to Eq. 1, resulting in the scaled parameters $\hat{\boldsymbol{\theta}}_j(\mathbf{x}_j)$. In total, the extracted features of the jth view are computed by $\mathbf{f}_j(\mathbf{x}_j; \hat{\boldsymbol{\Theta}}_j(\mathbf{x}_j)) = \mathbf{g}_j\big(\mathbf{z}_j; \hat{\boldsymbol{\theta}}_j(\mathbf{x}_j)\big)$, where the total parameterization is denoted by $\hat{\boldsymbol{\Theta}}_j(\mathbf{x}_j) = \big\{\tilde{\boldsymbol{\theta}}_j^c, \hat{\boldsymbol{\theta}}_j(\mathbf{x}_j)\big\}$ and $\mathbf{z}_j$ was omitted for a convenient notation. Finally, the total projection of the jth view to the shared space is computed by projecting the extracted features by $\mathbf{A}_j$ as follows:

$$\mathbf{f}_j^*(\mathbf{x}_j; \hat{\boldsymbol{\Theta}}_j(\mathbf{x}_j)) = \mathbf{A}_j^\top \mathbf{f}_j(\mathbf{x}_j; \hat{\boldsymbol{\Theta}}_j(\mathbf{x}_j)) \tag{3}$$

The projection matrices, $\mathbf{A}_1$ and $\mathbf{A}_2$, ensure that the projected vectors satisfy the constraint that all projections of the same domain are uncorrelated.

We present following extension of the Deep CCA optimization problem, which includes transformation functions with dynamically-scaled parameterization:

$$\begin{aligned} &\underset{\{\Theta_j^c, \theta_j^h, \mathbf{A}_j\}_{j=1}^2}{\text{argmax}} \quad tr\left(cov\Big(\mathbf{f}_1^*(\mathbf{x}_1; \hat{\Theta}_1(\mathbf{x}_1)), \mathbf{f}_2^*(\mathbf{x}_2; \hat{\Theta}_2(\mathbf{x}_2))\Big)\right) \\ &\text{s.t.} \\ &\qquad cov\Big(\mathbf{f}_1^*(\mathbf{x}_1; \hat{\Theta}_1(\mathbf{x}_1)), \mathbf{f}_1^*(\mathbf{x}_1; \hat{\Theta}_1(\mathbf{x}_1))\Big) = \mathbf{I} \\ &\qquad cov\Big(\mathbf{f}_2^*(\mathbf{x}_2; \hat{\Theta}_2(\mathbf{x}_2)), \mathbf{f}_2^*(\mathbf{x}_2; \hat{\Theta}_2(\mathbf{x}_2))\Big) = \mathbf{I} \end{aligned} \tag{4}$$

where tr is the trace operator and cov is the computation of the covariance matrix. The conventional static parameters of the entire $\mathbf{f}_j$ network, without the parameters of the scaling network, are denoted by $\Theta_j^c = \{\tilde{\theta}_j^c, \theta_j^c\}$.

Optimization. For training the Dynamically-Scaled Deep CCA, we follow the optimization process of the original Deep CCA [2], but with our extended formulation of dynamically scaled transformation functions, as described next. Suppose a set of paired instances of the two views is given, i.e. $\{(\mathbf{x}_1^i, \mathbf{x}_2^i)\}_{i=1}^N$, where $\mathbf{x}_1^i \in \mathbb{R}^{n_1}$ and $\mathbf{x}_2^i \in \mathbb{R}^{n_2}$. The obtained extracted features of the views are denoted by $\{\mathbf{f}_1(\mathbf{x}_1^i; \hat{\Theta}_1(\mathbf{x}_1^i))\}_{i=1}^N$ and $\{\mathbf{f}_2(\mathbf{x}_2^i; \hat{\Theta}_2(\mathbf{x}_2^i))\}_{i=1}^N$, respectively. These projections are normalized to have a zero mean with respect to averaging across the sample index i, i.e. $\bar{\mathbf{f}}_j(\mathbf{x}_j^i; \hat{\Theta}_j(\mathbf{x}_j^i)) = \mathbf{f}_j(\mathbf{x}_j^i; \hat{\Theta}_j(\mathbf{x}_j^i)) - \frac{1}{N}\sum_{n=1}^N \mathbf{f}_j(\mathbf{x}_j^n; \hat{\Theta}_j(\mathbf{x}_j^n))$. Let $\bar{\mathbf{F}}_1 \in \mathbb{R}^{d\times N}$ and $\bar{\mathbf{F}}_2 \in \mathbb{R}^{d\times N}$ be matrices, whose columns are $\{\bar{\mathbf{f}}_1(\mathbf{x}_1^i; \hat{\Theta}_1(\mathbf{x}_1^i))\}_{i=1}^N$ and $\{\bar{\mathbf{f}}_2(\mathbf{x}_2^i; \hat{\Theta}_2(\mathbf{x}_2^i))\}_{i=1}^N$ respectively.

The estimated cross-covariance matrix of the extracted features of each view is computed by $\boldsymbol{\Sigma}_{1,2} = \frac{1}{N-1}\bar{\mathbf{F}}_1\bar{\mathbf{F}}_2^T$. Similarly, define the estimated auto-covariance matrices of each view by $\boldsymbol{\Sigma}_{1,1} = \frac{1}{N-1}\bar{\mathbf{F}}_1\bar{\mathbf{F}}_1^T + r_1\mathbf{I}$ and $\boldsymbol{\Sigma}_{2,2} = \frac{1}{N-1}\bar{\mathbf{F}}_2\bar{\mathbf{F}}_2^T + r_2\mathbf{I}$, where $r_1 > 0$ and $r_2 > 0$ are fixed hyperparameters used to ensure that $\boldsymbol{\Sigma}_{1,1}$ and $\boldsymbol{\Sigma}_{2,2}$ are positive definite. The optimization problem from Eq. 4 in terms of the estimated covariance matrices becomes as follows:

$$\begin{aligned} &\underset{\{\Theta_j^c, \theta_j^h, \mathbf{A}_j\}_{j=1}^2}{\text{argmax}} \quad tr(\mathbf{A}_1^\top \boldsymbol{\Sigma}_{1,2}\mathbf{A}_2) \\ &\text{s.t.} \quad \mathbf{A}_1^\top \boldsymbol{\Sigma}_{1,1}\mathbf{A}_1 = \mathbf{I}, \quad \mathbf{A}_2^\top \boldsymbol{\Sigma}_{2,2}\mathbf{A}_2 = \mathbf{I} \end{aligned} \tag{5}$$

Let $\boldsymbol{\Psi} = \boldsymbol{\Sigma}_{1,1}^{-1/2}\boldsymbol{\Sigma}_{1,2}\boldsymbol{\Sigma}_{2,2}^{-1/2}$ and let $\{\epsilon_k\}_{k=1}^d$ be the set of the top d singular values of the matrix $\boldsymbol{\Psi}$. In order to find the optimal parameters of the feature extractors in Eq. 5, we minimize the same loss used in the original Deep CCA formulation [2], given by

$$\mathcal{L}_{dcca} = -\sum_{k=1}^d \epsilon_k \tag{6}$$

The Deep CCA loss can be optimized effectively by a gradient descent optimizer on sufficiently-large mini-batches [22,23], using the gradients that are detailed in [2]. We apply our suggested two-stage optimization scheme. The first stage is the warm-up period, for which our model consists of only conventional parameters ($\boldsymbol{\theta}_1^c$ and $\boldsymbol{\theta}_2^c$). In this stage, these parameters are optimized by minimizing the Deep CCA loss (Eq. 6) using a gradient descent optimizer. After a predefined number T of training iterations, the scaling networks ($\mathbf{h}_1$ and $\mathbf{h}_2$) are added to the model in order to scale the conventional parameters. The learnable parameters of the scaling networks ($\boldsymbol{\theta}_1^h$ and $\boldsymbol{\theta}_2^h$) are optimized simultaneously with the conventional parameters in order to minimize the same loss (Eq. 6).

The optimal projection matrices, $\mathbf{A}_1$ and $\mathbf{A}_2$, are computed only after optimizing the feature extractors, by $\mathbf{A}_1 = \boldsymbol{\Sigma}_{1,1}^{-1/2}\mathbf{U}$ and $\mathbf{A}_2 = \boldsymbol{\Sigma}_{2,2}^{-1/2}\mathbf{V}$, where the matrices $\mathbf{U}$ and $\mathbf{V}$ consist of the left and right singular vectors of $\boldsymbol{\Psi}$, which correspond to $\{\epsilon_k\}_{k=1}^d$.

Finally, we use dynamic scaling for the purpose of improving a CCA-based retrieval and present a dynamically-scaled variant of the Ranking-CCA model [7], which was specifically designed for the cross-modal retrieval task. A successful cross-modality retrieval model has to be able to distinguish well enough between mismatched samples. For this purpose, we propose to input the scaling networks of the DSL with the concatenation of both $\mathbf{z}$ and the original raw input to the network, $\mathbf{x}$. The additional input is able to add more context for learning how to dynamically scale the parameters, such that the resulting embedding vectors are distinguishable. In particular, $\mathbf{h}(\mathbf{z}(\mathbf{x}); \boldsymbol{\theta}^h)$ in Eq. 1 is replaced by $\mathbf{h}([\mathbf{z}(\mathbf{x}), \mathbf{x}]; \boldsymbol{\theta}^h)$. Let $\mathbf{f}_1^*$ and $\mathbf{f}_2^*$ be the dynamically-scaled transformation functions for the first and second views, respectively, as denoted in Eq. 3.

Following the original Ranking-CCA, the feature extractors and the projection matrices are not trained separately as done in Deep CCA, but they are trained concurrently at each training iteration. In particular, for each mini-batch of training samples, the projection matrices are explicitly computed as described above. In addition, instead of minimizing the Deep CCA loss $\mathcal{L}_{dcca}$ for optimizing the feature extractors, [7] suggests to optimize the entire architecture by end-to-end minimization of the pairwise ranking loss. This loss is tailored for the cross-modality retrieval task, by encouraging matching samples to be closer in the embedding space than mismatching samples. Each projection matrix serves as a differnetiable linear layer on top of the feature extractors and allows back-propagation of the gradients of the optimized loss to the previous layers. The pairwise ranking loss is computed as follows:

$$\sum_{i,j \neq i} \mathbb{1}\Big(m - s\big(\mathbf{f}_1^*(\mathbf{x}_1^i), \mathbf{f}_2^*(\mathbf{x}_2^i)\big) + s\big(\mathbf{f}_1^*(\mathbf{x}_1^i), \mathbf{f}_2^*(\mathbf{x}_2^j)\big)\Big) \tag{7}$$

where s is a scoring function of two input vectors, e.g. the cosine similarity, and $\mathbb{1}(u) = max(0, u)$. The hyperparameter m is the margin. Summing over j is done across all mismatching samples of $\mathbf{x}_1^i$ within the batch. A symmetric pairwise loss, is defined by switching each notation of 1 by 2 in Eq. 7, and vice versa. The parameterization is omitted in Eq. 7 in order to simplify the notations, but both $\mathbf{f}_1^*$ and $\mathbf{f}_2^*$ are dynamically-scaled in contrast to the original formulation.

For a given source instance of one view and a set of target instances of the other view, all instances are projected using the corresponding $\mathbf{f}_j^*$. The k closest projected target samples to the projected source sample in terms of the cosine-distance, are chosen as the top-k suggestions by the model.

3 Experiments

In order to evaluate the contribution of dynamic scaling, we perform two sets of experiments. One is focused on the total canonical correlation score, which is often used in the CCA literature. The second set focuses on the retrieval application.

Following the CCA literature, the neural networks consist of fully-connected layers. The number of layers and their widths are detailed in the supplementary material, where we follow [24] for MNIST and XRMB. For a given dataset, we train all neural network-based transformation functions with the same architecture for the same number of training epochs, after being initialized by the same seed. For a fair comparison, models that do not require a two-phase optimization process which includes a warm-up period, are trained for a number of iterations that is equal to the total number of iterations of models that do require two optimization phases. After each training epoch, the loss of the current checkpoint of the model is computed. The learnable parameters that achieve the lowest loss on the validation set are chosen as the parameters of the model. The hyperparameters of each model are tuned by evaluating, on the validation set, the total canonical correlation and recall measurements for the total canonical correlation and retrieval experiments, respectively. Regarding our proposed DS-DCCA and DS-Ranking CCA, we model the conventionally static layers of the mapping functions with neural networks, which are identical to the networks modelling the mapping functions in the network-based baselines. The scaling networks are fully connected layers followed by a batch normalization and a ReLU activation function. See supplementary material for more details in https://tomerfr.github.io/DynamicallyScaledDeepCCA/supplementary.pdf.

3.1 Total Canonical Correlation

We first evaluate our Dynamically-Scaled Deep CCA (DS-DCCA) model with respect to the total canonical correlation score on the common benchmarks of the CCA literature and compare it with several CCA-based models. These experiments directly test the performance of the model with respect to the main objective of the CCA formulation.

Baselines. We compare our proposed DS-DCCA model to well-known CCA-based models from the literature. In particular, methods modeled by classical approaches with no neural networks: (1) regularized CCA [20], (2) FKCCA [16], (3) NKCCA [16], (4) NCCA [17]. In addition, we compare our method to CCA methods, which are modelled by neural-networks: (5) DCCA [2], (6) DCCAE [23], (7) Soft-CCA [4], (8) Soft-HGR [21], (9) ℓ_0-CCA [15].

Datasets. Three datasets were used for evaluation: MNIST, Wisconsin X-Ray Microbeam (XRMB), and Flickr8k **MNIST** [14] consists of 28×28 grayscale images of handwritten digits. We employ a variant of MNIST that is used in the

Table 1. Total correlation of top d canonical components. Means and standard deviations of the test sets obtained in k-folds cross validation are reported, as well as the p-value computed via a paired t-test vs. our DS-DCCA.

Model	MNIST		XRMB		Flickr8k	
	MEAN±STD	P-VALUE	MEAN±STD	P-VALUE	MEAN±STD	P-VALUE
Up Bound (d)	50.00±0.00	-	112.00±0.00	-	128.00 ± 0.00	-
CCA	28.88 ± 0.27	1.76E-12	15.88 ± 0.07	3.34E-13	41.58 ± 0.61	2.07E-13
FKCCA	41.71 ± 0.07	1.78E-12	95.45 ± 0.18	1.04E-09	39.24 ± 0.60	7.31E-13
NKCCA	45.10 ± 0.04	6.45E-11	103.49 ± 0.14	1.10E-08	68.49 ± 1.00	4.31E-11
DCCA	46.76 ± 0.03	2.18E-08	108.73 ± 0.10	1.01E-07	67.65 ± 1.28	2.60E-10
DCCAE	46.75 ± 0.02	1.07E-08	108.71 ± 0.09	3.17E-08	67.75 ± 1.18	1.29E-10
NCCA	40.98 ± 0.12	1.37E-11	107.57 ± 0.18	5.11E-07	57.47 ± 4.07	1.27E-07
Soft-CCA	44.55 ± 0.18	1.38E-08	84.85 ± 1.43	2.50E-06	60.55 ± 1.05	5.92E-11
Soft-HGR	46.86 ± 0.04	1.54E-07	106.12 ± 0.11	5.68E-08	54.41 ± 4.35	1.42E-07
ℓ_0-CCA	47.17 ± 0.03	3.13E-06	108.72 ± 0.11	1.99E-07	72.83 ± 1.11	1.08E-09
DS-DCCA	**47.50 ± 0.03**	-	**110.88 ± 0.06**	-	**86.06 ± 1.28**	-

Table 2. Ablation study - The total correlation (TC) is reported as well as the percent of the remaining gap to the upper bound, which is improved by the full configuration of our DS-DCCA.

Dataset		Upper Bound	DCCA				DS-DCCA			
			Original	Wide 2	Wide 1&2	Global Scale	Scaling Out-puts	w/o θ^c	No Warm-up	Full Config.
MNIST	TC	50.00	46.71	46.90	46.86	46.70	47.22	47.42	47.36	**47.56**
	%	-	26	21	22	26	12	5	8	-
XRMB	TC	112.00	108.69	109.67	109.29	108.63	110.12	110.78	110.47	**110.84**
	%	-	65	50	57	66	39	5	24	-
Flickr8k	TC	128.00	67.51	76.74	76.15	67.65	79.11	80.25	80.65	**85.57**
	%	-	30	17	18	30	13	11	10	-

CCA literature [3,4], which defines the two different views of a given image to be its left and right halves, respectively. [3] splits the dataset to 50k/10k/10k for training/validating/testing. The dimension of the projections is selected to be $d = 50$ as used in the CCA literature. **XRMB** [25] consists of simultaneously recorded speech (112d) and articulatory measurements (273d). [16] splits the dataset to 30k/10k/10k for training/validating/testing. The dimension of the projections is selected to be $d = 112$ as used in the CCA literature. **Flickr8k** [13] is a dataset consisting of 8,000 images from Flickr.com and five textual captions per each image. We encode each image to a 2048d vector by the penultimate layer of ResNet50 [12] pre-trained on ImageNet [6]. After removing stop-words and applying the SpaCy lemmatization, each set of five captions is embedded to a single 300d vector by mean-pooling the Word2Vec [18] embedding vectors.

Table 3. Recall rates for Flickr8k, Flickr30k and IAPR TC-12. Bold and underlined results are first and second places, respectively.

Model	IMG→TXT			TXT→IMG		
	R_1	R_5	R_{10}	R_1	R_5	R_{10}
Flickr8k:						
CCA	27.3	57.0	68.3	24.5	55.5	68.4
FKCCA	17.4	42.9	56.5	17.3	43.6	55.4
NKCCA	27.0	58.5	73.8	26.5	57.8	71.2
DCCA	31.8	65.7	77.7	29.3	62.8	76.1
DCCAE	33.6	65.9	76.6	30.0	63.0	76.0
NCCA	8.0	25.5	34.6	8.5	24.3	34.7
Soft-CCA	28.8	61.6	73.5	27.1	60.1	72.3
Soft-HGR	25.4	55.1	69.0	21.5	54.3	68.8
ℓ_0-CCA	30.7	62.4	76.7	28.5	61.2	74.9
RCCA	33.6	66.9	78.0	31.9	64.6	78.6
DSRCCA						
z only	33.7	68.1	79.9	33.1	67.1	79.5
x only	33.9	69.8	79.8	**34.7**	66.9	80.0
z & **x**	**35.3**	**70.7**	**82.7**	34.4	**68.3**	**82.3**

Model	IMG→TXT			TXT→IMG		
	R_1	R_5	R_{10}	R_1	R_5	R_{10}
Flickr30k:						
RCCA	40.2	72.8	83.2	40.0	70.7	82.7
DSRCCA						
z only	41.3	74.2	82.8	40.7	**73.4**	84.7
x only	**44.1**	74.0	84.1	43.9	73.1	**85.4**
z & **x**	43.9	**75.0**	**84.5**	**44.2**	**73.4**	84.2
IAPR TC$_{12}$:						
RCCA	48.6	81.1	90.0	49.0	80.5	89.1
DSRCCA						
z only	48.9	80.3	90.2	49.5	80.3	89.0
x only	49.1	**82.7**	90.2	**51.1**	81.3	**91.1**
z & **x**	**49.6**	**82.7**	**91.4**	50.0	**82.0**	90.9

Total Canonical Correlation Results. The results for the total correlation scores are presented in Table 1. Each model is trained on the training set to learn a d-dimensional representation of each view. Following prior works [2,17], another regularized linear CCA [20] is trained on the projected training samples of each view, which ensures that the top d canonical components of each representation are extracted. The hyperparameters of each model are selected to be those that achieve the best total canonical correlation on the validation set. The total correlation is measured by summing the correlation coefficients between each pair of corresponding canonical components. Each model with its best perfomring hyperparameters is then evaluated using k-fold cross validation, which preserves the above mentioned subset ratios, i.e. 7, 5 and 8 folds for MNIST, XRMB and Flickr8k, respectively. For example, for each testing fold out of the 7 folds for MNIST, the remaining 6 folds are randomly split to 50k and 10k samples for training and validating.

For each model and each dataset, we report in Table 1 the mean and standard deviation of the model's results on the test set after performing k-fold cross validation. In addition, we report for each baseline the p-values computed via a paired t-test vs. our DS-DCCA. The mentioned upper bound of the total canonical correlation for each case is equal to the total number of canonical components (d). Evidently, the proposed DS-DCCA learns canonical components of the input views, which are more correlated on average in comparison to both classical and modern CCA-based models on the compared benchmarks. In particular, DS-DCCA improves the second best result by 11.7%, 65.9% and 24.0% of the remaining gap to the upper bound for MNIST, XRMB and Flickr8k, respectively. Moreover, the very low p-values emphasize that the leading results of our DS-DCCA are statistically significant.

Ablation. We compare our proposed DS-DCCA model (full config.) to: (1) The original variant of DCCA, (2) DCCA with a wider middle layer and (3) DCCA with wider first and second layers, reaching a comparable number of added parameters as the scaling networks, (4) DCCA, whose parameters of the last layer for each view are scaled by learnable scaling factors, which are independent of the inputs and remain static after the training (DCCA + Global Scaling). (5) A dynamically-scaled DCCA with a similar capacity to our proposed model, but its scaling networks scale the outputs of the last layer instead of its parameters (Scaling outputs), (6) DCCA model, whose last layer is parameterized by a hypernetwork, i.e. this layer is not parameterized by conventional parameters ($\boldsymbol{\theta}^c$) scaled by another network. Instead, the parameters are generated directly by the outputs of another network for the input $\mathbf{z}$ (w/o $\boldsymbol{\theta}^c$). (7) DS-DCCA, whose scaling networks were trained from the beginning of the process without a warm-up period (No warm-up).

The results are provided in Table 2. For each scenario, we report the total correlation, obtained on a single test set, and the percent of the remaining gap to the upper bound, which is improved by the full configuration of our DS-DCCA. As can be seen, the full configuration of our model, outperforms the compared variants. Our added scaling networks outperform the addition of neurons to each of the first layers of DCCA for reaching a comparable number of parameters. The added neurons have static and input-independent parameters, in contrast to our dynamically-scaled ones. The global scacling struggles to improve the original DCCA and even worsens the results on some datasets. These learnable scaling factors, which are independent of the inputs, are found not beneficial and make the training process more difficult. Similarly, our approach of dynamically scaling the parameters of the transformation functions is superior than dynamically scaling only their outputs. We note that scaling the parameters provide many more degrees of freedom than scaling the output vector. It is also evident that the warmup-period improves the results since it allows to initialize the conventional static parameters better, prior to optimizing the scaling networks. The lower result of the full hypernetwork model (no $\boldsymbol{\theta}^c$) also indicates that the scaling requires a network with weights that are not dynamic.

3.2 Cross-Modality Retrieval

We evaluate our dynamically-scaled variant of Ranking CCA (DS-R. CCA) in a cross-modality retrieval task on several image-text datasets. Given one sample of a modality, the goal of the cross-modality retrieval task is to retrieve the matching sample of the other modality from a set of samples. We evaluate the models on image annotation ($IMG \rightarrow TXT$), for which an image is given and the matching textual caption should be retrieved, and vice versa ($TXT \rightarrow IMG$), where a textual caption is given and the matching image is to be found. The performance is measured by recall statistics: the score R_k is computed as the percent of source samples for which the model retrieved the correct target sample as one of its top k suggestions from the total number of source samples.

Datasets. Three datasets were used for evaluating the retrieval task: (1) Flickr8k, (2) Flickr30k and (3) IAPR TC-12. All datasets were pre-processed and encoded to embedding vectors as described for Flickr8k in the total canonical correlation experiment. The projections' dimension of each view is selected to be $d = 128$ for all datasets. **Flickr8k** [13] was split as described in the previous section. **Flickr30k** [27] is an extended version of the Flickr8k dataset and consists of 31,784 images from Flickr.com. Following [26], random 1k and 1k samples are selected for validation and testing, respectively. **IAPR TC-12** [8] contains 20k images and one detailed caption per image. Following [7], random 2k and 1k samples are selected for validation and testing, respectively.

Baselines. For Flickr8k, the full configuration of our dynamically-scaled variant of Ranking CCA (DS-R.CCA) is compared to all CCA-based baselines from the total canonical correlation experiment. In addition, our variant is compared to the original non-dynamically scaled variant of Ranking CCA (R. CCA) [7], which was proposed specifically to learn representations for the purpose of the cross-modality retrieval task. For Flickr30k and IAPR TC-12, our model is compared to Ranking CCA (R. CCA). In order to demonstrate the effectiveness of inputting the concatenation of both $\mathbf{z}$ and $\mathbf{x}$ to the scaling networks, we compare this full configuration to variants of our DS-R.CCA, whose scaling networks are conditioned only on either $\mathbf{z}$ or $\mathbf{x}$.

Retrieval Results. The recall results in the cross-modality retrival task on Flickr8k, Flickr30k and IAPR TC-12 are presented in Table 3, where the best performing model and the second one are in bold and underlined, respectively. For each measurement of R_k, the reported results are the R_k obtained on the test set by the hyperparameters, which achieved the best R_k on the validation set. As expected, Ranking-CCA outperforms all CCA-based models, which are not designed specifically for the retrieval task. Evidently, the full configuration of our DS-R.CCA ($\mathbf{z}$ and $\mathbf{x}$) achieves the best recall retrieval rates in 13 measurements out of 18, and achieves the second best values in 4 out of the remaining 5 measurements. Evidently, both the embedding space of the deep CCA method ($\mathbf{z}$) and the input ($\mathbf{x}$) have relevancy to the scales. We note that the combination of the two can be seen as a form of a skip connection. The second best performing model among the compared ones is the $\mathbf{x}$-only variant of our DS-R.CCA, which emphasizes the importance of conditioning the dynamically scaled parameters on the raw inputs for learning distinguishable embedding vectors.

4 Conclusion

CCA methods learn orthogonal directions or, more generally, non-linear scalar mappings, in each of the views. In this work, we show the advantage of having computed weights, which vary based on the input pair, for scaling these scalar features so that matching between the two views is maximized. The method is based on a new type of layer that is applied as the last layer of the deep CCA method and that receives the activations of the previous layer. Our experiments demonstrate a clear advantage with regards to the total correlation score. The per-class behavior of the scales, as depicted in the visualization on MNIST, emphasizes the advantage of our method in cases where the input presents class-based variability that requires the ability to adjust the transformation function. Moreover, our approach is suited for the cross-modality retrieval task, since it adjusts the parameterization of the transformation functions conditioned on the given query. Correspondingly, our method achieves better recall rates for for the cross-modality retrieval task.

Acknowledgement. This project was supported by a grant from the Tel Aviv University Center for AI and Data Science (TAD).

References

1. Akaho, S.: A kernel method for canonical correlation analysis. In: Proceedings of the International Meeting of the Psychometric Society (2001)
2. Andrew, G., Arora, R., Bilmes, J., Livescu, K.: Deep canonical correlation analysis. In: ICML (2013)
3. Chandar, S., Khapra, M.M., Larochelle, H., Ravindran, B.: Correlational neural networks. Neural Comput. **28**(2) (2016)
4. Chang, X., Xiang, T., Hospedales, T.M.: Scalable and effective deep cca via soft decorrelation. In: CVPR (2018)
5. Chen, H., Chen, Z., Chai, Z., Jiang, B., Huang, B.: A single-side neural network-aided canonical correlation analysis with applications to fault diagnosis. IEEE Trans. Cybern. **52**(9), 1–13 (2021)
6. Deng, J., Dong, W., Socher, R., Li, L.J., Li, K., Fei-Fei, L.: Imagenet: A large-scale hierarchical image database. In: CVPR (2009)
7. Dorfer, M., Schlüter, J., Vall, A., Korzeniowski, F., Widmer, G.: End-to-end cross-modality retrieval with cca projections and pairwise ranking loss. Int. J. of Multimedia Inform. Retrieval (2018)
8. Grubinger, M., Clough, P., Müller, H., Deselaers, T.: The iapr tc-12 benchmark. In: International workshop ontoImage (2006)
9. Ha, D., Dai, A.M., Le, Q.V.: Hypernetworks. In: ICLR (2017)
10. Han, Y., Huang, G., Song, S., Yang, L., Wang, H., Wang, Y.: Dynamic neural networks: A survey. arXiv:2102.04906 (2021)
11. Harold, H.: Relations between two sets of variates. Biometrika 28(3/4) (1936)
12. He, K., Zhang, X., Ren, S., Sun, J.: Deep residual learning for image recognition. In: CVPR (2016)
13. Hodosh, M., Young, P., Hockenmaier, J.: Framing image description as a ranking task: Data, models and evaluation metrics. J. Artif. Intell. Res. **47**(1), 853–899 (2013)
14. LeCun, Y., Bottou, L., Bengio, Y., Haffner, P.: Gradient-based learning applied to document recognition. In: Proceedings of the IEEE (1998)
15. Lindenbaum, O., Salhov, M., Averbuch, A., Kluger, Y.: L0-sparse canonical correlation analysis. In: ICLR (2022)
16. Lopez-Paz, D., Sra, S., Smola, A., Ghahramani, Z., Schölkopf, B.: Randomized nonlinear component analysis. In: ICML (2014)

17. Michaeli, T., Wang, W., Livescu, K.: Nonparametric canonical correlation analysis. In: PMLR (2016)
18. Mikolov, T., Chen, K., Corrado, G., Dean, J.: Efficient estimation of word representations in vector space. ICLR 2013 (01 2013)
19. Sun, Q., Jia, X., Jing, X.Y.: Addressing contradiction between reconstruction and correlation maximization in deep canonical correlation autoencoders. In: ICANN (2022)
20. Vinod, H.D.: Canonical ridge and econometrics of joint production. J. Econ. **4**(2), 147–166 (1976)
21. Wang, L., Wu, J., et al.: An efficient approach to informative feature extraction from multimodal data. In: AAAI (2019)
22. Wang, W., Arora, R., Livescu, K., Bilmes, J.A.: Unsupervised learning of acoustic features via deep canonical correlation analysis. In: ICASSP (2015)
23. Wang, W., Arora, R., Livescu, K., Bilmes, J.: On deep multi-view representation learning. In: PMLR (2015)
24. Wang, W., Arora, R., Livescu, K., Srebro, N.: Stochastic optimization for deep cca via nonlinear orthogonal iterations. In: 2015 53rd Annual Allerton Conference on Communication, Control, and Computing (Allerton). IEEE (2015)
25. Westbury, J.R., Turner, G., Dembowski, J.: X-ray microbeam speech production database user's handbook. University of Wisconsin (1994)
26. Yan, F., Mikolajczyk, K.: Deep correlation for matching images and text. In: CVPR (2015)
27. Young, P., Lai, A., Hodosh, M., Hockenmaier, J.: From image descriptions to visual denotations: New similarity metrics for semantic inference over event descriptions. In: ACL (2014)

TCR: Short Video Title Generation and Cover Selection with Attention Refinement

Yakun Yu[1(✉)], Jiuding Yang[1], Weidong Guo[2], Hui Liu[2], Yu Xu[2], and Di Niu[1]

[1] University of Alberta, Edmonton, AB, Canada
{yakun2,jiuding,dniu}@ualberta.ca
[2] Tencent, Shenzhen, China

Abstract. With the widespread popularity of user-generated short videos, it becomes increasingly challenging for content creators to promote their content to potential viewers. Automatically generating appealing titles and covers for short videos can help grab viewers' attention. Existing studies on video captioning mostly focus on generating factual descriptions of actions, which do not conform to video titles intended for catching viewer attention. Furthermore, research for cover selection based on multimodal information is sparse. These problems motivate the need for tailored methods to specifically support the joint task of short video title generation and cover selection (TG-CS) as well as the demand for creating corresponding datasets to support the studies. In this paper, we first collect and present a real-world dataset named Short Video Title Generation (SVTG) that contains videos with appealing titles and covers. We then propose a Title generation and Cover selection with attention Refinement (TCR) method for TG-CS. The refinement procedure progressively selects high-quality samples and highly relevant frames and text tokens within each sample to refine model training. Extensive experiments show that our TCR method is superior to various existing video captioning methods in generating titles and is able to select better covers for noisy real-world short videos.

Keywords: Title generation · Cover selection · Multimodal learning

1 Introduction

Video titles and covers are two critical elements for capturing users' attention when using an app (e.g., TikTok). However, many short videos do not have well-edited titles beyond plain descriptions or a cover image other than the first frame. The sheer volume of daily video uploads further makes it hard for the platform to edit video covers in a timely manner. Therefore, there is an apparent demand for using artificial intelligence to improve the quality of video titling and cover image selection, especially with the objective of increasing user engagement.

Various video description generation techniques have been studied and developed recently, including video captioning [5,15], video-based summarization

Y. Yu, J. Yang, W. Guo—Equal contribution.

H. Kashima et al. (Eds.): PAKDD 2023, LNAI 13937, pp. 245–256, 2023.
https://doi.org/10.1007/978-3-031-33380-4_19

[6,7], etc. They are mainly designed for objectively describing actions and interactions of objects in a video. For example, for a video of a woman doing yoga, a generated caption is usually "a woman is exercising while her cat keeps disturbing her", which is correct but unvarnished. However, a video uploader would rather title the scene as "My cat sees me as his toy" in a hilarious way in order to attract viewers' attention. Therefore, there is still a gap in the literature to generate appealing and human-like video titles. However, a lack of open-sourced short video datasets from real-world social media platforms hinders the development of techniques for human-like video title generation. Furthermore, little work has been done on video cover selection [12], which is another challenge, again due to the lack of video cover selection dataset.

To address the issues mentioned above, we collect a Short Video Title Generation (SVTG) dataset that consists of 8,652 short video samples with a human-edited appealing title and a visual cover selected by the original content creator. To the best of our knowledge, SVTG is the first publicly available dataset that is designed for joint video title generation and cover selection (TG-CS) on real-world short videos, especially with a purpose of boosting click-throughs in mind rather than plainly describing the scene.

Furthermore, we propose a novel video Title generation and Cover selection with Refinement (TCR) approach that integrates a title-cover generator with cross-attention-based refinement learning for TG-CS task based on the multimodal information in short videos. Specifically, we use a multimodal title-cover generator to capture the temporal dependency between modalities. In order to handle noisy modality data extracted from real-life short videos, we further propose a refinement learning strategy to refine model training.

Our main contributions are summarized as follows: (1) We construct a new short video dataset, SVTG, which can facilitate research on short video title generation beyond generating the unvarnished factual description of objects, and research on video cover selection based on multimodal information. Our objective is to generate eye-catching video titles and covers, with the goal of engaging online user attention; (2) We propose the TCR method that can simultaneously generate appealing titles and covers by fully capturing the dependency between modalities and refinement training; (3) We evaluate TCR on the SVTG dataset as well as on the public How2 dataset. TCR has demonstrated its effectiveness on title generation across the two datasets and on cover selection under SVTG. We open-source the dataset and code along with supplementary materials to facilitate future research on generating titles and selecting covers for real-world social media videos[1].

2 Related Work

2.1 Text Generation

Text generation mainly includes text summarization and video captioning. The former aims to generate a short summary for a long document while preserv-

[1] https://github.com/PipiZong/TCR.git.

Table 1. The statistics of the SVTG dataset.

	#Videos	#Categories	#Authors	#Sentences	Vocab
Train	8,052	37	3,478	155,826	100,361
Valid	200	25	172	3,802	6,770
Test	400	27	288	7,352	10,494
All	8,652	37	3,618	166,980	105,487

ing the main idea of the document. The latter focuses on describing the visual content of a video using condensed text.

Conventional text summarization utilizes a single text modality to generate summaries [10,16]. Recent works [6,7] show promising results on summarization with the help of other modalities, e.g., the visual modality. For instance, Liu *et al.* [9] propose both RNN-based and Transformer-based encoder-decoder models with the input of videos and its ground-truth/ASR transcripts. Encoder-decoder networks are widely used in video captioning where the encoder converts the input into high-level contextual representations before the decoder takes the representations and generates words as the caption. For example, Tan *et al.* [15] propose a RNN-based encoder-decoder model with three visual reasoning modules to generate video captions.

The above methods perform well on generating factual descriptions of a video scene given the video and the corresponding well-edited transcripts, yet fail when they intend to generate titles beyond factual descriptions based on noisy multimodal information. In contrast, our method can handle noisy modality data and capture the subtle relations between modalities, thus being able to generate both factual descriptions and appealing titles.

2.2 Cover Selection

Cover selection aims to find a representative frame that conveys the main idea of a video. Song *et al.* [14] propose a thumbnail selection system using clustering methods. Ren *et al.* [12] select the best frame based on the prediction scores. They utilize only the visual modality, which is insufficient for real-world short videos where the keyframe should be determined by multiple modalities. Though Li *et al.* [7] propose VMSMO, to train a cover selection model based on multimodal information, i.e., news video and news article pairs, their model's performance is highly dependent on the quality of news articles that are well-edited by humans. However, the SVTG dataset does not provide such manually-created articles. Therefore, it is inappropriate to apply their method to our task for comparison. Moreover, we extract covers using a unified model instead of an independent cover selection model, which is more straightforward.

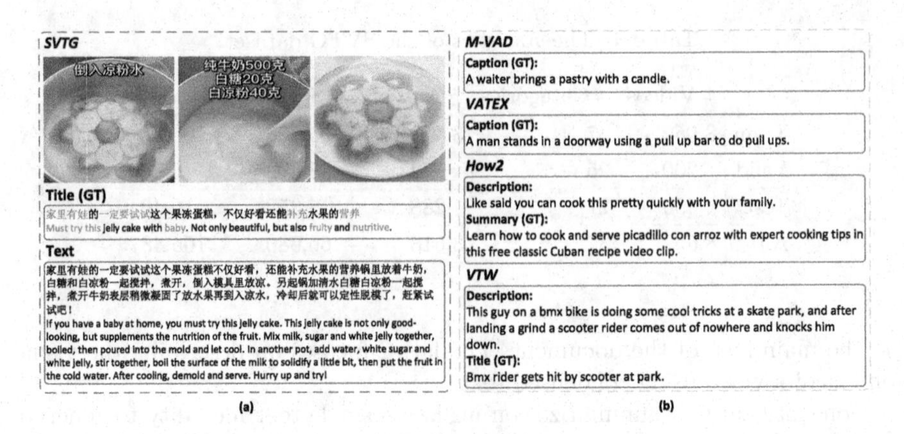

Fig. 1. Comparison between (a) SVTG data example and (b) Other benchmark dataset examples. SVTG example has the continuous frames on the top, title on the middle, and transcripts extracted by ASR/OCR on the bottom. Other examples show their ground truths (GT) and user-generated descriptions.

3 The SVTG Dataset

To the best of our knowledge, SVTG is the first publicly available Chinese dataset that designed for the joint TG-CS for short videos. We regard a title as appealing if it expresses the video maker's desire to attract viewers with some sentiment words in addition to describing the factual information (e.g., the interactions between objects). See Supplementary A for SVTG data collection. Table 1 shows the statistics of our dataset.

To show the uniqueness of SVTG, we compare it with four other benchmark datasets, as shown in Fig. 1. Compared with these datasets, SVTG has three unique characteristics. First, SVTG videos have more appealing titles, which can promote the study of deep learning models to generate short video titles beyond plain factual descriptions. As we observe in Fig. 1(a), SVTG videos usually contain strong sentiment words (e.g., "must" and "nutritive") to draw viewers' attention except simply describing the video scene. In contrast, from Fig. 1(b), we can see that the captions, summaries, and titles in other benchmark datasets usually emphasize on the actions of objects (e.g., "stand") and the interaction between objects (e.g., "a man using a bar"). These descriptions tend to be unvarnished and objective without any sentiments. If we annotated the video from Fig. 1(a) in the way as the other datasets do, the title would sound like "pour milk and add fruit to make cake" and would be less attractive to online viewers.

Second, SVTG is more challenging for video title generation, and closer to real-life videos. As demonstrated in Fig. 1(b), the existing benchmark datasets contain human-annotated video descriptions, subtitles or articles as the text modality. However, it is expensive or even infeasible to have such human-annotated video descriptions in real-world applications. In real-world settings,

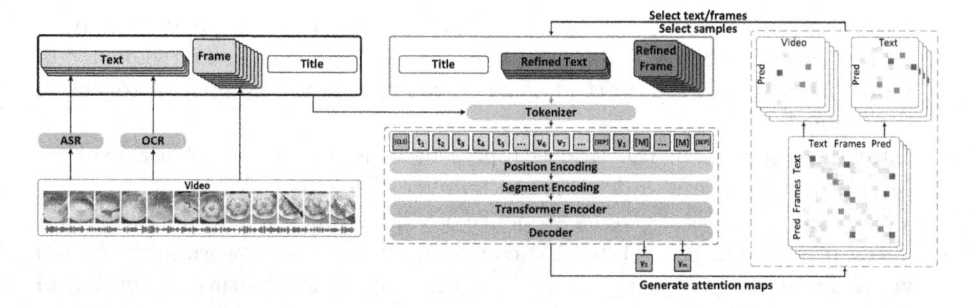

Fig. 2. The overall architecture of the proposed TCR method including data preprocessing in the black box, a title-cover generator in the blue dotted box, and a multimodal cross-attention-based refinement module in the orange dotted box. "Title" means the ground truth title. "Pred" in the attention map represents the predicted title. And "[M]" in the tokenized sequence stands for a masked token. (Color figure online)

when a content creator uploads a short video to the platform, the platform needs to select a cover and generate a title based on the video itself. All information in SVTG comes from the video itself without any other human annotations. Thus SVTG is closer to real-world applications. However, ASR- and OCR-detected text can be highly noisy in contrast to human-annotated summaries or descriptions, making SVTG more challenging than other datasets.

Third, to our best knowledge, SVTG is the first dataset that includes the video covers for facilitating the joint TG-CS on short videos. We argue that the combination of two tasks, title generation and cover selection, is crucial for attracting viewers' attention and thus should not be separated for consideration. Furthermore, having a unified dataset encourages the model to learn the shared connection of this joint task. See Supplementary B for more details about these benchmark datasets.

4 Method

We propose a self-attentive method named TCR for the joint TG-CS under the SVTG dataset. Figure 2 shows the structure of the TCR method. It consists of a multimodal title-cover generator and a refinement module. The generator generates titles and extracts covers by fully exploiting the dependency between modalities. The refinement module helps refine model training by selecting high-quality samples and highly relevant tokens/frames within each sample. We will explain the TCR method in detail below.

4.1 Title Generation and Cover Selection

Since the frames and the text inputs are both temporally ordered, we can directly combine them to explicitly learn the dependency between all pairs of text tokens and frames. We first preprocess the training data into the following formats:

- Text Representation: given the transcripts extracted by ASR and OCR from a video, the combined ASR and OCR text is tokenized into a sequence of WordPieces [17], i.e., $T = (t_1, t_2, ..., t_n)$, where n is the number of tokens in the sequence.
- Title Representation: the title is represented as a sequence of tokens $Y = (y_1, y_2, ..., y_m)$, where m denotes the number of tokens in Y.
- Video Representation: given a sequence of video frames of length L, we feed it into a pre-trained ResNet-101 2D convolutional neural network [3] to obtain visual features $V = \{v_l\}_{l=1}^{L} \in \mathbb{R}^{L \times d_v}$, where d_v is the feature dimension of the pretrained ResNet model.

The title-cover generator is composed of three embedding layers, 12-layer transformer encoders, and a language modeling (LM) head on top as the decoder. On the one hand, we regard the frames as L virtual tokens $\{v_i\}_{i=1}^{L}$ and concatenate them with text tokens T to form a new sequence of length $L + n$. We also add a special token [CLS] at the start of the sequence and another special token [SEP] at the sequence end. These tokens, including Y, are embedded by three embedding layers that are trainable: token-level embedding, position-level embedding, and segment-level embedding.

On the other hand, we feed the frame features V into a fully connected (FC) layer to project it into the same lower-dimensional space as the text token embeddings and regard the projected frame features as the frame token embeddings. To establish a connection between the text tokens and frames so that they could attend to each other, we combine their respective embeddings by simply replacing the virtual token embeddings with the frame embeddings. Therefore, the final representations for text tokens and video frames can be obtained by summing up their corresponding three embeddings. We then feed these representations as the input to the transformer encoders.

Assume the input is $X_{VT} \in \mathbb{R}^{L' \times d_h}$ where d_h is the hidden size of our model, the output of the i-th transformer encoder is denoted as $O_i \in \mathbb{R}^{L' \times d_h}$. Each encoder aggregates the output of the previous encoder using multiple self-attention heads, thus O_i is computed by:

$$O_i = \begin{cases} X_{VT}, & i = 0; \\ \text{Encoder}_i(O_{i-1}), & i > 0. \end{cases}$$

In this way, we can obtain visual-language fusion representations at different encoder layers. Let A_i denotes the attention score of each head in the i-th encoder.

$$\begin{aligned} A_i &= \text{Dropout}(\text{Softmax}(\frac{Q_i K_i^T}{\sqrt{d_h/12}} + \text{Attention Mask})), \\ Q_i,\ K_i &= O_{i-1} W_{Q_i},\ O_{i-1} W_{K_i}. \end{aligned} \tag{1}$$

Q and K are the queries and keys linearly transformed from the input of each encoder. Attention Mask $\in \mathbb{R}^{L' \times L'}$ (elements $\in \{0, -\infty\}$) is used to make sure the target token to be generated only attending to the leftward information including itself, the generated tokens, frames and text tokens [2]. The LM head consists of two dense layers and a layer normalization (LN) layer in between to calculate the vocabulary distribution for decoding. Once the generator is well-trained, we select a frame with the highest attention score obtained in Eq. 1 as the video cover.

4.2 Attention-Based Model Refinement

Considering redundant noise may exist in the ASR/OCR transcripts and continuous frames, noise filtering is intuitive to help improve the quality of the generated titles. To this end, we propose an attention-based refinement module, which will automatically select not only the key parts of the multimodal input within each sample but also select higher-quality samples. Moreover, the module can interact with the above generator to find the best combination of different input tokens for progressively refining the generator.

Token-Level Refinement. Given a title-cover generator, we could first generate titles for the training data. We then acquire the cross-attention scores between the generated titles and the inputs, i.e., the ASR/OCR tokens and the frame tokens, through Eq. 1. We assume the tokens that the generated titles attend to a lot are critical for building a generator with good performance. Therefore, we extract them as refined sentences/frames. For sentence refinement, we first locate the sentence that has the highest attention score with each generated token. Then, the top u sentences which the generated title attends to frequently are naturally selected as the key sentences.

Video titles always contain some words (e.g., "alley-oop" in a basketball game video) that frequently appear in the ASR/OCR text but are rarely related to video frames. Therefore, we need to select the related frames at a higher granularity than the way we select related sentences. To achieve this, we first assign a $weight \in [0, L]$ to each frame according to its attention scores with each generated title token, then calculate the total weights for each frame. Higher weights indicate greater contributions to title generation. We finally pick out the top v frames with the highest weights as the refined frames or keyframes for the video.

Sample-Level Refinement. In addition to refining the tokens within each sample, we denoise the data by selecting higher-quality data at the sample level. A training data will be deemed high-quality if the generated title by the generator M is similar to its ground truth title. The similarity is computed by the Rouge score [8].

Apparently, with refinement at both token-level and sample-level, we could refine the title-cover generator multiple times until we get the best test results. See the detailed refinement algorithm in Supplementary C.

5 Experimental Results

5.1 Baseline Models

We compare our proposed method with the following baseline models:

Lead-3 directly selects the first three sentences from the text as the title [10].

HSG constructs a heterogeneous graph with sentence nodes and word nodes, then selects sentences as the summary for a document by node classification [16].

MAST is a model consists of encoder layers, a hierarchical attention layer, and a decoder that generates a textual summary on multimodal inputs [6].

NMT is an RNN-based model [1] designed for sequence-to-sequence tasks and modified for video description tasks.

MFN generates a summary based on video and its ASR or ground-truth transcript through a multistage fusion model with a forget gate module to remove the redundant information [9].

Bert2Bert contains a text BERT encoder, a video transformer encoder and a BERT decoder, which is a popular framework used in recent video/image + language works [4].

5.2 Implementation Details

We evaluate the above baseline models and our TCR model on the SVTG dataset following the split criteria in Table 1. We implement our experiments using PyTorch [11] on an NVIDIA V100 GPU. To preprocess the videos, we sample 25 frames for each video and extract visual features from these sampled frames.

Our multimodal generator is adapted from a text pretrained model architecture [2]. The vocabulary size is 21,128. We use AdamW optimizer with a learning rate of $1e-5$. Linear warmup schedule is set for the first 10% of the total training steps with linear decay. Dropout is adopted for regularization. The batch size is set to 16, and the maximum input sequence length is set to 512. During training, we only mask 20% of the title tokens, and the maximum number of masked tokens is set to 20. We train up to 20 epochs with the cross-entropy loss computed by the predicted title tokens and the ground truth tokens. The validation set is used to select the best model with the highest mean score of the evaluation metric. For testing, we use beam search with beam size 5 to report the final test results. For refinement, we select the top 3 sentences/frames and perform one iteration.

In addition, we evaluate the TCR on the publicly available How2 dataset [13] to further validate its generation ability. The How2 dataset consists of 79,114 videos accompanied by corresponding user-generated descriptions and summaries. It has a train set of 73,993 samples, a validation set of 2,965 samples, and a test set of 2,156 samples. Since the TCR method is designed for handling real-world short videos with noisy multimodal information, we use the ASR transcripts instead of the provided descriptions following the work of Liu *et al.* [9] for a fair comparison.

5.3 Results

Here we first report the performance of the baseline models and our proposed TCR for title generation. Then we show the evaluation results of cover selection. Later, we present the ablation tests from different aspects.

Table 2. Rouge score comparison with the baselines for SVTG.

Model	R-1	R-2	R-L
Lead-3	29.44	18.40	37.35
HSG	41.06	27.45	40.59
NMT	41.20	28.74	41.91
MFN-rnn	16.64	1.47	14.36
MAST	39.73	27.42	41.00
MFN-transformer	19.24	3.99	18.40
Bert2Bert	24.84	5.75	22.58
TCR w/o Text	8.53	0.31	9.30
TCR w/o Visual	46.35	32.67	46.07
TCR	**47.62**	**33.98**	**47.08**

Title Generation Performance. We evaluate the quality of the generated titles by the standard Rouge F1 matric [8] following previous works [6,7,9,10,16]. R-1, R-2, and R-L refer to the unigram, bigram, and longest common subsequence overlap with the ground truth title, respectively.

We first compare the results of our method and the baseline models on SVTG, as listed in Table 2. We can observe from the table that TCR outperforms all the baseline models in terms of all Rouge scores. Specifically, TCR outperforms the best baseline model by 6.42 R-1 points, 5.24 R-2 points, and 5.17 R-L points, indicating its superior ability to learn video titles when given noisy ASR/OCR text and frames. Though these baseline models perform well in generating summaries and captions given user-annotated text descriptions, they are incapable of generating titles in social media language and handling noisy text information. In contrast, TCR can generate more accurate titles by fully exploiting the dependency between all text/frame tokens and denoised training.

Table 3. Rouge score comparison for How2.

Model	R-1	R-2	R-L
S2S	48.1	28.2	43.4
FT	51.1	31.0	45.8
HA-rnn	53.9	34.2	48.7
HA-transformer	55.1	36.0	50.1
MFN-rnn	60.0	43.6	56.1
MFN-transformer	59.3	42.1	55.0
TCR	**60.9**	**44.2**	**60.6**

To further examine the generation ability of TCR, we apply it to a public dataset, How2, that is designated to generate a factual summary for videos. Table 3 presents the results on How2. We use the dataset and baseline models in

Table 4. Evaluation scores on the selected frames and the default covers that come with the videos.

Pictures	*Superior*	*Informativeness*
Our selected frames	31.5%	15.5%
Default covers	25.5%	15.5%
Cannot tell	43.0%	69.0%

the work [9] for a fair comparison. See Supplementary D for model details about these baselines. We can observe from Table 3 that TCR outperforms the baseline models in all Rouge scores. Notably, it outperforms the best baseline MFN-rnn model by 4.5 R-L points. This finding again indicates that TCR is excellent at generation both within factual descriptions (e.g., summaries in How2) and beyond factual descriptions (e.g., titles in SVTG). Moreover, by comparing the performance of MFN in Table 2 and Table 3, we can conclude that MFN is unable to generalize to SVTG that is distinct from traditional video summary/caption datasets.

Cover Selection Performance. We compare the selected cover images with the default cover that comes with the videos by human judgments from the following aspects : *Informativeness* (which picture corresponds to the topic conveyed by the video title), and *Superior* (which picture is better to be a video cover that attracts you). See Supplementary E for more details about the human evaluation.

The results are reported in Table 4. We can observe from the table that participants cannot tell which one is better in most cases because the two pictures are almost identical. Moreover, our selected covers outperform the default covers by 23.5% in terms of *Superior* while they are equivalent in *Informativeness*, which demonstrates the effectiveness of our method to automatically select the cover image from a series of frames.

Ablation Studies. All ablation studies are performed on the SVTG dataset. The last three rows in Table 2 show the ablation results on the effect of different modalities. From this table, we can see that TCR outperforms TCR w/o Visual and TCR w/o Text, which demonstrates that multimodal information can benefit title generation more than a single modality. It can also be easily observed that the text modality is dominant while the visual modality has less contribution in title generation because ASR- and OCR-detected text usually contains words that describe the highlights of stories in short video cultures. Supplementary F further provides the ablation tests on the number of selected sentences/frames and refinement iterations.

5.4 Qualitative Analysis

Figure 3 shows examples including the generated titles by baseline models and our proposed model, as well as the cover selection comparison. The results from

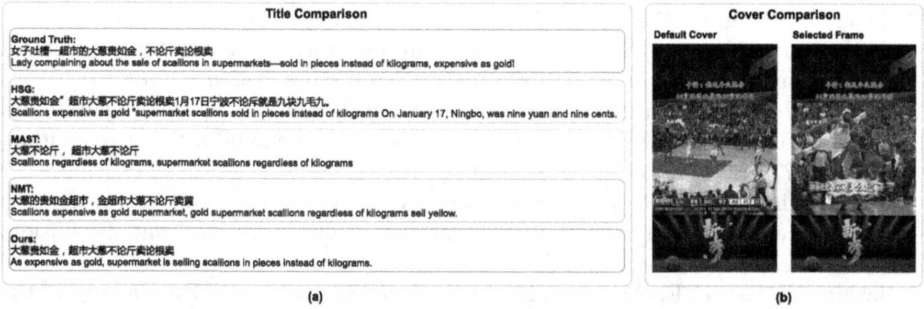

Fig. 3. Examples of (a) the generated titles by the baselines and TCR (b) the selected frame and its default cover that comes with the video. The video title is "Kobe Bryant pump fakes Vince Carter".

Fig. 3(a) demonstrate that our proposed model could generate a more accurate title than the baselines. Furthermore, as shown in Fig. 3(b), the selected frame is better than the default cover that comes with the video since the selected frame captures the jumping moment of the defence player while the default cover does not. To better understand the refinement process and what our model has learned, we also visualize the attention scores in Supplementary G.

6 Conclusion

In this paper, we first collect a real-world dataset named SVTG to specifically support the joint TG-CS task for short videos. To our best knowledge, SVTG is the first short video dataset that contains appealing titles instead of unvarnished factual captions or summaries like other benchmark datasets. Then we propose a TCR method that consists of a title-cover generator and a model refinement module. The title-cover generator generates the title and extracts the cover image by fully exploiting the dependency between modalities. The refinement module further selects high-quality training samples and relevant text/frames within each sample to refine the generator progressively. Experiments on the SVTG dataset and the public How2 dataset show the effectiveness of the proposed TCR method on title generation when given noisy data. Furthermore, the cover selection evaluation results suggest that our selected covers are preferred compared to the default covers. This further demonstrates the effectiveness of the TCR method on cover selection.

References

1. Caglayan, O., García-Martínez, M., Bardet, A., Aransa, W., Bougares, F., Barrault, L.: Nmtpy: a flexible toolkit for advanced neural machine translation systems. The Prague Bull. Math. Linguist. **109**(1), 15 (2017)

2. Dong, L., Yang, N., Wang, W., Wei, F., Liu, X., Wang, Y., Gao, J., Zhou, M., Hon, H.W.: Unified language model pre-training for natural language understanding and generation. Advances in Neural Information Processing Systems 32 (2019)
3. He, K., Zhang, X., Ren, S., Sun, J.: Deep residual learning for image recognition. In: Proceedings of the IEEE Conference on Computer Vision and Pattern Recognition, pp. 770–778 (2016)
4. Hu, R., Singh, A.: Unit: Multimodal multitask learning with a unified transformer. In: Proceedings of the IEEE/CVF International Conference on Computer Vision. pp. 1439–1449 (2021)
5. Iashin, V., Rahtu, E.: Multi-modal dense video captioning. In: Proceedings of the IEEE/CVF Conference on Computer Vision and Pattern Recognition Workshops, pp. 958–959 (2020)
6. Khullar, A., Arora, U.: Mast: Multimodal abstractive summarization with trimodal hierarchical attention. In: Proceedings of the First International Workshop on Natural Language Processing Beyond Text, pp. 60–69 (2020)
7. Li, M., Chen, X., Gao, S., Chan, Z., Zhao, D., Yan, R.: Vmsmo: Learning to generate multimodal summary for video-based news articles. In: Proceedings of the 2020 Conference on Empirical Methods in Natural Language Processing (EMNLP), pp. 9360–9369 (2020)
8. Lin, C.Y.: Rouge: A package for automatic evaluation of summaries. In: Text Summarization Bout, pp. 74–81 (2004)
9. Liu, N., Sun, X., Yu, H., Zhang, W., Xu, G.: Multistage fusion with forget gate for multimodal summarization in open-domain videos. In: Proceedings of the 2020 Conference on Empirical Methods in Natural Language Processing (EMNLP), pp. 1834–1845 (2020)
10. Nallapati, R., Zhai, F., Zhou, B.: Summarunner: A recurrent neural network based sequence model for extractive summarization of documents. In: Proceedings of the AAAI Conference on Artificial Intelligence, vol. 31 (2017)
11. Paszke, A., et al.: Pytorch: An imperative style, high-performance deep learning library. In: Advances in Neural Information Processing Systems, vol. 32 (2019)
12. Ren, J., Shen, X., Lin, Z., Mech, R.: Best frame selection in a short video. In: Proceedings of the IEEE/CVF Winter Conference on Applications of Computer Vision. pp. 3212–3221 (2020)
13. Sanabria, R., et al.: How2: A large-scale dataset for multimodal language understanding. In: NeurIPS (2018)
14. Song, Y., Redi, M., Vallmitjana, J., Jaimes, A.: To click or not to click: Automatic selection of beautiful thumbnails from videos. In: Proceedings of the 25th ACM International on Conference on Information and Knowledge Management, pp. 659–668 (2016)
15. Tan, G., Liu, D., Wang, M., Zha, Z.J.: Learning to discretely compose reasoning module networks for video captioning. In: Proceedings of the Twenty-Ninth International Conference on International Joint Conferences on Artificial Intelligence. pp. 745–752 (2021)
16. Wang, D., Liu, P., Zheng, Y., Qiu, X., Huang, X.J.: Heterogeneous graph neural networks for extractive document summarization. In: Proceedings of the 58th Annual Meeting of the Association for Computational Linguistics, pp. 6209–6219 (2020)
17. Wu, Y., et al.: Google's neural machine translation system: Bridging the gap between human and machine translation. arXiv preprint arXiv:1609.08144 (2016)

ItrievalKD: An Iterative Retrieval Framework Assisted with Knowledge Distillation for Noisy Text-to-Image Retrieval

Zhen Liu, Yongxin Zhu, Zhujin Gao, Xin Sheng, and Linli Xu(✉)

State Key Laboratory of Cognitive Intelligence,
University of Science and Technology of China, Hefei, China
{liuzhenz,zyx2016,gaozhujin,xins}@mail.ustc.edu.cn, linlixu@ustc.edu.cn

Abstract. Benefiting from the superiority of the pretraining paradigm on large-scale multi-modal data, current cross-modal pretrained models (such as CLIP) have shown excellent performance on text-to-image retrieval. However, the current research mainly focuses on the scenarios with strong matching of images and texts, which is not always available in practice. For example, in social media content or daily communication, the text is not always completely related to the image and may also contain some irrelevant content, which introduces non-negligible noise to text-to-image retrieval. The noisy multi-modal setting is significantly different from the current cross-modal pretraining corpus, which may lead to significant degradation of the retrieval performance of the general image-text retrieval models. In this paper, we focus on the task of noisy text-to-image retrieval and propose an iterative retrieval framework which firstly retrieves the key-semantic information from the noisy text with knowledge distillation, followed by retrieving the relevant image from the image pool with the key-semantic clue. Experiments on Noisy-MSCOCO and PhotoChat datasets confirm the superiority of the proposed iterative retrieval framework in the task of noisy text-to-image retrieval compared with the general retrieval models.

Keywords: Image-text retrieval · Knowledge distillation · Extractive summarization

1 Introduction

The task of cross-modal image-text retrieval is to retrieve samples from one modal with the guidance of the samples from the other modal. With the rapid growth of the data from various modalities, cross-modal image-text retrieval has a wide range of applications, helping users quickly locate data from a specific modal that is relevant to the current query. In general, cross-modal image-text retrieval consists of two sub-tasks, which are image-to-text (I2T) and text-to-image (T2I) retrieval respectively. In this paper, we focus on T2I retrieval which aims at retrieving the most relevant image according to the textual context.

H. Kashima et al. (Eds.): PAKDD 2023, LNAI 13937, pp. 257–268, 2023.
https://doi.org/10.1007/978-3-031-33380-4_20

As a cross-modal task, the major challenge of T2I retrieval is how to bridge the semantic gap between different modals. To tackle that, the transformer-based cross-modal pretraining models have been successfully applied to various tasks. By pretraining on large-scale image-text datasets, different modalities are encoded into a common semantic space, yielding modal-agnostic semantic representations. CLIP [1] is one of the most representative models, the zero-shot retrieval performance of which on the commonly-used image-text dataset MSCOCO [12] is competitive with the finetuning performance of the previous pretraining models.

Despite the outstanding performance of the methods based on large-scale cross-modal pretraining, it is often overlooked that most of the current research on T2I retrieval assumes that the query-key data is strongly correlated. Nevertheless, the image-text matching relationship may be weakly correlated in real scenarios. For example, on social media platforms such as Twitter, people tend to share their daily life in a combination of images and texts, where the text may involve some content that is irrelevant to the image. As a matter of fact, the image-text scenarios can be very noisy in practice. Figure 1 shows an example in the PhotoChat dataset from the photo sharing task [19]. The motivation of the photo sharing task is the popularity of photo-sharing in online chat, the goal of which is to retrieve the corresponding image of the conversational context, most of which is irrelevant chat. In this case, applying the CLIP model directly fails to retrieve the correct image, as the CLIP model obviously ignores the information of "strawberry" and "blueberry" in the dialogue.

Fig. 1. An example in the PhotoChat dataset. The top right image is the image relevant to the left dialog context, and the bottom right image is retrieved with the CLIP model.

To further analyze why the CLIP model fails to retrieve the correct image in the above example, we conduct a simple investigation. Specifically, we construct a simple noisy T2I scenario named Noisy-MSCOCO by injecting noise to the dataset MSCOCO. We generate some noisy sentences according to the captions from MSCOCO and then mix them to get the final noisy text. Figure 2 shows the

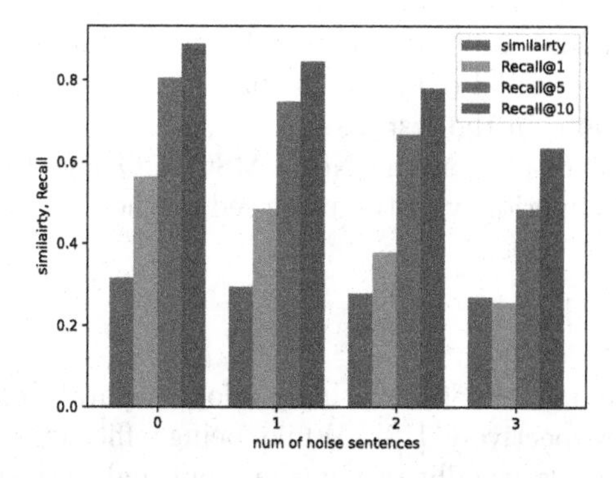

Fig. 2. The retrieval results and the similarity between image-text pairs on Noisy-MSCOCO given the number of noisy sentences. The results are reported in the zero-shot setting with CLIP.

retrieval performance based on CLIP as the number of noisy sentences grows. We also report the cosine similarity, i.e. the image-text alignment score between the representations of the two modalities calculated by CLIP. It can be seen that with the increasing noise, the image-text similarity between the text and the corresponding image decreases as expected, which results in a significant degradation in the retrieval performance. This poses an interesting problem in the retrieval task, which is motivated to capture the relevant information from raw data, whereas the noisy sentences are obviously harmful to the T2I retrieval performance.

In this paper, we focus on text-to-image retrieval where the text may contain a lot of noise, and define the problem as noisy text-to-image retrieval (NT2I). We propose an iterative retrieval framework assisted with knowledge distillation (ItrievalKD). Essentially, to alleviate the influence of irrelevant textual content on the retrieval performance, it is necessary to extract the image-related content from the noisy text as the key-semantic text. Unfortunately, the supervision information of key-semantic text is not available for training the extractor in most cases. Therefore in our iterative retrieval framework, we start with exploiting CLIP to obtain the key-semantic annotations, and then proceed to retrieve the relevant image from the image pool with the key-semantic clue. Furthermore, due to the lack of image annotations during testing, we propose to adopt knowledge distillation to distill the image-text matching knowledge from the cross-modal model CLIP to the plain-text model BERT [3], which can be used to obtain the key-semantic information in the testing phase. The relevant image can then be retrieved with the key-semantic clue from the image pool.

In summary, the main contributions of the work are:

- We propose an iterative retrieval framework for the noisy T2I task, where the text contains noise in text-to-image retrieval.

- We adopt knowledge distillation to transfer the image-text matching knowledge from the cross-modal model to the plain-text model to alleviate the lack of image annotations in the testing stage.
- The experimental results on the Noisy-MSCOCO and PhotoChat datasets demonstrate the superiority of the proposed method.

2 Related Work

Most early cross-modal retrieval methods adopt separate encoders to encode images and texts respectively [18]. While being efficient, these independent feature-encoding models usually produce sub-optimal performance due to the lack of interactions between modals. [11] is the first attempt to consider the dense pairwise cross-modal interactions which achieves tremendous accuracy improvements. After that, various cross-modal interaction methods [2,8] have been proposed to extract the features of both text and image. On the other hand, methods with only global cross-modal are restricted in the sense that text descriptions usually contain fine-grained correlations with images, which are easily smoothed by global alignment. To address that, some works [7,17] propose to explore the region (or patch) to word correspondences. An alternative solution is the pretrain-then-finetune paradigm driven by the global alignment method [1,9], which can achieve satisfactory results with improved robustness, with the help of the large-scale pretraining data.

3 Methodology

In this section, we elaborate on the iterative retrieval framework for the noisy text-to-image retrieval task. We start with the problem definition and a brief overview of the CLIP model which is employed in the iterative retrieval process. Then the architecture of the proposed model will be described in detail.

Problem Definition. Given a parallel image-text dataset (T, V), each sample pair consists of a noisy text t_i and a relevant image v_i, where $t_i = \{t_i^1, t_i^2, \ldots, t_i^k, \ldots, t_i^m\}$ is composed of multiple sentences and t_i^k represents the k-th sentence. The task is to retrieve the most relevant image v_i to the noisy text t_i from the image pool V with the proposed iterative retrieval model $R(t_i, V)$.

3.1 Preliminaries: CLIP

CLIP is trained to learn visual representations with natural language supervision. As shown in Fig. 3, it consists of a text encoder $\mathbb{T}$ which is a GPT [15] style Transformer model, and an image encoder $\mathbb{V}$ which can be either a Vision Transformer (ViT) [4] or a Residual Convolutional Neural Network (ResNet) [5]. Then the dot product between the two outputs of the above two encoders will be used as the alignment score of the input image and the text. The model is

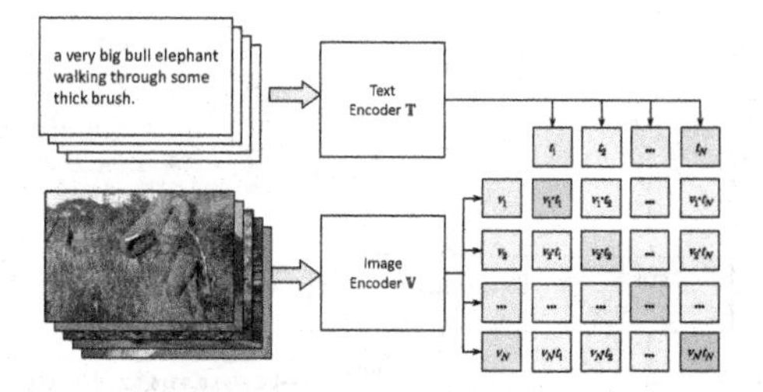

Fig. 3. The framework of the CLIP model.

pretrained to distinguish aligned image-text pairs from randomly combined ones with a contrastive loss,

$$\mathcal{L}_{\text{NCE}} = -\left(\log \frac{\exp(sim(v_i, t_i)/\alpha)}{\sum_j \exp(sim(v_i, t_j)/\alpha))} + \log \frac{\exp(sim(t_i, v_i)/\alpha)}{\sum_j \exp(sim(t_i, v_j)/\alpha))}\right) \tag{1}$$

where α is the temperature coefficient to be learned in CLIP. The image-text alignment score $sim(v_i, t_i)$, which is the similarity mentioned above, is calculated as follows,

$$sim(v_i, t_i) = \frac{\mathbb{T}(t_i) * \mathbb{V}(v_i)}{||\mathbb{T}(t_i)||^2 * ||\mathbb{V}(v_i)||^2} \tag{2}$$

3.2 Model Architecture

The overall framework of the proposed model is shown in Fig. 4, which is compared to the general method in the left panel. Instead of directly taking the noisy text as the query to retrieve the most relevant image from the image pool, which may degrade the retrieval performance as discussed above, our proposed method ItrievalKD first extracts the key-semantic information from the noisy text to alleviate the influence of irrelevant textual content on the retrieval performance, followed by retrieving the relevant image according to the key-semantic clue. Below we will describe the iterative retrieval framework in detail.

Retrieving the Key-Semantic Text in the Noisy Text. Due to the lack of ground truth regarding the key-semantic annotations in most NT2I cases as supervision, it is necessary to retrieve the key-semantic content in the noisy text at first. Here we propose a simple yet effective annotation strategy. We consider each sentence as a basic unit of semantic information in the noisy text. In the NT2I scenario, the key-semantic content in the noisy text should be highly-related to the corresponding image, and the rest should be irrelevant. Hence, the choice of key-semantic text heavily depends on the corresponding

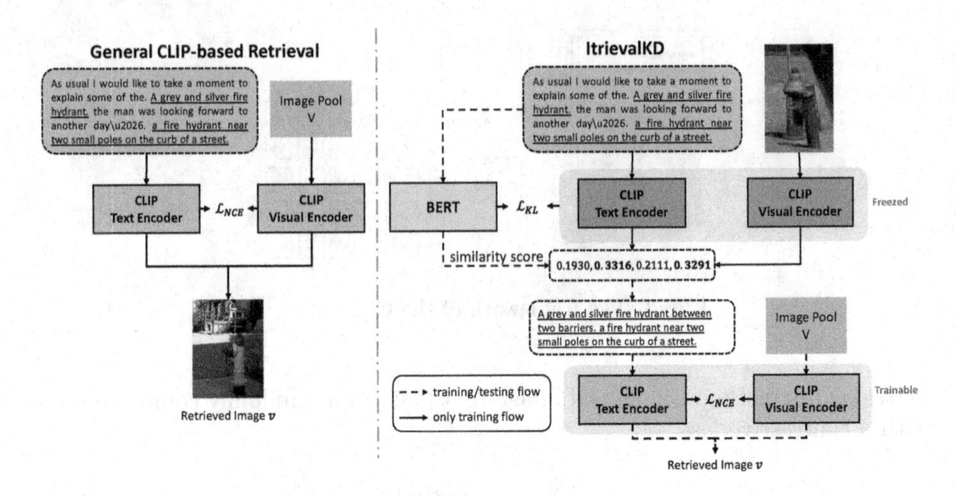

Fig. 4. An illustration of the general retrieval method with the CLIP model (the left panel) and the proposed ItrievalKD framework (the right panel). The underlined part of the input text corresponds to the key-semantic content, and the rest are noisy sentences.

image. Therefore, we calculate the score s_i^k for each sentence t_i^k using Eq. (2) as $s_i^k = sim(v_i, t_i^k)$, which represents the similarity score between the image v_i and the sentence t_i^k. The higher the score is, the more likely the sentence is key-semantic to the image. We then take κ sentences with the highest scores as the key-semantic sentences $\hat{t}_i$ of the noisy text t_i.

Knowledge Distillation for Key-Semantic Extraction. Nevertheless, the lack of the image information paired with the noisy text makes it impossible to directly apply the above strategy to select the key-semantic content during the testing stage, when only the noisy text is available. To resolve that, we need to transfer the image-text correlation knowledge of the CLIP model to a plain-text model, based on which the key-semantic text can be obtained from the plain-text model during the testing stage.

Specifically, we adopt the Knowledge Distillation (KD) technique [6] to distill the knowledge of the image-text content relevance from the teacher model (i.e., CLIP) to the student model (i.e., BERT). The student model BERT is required to mimic the behaviors of the teacher network CLIP when calculating the image-text content relevance scores, followed by ranking the sentences according to the scores.

Since the BERT model picks sentences in unit of sentence, we follow the same input form as BERTSUM [13], which is a method for the extractive summarization task. We insert a [CLS] token before each sentence and a [SEP] token after each sentence. Interval segment embedding is used to distinguish multiple sentences within a text. Finally, we obtain the input to the BERT model by combining the token embeddings, interval segment embeddings and position embeddings. The vector h_i^k, which is the corresponding vector of the k-th [CLS] token from the top BERT layer, will be used as the representation of the sentence t_i^k.

After obtaining the sentence representation h_i^k from BERT, we build a linear layer and a sigmoid layer on the top of the BERT outputs to learn the sentence-image matching scores,

$$\hat{s}_i^k = \sigma(Wh_i^k + b) \tag{3}$$

where σ is the activation function (sigmoid in this work).

We use the Kullback-Leibler (KL) divergence [10] to quantify the discrepancy between the ranking score distributions of the plain-text model BERT and the multi-modal model CLIP. Via knowledge distillation, the plain-text model BERT directly imitates the score distribution from the teacher model CLIP. Formally, the training objective is to minimize the following loss functions with temperature τ,

$$\begin{aligned} \mathcal{L}_{\text{KL}} &= -p \ln \frac{q}{p} \\ p(\hat{s}_i^k, \tau) &= \frac{\exp(\hat{s}_i^k/\tau)}{\sum_k \exp(\hat{s}_i^k/\tau)} \\ q(s_i^k, \tau) &= \frac{\exp(s_i^k/\tau)}{\sum_k \exp(s_i^k/\tau)} \end{aligned} \tag{4}$$

Image Retrieval with the Key-Semantic Clue. After obtaining the key-semantic text $\hat{t_i}$ according to the scores, we can take it instead of the noisy text t_i for cross-modal retrieval. We can adopt Eq. (1) to finetune the CLIP model to further augment the performance.

3.3 Training and Inference

Training. The BERT model is trained with the knowledge distilled from the CLIP model by minimizing $\mathcal{L}_{\text{KL}}$, while the parameters of the CLIP model are frozen. It is optional to finetune the CLIP model with $\mathcal{L}_{\text{NCE}}$ for further performance when retrieving the relevant image with the key-semantic clue. We report the performance both in the zero-shot setting and finetuning setting. As the CLIP model is prone to overfitting when finetuning, we use the noisy text for training to alleviate this problem in the finetuning setting.

Inference. We first retrieve the key-semantic text from the noisy text with the BERT model, and proceed to retrieve the relevant image from the image pool with the key-semantic clue.

4 Experiments

4.1 Datasets

Noisy-MSCOCO. Given the lack of available datasets in the NT2I scenario, we extend the MSCOCO dataset with additional noise to construct the

Noisy-MSCOCO dataset. Specifically, we randomly select 10,000, 1000 and 1000 image-text pairs from MSCOCO for training, validation and test respectively. Noisy sentences are generated with the GPT-2 [16] model by extending each caption with a prompt "and" where the maximum length is set as 35. Finally, we randomly sample n_{key} and n_{noise} sentences from the original captions and noisy sentences separately, followed by shuffling them to construct the Noisy-MSCOCO dataset. In practice, n_{key} is set to 3, and n_{noise} is selected from $\{0, 1, 2, 3\}$.

PhotoChat. PhotoChat is a multi-modal conversation dataset, where each dialogue is paired with an image that is shared during the conversation. Following previous works, we only consider the conversation content of the party who sends the image, because only this party can see the image before sending it.

4.2 Evaluation Metrics

We use Recall@K (R@K), computed as "the fraction of times a correct item was found among the top K results" as the evaluation metric. Specifically, we choose R@1, R@5, and R@10, as well as the sum of them which we denote as "SUM" as [19] to evaluate the proposed method.

4.3 Implementation Details

The proposed model mainly consists of modules based on BERT and CLIP. For BERT, we adopt the "bert-base-uncased" version. We set the batch size to 32, the maxinum input length to 256 and the temperature coefficient τ in Equation (4) to 1. During the validation and test stages, for the Noisy-MSCOCO dataset, we directly adopt n_{key} as κ which is the number of key sentences extracted; and for the PhotoChat dataset, we set κ to 3 in the zero-shot setting and 4 in the finetuning setting as this work best. The best BERT model is chosen according to the accuracy in predicting the key-semantic sentences on the validation set. We employ CLIP (ViT-B/32) and CLIP (RN50) from the series of the CLIP models, and set CLIP (ViT-B/32) as the default. During finetuning, the batch sizes of CLIP (ViT-B/32) and CLIP (RN50) are set as 128 and 64 respectively, and we scale the max input length of the CLIP model to 128 as the original CLIP model limits the text input length to 77 which may be exceeded by the text length in the PhotoChat dataset. The random seed is set to 1 and the Adam optimizer is employed with the learning rate of $1e-5$.

4.4 Baselines

We mainly compare the proposed framework with the general CLIP-based retrieval model. In addition, since the stage of extracting key sentences from the noisy text is similar to the extractive summarization task, we also select two classical unsupervised extractive summarization methods: 1) TF-IDF [10], a statistical method used to assess the importance of words in a document of a

Table 1. The zero-shot retrieval results on the Noisy-MSCOCO dataset. Key-CLIP, CLIP and ItrievalKD correspond to CLIP retrieval with the ground truth key-sentences, with the noisy text and iterative retrieval with the predicted key-sentences respectively.

n_{noise}	CLIP (ViT-B/32)	zero-shot for CLIP				CLIP (RN50)	zero-shot for CLIP			
		R@1	R@5	R@10	SUM		R@1	R@5	R@10	SUM
0	Key-CLIP	56.3	80.5	88.8	225.6	Key-CLIP	55.2	79.9	87.3	222.4
1	CLIP	48.4	74.7	84.5	207.6	CLIP	49.7	73.8	83.1	206.6
	ItrievalKD	52.9	80.4	88.8	222.1	ItrievalKD	53.7	79.0	87.4	220.1
2	CLIP	37.9	66.7	78.0	182.6	CLIP	44.6	70.8	80.3	195.7
	ItrievalKD	53.6	80.4	88.4	222.4	ItrievalKD	54.0	79.5	88.1	221.6
3	CLIP	25.7	48.4	63.4	137.5	CLIP	32.3	59.1	71.3	162.7
	ItrievalKD	53.9	79.8	88.4	222.1	ItrievalKD	54.0	77.7	86.4	218.1

Table 2. The zero-shot and finetuning retrieval results on the PhotoChat dataset.

CLIP version	model	zero-shot for CLIP				finetuning for CLIP			
		R@1	R@5	R@10	SUM	R@1	R@5	R@10	SUM
CLIP (Vit-B/32)	CLIP	23.3	42.9	52.3	118.5	38.5	64.0	72.3	174.8
	TF-IDF-CLIP	13.1	27.2	35.2	75.5	27.2	49.3	57.8	134.3
	TextRank-CLIP	12.8	27.9	35.7	76.4	22.5	42.5	50.9	115.9
	ItrievalKD	26.7	46.3	55.5	127.6	41.2	64.0	72.1	177.3
CLIP (RN50)	CLIP	25.8	43.6	52.0	121.4	31.6	58.7	67.6	157.9
	TF-IDF-CLIP	13.5	27.1	34.4	75	24.1	43.7	54.2	122.0
	TextRank-CLIP	10.6	20.6	27.2	58.4	19.0	37.1	47.9	104.0
	ItrievalKD	26.3	45.2	55.6	127.1	34.5	59.3	68.8	162.6

corpus. Specifically, we take the maximum TF-IDF value of the words in a sentence as the importance score of the sentence; 2) TextRank [14], a graph-based ranking algorithm, in which we construct the graph by treating each sentence as a node. We extract the key sentences from the noisy text with the above two extractive summarization methods, based on which we retrieve the relevant image, which are named as TF-IDF-CLIP and TextRank-CLIP.

4.5 Retrieval Results

The retrieval results on the Noisy-MSCOCO dataset are shown in Table 1. As CLIP is prone to overfitting on the MSCOCO dataset, we only report the results in the zero-shot setting. In the experiments, we compare the retrieval performance of the CLIP model with the ground truth annotations (Key-CLIP) to the one with the noisy text as the query, where we can observe that the retrieval performance of CLIP degrades significantly with the noise increases, compared to the model with no noise. In comparison, the proposed method ItrievalKD can effectively eliminate the influence of the noisy sentences by extracting the

Table 3. The accuarcy in retrieving key sentences in the noisy text with the sentence-image matching scores calculated by CLIP on the Noisy-MSCOCO dataset.

n_{noise}	CLIP (ViT-B/32)	CLIP (RN50)
1	0.9823	0.9843
2	0.9760	0.9793
3	0.9653	0.9643

key-semantic content from the noisy text and achieve comparable results with the noise-free performance of Key-CLIP. For example, R@1 drops from 56.3 to 25.7 when n_{nosie} increases to 3 in the zero-shot setting with CLIP (ViT-B/32), while reaching 53.9 when ItrievalKD is applied.

Table 2 shows the zero-shot and finetuning results on the PhotoChat dataset. The retrieval results of the proposed ItrievalKD surpasses the CLIP model in both zero-shot and finetuning settings, demonstrating its effectiveness. Especially, SUM increases from 118.5 to 127.6 in the zero-shot setting over CLIP (Vit-B/32). In addition, it can be observed that both of the two unsupervised summarization methods (i.e., TF-IDF-CLIP and TextRank-CLIP) even degrade the retrieval performance, which implies that the conventional unsupervised summarization methods are not suitable for key-semantic extraction in the NT2I task.

The results of the ItrievalKD framework based on CLIP (ViT-B/32) and CLIP (RN50) follow the similar trend, which demonstrates the effectiveness and robustness of the proposed method.

4.6 The Effectiveness of Retrieving the Key-Semantic Text in the Noisy Text with CLIP

We proceed to verify the effectiveness of retrieving the key-semantic sentences in the noisy text with CLIP. We show the performance of adopting CLIP to retrieve key sentences on the Noisy-MSCOCO dataset in Table 3. As the Noisy-MSCOCO dataset has key-sentence labels, we use accuracy to evaluate the performance of retrieving key sentences by CLIP. It is observed that, although the accuracy decreases slightly with the noise increases, the accuracy over CLIP (ViT-B/32) on the Noisy-MSCOCO dataset remains 96.53% even when n_{noise} is set to 3. It validates that the strong ability of retrieving the key sentences from the noisy text enables ItrievalKD to achieve comparable results with the noise-free performance of Key-CLIP as shown in Table 1.

4.7 Case Study

An example on the Noisy-MSCOCO dataset is given in Table 4. The general retrieval method given the entire noisy text as the query would return the wrong image, while the ItrievalKD method can retrieve the truly-relevant image. In this

Table 4. Case study on the Noisy-MSCOCO dataset in the zero-shot setting over CLIP. The underlined department of the text is the key-semantic clue.

Text	Negative Image
at 9:46 p.m. this morning, someone in Florida called the police to report that I am extremely excited to present the 6th edition of The Game, a collection of the a red fire hydrant near a dirt road with trees in the background A red fire hydrant in a forest setting. A close of a red fire hydrant next to a road In his final days in office, President Barack Obama has put his administration on a high alert. The	
	Positive Image

case, if the noisy text is used, the general retrieval model may pay attention to the key information "fire hydrant" while ignoring the details such as "forest", "tree", and "in the background". By capturing the key-semantic information in the noisy text, the proposed method can avoid this problem.

5 Conclusion

In this paper, we propose an iterative retrieval framework assisted with knowledge distillation ItrievalKD for the text-to-image retrieval task when the query text contains noise unrelated to the relevant image. As the irrelevant information in the text is harmful to the capturing of key-semantic part for the general retrieval model, the proposed method ItrievalKD first obtains the key-semantic information from the noisy text, followed by retrieving the relevant image from the image pool based on the key-semantic clue. We verify the effectiveness of the proposed method on the Noisy-MSCOCO and PhotoChat datasets.

Acknowledgments. This research was supported by the National Key Research and Development Program of China (Grant No. 2022YFB3103100), the National Natural Science Foundation of China (Grant No. 62276245), and Anhui Provincial Natural Science Foundation (Grant No. 2008085J31).

References

1. Learning transferable visual models from natural language supervision. In: International Conference on Machine Learning, pp. 8748–8763. PMLR (2021)
2. Cui, Y., et al.: Rosita: Enhancing vision-and-language semantic alignments via cross-and intra-modal knowledge integration. In: Proceedings of the 29th ACM International Conference on Multimedia, pp. 797–806 (2021)

3. Devlin, J., Chang, M.W., Lee, K., Toutanova, K.: Bert: Pre-training of deep bidirectional transformers for language understanding. arXiv preprint arXiv:1810.04805 (2018)
4. Dosovitskiy, A., et al.: An image is worth 16x16 words: Transformers for image recognition at scale. arXiv preprint arXiv:2010.11929 (2020)
5. He, K., Zhang, X., Ren, S., Sun, J.: Deep residual learning for image recognition. In: Proceedings of the IEEE Conference on Computer Vision and Pattern Recognition, pp. 770–778 (2016)
6. Hinton, G., Vinyals, O., Dean, J.: Distilling the knowledge in a neural network. Comput. Sci. **14**(7), 38–39 (2015)
7. Ji, Z., Chen, K., Wang, H.: Step-wise hierarchical alignment network for image-text matching. arXiv preprint arXiv:2106.06509 (2021)
8. Ji, Z., Wang, H., Han, J., Pang, Y.: Saliency-guided attention network for image-sentence matching. In: Proceedings of the IEEE/CVF International Conference on Computer Vision, pp. 5754–5763 (2019)
9. Jia, C., et al.: Scaling up visual and vision-language representation learning with noisy text supervision. In: International Conference on Machine Learning, pp. 4904–4916. PMLR (2021)
10. Kullback, S., Leibler, R.A.: On information and sufficiency. Ann. Math. Stat. **22**(1), 79–86 (1951)
11. Lee, K.H., Chen, X., Hua, G., Hu, H., He, X.: Stacked cross attention for image-text matching. In: Proceedings of the European Conference computer vision (ECCV), pp. 201–216 (2018)
12. Lin, T.-Y., et al.: Microsoft COCO: Common Objects in Context. In: Fleet, D., Pajdla, T., Schiele, B., Tuytelaars, T. (eds.) ECCV 2014. LNCS, vol. 8693, pp. 740–755. Springer, Cham (2014). https://doi.org/10.1007/978-3-319-10602-1_48
13. Liu, Y.: Fine-tune bert for extractive summarization (2019)
14. Mihalcea, R., Tarau, P.: Textrank: Bringing order into text (2004)
15. Radford, A., Narasimhan, K., Salimans, T., Sutskever, I., et al.: Improving language understanding by generative pre-training (2018)
16. Radford, A., Wu, J., Child, R., Luan, D., Amodei, D., Sutskever, I., et al.: Language models are unsupervised multitask learners. OpenAI blog **1**(8), 9 (2019)
17. Wu, H., et al.: Unified visual-semantic embeddings: Bridging vision and language with structured meaning representations. In: Proceedings of the IEEE/CVF Conference on Computer Vision and Pattern Recognition, pp. 6609–6618 (2019)
18. Wu, Y., Wang, S., Song, G., Huang, Q.: Learning fragment self-attention embeddings for image-text matching. In: Proceedings of the 27th ACM International Conference on Multimedia, pp. 2088–2096 (2019)
19. Zang, X., Liu, L., Wang, M., Song, Y., Zhang, H., Chen, J.: Photochat: A human-human dialogue dataset with photo sharing behavior for joint image-text modeling. arXiv preprint arXiv:2108.01453 (2021)

Recommender Systems

Semantic Relation Transfer for Non-overlapped Cross-domain Recommendations

Zhi Li[1(✉)], Daichi Amagata[1], Yihong Zhang[1], Takahiro Hara[1], Shuichiro Haruta[2], Kei Yonekawa[2], and Mori Kurokawa[2]

[1] Osaka University, Suita, Japan
{li.zhi,amagata.daichi,hang.yihong,hara}@ist.osaka-u.ac.jp
[2] KDDI Research, Inc., Fujimino, Japan
{sh-haruta,ke-yonekawa,mo-kurokawa}@kddi-research.jp

Abstract. Although cross-domain recommender systems (CDRSs) are promising approaches to solving the cold-start problem, most CDRSs require overlapped users, which significantly limits their applications. To remove the overlap limitation, researchers introduced domain adversarial learning and embedding attribution alignment to develop non-overlapped CDRSs. Existing non-overlapped CDRSs, however, have several drawbacks. They ignore the semantic relations between source and target items, leading to noisy knowledge transfer. Moreover, they learn knowledge from both domain-shared and domain-specific preferences and are hence easily misled by the source-domain-specific preferences. To overcome these drawbacks, we propose a novel semantic relation-based knowledge transfer framework (SRTrans). We semantically cluster the source and the target items and calculate their similarities to extract relational knowledge between domains. To transfer the relational knowledge, we develop a new two-tier graph transfer network. Last, we introduce a task-oriented knowledge distillation supervision and combine it with a prediction loss to alleviate the negative impact of the source-domain-specific preferences. Our experimental results on real-world datasets demonstrate that SRTrans significantly outperforms state-of-the-art models.

Keywords: non-overlapped cross-domain recommendation · graph neural network · knowledge distillation

1 Introduction

The cold-start problem is a general challenge in practical recommender systems [8,14,15]. Cross-domain recommendation (CDR) is a promising solution to alleviate the cold-start problem, because it utilizes sufficient data from a source domain as prior knowledge to support the recommendation in the sparse target domain [7,10,17]. Most cross-domain recommender systems (CDRSs) require overlapped users to transfer individual-level knowledge across domains. These

H. Kashima et al. (Eds.): PAKDD 2023, LNAI 13937, pp. 271–283, 2023.
https://doi.org/10.1007/978-3-031-33380-4_21

methods consider different categories of items in one service (e.g., books and movies on Amazon's e-commerce platform) as different domains [17]. In this scenario, overlapped users are easily collected and can serve as a bridge to connect different domains and also to transfer knowledge. However, matching users in different services is infeasible, particularly when services are provided by different companies. The aforementioned CDRSs cannot make use of source data from external services because of a lack of overlapped users.

To make use of data from external services, non-overlapped CDRSs were developed. These systems transfer statistic-level knowledge and hence can remove the requirement for overlapped users. Domain adversarial learning (DAL) and embedding attribution alignment (EAA) are two mainstream approaches to constructing these CDRSs. DAL basically introduces a domain discriminator to extract domain-independent user and item embeddings [13]. By doing so, users' preferences are better modeled by utilizing both source and target interactions. EAA aligns the attribution of user and item embeddings between source and target domains to guide the target embedding learning with the source knowledge [1,6,10]. However, the state-of-the-art DAL and EAA methods [1,10,13] ignore semantic relations between source and target items and transfer knowledge from *all* source data. This introduces noisy source knowledge and results in a negative transfer and performance degradation, because user preferences contain both domain-shared and domain-*specific* parts [9].

Motivated by the above observations, we propose a novel semantic relation-based knowledge transfer framework (SRTrans) for non-overlapped CDRSs. Different from the state-of-the-art DAL and EAA methods, which extract and transfer knowledge from *all* source data including irrelevant noise, SRTrans introduces a new two-tier graph transfer network to transfer only the relational knowledge from the source to target domains (i.e., to alleviate the negative impact of domain-specific preferences). SRTrans combines a prediction loss with a new task-oriented knowledge distillation supervision. This distills the domain-shared preferences from the knowledge learned by the two-tier graph transfer network. By transferring only the source-domain-*shared* preferences, SRTrans can mitigate the negative impact of the source-domain-specific preferences and can transfer knowledge better than the state-of-the-art methods. In summary, our contributions are three-fold:

1 We propose a semantic relation-based graph transfer framework, namely SRTrans, for non-overlapped CDRSs. SRTrans is robust against noisy source data and can extract useful source knowledge.
2 To mitigate the negative impact of source-domain-specific preferences, we introduce a new task-oriented knowledge distillation supervision and combine it with a prediction loss. This approach improves the final performance of SRTrans.
3 We conduct extensive experiments on real-world datasets, and the results demonstrate that SRTrans outperforms the state-of-the-art methods.

2 Related Work

User-overlapped CDR. Most CDRSs bridge source and target domains through overlapped users. These systems transfer individual-level knowledge learned in the source domain to the target domain. For example, BiTGCF [9] and GA-DTCDR [16] respectively introduced a graph neural network and an attention mechanism as a cross-domain feature transfer layer to fuse overlapped users' source and target latent features. However, the assumption of overlapped users limits their applications.

Another research direction is to learn mappings between source and target user embeddings. EMCDR [11] proposed a multi-layer perceptron to map source user embeddings to target embeddings. PTUPCDR [17] further improved EMCDR by learning personalized meta-transfer mappings.

Non-overlapped CDR. To remove the limitation of overlapped users, non-overlapped CDRSs were developed, and domain adversarial learning (DAL) for non-overlapped CDRSs was introduced. For example, RecSys-Dan [13] devised a discriminator and minimized the divergence of the predictions between source and target domains. Some studies proposed embedding attribution alignment (EAA), which aligns embedding attributions between source and target domains. MMT-Net [6] developed a contextual CDRS and regularized the contextual-jointed target user and item embedding learning with learned source embedding distributions. ESAM [1] and CFAA [10] removed the requirement for domain-shared contextual features and aligned the attribution distribution and correlation between source and target domains.

However, existing DAL and EAA approaches ignore the relations between source and target and leverage all source data to transfer knowledge, resulting in an impaired transfer or even a negative transfer. These approaches, moreover, learn from both domain-shared and domain-specific preferences and yield sub-optimal performances. The above drawbacks motivate us to propose SRTrans, which extracts relational knowledge and alleviates the negative impact of the domain-specific preferences in the source domain.

3 Method

Problem Formulation. We first formulate the cross-domain recommendation task as the top-K recommendation in a sparse target domain $\mathcal{D}_t$, under the assumption of the existence of an auxiliary dense domain $\mathcal{D}_s$ that is considered to be the source. Let $\mathcal{U}_t$ and $\mathcal{V}_t$ denote sets of users and items in $\mathcal{D}_t$, respectively. The interaction set between $\mathcal{U}_t$ and $\mathcal{V}_t$ is denoted as $\mathcal{R}_t = \{(u,v)|u \in \mathcal{U}_t, v \in \mathcal{V}_t\}$. Analogously, we denote the user set, item set, and interaction set in $\mathcal{D}_s$ as $\mathcal{U}_s$, $\mathcal{V}_s$, and $\mathcal{R}_s$, respectively. It is important to note that there is no overlap of user and item between $\mathcal{D}_s$ and $\mathcal{D}_t$. Given a user $u \in \mathcal{U}_t$, the top-K recommendations aim to predict a preference score $\hat{y}_{u,v}$ for each item $v \in \bar{\mathcal{V}}_u = \{v|v \in \mathcal{V}_t, v \notin \mathcal{V}_u\}$, where $\mathcal{V}_u$ is the set of items that interacted with u. The preference score $\hat{y}_{u,v}$ is defined as $\hat{y}_{u,v} = f(\mathbf{e}_u, \mathbf{e}_v)$, where $f(\cdot)$ is a user preference estimator, $\mathbf{e}_u$ is the

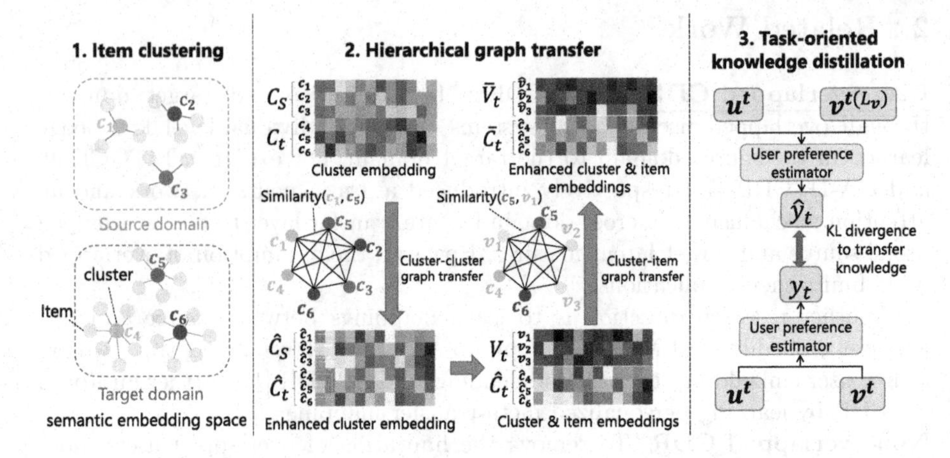

Fig. 1. Overview of our proposed SRTrans

user embedding, and $\mathbf{e}_v$ is the item embedding. The K items with the largest scores are recommended to u.

Overview. Motivated by the previous non-overlapped CDRSs that ignore the relations among interactions and are easily misled by the source-domain-specific preferences, we propose a novel semantic relation-based graph transfer framework (SRTrans) that extracts and transfers cluster-based relational knowledge while alleviating the misleading source-domain-specific preferences. SRTrans is depicted in Fig. 1. The numbers below correspond to the ones in Fig. 1.

(1) We first semantically cluster items and calculate cluster embeddings through an adaptive cluster approach. (2) Then, we calculate similarities of these clusters to measure their relations. We also propose a novel two-tier graph transfer network that extracts and transfers relational knowledge based on the cluster relations, where the relational knowledge is finally aggregated into target items by graph neural networks. (3) After that, we design a new task-oriented knowledge distillation supervision that can be combined with a prediction loss to avoid learning the source-domain-specific preferences.

3.1 Adaptive Semantic Item Cluster

To learn semantic item clusters, we first compute the semantic embeddings of items. Given the item texts, i.e., item descriptions, we apply the semantic token embeddings from a pre-trained BERT model [2] to represent the tokens in the item texts. Let T_v denote the text of an item v, and the semantic embedding of v is the *tf-idf* weighted aggregation of token embeddings:

$$\mathbf{v}_{txt} = \sum_{w \in T_v} tf\text{-}idf(w) \cdot \mathrm{BERT}(w), \tag{1}$$

where $\mathrm{BERT}(w)$ and $tf\text{-}idf(w)$ are respectively the embedding and the *tf-idf* score for token $w \in T_v$. The *tf-idf* score is calculated with combination of the source and target item text corpora. By doing so, we can focus on important tokens and alleviate the negative impact of noisy tokens. The dimension of the BERT semantic embeddings (768) is so high and may lead to tremendous computing costs and over-fitting issues in our sparse target domain. We hence assign a dense layer ϕ_{emb} with output dimension $d \ll 768$ to encode $\mathbf{v}_{txt}$. The encoded item embedding $\mathbf{v}$ is given by

$$\mathbf{v} = \phi_{emb}(\mathbf{v}_{txt}). \tag{2}$$

After that, we calculate semantic embeddings for all source and target items, which are denoted as $\mathbf{V}^s$ and $\mathbf{V}^t$, respectively.

To get semantic item clusters that can facilitate the final recommendation task, we borrow the idea of DE-RRD [4]. Formally, given the cluster number $\mathcal{N}_c$, we calculate the cluster assignment probability vector $\mathbf{p} \in \mathbb{R}^{\mathcal{N}_c}$ with a small network ϕ_{ca}, where the vector of an item v is given by

$$\mathbf{p}_v = \phi_{ca}(\mathbf{v}). \tag{3}$$

Each element $p_{v,j}$ of $\mathbf{p}_v$ represents the probability that the item v is assigned to the cluster j. With the help of $\mathbf{p}_v$, we assign a binary gradient vector $\mathbf{m}_v \in \mathbb{R}^{\mathcal{N}_c}$ to indicate the cluster of item v, where the element j of $\mathbf{m}_v$ is given by

$$m_{v,j} \sim \text{Bern}\left(\frac{\exp(p_{v,j})}{\sum_{k\in\mathcal{N}_c}\exp(p_{v,k})}\right), \tag{4}$$

where Bern$(\cdot)$ represents the Bernoulli distribution. We use the Gumbel-Softmax reparameterization trick to differentiate through the Bernoulli sampling process:

$$m_{v,j} = \frac{\exp((p_{v,j}) + g_j/\tau)}{\sum_{k\in\mathcal{N}_c}\exp((p_{v,k} + g_k)/\tau)}, \quad \mathbf{g} \sim \text{Gumbel}(0,\mathbf{1}) \tag{5}$$

where $\mathbf{g} \in \mathbb{R}^{\mathcal{N}_c}$ is the Gumbel noise drawn from Gumbel$(0, \mathbf{1})$ distribution. Note that g_j is the j-th element of $\mathbf{g}$ and τ is the temperature parameter. The sampling process is separated from $\mathbf{p}_v$, so the cluster assigning network ϕ_{ca} can be end-to-end updated through backpropagations. Let $\mathbf{M^s} \in \mathbb{R}^{|\mathcal{V}^s|\times\mathcal{N}_c}$ ($\mathbf{M^t} \in \mathbb{R}^{|\mathcal{V}^t|\times\mathcal{N}_c}$) denote the binary gradient matrix for the source (target) domain, where each row in $\mathbf{M^s}$ ($\mathbf{M^t}$) indicates the cluster of a source (target) item and can be computed by Eqs.s (3) and (5).

We next declare source cluster embeddings $\mathbf{C}^s$ and target cluster embeddings $\mathbf{C}^t$ as the average semantic embeddings of the items assigned to the corresponding cluster:

$$\mathbf{C}^s = (\tilde{\mathbf{M}}^s)^T\mathbf{V}^s \quad \text{and} \quad \mathbf{C}^t = (\tilde{\mathbf{M}}^t)^T\mathbf{V}^t, \tag{6}$$

where $\tilde{\mathbf{M}}^s$ and $\tilde{\mathbf{M}}^t$ respectively are assignment matrices normalized by the number of items in each cluster, i.e., $\tilde{\mathbf{M}}^s_{[:,i]} = \mathbf{M}^s_{[:,i]}/\sum_i \mathbf{M}^s_{[:,i]}$.

3.2 Two-tier Graph Transfer

We propose a novel two-tier graph transfer network to extract and transfer cluster-based relational knowledge from the source to the target. This network consists of a cluster-level relational graph transfer and a cluster-item relational graph transfer. The former transfers knowledge from the source to the target clusters by constructing a relational graph whose adjacency matrix is denoted as the similarities between the source and target clusters. Analogously, the latter transfers knowledge from the target clusters to the target items by building a graph whose adjacency matrix is denoted as the similarities between the target clusters and items. With these similarity-based adjacency matrices, the graph transfer network can transfer the most relevant source knowledge from source clusters into the target items and thus alleviate the negative impact of irrelevant noises. Therefore, we can remove the drawbacks of the state-of-the-art methods (transferring all source knowledge, including domain-specific knowledge).

Cluster-level Relational Graph Transfer. To extract and transfer cluster-level relational knowledge from the source to the target, we first define a cluster-based relational graph $\mathcal{G}_c$ to represent the relations between clusters from different domains. Let $\mathbf{C} = (\mathbf{C}^s; \mathbf{C}^t) \in \mathbb{R}^{(2\mathcal{N}_c)\times d}$ denote the fused source and target cluster representations and serve as vertices in graph $\mathcal{G}_c$. We further define the edges and the corresponding edge weights. The edge weight $\mathcal{S}(\mathbf{c}_i, \mathbf{c}_j)$ between clusters $\mathbf{c}_i$ and $\mathbf{c}_j$ is measured by the cosine similarity with softmax and is formulated by:

$$\mathcal{S}(\mathbf{c}_i, \mathbf{c}_j) = \text{softmax}(\frac{\mathbf{c}_i \cdot \mathbf{c}_j}{\|\mathbf{c}_i\| \|\mathbf{c}_j\|}). \tag{7}$$

For simplicity, we denote the cluster-based relational graph as $\mathcal{G}_c = (\mathbf{C}, \mathcal{A})$, where $\mathbf{C} = \{\mathbf{c}_i | i \in [1, 2\mathcal{N}_c]\} \in \mathbb{R}^{2\mathcal{N}_c \times d}$ represents the set of vertices and each vertex corresponds to a cluster, whereas $\mathcal{A} = \{\mathcal{S}(\mathbf{c}_i, \mathbf{c}_j) | \mathbf{c}_i \in \mathbf{C}, \mathbf{c}_j \in \mathbf{C}\}$ is the adjacency matrix, which indicates the relations between clusters. Note that $\mathcal{A} = \{\mathcal{S}(\mathbf{C}^s, \mathbf{C}^s), \mathcal{S}(\mathbf{C}^s, \mathbf{C}^t); \mathcal{S}^T(\mathbf{C}^s, \mathbf{C}^t), \mathcal{S}(\mathbf{C}^t, \mathbf{C}^t)\}$ contains cluster relations within and across domains.

After constructing the cluster-based relational graph $\mathcal{G}_c$, we use the graph convolutional network (GCN) [5] to transfer relational knowledge from source clusters $\mathbf{C}^s$ to target clusters $\mathbf{C}^t$, where the GCN is formulated as:

$$\mathbf{C}^{(l+1)} = \text{ReLU}\left(\tilde{\mathbf{D}}^{-\frac{1}{2}} \tilde{\mathcal{A}} \tilde{\mathbf{D}}^{-\frac{1}{2}} \mathbf{C}^{(l)} \mathbf{W}_c^{(l)}\right). \tag{8}$$

Here $\tilde{\mathcal{A}} = \mathcal{A} + \mathbf{I}$, where $\mathbf{I}$ is the identity matrix, corresponds to adding self loops to the graph. Also, $\tilde{\mathbf{D}}$ is the degree matrix with elements $\tilde{D}_{ii} = \sum_j \tilde{A}_{ij}$, and $\mathbf{W}_c^{(l)}$ is a trainable weight matrix for layer l. The input $\mathbf{C}^{(0)} = \mathbf{C}$. After aggregating L_c GCN layers, we get the relational enhanced target cluster representations as the last $\mathcal{N}_c$ rows of $\mathbf{C}^{(L_c)}$, which is denoted as $\mathbf{C}^{t(L_c)} = \{\mathbf{c}_i^{t(L_c)} | i \in [1, \mathcal{N}_c]\}$. By fusing and aggregating source and target clusters together, the enhanced target clusters can refine relational knowledge within and across domains.

Cluster-item Relational Graph Transfer. After calculating the enhanced target clusters $\mathbf{C}^{t(L_c)}$, we further construct a cluster-item relational graph $\mathcal{G}_v$ to measure the relations between enhanced target clusters $\mathbf{C}^{t(L_c)}$ and the target items $\mathbf{V}^t$. With $\mathcal{G}_v$ and cluster-item relations, the useful knowledge from the most relevant $\mathbf{C}^{t(L_c)}$ is aggregated into the target items, while the irrelevant noise is alleviated. To fuse knowledge from both target clusters and items, we concatenate $\mathbf{C}^{t(L_c)}$ and $\mathbf{V}^t$ to serve as vertices in the graph $\mathcal{G}_v$, where the vertex set is denoted as $\mathbf{H} = (\mathbf{C}^{t(L_c)}; \mathbf{V}^t) \in \mathbb{R}^{(\mathcal{N}_c+|\mathcal{V}_t|)\times d}$. Analogously, the edge weight $\mathcal{S}(\mathbf{h}_i, \mathbf{h}_j)$ between vertex $\mathbf{h}_i$ and $\mathbf{h}_j$ is defined as the cosine similarity with softmax:

$$\mathcal{S}(\mathbf{h}_i, \mathbf{h}_j) = \text{softmax}(\frac{\mathbf{h}_i \cdot \mathbf{h}_j}{\|\mathbf{h}_i\|\|\mathbf{h}_j\|}). \tag{9}$$

The cluster-item relational graph $\mathcal{G}_v$ is denoted as $\mathcal{G}_v = (\mathbf{H}, \mathcal{B})$, where the vertex set and the adjacency matrix are $\mathbf{H} = \{\mathbf{h}_i | i \in [1, \mathcal{N}_c + |\mathcal{V}_t|]\}$ and $\mathcal{B} = \{\mathcal{S}(\mathbf{h}_i, \mathbf{h}_j) | \mathbf{h}_i \in \mathbf{H}, \mathbf{h}_j \in \mathbf{H}\}$, respectively.

By leveraging $\mathcal{G}_v$, we next transfer relational knowledge via a GCN, which is formulated as:

$$\mathbf{H}^{(l+1)} = \text{ReLU}\left(\tilde{\mathbf{D}}_v^{-\frac{1}{2}} \tilde{\mathcal{B}} \tilde{\mathbf{D}}_v^{-\frac{1}{2}} \mathbf{H}^{(l)} \mathbf{W}_v^{(l)}\right). \tag{10}$$

Here, $\tilde{\mathcal{B}}$ is the adjacency matrix with self loops, and $\tilde{\mathbf{D}}_v$ is the degree matrix on $\tilde{\mathcal{B}}$. $\mathbf{W}_c^{(l)}$ is a trainable weight matrix for layer l. Also, we define $\mathbf{H}^{(0)} = \mathbf{H}$. After aggregating L_v GCN layers, we get the knowledge-enhanced target item embeddings $\mathbf{V}^{t(L_v)}$ as the last $|\mathcal{V}_t|$ rows of $\mathbf{W}_c^{(L_v)}$.

3.3 Task-oriented Knowledge Distillation

Existing non-overlapped CDRSs [1,10,13] are easily misled by domain-specific preferences, because the source and target embedding space is directly aligned. Motivated by this finding, we combine a knowledge distillation with the target prediction by introducing a new task-oriented knowledge distillation supervision. In this way, the target prediction loss can supervise the knowledge distillation to learn the domain-*share* preferences. As a result, the negative transfer issues incurred by the misleading source-domain-specific preferences can be alleviated.

Formally, we first define the user embeddings of source and target domains as trainable parameter matrices $\mathbf{U}^s$ and $\mathbf{U}^t$, respectively. Each row in $\mathbf{U}^s$ ($\mathbf{U}^t$) corresponds to a source (target) user and can be retrieved with the user ID. The prediction score of the target user u^t for the target item v^t before and after knowledge enhancement in Sect. 3.2 are then given by:

$$\hat{y}_{u,v}^t = f(\mathbf{u}^t, \mathbf{v}^t) \quad \text{and} \quad \tilde{y}_{u,v}^t = f(\mathbf{u}^t, \mathbf{v}^{t(L_v)}), \tag{11}$$

respectively, where $\mathbf{v}^t$ and $\mathbf{v}^{t(L_v)}$ are respectively the item embedding before and after knowledge enhancement. For a user preference estimator $f(\cdot)$, we adopt inner product [12] and LGC [3] to evaluate the performance of our SRTrans on different models. After that, the task-oriented knowledge distillation is defined as the KL-divergence between $\hat{y}_{u,v}^t$ and $\tilde{y}_{u,v}^t$, which is formulated by:

$$\mathcal{L}_{KD}(\hat{y}^t | \tilde{y}^t) = \sum_{(u,v)\in\mathcal{R}^t} \left(\hat{y}_{u,v}^t \log \frac{\hat{y}_{u,v}^t}{\tilde{y}_{u,v}^t} + (1 - \hat{y}_{u,v}^t) \log \frac{(1 - \hat{y}_{u,v}^t)}{(1 - \tilde{y}_{u,v}^t)}\right). \tag{12}$$

We use the pair-wise BPR loss [3] to measure the loss of predictions. To achieve this, we randomly sample a negative item for each source and each target interaction. Taking the target domain as an example, a new interaction $r^t \in \mathcal{R}_t$ is a triplet $r^t = (u, v, v')$, where $u \in \mathcal{U}_t$, $v \in \mathcal{V}_t$, and $v' \in \bar{\mathcal{V}}_u$. Then, the pair-wise BPR loss is given by

$$\mathcal{L}_{BPR} = -\left(\sum_{(u,v,v') \in \mathcal{R}_t} \ln \sigma \left(\hat{y}_{uv} - \hat{y}_{uv'} \right) + \sum_{(u,v,v') \in \mathcal{R}_s} \ln \sigma \left(\hat{y}_{uv} - \hat{y}_{uv'} \right) \right). \quad (13)$$

The total loss is measured by combining the knowledge distillation loss $\mathcal{L}_{KD}$ and the prediction loss $\mathcal{L}_{BPR}$, that is

$$\mathcal{L} = \mathcal{L}_{BPR} + \lambda \mathcal{L}_{KD}, \quad (14)$$

where λ is a hyper-parameter used to balance the weights of different losses. By combining $\mathcal{L}_{KD}$ and $\mathcal{L}_{BPR}$, the negative impact from the source-domain-specific preferences can be alleviated under the supervision of the prediction loss.

4 Experiment

4.1 Experiment Setting

Datasets. We used two public and two private datasets to investigate the recommendation performance of SRTrans in practical applications and for benchmarking purposes.

The public datasets were MovieLens25M[1] and Amazon[2]. For MovieLens25M (ML), we used the movie ratings from *30/9/2016* to *1/10/2018*, where the movie descriptions were collected from TMDB[3]. For Amazon, we built the Amazon-Book (AB) subset containing ratings for books from *30/9/2016* to *3/10/2018.*

The private datasets have an online advertisement dataset (AD) and an e-commerce dataset (E-com). AD contains Web browsing records from *1/8/2017* to *31/8/2017* on an ads platform and the textual content of Web pages. E-com has the purchase records on an e-commerce platform and the textual descriptions of products; purchase records in E-com have the same period as that of AD.

Each CDR scenario consisted of a source domain and a relatively sparse target domain. The CDR scenario from domain A to domain B is denoted as A→B, where A and B are the source domain and the target domain, respectively. All the CDR scenarios included: (1) AD→E-com and E-com→AD; (2) ML→AB and AB→ML. For each source domain, we selected users with 3 to 10 interaction records and items with 10 to 15 interaction records to adapt to a dense setting. Inversely, for each target domain, we selected users with 3 to 5 interactions and items with 5 to 15 interactions to form a relatively sparse environment. Some basic information on the pre-processed datasets is summarized in Table 1.

[1] grouplens.org/datasets/movielens/25m/.
[2] jmcauley.ucsd.edu/data/amazon/.
[3] www.themoviedb.org/documentation/api.

Table 1. Basic information on the datasets we used

	Dataset	#users	#items	#interactions	avg. #interactions per user
As source	ML	18,232	14,435	421,803	23.14
	AB	27,662	12,708	129,899	4.70
	AD	18,829	12,253	360,880	19.17
	E-com	17,418	6,142	81,499	4.68
As target	ML	6,298	9,873	31,445	4.99
	AB	13,350	10,477	61,004	4.57
	AD	11,010	12,031	55,050	5.00
	E-com	12,558	5,118	46,871	3.73

Evaluation Criteria. For each user in the target domains, we took her last and second-last interactions to form the test and validation sets, respectively. The remaining interactions were used as the training set. Then, we ranked her test item with 99 randomly sampled negative items as the candidate recommendations, where negative items are items that have no interaction with this user. We employed widely used *Hit Ratio* (HR) and *Normalized Discounted Cumulative Gain* (NDCG) to measure the accuracy of the top-K recommendation.

Baselines. We compared *SRTrans*[4] with the single-domain baseline (*BASE*) and the following state-of-the-art non-overlapped CDRSs: *RecSys-DAN* [13], *ESAM* [1], and *CFAA* [10]. Because these baselines and our SRTrans are model-agnostic frameworks, we adopted two base models for them: *BPR-MF* [12] (an MF-based model) and *LightGCN* [3] (a GNN-based model). For fair comparison, the embedding module of the baselines was replaced with our semantic encoded item embedding that can be calculated by Eq. (2).

Implementation Details. We adopted the same L2 penalty and mini-batch trick for the evaluated methods and set them to 0.01 and 2048, respectively. The GCN layer for the base model LightGCN was set to 3. The embedding dimension d was 32 and 16 for public and private datasets, respectively. For SRTrans, the temperature parameter τ was 0.0001 and the weight of the knowledge distillation loss λ was 0.1. The number of graph transfer layers L_c and L_v was 1. The number of clusters $\mathcal{N}_c$ was 32. We implemented MF- and LightGCN-based SRTrans, RecSys-DAN, ESAM, and CFAA with PyTorch framework, using the Adam optimizer where the learning rate was set to 0.01. The above hyper-parameters were fine-tuned according to the performance on the validation set.

4.2 Experiment Results

Comparison. Table 2 shows the comparison results. We find that: (1) SRTrans outperforms the baselines w.r.t. HR@5 and NDCG@5 in most cases, especially when the base model is LightGCN. This is because SRTrans transfers relational knowledge into the individual item, and GNN-based models can further fuse knowledge in these items by structurally aggregating them. (2) Although

[4] https://github.com/ZL6298/SRTrans/.

Table 2. Comparison between our proposal and state-of-the-art. Performances ± 95% confidence intervals are reported. Bold shows the winner.

		AD→E-com		E-com→AD	
	Method	HR@1	NDCG@5	HR@1	NDCG@5
MF	BASE	0.224 ± 0.033	0.310 ± 0.014	0.180 ± 0.008	0.299 ± 0.006
	RecSys-DAN	0.269 ± 0.025	0.356 ± 0.015	0.046 ± 0.010	0.102 ± 0.014
	ESAM	**0.283 ± 0.035**	**0.365 ± 0.016**	0.125 ± 0.012	0.239 ± 0.013
	CFAA	0.251 ± 0.036	0.336 ± 0.014	0.140 ± 0.008	0.257 ± 0.007
	SRTrans (ours)	0.243 ± 0.028	0.331 ± 0.011	**0.187 ± 0.005**	**0.307 ± 0.004**
LightGCN	BASE	0.297 ± 0.012	0.372 ± 0.009	0.190 ± 0.008	**0.315 ± 0.008**
	RecSys-DAN	0.254 ± 0.016	0.277 ± 0.016	0.043 ± 0.006	0.065 ± 0.007
	ESAM	0.216 ± 0.033	0.314 ± 0.023	0.114 ± 0.010	0.221 ± 0.012
	CFAA	0.282 ± 0.035	0.375 ± 0.021	0.042 ± 0.004	0.114 ± 0.007
	SRTrans (ours)	**0.312 ± 0.011**	**0.384 ± 0.009**	**0.191 ± 0.010**	0.306 ± 0.009
		ML→AB		AB→ML	
	Method	HR@1	NDCG@5	HR@1	NDCG@5
MF	BASE	0.072 ± 0.003	0.136 ± 0.004	0.031 ± 0.002	0.076 ± 0.002
	RecSys-DAN	0.021 ± 0.006	0.048 ± 0.007	0.005 ± 0.001	0.019 ± 0.002
	ESAM	0.034 ± 0.006	0.079 ± 0.008	**0.032 ± 0.003**	**0.083 ± 0.004**
	CFAA	0.055 ± 0.004	0.109 ± 0.006	0.029 ± 0.003	0.076 ± 0.004
	SRTrans (ours)	**0.075 ± 0.005**	**0.142 ± 0.005**	0.030 ± 0.002	0.075 ± 0.003
LightGCN	BASE	0.122 ± 0.008	0.194 ± 0.006	0.085 ± 0.007	0.165 ± 0.008
	RecSys-DAN	0.042 ± 0.009	0.084 ± 0.009	0.013 ± 0.001	0.041 ± 0.004
	ESAM	0.046 ± 0.010	0.119 ± 0.011	0.070 ± 0.011	0.166 ± 0.018
	CFAA	0.049 ± 0.010	0.121 ± 0.012	0.016 ± 0.001	0.049 ± 0.002
	SRTrans (ours)	**0.129 ± 0.011**	**0.204 ± 0.009**	**0.107 ± 0.008**	**0.202 ± 0.009**

cross-domain baselines transfer user preferences (RecSys-DAN) or align embeddings spaces (ESAM and CFAA) from the source domain to the target domain, they often perform worse than the single-domain method (BASE). This result indicates that domain-specific preferences and noisy source data incur negative transfer issues. (3) SRTrans achieves comparable performance or outperforms the single-domain model (BASE) in most scenarios. This observation confirms that SRTrans remarkably alleviates the negative transfer issue.

Visualization. To better show the knowledge transfer process and explain the learned cluster relations, we visualize the cluster-based relational graph by sampling a mini-batch of interactions from the training data. The result of ML→AB is shown in Fig. 2.

Figure 2(a) depicts a heatmap of an adjacency matrix that describes the relation between the source and target clusters. The colors indicate cosine similarities between clusters. From this figure, we can find that highly related source and target clusters (indicated by light colors) and irrelevant ones (indicated by dark colors) are identified. Figure 2(b) gives a sub-graph constructed by using 10 most related source-target cluster pairs. This figure shows how the knowledge in source clusters is transferred to target clusters. For example, target cluster 23 receives more knowledge from its highly relevant source clusters 13, 30, 7, and 14, compared to other irrelevant clusters. This example highlights the mechanism of alleviating negative transfer issues.

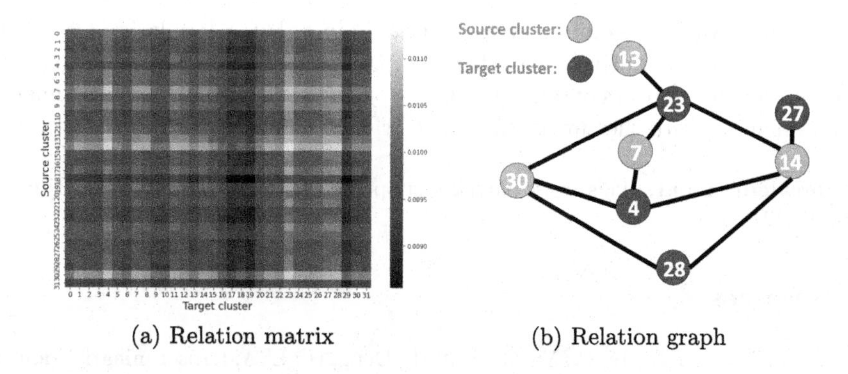

(a) Relation matrix (b) Relation graph

Fig. 2. Visualization result on a mini-batch of data from the ML $\rightarrow$ AB scenario

Table 3. Performances of variants of SRTrans

Method	AD→E-com		E-com→AD	
	HR@1	NDCG@5	HR@1	NDCG@5
w/o SI	0.292 ± 0.008	0.351 ± 0.006	0.090 ± 0.005	0.140 ± 0.005
w/o AC	0.310 ± 0.012	0.376 ± 0.009	**0.194 ± 0.009**	0.310 ± 0.008
w/o KD	0.297 ± 0.012	0.372 ± 0.009	0.190 ± 0.008	**0.315 ± 0.008**
SRTrans	**0.312 ± 0.011**	**0.384 ± 0.009**	0.191 ± 0.01	0.306 ± 0.009
	ML→AB		AB→ML	
w/o SI	0.101 ± 0.008	0.159 ± 0.007	0.060±0.004	0.115±0.005
w/o AC	**0.130 ± 0.011**	**0.207 ± 0.009**	0.097±0.007	0.188±0.009
w/o KD	0.122 ± 0.008	0.194 ± 0.006	0.085±0.007	0.165±0.008
SRTrans	0.128 ± 0.011	0.201 ± 0.009	**0.107 ± 0.008**	**0.202 ± 0.009**

Ablation Study. To study how each module of SRTrans contributes to the final performance, we compared SRTrans with its several variants, namely (1) w/o SI, which replaces semantic item embeddings with randomly initialized ones, (2) w/o AC, which removes adaptive cluster module and directly calculates similarities between single source and target items to build knowledge transfer graph, and (3) w/o KD, which is SRTrans without knowledge distillation and is equal to the single-domain BASE model. Table 3 reports the result.

We see that w/o SI has the lowest performance. This result indicates that the semantic features are essential to extracting the relational knowledge. Moreover, w/o AC shows a comparable performance to SRTrans in some cases, suggesting that transferring item-based relational knowledge can also alleviate the performance degradation caused by source noise data.

5 Conclusion

This work proposed a novel semantic relation-based knowledge transfer framework (SRTrans) for non-overlapped cross-domain recommendations. SRTrans introduces a new two-tier graph transfer network that extracts relational knowledge from a source domain to enhance the target item embeddings. With these embeddings, SRTrans combines a task-oriented knowledge distillation loss with a prediction loss to adaptively learn from domain-shared preferences and to alleviate the negative impacts of source-domain-specific preferences. Our experimental results demonstrate the superiority of SRTrans.

Acknowledgement. This work partially supported by JST CREST Grant Number JPMJCR21F2.

References

1. Chen, Z., Xiao, R., Li, C., Ye, G., Sun, H., Deng, H.: ESAM: discriminative domain adaptation with non-displayed items to improve long-tail performance. In: SIGIR, pp. 579–588 (2020)
2. Devlin, J., Chang, M., Lee, K., Toutanova, K.: BERT: pre-training of deep bidirectional transformers for language understanding. In: NAACL-HLT, pp. 4171–4186 (2019)
3. He, X., Deng, K., Wang, X., Li, Y., Zhang, Y., Wang, M.: Lightgcn: Simplifying and powering graph convolution network for recommendation. In: SIGIR, pp. 639–648 (2020)
4. Kang, S., Hwang, J., Kweon, W., Yu, H.: DE-RRD: A knowledge distillation framework for recommender system. In: CIKM, pp. 605–614 (2020)
5. Kipf, T.N., Welling, M.: Semi-supervised classification with graph convolutional networks. In: ICLR (2017)
6. Krishnan, A., Das, M., Bendre, M., Yang, H., Sundaram, H.: Transfer learning via contextual invariants for one-to-many cross-domain recommendation. In: SIGIR, pp. 1081–1090 (2020)
7. Li, Z., et al.: Debiasing graph transfer learning via item semantic clustering for cross-domain recommendations. In: IEEE Big Data, pp. 762–769 (2022)
8. Li, Z., Amagata, D., Zhang, Y., Maekawa, T., Hara, T., Yonekawa, K., Kurokawa, M.: Hml4rec: hierarchical meta-learning for cold-start recommendation in flash sale e-commerce. Knowl.-Based Syst. **255**, 109674 (2022)
9. Liu, M., Li, J., Li, G., Pan, P.: Cross domain recommendation via bi-directional transfer graph collaborative filtering networks. In: CIKM, pp. 885–894 (2020)
10. Liu, W., Zheng, X., Hu, M., Chen, C.: Collaborative filtering with attribution alignment for review-based non-overlapped cross domain recommendation. In: Web Conference, pp. 1181–1190 (2022)
11. Man, T., Shen, H., Jin, X., Cheng, X.: Cross-domain recommendation: An embedding and mapping approach. In: Sierra, C. (ed.) IJCAI, pp. 2464–2470 (2017)
12. Rendle, S., Freudenthaler, C., Gantner, Z., Schmidt-Thieme, L.: BPR: bayesian personalized ranking from implicit feedback. In: UAI, pp. 452–461 (2009)
13. Wang, C., Niepert, M., Li, H.: Recsys-dan: discriminative adversarial networks for cross-domain recommender systems. IEEE Trans. Neural Netw. Learn. Syst. **31**(8), 2731–2740 (2020)
14. Wang, H., et al.: Preliminary investigation of alleviating user cold-start problem in e-commerce with deep cross-domain recommender system. In: ECNLP, pp. 398–403 (2019)
15. Wang, H., et al.: A dnn-based cross-domain recommender system for alleviating cold-start problem in e-commerce. IEEE Open J. Indust. Electron. Society **1**, 194–206 (2020)

16. Zhu, F., Wang, Y., Chen, C., Liu, G., Zheng, X.: A graphical and attentional framework for dual-target cross-domain recommendation. In: IJCAI, pp. 3001–3008 (2020)
17. Zhu, Y., et al..: Personalized transfer of user preferences for cross-domain recommendation. In: WSDM, pp. 1507–1515 (2022)

Interest Driven Graph Structure Learning for Session-Based Recommendation

Huachi Zhou[1], Shuang Zhou[1], Keyu Duan[2], Xiao Huang[1(✉)], Qiaoyu Tan[3], and Zailiang Yu[4]

[1] The Hong Kong Polytechnic University, Hung Hom, Hong Kong
huachi.zhou@connect.polyu.hk, {csszhou,xiaohuang}@comp.polyu.edu.hk
[2] National University of Singapore, Singapore, Singapore
k.duan@u.nus.edu
[3] Texas A&M University, College Station, USA
qytan@tamu.edu
[4] Zhejiang Lab, Hangzhou, China
yuzl@zhejianglab.com

Abstract. In session-based recommendations, to capture user interests, traditional studies often directly embed item sequences. Recent efforts explore converting a session into a graph and applying graph neural networks to learn representations of user interests. They rely on predefined principles to create edges, e.g., co-occurrence of item pairs in the sequence. However, in practice, user interests are more complicated and diverse than manually predefined principles. Adjacent items in the sequences may not be related to the same interest, while items far away from each other could be related in some scenarios. For example, at the end of shopping, the user remembers to purchase items associated with the one purchased at the beginning. While using predefined rules may undermine the quality of the session graph, it is challenging to learn a reasonable one that is in line with the user interest. Sessions are diverse in length, the total number of interests, etc. Signals for supervision are not available to support graph construction. To this end, we explore coupling the session graph construction with user-interest learning, and propose a novel framework - PIGR. It recognizes items with similar representations learned based on sequential behavior and preserves their interactions. Related items reside in the same induced subgraph and are clustered into one interest. A unified session-level vector is retrieved from the different granularity of interests to guide the next-item recommendation. Empirical experiments on real-world datasets demonstrate that PIGR significantly outperforms state-of-the-art baselines.

1 Introduction

Session-based recommendation has received considerable attention [17] because online users may not log in for fear of breach of privacy, making tracing the historical behaviors of users infeasible. Conventional sequential recommendation is based on rich explicit user-item interactions to reveal user preferences [5].

H. Kashima et al. (Eds.): PAKDD 2023, LNAI 13937, pp. 284–296, 2023.
https://doi.org/10.1007/978-3-031-33380-4_22

Session-based recommendation targets at predicting the next item choice given an anonymous sequence clicked in one session [6]. Early studies on this emerging domain mainly focus on mining actionable patterns from the chronologically ordered items. Multi-layer recurrent neural networks [8] and co-attention mechanism [3] are designed to process consecutive clicks.

Graph neural networks (GNNs) have been under broad research in session-based recommendation. One-way sequence modeling only captures adjacent dependency among consecutive items. The transition may be too sparse to effectively derive user preferences. In many scenarios, distant items might be relevant and nonadjacent dependency could reduce the overfitting brought by the sparsity of sessions. GNNs have been intensively explored to resolve mentioned problems [19]. The basic idea is to convert each click sequence into graphs to enable message passing between distant items. Along this line, advanced models are proposed to better capture collaborative signals. For example, researchers develop dual graph neural networks to exploit both global-level and local-level item transitions [18] or hypergraphs to learn the inherent dependency of items across all sessions [20]. Unanimously, these GNN-based methods express the connectivity of graph structure by manually predefined principles. A dominant heuristic principle is to use co-occurrences of item pairs as edges [2,7,22].

Despite the effectiveness, we argue that user interests are naturally far more complicated and diverse than manually predefined principles. Adjacent items may not have a strong semantic relation, while distant items not adjacent to the same pivot item might still be semantically related in some scenarios. As illustrated in Fig. 1, because the user casually clicks Airpods after wok, there is an edge from wok to Airpods in the graph based on the principle. But wok and pot are more related compared with Airpods. When GNNs recursively aggregate representations of connected items, features not in the same interest will propagate to the same node. Consequently, it may generate inaccurate summary of user interests and lead to suboptimal model performance.

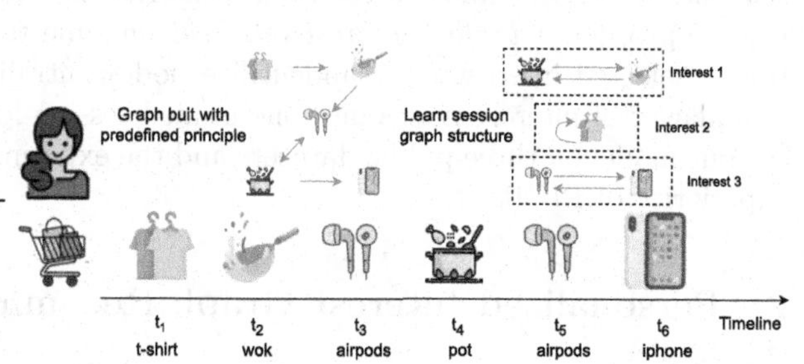

Fig. 1. A running example of two approaches of constructing the session graph, where dashed rectangle indicates one interest.

Motivated by the aforementioned issues, we explore the viability of coupling the session graph construction from scratch with distilling user interests. The items yearn for beneficial information from proximal nodes that share similar features and are clustered into different granularity of interests. The larger magnitude of interest is more possibly being the reason that drives the user to consume next item. However, it is a non-trivial and challenging task. First, the supervision signal indicating item node linkage is unobservable. Items belonging to the same interest are expected to be recognized and resided into the same induced subgraph. But sequences do not have underlying intrinsic graphs that discriminate whether two items in the session should interact or not. The graph structure modification has to take on the opportunity of maximizing the prediction performance. Second, user interests are diverse. Sessions have differentiated sequence structures in terms of session length, distinct items, and interest number. Moreover, user personalized interests evolve from historical actions. The proposed solution is agnostic regarding the specifics of interest distribution for each session in advance. It is demanding to specify an appropriate number of interests for each session ahead of training.

To this end, we propose a novel method dubbed Personalized Interest Graph Recommender (PIGR) to construct session graph structure driven by distilling user interests. The model finds a reasonable graph in the absence of side information with only user-item interactions available. Our main contributions are summarized below: (1) Instead of using predefined principles, we propose a differentiable framework PIGR to enable session graph construction and adaptive interest extraction simultaneously. (2) We propose a *session graph structure learning* module to preserve the connection between similar items inferred by sequential behavior and cluster them centered around the same interest node. (3) We propose a *unified interest retrieval* module to propagate item features to the selected interest node and utilize node centrality to aggregate different granularity of interest nodes into one unified session-level vector. (4) We evaluate our model on three public datasets and the experimental results validate the superiority of PIGR.

2 Personalized Interest Graph Recommender - PIGR

Problem Formulation. Session-based recommender system aims to predict the next item based on an anonymous session. A session contains a series of consecutive items sorted by clicked timestamps in ascending order. Gathering items from all sessions forms the item set $\mathcal{I}$, where $|\mathcal{I}|$ represents cardinality size. For inference, given a session s with m present items, session-based recommender system predicts the probability of item q being picked as the next item, i.e., $p(q|s)$. Among the candidate set $\mathcal{I}$, the item with the highest probability will then be selected as the next one.

Our Solution. Figure 2 provides a pipeline illustration of the PIGR framework. In detail, it is composed of two modules as follows: (i) Session graph structure learning module: It infers item similarity by taking sequential behavior into consideration and explicitly guides the session graph construction by clustering similar items centered around the same interest node; (ii) Unified interest retrieval module: It propagates features to the interest node and formulates an adaptive number of interests. Then it encodes interest nodes over the entire graph to a unified session-level vector.

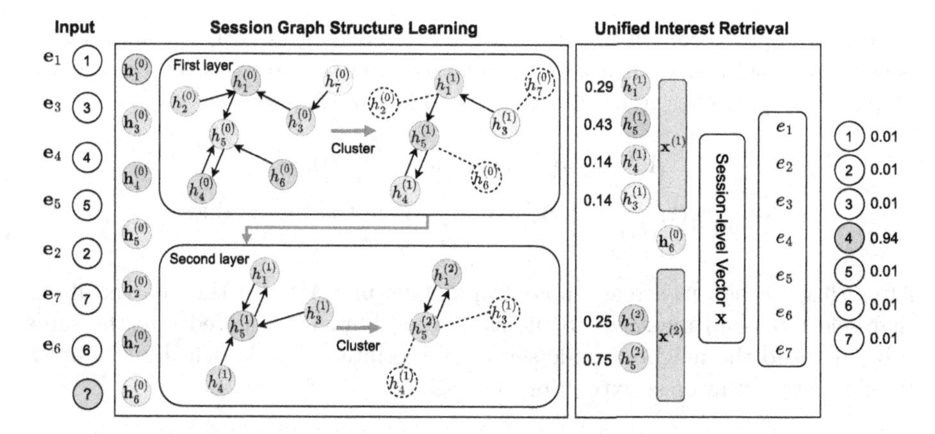

Fig. 2. The circles of the same color belong to the same interest. The first module evaluates the node similarities and converts the session to a graph. The second module propagates neighbor features along learned edges and outputs a unified session-level interest embedding.

2.1 Session Graph Structure Learning Module

At the beginning, the model may not accurately estimate node pairwise strength with only initial embedding. The item order underlying the sequence acts as prior knowledge of graph construction and is beneficial to node connection exploration. The absolute order is designed to depict the dependency contained in the absolute position of items in the sequence while the relative pattern emphasizes the correlation between the current item and prefix items. We use the position embeddings and Gated Recurrent Units (GRU) to model the absolute and relative order of items. The hidden representation of item j processed by GRU is as follows:

$$\boldsymbol{h}_j^{(0)} = GRU(\boldsymbol{h}_{j-1}^{(0)}, [\boldsymbol{e}_j \oplus \boldsymbol{p}_j]), \tag{1}$$

where $\oplus$ is the concatenation function; $\boldsymbol{p}_j \in \mathbb{R}^d$ denotes as a trainable position vector; embedding $\boldsymbol{e}_j$ is from learnable parameter matrix $\boldsymbol{E} \in \mathbb{R}^{|\mathcal{I}| \times d}$. Then we leverage items that integrate sequential information to learn the edges. Given the hidden representations of m nodes at the k-th layer $\left[\boldsymbol{h}_1^{(k)}, \boldsymbol{h}_2^{(k)}, \ldots, \boldsymbol{h}_m^{(k)}\right]$, we measure edge strengths by the cosine similarity metric:

$$\tilde{\boldsymbol{A}}_{ij}^{(k)} = cos(\boldsymbol{h}_i^{(k)}, \boldsymbol{h}_j^{(k)}) + \epsilon \cdot \tau(\boldsymbol{w}_n \boldsymbol{h}_i^{(k)}), \tag{2}$$

where $\epsilon \sim \mathcal{N}(0, 1)$ is a scalar independently sampled from the standard normal distribution, τ is the softplus activation function, and $\boldsymbol{w}_n \in \mathbb{R}^d$ is a learnable vector shared across layers. At the early training stage, it struggles to yield satisfactory item hidden representations. Each node may not selectively determine the optimal neighbors. Therefore, we add the trainable noise which slightly disturbs neighbor weights. Empirically, a node may interact with only a sparse set of nodes. To improve computing efficiency and remove edges with low information density, we pool a sparsified adjacent matrix from the previous fully connected graph by keeping each item with top t neighbors as follows:

$$\begin{aligned} \boldsymbol{A}_{i:}^{(k)} &= Softmax(Topt(\tilde{\boldsymbol{A}}_{i:}^{(k)}, t)), \\ Topt(\tilde{\boldsymbol{A}}_{i:}^{(k)}, t)_j &= \begin{cases} \tilde{\boldsymbol{A}}_{ij}^{(k)}, \text{ if } \tilde{\boldsymbol{A}}_{ij}^{(k)} \text{ is in the top } t \text{ values} \\ -\infty, \text{ otherwise} \end{cases}. \end{aligned} \tag{3}$$

After this, we obtain a reasonable graph structure $\boldsymbol{A}^{(k)}$. In this matrix, items that select to send messages to the same neighbor are clustered into the same subgraph and the neighbor node serves as the interest node, which will inform the downstream interest extraction process.

2.2 Unified Interest Retrieval Module

The former step finds related nodes for each selected interest node. Then we aggregate neighbor representations by performing message passing strategy on the built adjacent matrix $\boldsymbol{A}^{(k)}$ and the hidden representations $\boldsymbol{H}^{(k)}$. The $(k+1)$-th step message passing is computed by:

$$\boldsymbol{H}^{(k+1)} = MLP(\boldsymbol{A}^{(k)}\boldsymbol{H}^{(k)}) + \boldsymbol{H}^{(k)}, \tag{4}$$

where $MLP(\cdot)$ represents a two-layer perceptron network to integrate non-linear signal to each node and generate more expressive hidden representations. In such a manner, semantically similar node features are fused into the same coarsened node along learned edges. These coarsened nodes implicitly denote a set of clusters of multiple scales and propagation operation actually forces each node mapping to one interest then aggregate each interest. In particular, compared with efforts assigning soft cluster assignment matrix to nodes [23], we provide a general recipe to extract an adaptive number of interests without explicitly claiming ahead of training. Then we encode all interest nodes over the entire graph to the output and obtain a composite interest vector. First, the node centrality, i.e., sum of weighted in-degrees indicates the importance of each interest node in the dynamically learned graph structure formulated as $o_i^{(k+1)} = \sum_{j\in\mathcal{N}(i)} \boldsymbol{A}_{ij}^{(k+1)}$. Second, we attend interest nodes obtained from k layers with pooling to preserve the varying locality. The graph-level representation is then expressed by:

$$\boldsymbol{h} = \frac{1}{k+1}\sum_{i=0}^{k}\sum_{j=1}^{m} o_j^{(i+1)} \cdot \boldsymbol{h}_j^{(i+1)}. \tag{5}$$

The summarized interest ignores dependency contained in the linear order of items along time step. So we refine the graph-level representation with sequential information $\widetilde{\boldsymbol{h}}$, which is the last item output from Eq. (1) to learn the unified session-level vector as follows:

$$\begin{aligned} \boldsymbol{\alpha} &= \sigma(\boldsymbol{W}_\alpha[\widetilde{\boldsymbol{h}} \oplus \boldsymbol{h}]), \\ \boldsymbol{x} &= \boldsymbol{\alpha} \odot \widetilde{\boldsymbol{h}} + (\boldsymbol{1} - \boldsymbol{\alpha}) \odot \boldsymbol{h}, \end{aligned} \tag{6}$$

where σ is the sigmoid activation function; $\boldsymbol{W}_\alpha \in \mathbb{R}^{2d \times d}$ is a transformation matrix and $\boldsymbol{\alpha}$ balances the relative importance.

2.3 Training Objective

Our training target for session s is to minimize the following learning objective:

$$\mathcal{L} = \mathcal{L}_c(\hat{y}, y) + \beta \mathcal{L}_{reg}(\boldsymbol{h}, \widetilde{\boldsymbol{h}}), \tag{7}$$

where β controls the magnitude of the second loss. The $\mathcal{L}_c$ is the cross-entropy loss where $y \in \mathbb{R}^{|\mathcal{I}|}$ is the ground truth vector of session s and $\hat{y} \in \mathbb{R}^{|\mathcal{I}|}$ represents the estimated next item clicked probability. The next clicked probability concerning all items is given by: $\hat{y} = softmax(\boldsymbol{x}\boldsymbol{E}^\top)$. And we treat the recommendation task as a classification problem:

$$\mathcal{L}_c = -\sum_{i=1}^{|\mathcal{I}|} y_i \log(\hat{y}_i) + (1 - y_i)\log(1 - \hat{y}_i). \tag{8}$$

The $\mathcal{L}_{reg}$ acts as a regularization loss with data augmentation following [20]:

$$\mathcal{L}_{reg} = -\frac{1}{|\mathcal{B}|}\left(\log \sigma(\boldsymbol{h} \odot \widetilde{\boldsymbol{h}}) + \mathbb{E}_{\hat{\boldsymbol{h}} \sim \mathbb{P}}\big(\log \sigma(1 - \boldsymbol{h} \odot \hat{\boldsymbol{h}})\big) \right), \tag{9}$$

where σ is the sigmoid function, $|\mathcal{B}|$ is the batch size and $\hat{\boldsymbol{h}}$ is the derived embedding from $\widetilde{\boldsymbol{h}}$ with random permutation $\mathbb{P}$. The $\widetilde{\boldsymbol{h}}$ is regarded as congruent linear view of the graph. Maximizing the mutual information through regularization loss provides additional supervision signal and guarantees the interest in compliance with the sequential behaviors.

3 Experiments

In this section, we aim to answer five research questions: **RQ1**: How effective is PIGR compared with the state-of-the-art baselines? **RQ2**: How much do different components utilized by PIGR contribute to the whole model performance? **RQ3**: How is the capability of the methods in handling sessions with different lengths? **RQ4**: What is the influence of hyper-parameters on the PIGR? **RQ5**: What is the distribution of the interest number learned by PIGR?

Table 1. Detailed datasets statistics.

Datasets	Items num	Training num	Test num	Avg length	Length range
LastFM	24,699	799,884	206,723	17.26	Long
Gowalla	57,995	1,064,565	323,593	7.13	Medium
Yoochoose	17,390	312,527	91,428	4.24	Short

3.1 Experimental Setup

Datasets Processing. To study the property of the proposed framework PIGR, we conduct experiments on three real-world datasets LastFM[1], Gowalla[2] and Yoochoose[3] with different average length ranges. We summarize detailed dataset statistics in Table 1. Following previous experimental protocol [24], we filter sequences whose lengths are smaller than 2 in each dataset. Similar to [2], we use the most recent 20% of the original sequences as test sets and leave the rest as training set. And we split the last 20% subset of training set to tune hyper-parameters. Moreover, we apply the segmentation preprocessing technique to each sequence. For an anonymous sequence with elements $[s_1, s_2, \ldots, s_l]$, we generate a series of subsequence and label pairs for model input, i.e., $[[s_1], [s_2]]$, $[[s_1, s_2], [s_3]]$, ..., $[[s_1, s_2, \ldots, s_{l-1}], [s_l]]$.

Table 2. Overall performance comparison w.r.t. Recall@N and NDCG@N scores on the three benchmark datasets where p-value <0.01.

Methods	LastFM		Gowalla		Yoochoose	
	Recall@20	NDCG@20	Recall@20	NDCG@20	Recall@20	NDCG@20
STAMP	22.53	9.95	35.38	19.57	59.35	32.10
MIND	25.73	15.74	25.13	13.55	54.86	27.30
Comirec-SA	16.48	6.95	27.06	16.81	53.46	26.24
CO-SAN	25.92	11.59	48.25	28.32	<u>71.02</u>	<u>40.20</u>
FGNN	25.59	11.65	47.19	27.76	67.14	36.72
GC-SAN	28.27	13.89	51.67	31.22	69.11	38.41
SR-GNN	25.47	12.20	49.29	29.51	68.51	38.32
LESSR	<u>28.33</u>	<u>13.93</u>	52.50	<u>32.82</u>	70.04	40.18
DHCN	27.35	12.47	<u>52.79</u>	31.04	69.52	38.73
PIGR w/ SI	29.02	13.86	52.70	31.78	71.21	40.64
PIGR w/ FG	28.75	13.18	51.18	31.65	70.25	39.20
PIGR	**31.07**	**14.53**	**54.73**	**33.54**	**71.81**	**40.92**
Improvement(%)	**+9.7%**	**+4.3%**	**+3.67%**	**+2.19%**	**+1.11%**	**+1.79%**

[1] http://ocelma.net/MusicRecommendationDataset/index.html.
[2] https://snap.stanford.edu/data/loc-gowalla.html.
[3] https://www.kaggle.com/datasets/chadgostopp/recsys-challenge-2015.

Experimental Settings. We consider the following representative methods to compare with PIGR. (i) To verify the usefulness of modeling distant item transition, two sequential models (STAMP [11], COSAN [12]) and two multi-interest models (MIND [9], Comirec-SA [1]) are included; (ii) To prove the superiority of learning personalized interest graph structure, GNN-based models (FGNN [14], GC-SAN [22], SR-GNN [19], LESSR [2], DHCN [20]) with predefined principles are included. Besides, we also incorporate two variants of PIGR to verify our motivation. PIGR with fixed-graph (PIGR w/ FG) removes Eqs. (1), (2), (3) and constructs a fixed graph using popular co-occurrence rules. PIGR with static interest (PIGR w/ SI) excludes Eqs. (4), (5), (6) and employs the self-attention technique to capture static long-term and short-term interests. For the implementation details, we implement PIGR with Pytorch, where the learning rate is set to 0.0005 and the batch size is set to 512. The Adam optimizer is adopted. We apply the grid search strategy following [20] to tune hyper-parameters based on the validation performance. Each method is independently run five times and reported the average performance. And we adopt two standard evaluation metrics **Recall@N** and **NDCG@N** to measure model performance.

3.2 Comparison with Baselines (RQ1 & RQ2)

Table 2 summarizes all methods performance in terms of Recall@20 and NDCG@20 scores on three datasets. We have the following observations. First, the traditional sequential methods generally behave worse than the GNN models. These cases confirm the necessity of modeling distant item transition in sessions and the power of graph neural networks. Second, the performance of graph neural network competitors is inferior to PIGR. These methods construct the session graph based on the manually predefined principles and may easily introduce unnecessary edges in the sequence. As the session length extends, the relationships among items are more complex than predefined principles. Therefore, PIGR outperforms GNN models by a large margin. Besides, PIGR extracts diverse interests from the learned session graph and consistently outperforms PIGR w/ SI. And PIGR takes advantage of the learnable graph-structured information and achieves better performance than PIGR w/ FG. These results suggest that a promising direction is to learn a personalized graph structure from the session to extract an adaptive number of interests.

Table 3. Performance comparison with different session length ranges in terms of Recall@N and NDCG@N scores on Gowalla dataset.

Methods	Long		Medium		Short	
	Recall@20	NDCG@20	Recall@20	NDCG@20	Recall@20	NDCG@20
STAMP	34.91	19.83	39.08	22.91	33.62	17.34
MIND	20.72	10.44	28.36	15.84	27.86	15.45
Comirec-SA	26.79	16.19	27.85	17.49	26.85	17.04
CO-SAN	51.80	29.80	49.98	29.06	45.36	27.10
FGNN	45.03	24.65	46.98	26.92	44.41	26.77
GC-SAN	54.54	32.07	54.03	32.87	47.70	29.47
SR-GNN	51.81	30.22	51.01	30.62	45.95	28.15
LESSR	55.73	33.97	55.26	34.90	47.91	30.48
DHCN	56.35	32.10	54.52	31.93	48.68	29.51
PIGR	**58.43**	**35.26**	**57.30**	**35.37**	**49.82**	**30.89**

3.3 The Influence of Session Length (RQ3)

Table 4. Length range definition on Gowalla dataset.

Length range	Min length	Max length	Number
Long	15	200	114,540
Medium	6	14	86,339
Short	1	5	122,714

To specifically explore previous baselines performance on sessions in different length ranges, we partition the prediction results of test sessions on Gowalla dataset following the definition in Table 4 in line with each session length. There is a relatively balanced sequence length distribution on Gowalla dataset and it could fairly manifest all models capability of coping with different length ranges. The separate results are reported in Table 3. First, graph neural network models perform equally well or considerably better than the sequential models. It proves the importance of depicting item topological dependency. Second, PIGR has better improvement in the long sessions than in the short sessions and medium sessions. Since the user's complete preference is much more diverse in the long sessions, manually designing the graph structure is not an appropriate choice. It demonstrates the superiority of PIGR handing session lengths in different ranges.

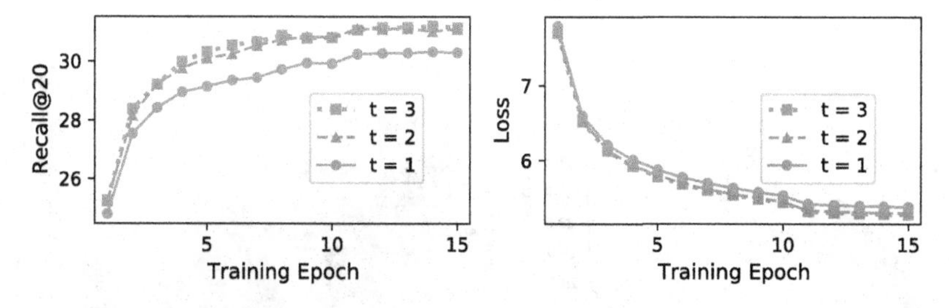

Fig. 3. Empirical training loss and corresponding Recall@20 score with different neighbor number t.

3.4 Hyper-parameter Sensitivities (RQ4)

To evaluate the impact of hyper-parameters, we conduct two groups of hyper-parameter sensitivity experiments. In the first group, to study the impact of the neighbor number t on the training convergence rate, we draw Fig. 3, which characterizes the empirical training loss and performance in terms of Recall@20 scores curve over epoch on LastFM dataset. We have the following observations. First, they do not exhibit a faster convergence rate considering Recall scores ranging t from 1 to 3. Remarkably, the loss curve illustrates that growing neighbor number does not turn the model to converge to a better global minimum. Overall, these observations collectively indicate that increasing neighbor number enhances the model generalization ability and does not impact the convergence rate.

In the second group, we examine joint effects of two hyper-parameters: neighbor number t and embedding dimension d. The prediction results on LastFM dataset are drawn in three-dimensional map in Fig. 4(a). We observe that the performance is consistently better while continuously increasing embedding dimensions. And the model achieves substantial improvement when t and d equals 2 and 150 respectively and marginal improvement with larger values.

3.5 Adaptability Analysis (RQ5)

To prove that PIGR extracts an adaptive number of interests, we select all sessions whose lengths are longer than ten on Yoochoose dataset and calculate the extracted interest number. The visualization is shown in Fig. 4(b), where each cell represents the occurrence ratio of different interest numbers among the same session length. We observe that PIGR extracts different interest numbers within the same session length. And the assigned probability mass is not all concentrated in one cell, proving that PIGR learns an adaptive number of interests.

4 Related Work

Session-based Recommendations. Li et al. [10] incorporate an attention mechanism to calculate each item score to user current interest. Then Yuan

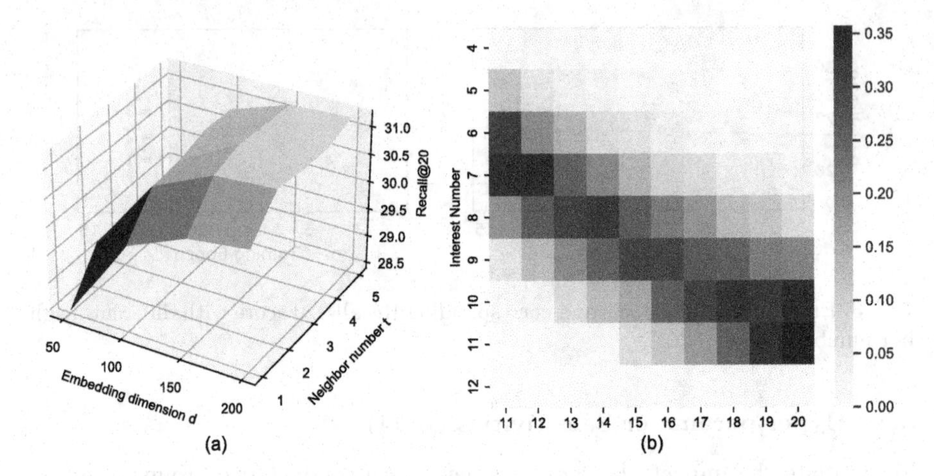

Fig. 4. (a) Test performance under joint impacts of embedding dimension d and neighbor number t. (b) Heat map of extracted interest probability distribution among session lengths.

et al. [24] incorporate α-entmax technologies to filter redundant items. Wu et al. [19] introduce graph gated neural networks to fully explore topological properties of the sequence. Pan et al. [13] add star nodes to link nonadjacent items. However, several works think that current sequences only focus on explicit item dependencies in a single session, ignoring implicit global information between sessions. Wang et al. [18] propose to construct a global item-item graph based on each pair of item occurrences in all training sessions. To capture dynamic user preferences. Qiu et al. [15] propose to utilize a sample reservoir to store valuable samples while Zhou et al. [25] propose to capture temporal information.

Multi-interest Recommendations. A user's sequence may display different users' intents. A next item choice may be due to the influence of multiple interest factors. Cen et al. [1] propose two multi-interest extraction mechanisms: self-attention and dynamic routing. Similarly, Li et al. [9] leverage capsule network to model multi-interests of users at Tmall. Xiao et al. [21] propose a Transformer-based framework to capture diverse interests expressed by the user behaviors. Tan et al. [16] only select the most related k interests from the prototype pool for each user. Cho et al. [4] set up K general proxy to encode general interests shared by multiple sessions.

5 Conclusions and Future Work

In this paper, we propose a novel framework named PIGR for session-based recommendations. Compared with existing GNN-based solutions, PIGR learns a reasonable graph structure from the session instead of directly extracting the co-occurrence of item pairs as edges. This learning process is driven by extracting adaptive interest for each session. Items with similar representations learned by sequential behavior will be clustered to center around the same interest node. The different granularity of interests at each layer can be retrieved as one unified session-level vector of the user. Extensive experiments demonstrate that PIGR achieves significant performance improvement over the state-of-the-art baselines on real-world datasets. Our future work is to explore the applicability of embedding item multiple attributes to graph structure learning.

Acknowledgments. The work described in this paper was partially supported by a grant from the Research Grants Council of the Hong Kong Special Administrative Region, China (Project No. PolyU 25208322) and Key Research Project of Zhejiang Lab (No.2022PG0AC03).

References

1. Cen, Y., Zhang, J., Zou, X., Zhou, C., Yang, H., Tang, J.: Controllable multi-interest framework for recommendation. In: KDD, pp. 2942–2951 (2020)
2. Chen, T., Wong, R.C.W.: Handling information loss of graph neural networks for session-based recommendation. In: KDD, pp. 1172–1180 (2020)
3. Chen, W., Cai, F., Chen, H., de Rijke, M.: A dynamic co-attention network for session-based recommendation. In: CIKM, pp. 1461–1470 (2019)
4. Cho, J., Kang, S., Hyun, D., Yu, H.: Unsupervised proxy selection for session-based recommender systems. In: SIGIR, pp. 327–336 (2021)
5. Fang, H., Zhang, D., Shu, Y., Guo, G.: Deep learning for sequential recommendation: Algorithms, influential factors, and evaluations. TOIS, pp. 1–42 (2020)
6. Hidasi, B., Karatzoglou, A., Baltrunas, L., Tikk, D.: Session-based recommendations with recurrent neural networks. arXiv preprint arXiv:1511.06939 (2015)
7. Huang, C., et al.: Graph-enhanced multi-task learning of multi-level transition dynamics for session-based recommendation. In: AAAI, pp. 4123–4130 (2021)
8. Jannach, D., Ludewig, M.: When recurrent neural networks meet the neighborhood for session-based recommendation. In: RecSys, pp. 306–310 (2017)
9. Li, C., et al.: Multi-interest network with dynamic routing for recommendation at tmall. In: CIKM, pp. 2615–2623 (2019)
10. Li, J., Ren, P., Chen, Z., Ren, Z., Lian, T., Ma, J.: Neural attentive session-based recommendation. In: CIKM, pp. 1419–1428 (2017)
11. Liu, Q., Zeng, Y., Mokhosi, R., Zhang, H.: Stamp: short-term attention/memory priority model for session-based recommendation. In: KDD, pp. 1831–1839 (2018)
12. Luo, A., et al.: Collaborative self-attention network for session-based recommendation. In: IJCAI, pp. 2591–2597 (2020)
13. Pan, Z., Cai, F., Chen, W., Chen, H., de Rijke, M.: Star graph neural networks for session-based recommendation. In: CIKM, pp. 1195–1204 (2020)
14. Qiu, R., Li, J., Huang, Z., Yin, H.: Rethinking the item order in session-based recommendation with graph neural networks. In: CIKM, pp. 579–588 (2019)
15. Qiu, R., Yin, H., Huang, Z., Chen, T.: Gag: Global attributed graph neural network for streaming session-based recommendation. In: SIGIR, pp. 669–678 (2020)
16. Tan, Q., et al.: Sparse-interest network for sequential recommendation. In: WSDM, pp. 598–606 (2021)
17. Wang, S., Cao, L., Wang, Y., Sheng, Q.Z., Orgun, M.A., Lian, D.: A survey on session-based recommender systems. In: CSUR, pp. 1–38 (2021)

18. Wang, Z., Wei, W., Cong, G., Li, X.L., Mao, X.L., Qiu, M.: Global context enhanced graph neural networks for session-based recommendation. In: SIGIR, pp. 169–178 (2020)
19. Wu, S., Tang, Y., Zhu, Y., Wang, L., Xie, X., Tan, T.: Session-based recommendation with graph neural networks. In: AAAI, pp. 346–353 (2019)
20. Xia, X., Yin, H., Yu, J., Wang, Q., Cui, L., Zhang, X.: Self-supervised hypergraph convolutional networks for session-based recommendation. In: AAAI, pp. 4503–4511 (2021)
21. Xiao, Z., Yang, L., Jiang, W., Wei, Y., Hu, Y., Wang, H.: Deep multi-interest network for click-through rate prediction. In: CIKM, pp. 2265–2268 (2020)
22. Xu, C., et al.: Graph contextualized self-attention network for session-based recommendation. In: IJCAI, pp. 3940–3946 (2019)
23. Ying, Z., You, J., Morris, C., Ren, X., Hamilton, W., Leskovec, J.: Hierarchical graph representation learning with differentiable pooling (2018)
24. Yuan, J., Song, Z., Sun, M., Wang, X., Zhao, W.X.: Dual sparse attention network for session-based recommendation. In: AAAI, pp. 4635–4643 (2021)
25. Zhou, H., Tan, Q., Huang, X., Zhou, K., Wang, X.: Temporal augmented graph neural networks for session-based recommendations. In: SIGIR, pp. 1798–1802 (2021)

Multi-behavior Guided Temporal Graph Attention Network for Recommendation

Weijun Xu, Han Li, and Meihong Wang(✉)

Department of Software Engineering, Xiamen University, Xiamen, China
{xuweijun,lihan}@stu.xmu.edu.cn, wangmh@xmu.edu.cn

Abstract. The traditional recommendation approaches learn the representations of users and items utilizing only a single type of behavior data, which results in them facing the data sparsity issue. To alleviate the dilemma, multi-behavior recommendations leverage different types of behaviors to assist in modeling users' preferences. Despite their remarkable effectiveness, two significant challenges have remained less explored: 1) Effectively distinguishing the contributions of different types of behaviors during capturing users' preferences; 2) Sufficiently exploiting the temporal information of user-item interactions. To tackle these challenges, we develop a new model named *Multi-behavior Guided Temporal Graph Attention Network* (MB-TGAT) to discriminate the diverse influence of various behaviors and to explore the evolutionary tendencies of users' recent preferences. In particular, we propose a behavior-aware attention mechanism to differentiate the strengths of different behaviors in the user-item aggregation phase. Furthermore, we tailor a phased message passing mechanism based on GNNs and design an evolution sequence self-attention to extract the users' preferences from static and dynamic perspectives, respectively. Extensive experiments on three real-world datasets demonstrate the superiority of our model, noticeably with 37.27%, 37.31% and 14.63% performance gain over the state-of-the-art baselines on the Taobao, IJCAI-15 and YooChoose datasets, respectively.

Keywords: Multi-behavior Recommendation · Collaborative Filtering · Graph Neural Network

1 Introduction

In the era of information clutter, recommender systems have emerged as effective approaches to alleviate the information overloading issue, which aim to recommend as precisely as possible the user's preferred contents. Collaborative filtering (CF) is a fundamental recommendation method that models users' preferences by capturing the collaborative signals hide in user-item interactions. Early CF methods [6,12] utilize matrix factorization to learn the latent feature representations of users and items. With the flourishing of graph neural networks (GNNs) [9,14], they are introduced into recommendation [5,10,15,16] to dig the high-order relationships among nodes in the user-item interaction graph, which efficiently capture the hidden collaborative signals and obtain excellent node

H. Kashima et al. (Eds.): PAKDD 2023, LNAI 13937, pp. 297–309, 2023.
https://doi.org/10.1007/978-3-031-33380-4_23

representations. However, these methods often face the data sparsity problem since they use user-item interactions limited to a single type of behavior (e.g., purchase).

In fact, in the real-world scenarios, we can effortlessly collect various types of interaction behaviors data, such as click, add-to-cart and purchase in e-commerce platforms. Different types of behaviors express users' diverse intentions and preference levels, hence it is valuable to consider behavior multiplicity during learning the representations of users and items. Early effort [11] expands from BPR to discriminate the contributions of multiple types of feedback. NMTR [4] artificially constrains the cascading relationship among different types of behaviors. EHCF [3] correlates the prediction of each behavior in a transfer way for multi-relational recommendation. Newer approaches [2,8,18] are designed based on GNNs, which utilize the message passing mechanism to capture the high-order connectivity in the user-item multi-behavior interaction graph and adopt attention weight to differentiate the influence of various behaviors.

Despite their remarkable success, we observe that these studies still face two challenges: *First*, distinguishing the contributions of different types of behaviors is challenging but worthwhile for multi-behavior recommendation [8], which ensures that multiple types of behaviors can be sufficiently leveraged. *Second*, we should consider the changes in the temporal dimension of the items interacted with users in a short-term, which reveals the evolutionary trends of their recent preferences. In other words, when we model the complicated user-item multi-behavior relationships, the fact that users' preferences evolve over time should be taken into account. However, this is not an effortless challenge.

To address the above challenges, we propose a novel *Multi-behavior Guided Temporal Graph Attention Network* (MB-TGAT) to differentiate the contributions of different behaviors and to explore the short-term preferences evolution of users. Specifically, we devise a behavior-aware attention mechanism to discriminate the importance of various behaviors in the user-item aggregation phase. On this basis, we distill the users' preferences from both static and dynamic perspectives, as we consider that all items a user interacting with express his/her long-term static intrinsic preferences, while the diversity of items interacted with the user in a short-term implies his/her recent dynamic preferences, and both perspectives mutually complement with each other. We first customize a phased message passing mechanism based on GNNs to learn users' static preference representations. Then we construct an evolution sequence based on the temporal order of user-item interactions and design an evolution sequence self-attention to capture the potential dependencies between items within the evolution sequence to uncover the evolution propensity of users' short-term preferences.

To sum up, the main contributions of this work are summarized as follows:

- We propose a novel framework named MB-TGAT for multi-behavior recommendation, which emphasizes the importance of distinguishing the contributions of different types of behaviors and effectively extracts users' preferences from both static and dynamic perspectives.

- In MB-TGAT, we first design a behavior-aware attention mechanism to discriminate the influence of diverse behaviors, and then a tailored phased message passing pattern and an evolution sequence self-attention are devised to distill user preferences from static and dynamic views, respectively.
- We conduct extensive experiments on three real-world datasets to demonstrate the effectiveness of our model. Comparing with various state-of-the-art single-behavior and multi-behavior recommendations, our MB-TGAT achieves remarkable improvements. Furthermore, the ablation studies are performed to illustrate the indispensability of each sub-module.

2 Related Work

2.1 Graph-Based Recommendation

In recent years, the superior performance of GNNs [9,14] for graph representation learning has attracted widespread focus. In recommender systems, the user-item interactions constitute a bipartite graph, and a number of works [5,10,15,16,19] have shown bright promise in applying GNNs to recommendation tasks. For instance, NGCF [15] utilizes the high-order connectivity in the graph and a message passing mechanism to encode the collaborative signals. LightGCN [5] simplifies the message passing mechanism of GCN [9] to make it more concise and appropriate for recommendation. Additionally, STAM [19] encodes temporal information into the representation of items to express the order of items interacted with users. Recently, SGL [16] and NCL [10] develop a class of model-agnostic contrastive learning frameworks for GNN-based recommendation, which markedly alleviate the data sparsity issue. However, all of these methods only leverage a single type behavior to learn the representations of user and item.

2.2 Multi-behavior Recommendation

Multi-behavior recommendations [2–4,8,17] are devised to leverage multiple types of user-item interactions to remedy the limitations caused by sparse target behavior data in the representation learning of users and items. To be specific, MC-BPR [11] assumes that different levels of user feedback reflect different degrees of preference. NMTR [4] proposes a prior constraint on the preference strength by defining the cascading relationship among multiple behaviors. EHCF [3] transfers the prediction of different behaviors from low-level behavior to high-level behavior to exploit the complicated relations among them. GHCF [2] encodes both nodes and relations in the graph and utilizes GCN propagation layer to capture the collaborative high-hop signals. Furthermore, recently emerged numerous methods [8,17,18] adopt behavior-aware attention mechanisms to distinguish the strengths of different types of behaviors. Specifically, they first learn the representations of users and items under the specific behavior, then use attention weights to integrate the representations learned from each behavior space to express the final user preferences. Nevertheless, none of those approaches involve the temporal information of user-item interactions.

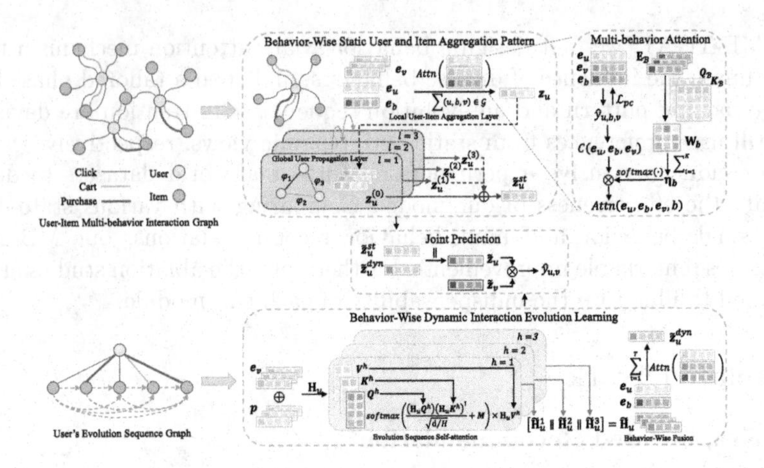

Fig. 1. The overall framework of MB-TGAT.

3 Preliminaries

Different from single-behavior recommendations, which contain only one type of interaction behavior, multi-behavior recommendations involve multiple types of behaviors, such as click, add-to-cart, purchase, etc. in e-commerce platforms. Specifically, among the various behaviors, as the purchase behavior is highly related to the revenue of platforms, which is regarded as the *target behavior* that we aim to predict, while other behaviors are considered as the *auxiliary behavior*.

Suppose there exist N users $\mathcal{U} = \{u_1, u_2, \ldots, u_N\}$, M items $\mathcal{V} = \{v_1, v_2, \ldots, v_M\}$ and K behaviors $\mathcal{B} = \{b_1, b_2, \ldots, b_K\}$, among which b_K is presented for the target behavior and $\{b_1, b_2, \ldots, b_{K-1}\}$ are expressed for the auxiliary behaviors. Here we utilize the triplet (u, b, v) to represent user-item multi-behavior interaction, and the multi-behavior interaction graph can be denoted as $\mathcal{G} = \{(u, b, v)|u \in \mathcal{U}, v \in \mathcal{V}, b \in \mathcal{B}\}$, where each triplet indicates that the user u interacts with item v under the behavior b. Note that $\mathcal{G}$ is an undirected graph, which means that (u, b, v) and (v, b, u) represent the same user-item interaction with behavior b. In addition, from the perspective of user-item interaction sequences, all interactions of each user (regardless of behavior types) can constitute a sequence in chronological order, as the items can. Formally, they are presented as $S_u = \{v_1, v_2, \ldots, v_{|S_u|}\}$, $S_v = \{u_1, u_2, \ldots, u_{|S_v|}\}$, respectively. Finally, we define three matrices $\mathbf{E}_{\mathcal{U}} \in \mathbb{R}^{N \times d}$, $\mathbf{E}_{\mathcal{V}} \in \mathbb{R}^{M \times d}$ and $\mathbf{E}_{\mathcal{B}} \in \mathbb{R}^{K \times d}$ to represent the initial embeddings of all users, items and behaviors separately.

4 Methodology

In this section, we introduce the proposed MB-TGAT in detail, the overall framework of which is shown in Fig. 1, and it is composed of four vital parts: 1) Multi-behavior attention; 2) Behavior-wise static user and item aggregation pattern; 3) Behavior-wise dynamic interaction evolution learning; 4) Joint prediction.

4.1 Multi-behavior Attention

Here, we design a behavior-aware attention mechanism to distinguish the contributions of various behaviors during learning representations of users and items.

Since the user-item interaction is described by a triplet (u, b, v), we can draw on the idea of TransE [1] to calculate the triplet's plausibility score $\hat{y}_{u,b,v}$ as follows:

$$\hat{y}_{u,b,v} = \|\mathbf{e}_u + \mathbf{e}_b - \mathbf{e}_v\|_2^2, \tag{1}$$

where $\mathbf{e}_u \in \mathbb{R}^d, \mathbf{e}_b \in \mathbb{R}^d$ and $\mathbf{e}_v \in \mathbb{R}^d$ represent the initialized embedding of user, behavior type and item, respectively. A smaller $\hat{y}_{u,b,v}$ indicates the triplets more likely to exist in the user-item multi-behavior interaction graph (as shown in Fig. 1), i.e., a higher likelihood that user u interact with item v under the behavior b, and vice versa. Therefore, we cleverly take the reciprocal of $\hat{y}_{u,b,v}$ as the behavior-aware user-item correlation, i.e., $C(\mathbf{e}_u, \mathbf{e}_b, \mathbf{e}_v) = 1/\hat{y}_{u,b,v}$.

To enhance the expression of $\hat{y}_{u,b,v}$, we adopt a pairwise ranking loss:

$$\mathcal{L}_{pc} = \sum_{(u,b,v,v')\in\widetilde{\mathcal{G}}} - \ln \sigma\left(\hat{y}_{u,b,v'} - \hat{y}_{u,b,v}\right), \tag{2}$$

where $\widetilde{\mathcal{G}} = \{(u, b, v, v') \mid (u, b, v) \in \mathcal{G}, (u, b, v') \notin \mathcal{G}\}$, and v' is a randomly sampled item that user u not interacts with under b and $\sigma(\cdot)$ is the sigmoid function.

However, $C(\mathbf{e}_u, \mathbf{e}_b, \mathbf{e}_v)$ can only describe the correlation between users and items under the specific behavior, but cannot express the distinction between multiple behaviors. Hence, we exploit the self-attention mechanism to explore the dependencies among all types of behaviors, which is formally presented:

$$\mathbf{W}_b = f\left(\frac{(\mathbf{E}_{\mathcal{B}}\mathbf{Q}_{\mathcal{B}})(\mathbf{E}_{\mathcal{B}}\mathbf{K}_{\mathcal{B}})^\top}{\sqrt{d}}\right), \quad \eta_{b_j} = f\left(\sum_{i=1}^{K}\mathbf{W}_{b_{i,j}}\right), \tag{3}$$

where $\mathbf{Q}_{\mathcal{B}} \in \mathbb{R}^{d\times d}$ and $\mathbf{K}_{\mathcal{B}} \in \mathbb{R}^{d\times d}$ are learnable transformation matrices for embedding projection and $f(\cdot)$ denotes the $softmax(\cdot)$ function. Since the j-th column elements in $\mathbf{W}_b \in \mathbb{R}^{K\times K}$ indicate the impact of behavior $b_j \in \mathcal{B}$ on other behaviors, we regard their summation as the importance of b_j.

Further, we incorporate the importance of each behavior with the user-item correlation to indicate the contributions of behavior b_j for user u interacting with item v, the behavior-aware attention function as follows:

$$Attn(\mathbf{e}_u, \mathbf{e}_b, \mathbf{e}_v, b_j) = \eta_{b_j} \cdot C(\mathbf{e}_u, \mathbf{e}_b, \mathbf{e}_v). \tag{4}$$

4.2 Behavior-Wise Static User and Item Aggregation Pattern

The general GNN-based recommendations [5,15] are prone to overfitting as the layers of GNNs stack up and the actual number of neighbors aggregated is restricted. Therefore, we split the propagation process into two phases and propose a customized phased message passing framework. To be specific, it leverages a local user-item aggregation layer to aggregate the first-order neighbors of the users/items, followed by accessing higher-hop neighbors via a global user/item propagation layer. In this way, more neighbors of users/items are explored.

Behavior-Wise Local User-Item Aggregation Layer. We adopt the behavior-aware attention to aggregate the first-order neighbors of users and items in the user-item multi-behavior interaction graph, which can be represented as:

$$\mathbf{z}_u = \mathbf{e}_u + \sum_{(u,b,v)\in\mathcal{G}} \xi_{(u,b,v)}\mathbf{e}_v, \quad \mathbf{z}_v = \mathbf{e}_v + \sum_{(u,b,v)\in\mathcal{G}} \xi_{(v,b,u)}\mathbf{e}_u, \tag{5}$$

where $\mathbf{z}_u \in \mathbb{R}^d$ and $\mathbf{z}_v \in \mathbb{R}^d$ denote the local preference representations of users and items separately. In addition, $\xi_{(u,b,v)}$ and $\xi_{(v,b,u)}$ express the behavior-wise attentive aggregation weights, which are formally calculated as:

$$\xi_{(u,b,v)} = f(\bar{\xi}_{(u,b,v)}), \quad \xi_{(v,b,u)} = f(\bar{\xi}_{(v,b,u)}), \quad \bar{\xi}_{(u,b,v)} = Attn(\mathbf{e}_u, \mathbf{e}_b, \mathbf{e}_v, b), \tag{6}$$

where $Attn(\cdot)$ denotes the behavior-aware attention function defined in Eqn (4). Note that we consider the multi-behavior interaction graph to be an undirected graph, hence we present $\bar{\xi}_{(v,b,u)} = \bar{\xi}_{(u,b,v)}$.

Global User and Item Propagation Layer. As mentioned above, all of a user's interactions can constitute an item sequence S_u according to the chronological order of user-item interactions. Thus, all users' interaction sequences naturally form a global user context graph $\mathcal{G}_u^{global} = \{(u,e)|u \in \mathcal{U}, e \in \mathcal{E}_u\}$, where $\mathcal{E}_u$ is the set of edges and edge $e_{i,j}$ exists if S_{u_i} and S_{u_j} share at least one item, and the edge's weight is $\varphi_{u_{i,j}} = \left|S_{u_i} \cap S_{u_j}\right| / \left|S_{u_i} \cup S_{u_j}\right|$. Next, we follow LightGCN [5] to get the weight matrix $\boldsymbol{\varphi}_u' = \mathbf{D}^{\frac{1}{2}}\mathbf{A}\mathbf{D}^{\frac{1}{2}}$, where $A_{i,j} = \varphi_{u_{i,j}}$, $D_{i,i} = \sum_j A_{i,j}$, and to yield the static preference representation $\tilde{\mathbf{z}}_u^{st}$ of user as follows:

$$\mathbf{z}_{u_i}^{(l+1)} = \sum_{u_j \in \mathcal{N}_{u_i}} \varphi_{u_{i,j}}' \mathbf{z}_{u_j}^{(l)}, \quad \tilde{\mathbf{z}}_u^{st} = \frac{1}{L+1}\sum_{l=0}^{L} \mathbf{z}_u^{(l)}, \tag{7}$$

where $\mathbf{z}_{u_i}^{(l+1)}$, $\mathcal{N}_{u_i}$ denote the embedding of user u_i at the $(l+1)$-th propagation layer and the set of its neighbors in $\mathcal{G}_u^{global}$, respectively. Here we let $\mathbf{z}_u^{(0)} = \mathbf{z}_u$. L denotes the total number of GNNs propagation layers. Similarly, the item's representation $\tilde{\mathbf{z}}_v^{st}$ can be obtained in the same approaches.

4.3 Behavior-Wise Dynamic Interaction Evolution Learning

To capture the dynamic preference variations of users, we construct a short-term user-item interaction evolution sequence to express the evolution of users' recent preferences. Specifically, based on the user-item interaction sequence S_u, we can construct the user's short-term interaction evolution sequence $H_u = \{\mathbf{e}_v^1 + \mathbf{p}^1, \ldots, \mathbf{e}_v^t + \mathbf{p}^t, \ldots, \mathbf{e}_v^T + \mathbf{p}^T\}$, where T is the length of H_u. If the length of sequence S_u is longer than T, we truncate S_u to the last T items to build H_u. Otherwise, we repeatedly add a 'padding item', a zero embedding, to the left until the length is T. $\mathbf{p}^t \in \mathbb{R}^d$ indicates the untrainable positional embedding of the t-th position in H_u, which is encoded by the positional encoding in [13]. Corresponding to H_u, there exists a behavior sequence $H_u^b = \{\mathbf{e}_b^1, \ldots, \mathbf{e}_b^t, \ldots, \mathbf{e}_b^T\}$, where $\mathbf{e}_b^i$ expresses the behavior type of user u interacting with the i-th item in H_u. Finally, we pack H_u together into an embedding matrix $\mathbf{H}_u \in \mathbb{R}^{T \times d}$.

Evolution Sequence Self-Attention. Inspired by [7,13], we design an unidirectional multi-head self-attention for evolution sequence to mine the potential relationships between items. In the user's evolution sequence graph (shown in Fig. 1), we specify that the t-th item in the evolution sequence can only attend over the items before and including t, and cannot attend to the future items ($> t$), as the future interacting items should not be of reference value to it.

We use a multi-head attention equipped with a mask matrix $\mathbf{M} \in \mathbb{R}^{T \times T}$ to implement the above unidirectional attention paradigm. Here, we take $\mathbf{H}_u$ as the keys, values and queries in [13] and project them into different spaces through the linear projection matrices $\mathbf{K}^h \in \mathbb{R}^{d \times \frac{d}{H}}$, $\mathbf{V}^h \in \mathbb{R}^{d \times \frac{d}{H}}$ and $\mathbf{Q}^h \in \mathbb{R}^{d \times \frac{d}{H}}$ separately, where H denotes the number of attention heads. Further, the *Scaled Dot-Product Attention* is adopted to get the attention weight matrix $\boldsymbol{\omega}^h$, formalized as follows:

$$\widetilde{\mathbf{H}}_u = \mathop{\Big\|}_{h=1}^{H} \boldsymbol{\omega}^h(\mathbf{H}_u \mathbf{V}^h), \quad \omega_{i,j}^h = f\left(\frac{\left((\mathbf{H}_u \mathbf{Q}^h)(\mathbf{H}_u \mathbf{K}^h)^\top\right)_{i,j}}{\sqrt{d/H}} + m_{i,j}\right), \tag{8}$$

where $\widetilde{\mathbf{H}}_u \in \mathbb{R}^{T \times d}$ is the new embedding matrix of users' evolution sequence and $\|$ denotes concatenation operation. $\omega_{i,j}^h$ is the element of attention weight matrix $\boldsymbol{\omega}^h$ and $m_{i,j} \in \{0, -\infty\}$ is the entry of mask matrix $\mathbf{M}$. When $m_{i,j} = -\infty$, we get a zero attention weight $\omega_{i,j}^h = 0$, which guarantees that the j-th item cannot attend to the i-th item behind it in the evolution sequences (i.e. $j < i$).

Behavior-Wise Fusion. In the previous statements, each item in the users' evolution sequence has a corresponding behavior type. We dexterously leverage the behavior-aware attention to yield the users' dynamic preference representation $\tilde{\mathbf{z}}_u^{dyn}$ through integrating the items' embedding into the embedding of specific user interacting with them, which is formulated as:

$$\tilde{\mathbf{z}}_u^{dyn} = \mathbf{e}_u + \sum_{i=1}^{T} \alpha_i \tilde{\mathbf{h}}_u^i, \quad \alpha_i = \frac{exp\,(\bar{\alpha}_i)}{\sum_{t=1}^{T} exp\,(\bar{\alpha}_t)}, \quad \bar{\alpha}_i = Attn\left(\mathbf{e}_u, \mathbf{e}_b^i, \tilde{\mathbf{h}}_u^i, b\right), \tag{9}$$

where $\tilde{\mathbf{h}}_u^i \in \mathbb{R}^d$ represents the i-th item's new representation in the evolution sequence of user u and $\mathbf{e}_b^i \in \mathbb{R}^d$ is the embedding of corresponding behavior type.

Similarly, the items' dynamic preference representation $\tilde{\mathbf{z}}_v^{dyn}$ can be derived by performing the above operations on the evolution sequence of items.

Table 1. Statistical information of datasets.

Datasets	#Users	#Items	#Click	#Cart	#Purchase	#Avg.length
Taobao	28,201	15,885	549,124	59,830	154,724	27.08
IJCAI-15	34,193	14,102	1,124,515	817	249,536	40.21
YooChoose	42,476	7,868	550,078	–	284,628	19.65

4.4 Joint Prediction and Model Training

We concatenate the static and dynamic preference representations and use a nonlinear transformation to yield the final user/item embedding $\tilde{\mathbf{z}}_u/\tilde{\mathbf{z}}_v$ as follows:

$$\tilde{\mathbf{z}}_u = LeakyReLU\left(\left(\tilde{\mathbf{z}}_u^{st} \| \tilde{\mathbf{z}}_u^{dyn}\right)\mathbf{W}_1\right), \quad \tilde{\mathbf{z}}_v = LeakyReLU\left(\left(\tilde{\mathbf{z}}_v^{st} \| \tilde{\mathbf{z}}_v^{dyn}\right)\mathbf{W}_2\right), \tag{10}$$

where $\mathbf{W}_1 \in \mathbb{R}^{2d \times d}$, $\mathbf{W}_2 \in \mathbb{R}^{2d \times d}$ are trainable parameters and $LeakyReLU(\cdot)$ is an activation function. $\|$ is the concatenation operation. Thereafter, we apply the inner product to predict user and item matching score $\hat{y}_{u,v} = \tilde{\mathbf{z}}_u^\top \tilde{\mathbf{z}}_v$.

Finally, We integrate the $\mathcal{L}_{pc}$ with the BPR loss [12] to optimize our model:

$$\mathcal{L} = \mathcal{L}_{pc} + \sum_{(u,v,v') \in \mathcal{O}} - \ln \sigma\left(\hat{y}_{u,v} - \hat{y}_{u,v'}\right) + \lambda \|\Theta\|_2^2, \tag{11}$$

where $\mathcal{O} = \{(u, v, v') \mid (u, b, v) \in \mathcal{G}, (u, b, v') \notin \mathcal{G}\}$ denotes the training data, v' is a randomly sampled item that user u not interacts with (regardless of behavior type) and $\sigma(\cdot)$ is the sigmoid function. Θ is all trainable parameters in the model and λ controls the weight of L_2 regularization to prevent model overfitting.

In addition, our model could achieve comparable time complexity with the GNN-based multi-behavior recommendations (details in supplement[1]).

5 Experiments

5.1 Experimental Settings

Datasets. We conduct experiments on three real-world datasets and the detailed statistical information of them is summarized in Table 1. i) **Taobao**[2]. This dataset is collected from the e-commerce platform Taobao. ii) **IJCAI-15**[3]. This dataset is from IJCAI 2015 Contest. iii) **YooChoose.** This dataset is released in the RecSys Challenge 2015. Following [2,8], we regard purchase as the *target behavior* and other types of behaviors are considered as *auxiliary behavior*. We adopt the widely used leave-one-out strategy [2,3] to split the datasets.

---

[1] https://github.com/XiaoLangLangY/MB-TGAT.
[2] https://tianchi.aliyun.com/dataset/649.
[3] https://tianchi.aliyun.com/dataset/42.

Table 2. The performance comparison of all models. All the numbers are percentage with '%' omitted. The best result is **bolded** and the runner-up is underlined.

Model	Taobao				IJCAI-15				YooChoose			
	H@20	H@50	N@20	N@50	H@20	H@50	N@20	N@50	H@20	H@50	N@20	N@50
BPR	5.33	8.00	2.55	3.07	2.59	3.67	1.32	1.54	60.01	69.00	33.97	35.77
NeuMF	4.99	7.38	2.40	2.87	2.33	3.40	1.18	1.39	60.74	68.74	34.08	35.68
LightGCN	7.11	10.72	3.35	4.07	4.21	6.10	2.03	2.40	62.36	72.76	34.46	36.54
SGL	7.29	10.78	3.43	4.12	4.52	6.59	2.16	2.57	61.35	73.01	33.00	35.33
NCL	6.92	10.33	3.31	3.98	3.84	5.74	1.82	2.20	62.61	72.78	36.60	38.63
STAM	6.85	9.99	3.20	3.83	4.55	6.37	2.19	2.55	43.89	54.12	22.54	24.58
MC-BPR	7.72	13.05	3.05	4.10	4.05	8.54	1.41	2.29	46.18	62.80	19.17	22.49
NMTR	15.52	23.23	5.66	7.19	11.05	20.46	3.91	5.77	52.82	70.05	20.70	24.14
MBGCN	7.48	11.05	3.53	4.24	4.12	6.14	1.94	2.34	54.34	62.94	31.58	33.30
EHCF	13.07	19.58	6.18	7.47	10.53	17.42	4.48	5.84	57.46	70.64	28.58	31.23
GHCF	14.81	21.47	6.95	8.27	11.66	19.43	4.85	6.38	63.70	77.06	28.86	31.57
MB-TGAT	**20.58**	**27.11**	**9.54**	**10.85**	**16.01**	**26.90**	**6.45**	**8.61**	**73.02**	**80.72**	**39.70**	**41.27**
%Improv	32.60	16.70	37.27	31.20	37.31	31.48	32.99	34.95	14.63	4.75	8.47	6.83

Baselines. To demonstrate the superiority of our MB-TGAT, we compare it with several state-of-the-art models, which are divided into two categories: **Single-behavior Recommendation:** BPR [12] optimizes the pairwise BPR loss to learn the latent features of users and items. NeuMF [6] combines traditional MF and MLP to capture user-item interaction signals. LightGCN [5] simplifies GCN to make it more concise and appropriate for recommendation. SGL [16] introduces self-supervised learning to enhance GNN-based collaborative filtering. NCL [10] constructs node-level contrastive objectives based on two types of neighbors from graph structure and semantic space. STAM [19] incorporates temporal information into neighbor embedding learning. **Multi-behavior Recommendation:** MC-BPR [11] extends from BPR to account for the levels of user feedback in multi-behavior data. NMTR [4] constrains the cascading relationship among various behaviors. MBGCN [8] distinguishes behavior strength by an user-item propagation layer. EHCF [3] links the prediction of each behavior in a transfer manner and applies non-sampling optimization. GHCF [2] encodes both nodes and relations for multi-relational prediction.

Evaluation Metrics and Implementation Details. We adopt two metrics called HR@K and NDCG@K, where K is set to 20 and 50. Following [5,10], we apply the full-ranking strategy to report the metrics. We implement our MB-TGAT in PyTorch and the learning rate is set $1e^{-3}$. We set the embedding dimension to 64 and tune the length of evolution sequence T in $\{5, 10, 20, 30, 40\}$. L is searched in the range of $\{1, 2, 3, 4\}$ and H in $\{1, 2, 4, 8\}$. For the baselines, we refer to the setting of original papers and follow their tuning strategies.

5.2 Performance Comparison

We show the performance of all models on all datasets in Table 2, from which we summarize the following observations:

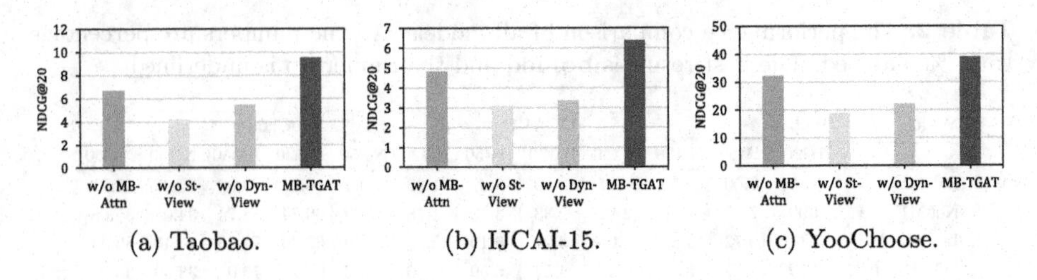

(a) Taobao. (b) IJCAI-15. (c) YooChoose.

Fig. 2. Performance of MB-TGAT and three variants on three datasets.

First, our MB-TGAT shows overwhelming superiority over all state-of-the-art single-behavior and multi-behavior recommendations and achieves remarkable improvements of 32.60% on Taobao, 37.31% on IJCAI-15 and 14.63% on YooChoose in terms of the HR@20. We attribute the outstanding performance to the following reasons: 1) Benefiting from the behavior-aware attention mechanism, we effectively distinguish the contributions of different behaviors during modeling users' preferences. 2) The evolution sequence self-attention adequately exploits the temporal information of user-item interactions to explore the evolutionary trends of users' short-term preferences. *Second*, most of multi-behavior recommendations outperform single-behavior recommendations, since they leverage auxiliary behaviors to assist the target behavior in characterizing users' preferences from different behavior dimensions to achieve better recommendation performance. However, the opposite is true on the YooChoose. The possible reasons are that there are fewer auxiliary behavior interactions in YooChoose and the type of auxiliary behavior is only *Click*, which is noisy and weakens the recommendation performance. *Third*, we carefully observe that GNN-based methods (e.g., LightGCN, NCL) work well than the traditional CF models (e.g., BPR, NeuMF) in single-behavior recommendations. This phenomenon reveals the helpfulness that GNNs capture the potential collaborative signals through mining the high-order connectivity in the user-item interaction graph.

In addition, experiments indicate that our model alleviates the data sparsity issue of target behavior by using the auxiliary behaviors (details in supplement[4]).

5.3 Ablation Study

To verify the effectiveness of each essential component, we design three variants for comparing with our MB-TGAT: **w/o MB-Attn** replaces the $\xi_{(u,b,v)}$ and $\xi_{(v,b,u)}$ in Eq. (5) with the degree normalization weight $1/\sqrt{|\mathcal{N}_u||\mathcal{N}_v|}$ and the mean pooling is adopted in the Eq. (9) to integrate the items within the evolution sequence and removes the loss $\mathcal{L}_{pc}$. **w/o St-View** discards the *static user and item aggregation pattern* and just characterize users' preferences from the dynamic view. **w/o Dyn-View** does not include the *dynamic interaction evolution learning* and models users' preferences from a single static perspective.

[4] https://github.com/XiaoLangLangY/MB-TGAT.

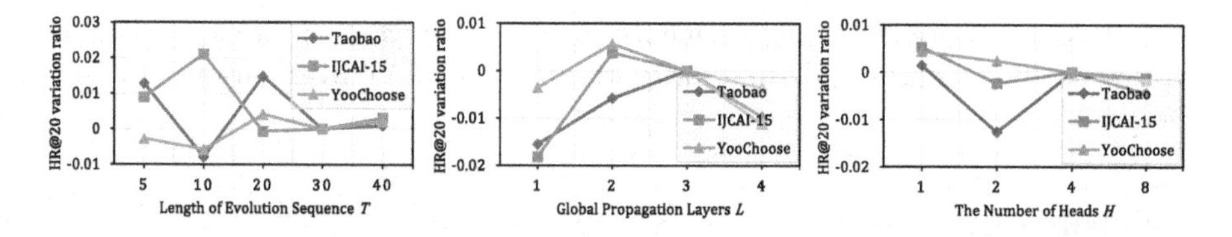

Fig. 3. Performance comparison w.r.t. different T, L and H on three datasets.

As shown in Fig. 2, we can observe that all three variants perform worse than vanilla MB-TGAT, with **w/o St-View** performing the worst, which indicates that each part of the model is indispensable and the long-term static intrinsic preference of user is the dominant factor in determining user's subsequent interactions. In addition, the fact that **w/o MB-Attn** performs slightly inferior to MB-TGAT confirms that we successfully distinguish the contributions of different behaviors during modeling user's preferences.

5.4 Hyper-parameter Analysis

To explore the impact of different hyper-parameter settings on the performance of MB-TGAT, we select three significant hyper-parameters for demonstration and the results are reported in Fig. 3. 1) The length of evolution sequence T. We observe that a suitable T is beneficial to improve the performance of MB-TGAT, as it implies the evolutionary trends of users' short-term preferences. Specifically, the best result is achievable when T is 20 on Taobao and YooChoose, and 10 on IJCAI-15. 2) The number of global propagation layers L. The result shows that our model achieves excellent performance with layers L in $\{2, 3\}$ on all datasets. Nevertheless, stacking too many layers leads to performance degradation, because it introduces noise to the representations of user and item and causes the over-smoothing issue. 3) The number of heads H for multi-head self-attention. We are surprised to observe that $H = 1$ achieves a slightly better performance. We speculate on the primary reason is that the embedding dimension d is only 64 in our model, which is not suitable for splitting into smaller subspaces.

6 Conclusion

In this paper, we propose a new multi-behavior guided temporal graph attention network for recommendation. To be specific, we first design a behavior-aware attention to discriminate the contributions of multiple behaviors during learning representations of users and items. Further, we fully consider both static and dynamic perspectives of users' preferences, and then develop a customized phased message passing mechanism and an evolution sequence self-attention to capture users' intrinsic preferences and recent dynamic preferences, respectively.

We conduct comprehensive experiments on three real-world datasets to demonstrate the effectiveness and superiority of our MB-TGAT over various state-of-the-arts.

References

1. Bordes, A., Usunier, N., Garcia-Duran, A., Weston, J., Yakhnenko, O.: Translating embeddings for modeling multi-relational data. In: Advances in Neural Information Processing Systems, vol. 26 (2013)
2. Chen, C., et al.: Graph heterogeneous multi-relational recommendation. In: Proceedings of the AAAI Conference on Artificial Intelligence, vol. 35, pp. 3958–3966 (2021)
3. Chen, C., Zhang, M., Zhang, Y., Ma, W., Liu, Y., Ma, S.: Efficient heterogeneous collaborative filtering without negative sampling for recommendation. In: Proceedings of the AAAI Conference on Artificial Intelligence, vol. 34, pp. 19–26 (2020)
4. Gao, C., et al.: Learning to recommend with multiple cascading behaviors. IEEE Trans. Knowl. Data Eng. **33**(6), 2588–2601 (2019)
5. He, X., Deng, K., Wang, X., Li, Y., Zhang, Y., Wang, M.: Lightgcn: Simplifying and powering graph convolution network for recommendation. In: Proceedings of the 43rd International ACM SIGIR conference on research and development in Information Retrieval, pp. 639–648 (2020)
6. He, X., Liao, L., Zhang, H., Nie, L., Hu, X., Chua, T.S.: Neural collaborative filtering. In: Proceedings of the 26th International Conference on World Wide Web, pp. 173–182 (2017)
7. Hsu, C., Li, C.T.: Retagnn: Relational temporal attentive graph neural networks for holistic sequential recommendation. In: Proceedings of the Web Conference 2021, pp. 2968–2979 (2021)
8. Jin, B., Gao, C., He, X., Jin, D., Li, Y.: Multi-behavior recommendation with graph convolutional networks. In: Proceedings of the 43rd International ACM SIGIR Conference on Research and Development in Information Retrieval, pp. 659–668 (2020)
9. Kipf, T.N., Welling, M.: Semi-supervised classification with graph convolutional networks. arXiv preprint arXiv:1609.02907 (2016)
10. Lin, Z., Tian, C., Hou, Y., Zhao, W.X.: Improving graph collaborative filtering with neighborhood-enriched contrastive learning. In: Proceedings of the ACM Web Conference 2022, pp. 2320–2329 (2022)
11. Loni, B., Pagano, R., Larson, M., Hanjalic, A.: Bayesian personalized ranking with multi-channel user feedback. In: Proceedings of the 10th ACM Conference on Recommender Systems, pp. 361–364 (2016)
12. Rendle, S., Freudenthaler, C., Gantner, Z., Schmidt-Thieme, L.: Bpr: Bayesian personalized ranking from implicit feedback. In: Proceedings of the Twenty-Fifth Conference on Uncertainty in Artificial Intelligence, pp. 452–461 (2009)
13. Vaswani, A., et al.: Attention is all you need. In: Advances in Neural Information Processing Systems, vol. 30 (2017)
14. Veličković, P., Cucurull, G., Casanova, A., Romero, A., Lio, P., Bengio, Y.: Graph attention networks. arXiv preprint arXiv:1710.10903 (2017)
15. Wang, X., He, X., Wang, M., Feng, F., Chua, T.S.: Neural graph collaborative filtering. In: Proceedings of the 42nd international ACM SIGIR conference on Research and development in Information Retrieval, pp. 165–174 (2019)

16. Wu, J., et al.: Self-supervised graph learning for recommendation. In: Proceedings of the 44th International ACM SIGIR Conference on Research and Development in Information Retrieval, pp. 726–735 (2021)
17. Xia, L., Huang, C., Xu, Y., Dai, P., Lu, M., Bo, L.: Multi-behavior enhanced recommendation with cross-interaction collaborative relation modeling. In: 2021 IEEE 37th International Conference on Data Engineering (ICDE), pp. 1931–1936. IEEE (2021)
18. Xia, L., et al.: Knowledge-enhanced hierarchical graph transformer network for multi-behavior recommendation. In: Proceedings of the AAAI Conference on Artificial Intelligence. vol. 35, pp. 4486–4493 (2021)
19. Yang, Z., Ding, M., Xu, B., Yang, H., Tang, J.: Stam: A spatiotemporal aggregation method for graph neural network-based recommendation. In: Proceedings of the ACM Web Conference 2022, pp. 3217–3228 (2022)

Pure Spectral Graph Embeddings: Reinterpreting Graph Convolution for Top-N Recommendation

Edoardo D'Amico[1,2](✉), Aonghus Lawlor[1,2], and Neil Hurley[1,2]

[1] Insight Centre for Data Analytics, Dublin, Ireland
{edoardo.damico,aonghus.lawlor,neil.hurley}@insight-centre.org
[2] University College Dublin, Dublin, Ireland

Abstract. The use of graph convolution in the development of recommender system algorithms has recently achieved state-of-the-art results in the collaborative filtering task (CF). While it has been demonstrated that the graph convolution operation is connected to a filtering operation on the graph spectral domain, the theoretical rationale for why this leads to higher performance on the collaborative filtering problem remains unknown. The presented work makes two contributions. First, we investigate the effect of using graph convolution throughout the user and item representation learning processes, demonstrating how the latent features learned are pushed from the filtering operation into the subspace spanned by the eigenvectors associated with the highest eigenvalues of the normalised adjacency matrix, and how vectors lying on this subspace are the optimal solutions for an objective function related to the sum of the prediction function over the training data. Then, we present an approach that directly leverages the eigenvectors to emulate the solution obtained through graph convolution, eliminating the requirement for a time-consuming gradient descent training procedure while also delivering higher performance on three real-world datasets.

Keywords: Collaborative filtering · Graph convolution · Spectral methods

1 Introduction

Graph convolutional networks (GCN) are a form of deep learning network which leverages the structural information in a graph representation of the training data [9]. The convolutional layers of the network aggregate each nodal feature with those of its neighbours in the graph. By constructing a network of h convolutional layers, the node embedding becomes dependent on the features of nodes that are h-hops away from it in the network. We focus on the Light-GCN algorithm [8] that has received a lot of attention recently due to the fact that it has demonstrated that, given only user-item interaction data without rich user and item features, the convolutional layers can be greatly simplified.

H. Kashima et al. (Eds.): PAKDD 2023, LNAI 13937, pp. 310–321, 2023.
https://doi.org/10.1007/978-3-031-33380-4_24

In particular, it argues that the non-linear activation and the trainable weights of the full GCN can be removed from the convolution without any degradation to the accuracy of the model and a substantial saving in the complexity of training. LightGCN has been shown to obtain state-of-the-art performance in terms of top-N performance measures on a number of recommender datasets. In this paper, we address the question of why LightGCN achieves good performance, despite its simple convolutional layers. While it is surely true that LightGCN is less complex than a regular GCN [9], it is also true that it requires training by gradient descent and that each update to the model parameters is much more complex than the updates of standard matrix factorisation algorithms, such as BPR. Hence, we ask if LightGCN is fundamentally better at capturing features in the dataset that a standard matrix factorisation model will miss. We show that, without the non-linear activation functions, the convolutions of LightGCN can be understood as graph filters that have the effect of generating features that are largely embedded in a subspace spanned by the eigenvalues of the normalised interaction matrix corresponding to its largest eigenvalues. We show why this is a suitable subspace in which to find quality solutions to solve the top-N recommendation problem. With this spectral interpretation of LightGCN, we proceed to build spectral recommender model, which we call *Pure Spectral Graph Embeddings* (PSGE) that leverages the principles behind LightGCN, while having a closed-form solution that can be found through an eigen-decomposition of the interaction matrix, rather than through a gradient descent algorithm. Given that fast algorithms for eigen-decomposition of sparse matrices are available [5], PSGE can be learned in a fraction of the time that it takes to train LightGCN. We demonstrate that PSGE out-performs LightGCN on a number of recommendation datasets. We also test its performance against the other leading linear algorithms in the literature and show that PSGE can be configured to achieve high recommendation performance while reducing the popularity bias that is evident in these other similar algorithms.

2 Preliminaries

Let $\mathcal{U}$ be a set of users of size $|\mathcal{U}| = U$ and $\mathcal{I}$ be a set of items of size $|\mathcal{I}| = I$. Given a $U \times I$ interaction dataset $\mathrm{R} = \{r_{ui}\}$ where r_{ui} represents implicit feedback given by user u on item i, the top-N recommendation problem is to recommend a set of $N > 0$ items that the system predicts are relevant to a given user u. Typically, a prediction function computes a relevance score $\hat{r}_{ui}$, the items are ranked according to $\hat{r}_{ui}$ and the top items in this ordering are recommended. We focus on latent space methods, where, for each user and item, a f-dimensional embedding, denoted respectively as $\mathbf{p}_u$ and $\mathbf{q}_i$ is learned from the interaction data, and the prediction function is the inner product of the user and item embeddings. Write P for the $U \times f$ matrix whose rows are the user embeddings $\mathbf{p}_u$ and Q for the $I \times f$ matrix whose rows are the item embeddings $\mathbf{q}_i$.

Graph convolution methods interpret user-item interactions as edges of a graph. More formally, the interaction data R can be represented as an undirected bipartite graph $\mathcal{G}_R = (\mathcal{V}, \mathcal{E})$ with nodes $\mathcal{V} = \mathcal{U} \cup \mathcal{I}$ and edges $\mathcal{E} = \{(u,i) | u \in \mathcal{U}, i \in \mathcal{I}, r_{ui} \neq 0\}$ connect user nodes u to item nodes i whenever there is an interaction between them in R.

2.1 Graph Signals and the Graph Fourier Transform

Given an undirected weighted graph of order n, with adjacency matrix $\mathrm{A} = \{a_{ij}\} \in \mathbb{R}$, a signal over the graph is a function $f : \mathcal{V} \rightarrow \mathbb{R}$. For any signal, we can form the n-dimensional vector $\mathbf{x} \in \mathbb{R}^n$ such that the ith component of $\mathbf{x}$ represents the value of the signal at the ith vertex of $\mathcal{V}$. Notice how concatenating the embedding matrices, P and Q into a single $(U+I) \times f$ dimensional matrix X, we obtain a matrix in which each *column* represents a signal over the bipartite graph $\mathcal{G}_R$. From this alternative point of view, the learning process involves the learning of f different signals over the graph which can be thought as *latent features* constructing the user and item representations.

A graph convolution operation on a signal is a weighted sum of the signal at a node with its values in a neighbourhood of up to $(n-1)$-hops from the node and can be represented as a polynomial over a propagation matrix S with weights g_i:

$$conv(\mathbf{x}, \mathbf{g}) = \sum_{i=0}^{n-1} g_i \mathrm{S}^i \mathbf{x} \equiv \mathbf{g} * \mathbf{x}.$$

A common choice for the propagation matrix S is the normalised Laplacian Δ, of a graph with adjacency A, defined as $\Delta = \mathrm{I} - \mathrm{D}^{-1/2}\mathrm{A}\mathrm{D}^{-1/2}$, where D is the diagonal matrix of node degrees with diagonal elements $d_{ii} = \sum_j a_{ij}$. Δ is a symmetric positive semi-definite matrix, so that its eigenvalues λ_i are non-negative and its eigenvectors form an orthogonal basis, allowing the decomposition, $\Delta = \mathrm{U}\Lambda\mathrm{U}^\top$, where U is the matrix whose columns are the n orthonormal eigenvectors and $\Lambda = \mathrm{diag}(\lambda_1, \ldots, \lambda_n)$ is the diagonal matrix of the eigenvalues, assumed ordered such that $0 = \lambda_1 \leq \lambda_2 \leq \cdots \leq \lambda_n$. Note that, while Δ is a common choice, any real symmetric matrix associated with the graph can be chosen to define the graph spectrum.

The *Graph Fourier Transform* [16] $\hat{\mathbf{x}}$ of a graph signal $\mathbf{x}$ is defined as its projection into the eigenvector basis U, i.e., $\hat{\mathbf{x}} = \mathrm{U}^\top \mathbf{x}$, with inverse operation defined as $\mathbf{x} = \mathrm{U}\hat{\mathbf{x}}$. In the Fourier domain of the eigenvector basis, a convolution is a simple element-wise multiplication, such that $\widehat{\mathbf{g} * \mathbf{x}} = \hat{\mathbf{g}}\hat{\mathbf{x}}$, where[1],

$$\hat{g}_i = \hat{g}_i(\lambda_i) = \sum_{j=0}^{n-1} g_j \mathrm{U}_j^i$$

This shows that the spectral coefficient $\hat{x}_i(\lambda_i)$ reflecting the correlation of the signal $\mathbf{x}$ with the ith eigenvector, is scaled by $\hat{g}_i(\lambda_i)$ and hence $\hat{\mathbf{g}}$ can be thought of as a spectral filter, which can enhance or diminish certain frequencies of the

[1] A polynomial $p(\mathrm{A})$ has the same eigenvectors as A, with eigenvalues given by $p(\lambda)$, where λ is an eigenvalue of A.

signal. We can understand the impact of the convolution on a signal most easily by studying $\hat{g}_i(\lambda_i)$ in the Fourier domain.

In the collaborative filtering problem only the historical user-item interactions are available, making impossible to seed the initial representations (signals) for the user and item nodes. In [20], it is proposed to initialise the node embeddings as free parameters and learn jointly the representations and filters from the training data. The complexity of this approach has been shown to downgrade the quality of the user and item representations learnt [3,8]. To overcome this problem, in [8], the "light convolution" method, LightGCN, is proposed, where the only free parameters correspond to the user and item representations. In the following, we show that the chosen propagation matrix corresponds to a fixed high-pass filter in the spectral domain defined by the normalised adjacency matrix.

3 LightGCN as a High-Pass Filter

LightGCN [8] is a state-of-the-art graph convolution model for the top-N recommendation task. It uses $\mathrm{S} = \mathrm{D}^{-1/2}\mathrm{A}\mathrm{D}^{-1/2}$ as a propagation matrix to exchange information along the edges of the graph, where A is the adjacency matrix of the user-item interaction graph $\mathcal{G}_R$. At the first step, the latent features of user and items (signals) $\mathrm{X}^{(0)} = [\mathrm{P}^{(0)}; \mathrm{Q}^{(0)}] \in \mathbb{R}^{(U+I)\times f}$ are randomly initialised and then updated at every convolution step as $\mathrm{X}^{(k)} = \mathrm{S}\mathrm{X}^{(k-1)}$. The final user and item latent features are then computed as a weighted combination of the signals at each convolution step:

$$\mathrm{X} = \alpha_0\mathrm{X}^{(0)} + \ldots + \alpha_k\mathrm{X}^{(k)} = \left(\alpha_0\mathrm{I} + \alpha_1\mathrm{S} + \ldots + \alpha_k\mathrm{S}^k\right)\mathrm{X}^{(0)}$$

The authors reported that learning the signals and the coefficients α_i jointly lead to worse results than assigning the uniform weights, $\alpha_0 = \alpha_1 = \ldots = \alpha_k = 1/(k+1)$.

By carrying out an analysis over the spectrum defined from the symmetric normalised Laplacian Δ, in [15], the convolution operation is shown to correspond to a low-pass filter. If we instead consider the spectrum defined by the propagation matrix $\mathrm{S} = \mathrm{I} - \Delta$ -i.e. the symmetric normalised adjacency matrix- we show how the convolution correspond to an high-pass filter.

$$\hat{g}_i(\lambda_i) = \frac{1}{k+1}(1 + \lambda_i + \lambda_i^2 + \ldots + \lambda_i^k)$$

where λ_i are the eigenvalues of S. Applying the sum of the geometric series, we can express it more concisely as:

$$\hat{g}(\lambda) = \begin{cases} \frac{1}{k+1}\frac{1-\lambda^{k+1}}{1-\lambda} & \lambda < 1 \\ 1 & \lambda = 1 \end{cases} \tag{1}$$

In Fig. 1, $\hat{g}_i$ is plotted against λ_i for different values of the convolution depth k. It illustrates that the convolution acts as a high-pass filter over the spectrum

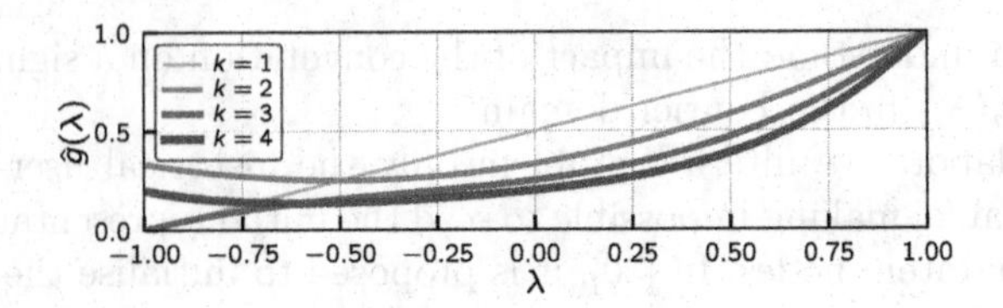

Fig. 1. LightGCN spectral filter for different values of k.

defined by S, reducing the strength of the lower frequencies, with stronger filtering as k increases. This implies that the convolution operation transforms the input signals so that they are focused in a subspace spanned by the eigenvectors corresponding to the high eigenvalues of the normalised adjacency S.

4 A Spectral Interpretation of LightGCN

We have highlighted how the final latent features learnt from LightGCN are substantially contained in the span of the largest eigenvectors of the normalised adjacency matrix. The theoretical underpinnings for why this is beneficial for the recommendation task are discussed in this section. We show that leveraging vectors lying in this subspace as latent features, leads to the optimisation of a target function which is a weighted summation of the prediction function over the training data.

When the problem is rating prediction, the goal of a recommendation algorithm is to learn a prediction matrix, $\hat{\mathrm{R}}$ which well approximates the rating matrix R. This can be formulated in terms of finding $\hat{\mathrm{R}}$ which is close to R in the Frobenius norm:

$$\min_{\hat{R}} \left(\|\mathrm{R} - \hat{\mathrm{R}}\|_F^2\right) = \min_{\hat{R}} \left(\mathrm{Tr}((\mathrm{R} - \hat{\mathrm{R}})(\mathrm{R} - \hat{\mathrm{R}})^T)\right)$$
$$= \min_{\hat{R}} \left(\mathrm{Tr}(\mathrm{R}\mathrm{R}^T) - 2\,\mathrm{Tr}(\mathrm{R}\hat{\mathrm{R}}^T) + \mathrm{Tr}(\hat{\mathrm{R}}\hat{\mathrm{R}}^{\mathrm{T}})\right) = \min_{\hat{R}} \left(-2\,\mathrm{Tr}(\mathrm{R}\hat{\mathrm{R}}^T) + \|\hat{\mathrm{R}}\|_F^2\right).$$

However, when the problem is top-N recommendation where the requirement is to learn a score to sort the items in order of preference, the scale of the prediction function is irrelevant to the order and can be fixed to any arbitrary value. Hence, we can write the target objective as:

$$\max_{\hat{\mathrm{R}}} \left(\mathrm{Tr}(\mathrm{R}\hat{\mathrm{R}}^T)\right) \quad \text{s.t. } \|\hat{\mathrm{R}}\|_F^2 \text{ is fixed.}$$

In fact, for implicit binary datasets, the trace has a natural interpretation as the sum of the predictions over the positive interaction data. Furthermore, note that the trace can be written as a quadratic form over the adjacency matrix of the user-item interaction graph.

Proposition 1. *The quadratic form $\mathcal{Q}_A(\boldsymbol{x})$ induced by the adjacency matrix of the user-item interaction graph, on a signal $\boldsymbol{x} = [\boldsymbol{p}; \boldsymbol{q}]$, corresponds to twice the sum of the prediction function over all positive interactions in the training dataset.*

$$\mathcal{Q}_A(\boldsymbol{x}) = \boldsymbol{x}^T \mathrm{A}\boldsymbol{x} = \sum_{\ell,k} a_{\ell k} x_\ell x_k = 2 \sum_{\{(u,i)|r_{ui}\neq 0\}} p_u q_i \,.$$

This can be generalised to a $(U + I) \times f$ matrix X *of f signals as:*

$$\mathcal{Q}_A(\mathrm{X}) = \mathrm{Tr}\left(\mathrm{X}^T \mathrm{A}\mathrm{X}\right) = \mathrm{Tr}\left(\mathrm{A}\mathrm{X}\mathrm{X}^T\right) = 2\,\mathrm{Tr}\left(\mathrm{P}^T \mathrm{R}\mathrm{Q}\right) = 2 \sum_{\{(u,i)|r_{ui}\neq 0\}} \boldsymbol{p}_u^T \boldsymbol{q}_i \,.$$

such that the sum of the prediction functions over the training data positive interactions is the trace of a quadratic form on the interaction data.

This is an intuitive objective for the top-N recommendation task,as opposed to the rating prediction task. Note that any rank f, symmetric matrix $\mathrm{X}\mathrm{X}^T$ can be written as $\mathrm{Y}\Sigma\mathrm{Y}^T$ where Y is orthogonal (i.e. $\mathrm{Y}^T\mathrm{Y} = \mathrm{I}_f$) and Σ is a $f \times f$ diagonal matrix. So we can equivalently write the trace as $\mathcal{Q}_A(\mathrm{X}, \Sigma) = \mathrm{Tr}\left(\mathrm{X}^T \mathrm{A}\mathrm{X}\Sigma\right)$, for orthogonal X. As such, we recognise that the problem of learning f signals (latent features) to construct the user and item embeddings which maximises the sum of the prediction function over the training data is solved by the generalised Rayleigh-Ritz theorem [12].

Theorem 1 (Rayleigh-Ritz). *For a real symmetric $n \times n$ matrix* A*:*

$$\max_{\mathrm{X}}\{\mathrm{Tr}\left(\mathrm{X}^T \mathrm{A}\mathrm{X}\right) \text{ s.t. } \mathrm{X}^T\mathrm{X} = \mathrm{I}_f\} = \lambda_1 + \cdots + \lambda_f$$

and the maximising matrix is $\mathrm{X} = [\boldsymbol{v}_1, \ldots, \boldsymbol{v}_f]$ where λ_i are the f largest eigenvalues of A *and $\boldsymbol{v}_i$ the corresponding orthonormal eigenvectors. Furthermore [19], the quadratic form $\mathrm{Tr}\left(\mathrm{X}^T \mathrm{A}\mathrm{X}\Sigma\right)$, where $\Sigma = \mathrm{diag}(\sigma_i)$ is a fixed diagonal matrix, is optimised by the same matrix of orthonormal eigenvectors, such that*

$$\mathrm{Tr}\left(\mathrm{X}^T \mathrm{A}\mathrm{X}\Sigma\right) = \sum_{i=1}^{f} \lambda_i \sigma_i \,. \tag{2}$$

4.1 Inverse Propensity Control

It is well recognised that recommender system datasets tend to exhibit biases in the manner in which the interaction data is observed [2]. Propensity scoring provides one means of taking such biases into account during model learning [14,21]. The propensity score is an estimate of the probability that any particular interaction is observed. The contribution of each observed interaction to the loss function is multiplied by its inverse propensity score, prior to model learning. The symmetric normalised adjacency matrix of the LightGCN method can be viewed as an inverse propensity weighted adjacency. In particular, each observed

a_{ui} is weighted by a term depending on the user and item degrees: $d_u^{-1/2} d_i^{-1/2}$. The quadratic form over this normalised matrix, $\tilde{A}$ is a weighted sum of the predictions on the training data, where each prediction is down-weighted according to the user and item degree:

$$\mathcal{Q}_{\tilde{A}}(\mathbf{x}) = \mathrm{Tr}\left(X^T \tilde{A} X\right) = \mathrm{Tr}\left(X^T D^{-1/2} A D^{-1/2} X\right) = 2 \sum_{\{(u,i)|r_{ui} \neq 0\}} \frac{1}{d_u^{1/2} d_i^{1/2}} \mathbf{p}_u^T \mathbf{q}_i \,.$$

Without such normalisation, the target function can be trivially maximised by giving larger embedding weights to the users and items with many interactions in the training set. The normalisation should therefore have the effect of increasing the embedding weights of unpopular items and users with short profiles.

The eigenvectors with largest eigenvalues of the normalised adjacency provide the optimal solution for this modified target objective. Hence, we can conclude that to a large extent the LightGCN method is effective because the convolution focuses on embeddings that are largely contained in the subspace spanned by the eigenvectors of largest eigenvalues of the normalised adjacency; and that these eigenvectors provide an optimal solution to the target objective of maximising the propensity-weighted sum of the predictions over the training data. It is noteworthy that, in the maximisation of the quadratic form, each eigenvector contributes proportionally to its associated eigenvalue (Theorem 1), meaning that is reasonable to assume that the latent features associated to higher eigenvectors should have more weight with respect to those associated to lower eigenvectors. The shape of the high pass filter employed by LightGCN throughout the learning process, Fig. 1, can deliver such a spectrum.

5 Pure Spectral Graph Embeddings Model

Given the interpretation of LightGCN in terms of the spectrum of the adjacency matrix, it is worth asking if spectral methods can be developed that are competitive with LightGCN on accuracy. Firstly, we show how the PureSVD [4] can be interpreted under a trace maximisation problem. Explaining the doubts presented in the original paper regarding how a method devised for rating prediction is performing so well with implicit feedbacks.

5.1 PureSVD

$\mathcal{Q}_A(X, \Sigma)$ is maximised when $X = [P; Q]$ are the eigenvectors of A and the prediction function is then $\hat{R} = P\Sigma Q^T$, where $\Sigma = \mathrm{diag}(\sigma_i)$. Writing $\mathbf{u}$ for a $U \times 1$-dimensional eigenvector of RR^T, and $\mathbf{v}$ for the corresponding $I \times 1$-dimensional eigenvector of $R^T R$, with eigenvalue $\lambda^2 \geq 0$, the eigenvectors of A are $\mathbf{x} = \frac{1}{\sqrt{2}}[\mathbf{u}; \mathbf{v}]$ and $\mathbf{x} = \frac{1}{\sqrt{2}}[\mathbf{u}; -\mathbf{v}]$ with eigenvalues $\pm\lambda$. Moreover, $\mathbf{u} = R\mathbf{v}/\lambda$ or, gathering all f eigenvectors into the columns of P and Q, we have $P = RQ\Lambda^{-1}$ where $\Lambda = \mathrm{diag}(\lambda_i)$. The eigenvectors $\mathbf{u}$ and $\mathbf{v}$ can be obtained from a singular value decomposition of R [10].

The coefficients σ_i in the above expression can be interpreted as a weight given to each of the f signals from which the embedding is formed. Now, $\|\hat{R}\|^2 = \|\mathrm{P}\Sigma\mathrm{Q}^T\|^2 = \sum_{i=1}^{f} \sigma_i^2$ and it follows that the best choice of σ_i to maximise $\sum_{i=1}^{f} \lambda_i \sigma_i$, under a fixed constraint on its norm is $\sigma_i \propto \lambda_i$. The prediction function is then $\mathrm{PQ}^T\Sigma = \mathrm{RQ}\Lambda^{-1}\Lambda\mathrm{Q}^T = \mathrm{RQQ}^T$ which is exactly the PureSVD method. We have arrived at this method through trace maximisation under a norm constraint on the prediction function, as opposed to the Frobenius norm minimisation approach, appropriate for rating prediction. The trace maximisation perspective allows for the development of other methods, which differ from PureSVD in the manner in which the norm of the prediction matrix is constrained.

5.2 Propensity Weighted Norm Constraint

Given that we wish to control the size of the embeddings of highly active users or popular items, it is useful to consider a constraint on the prediction function that controls the embedding size in proportion to the degree in the interaction dataset. In particular, we consider the following trace maximisation problem:

$$\max_{\hat{\mathrm{R}}} \left(\mathrm{Tr}(\mathrm{R}\hat{\mathrm{R}}^T) \right) \quad \text{s.t. } \|\mathrm{D}_U^{\alpha} \hat{\mathrm{R}} \mathrm{D}_I^{\beta}\|_F^2 \quad \text{is fixed}.$$

where D_U is the diagonal matrix of user degrees and D_I the diagonal matrix of item degrees. By scaling the contribution of each prediction $\hat{r}_{ui} = \mathbf{p}_u^T \mathbf{q}_i$ in this fixed norm by the degrees $d_u^{\alpha} d_i^{\beta}$, the effect, as α and β get larger will be that the size of high degree embeddings gets smaller. Here we have generalised the exponent used in LightGCN to allow for two tuneable parameters α and β such that we can explicitly control the propensity score attributed to the users and items. With a change of variables $\tilde{\mathrm{P}} = \mathrm{D}_U^{\alpha}\mathrm{P}$ and $\tilde{\mathrm{Q}} = \mathrm{D}_I^{\beta}\mathrm{Q}$, we have

$$\begin{aligned} \mathrm{Tr}(\mathrm{R}\hat{\mathrm{R}}^T) &= \mathrm{Tr}(\mathrm{RQP}^T) = \mathrm{Tr}(\mathrm{RD}_I^{-\beta}\tilde{\mathrm{Q}}\tilde{\mathrm{P}}^T\mathrm{D}_U^{-\alpha}) \\ &= \mathrm{Tr}((\mathrm{D}_U^{-\alpha}\mathrm{RD}_I^{-\beta})\tilde{\mathrm{Q}}\tilde{\mathrm{P}}^T) = \frac{1}{2}\mathrm{Tr}(\tilde{\mathrm{X}}^T(\mathrm{DAD})\tilde{\mathrm{X}}) \end{aligned}$$

where D is the $(U+I) \times (U+I)$ diagonal matrix $diag(\mathrm{D}_{\mathrm{U}}^{-\alpha}, \mathrm{D}_{\mathrm{I}}^{-\beta})$. Writing $\tilde{\mathrm{A}} = \mathrm{DAD}$ as the normalised adjacency, the trace is maximised by choosing $\tilde{\mathrm{P}}$ and $\tilde{\mathrm{Q}}$ from the PureSVD solution on the normalised interaction matrix $\tilde{\mathrm{R}} = \mathrm{D}_U^{-\alpha}\mathrm{RD}_I^{-\beta}$.

Having found $\tilde{\mathrm{P}}$ and $\tilde{\mathrm{Q}}$, one way to proceed is to rescale them back to the required factors P and Q, to obtain a fully popularity-controlled prediction matrix. On the other hand, it is well known that to achieve high recommendation accuracy, some popularity bias in the model's predictions is required [17]. So, instead, we complete the prediction function by noting that, since the embeddings are produced from the SVD of $\tilde{\mathrm{R}}$:

Table 1. Recommendation performance. bold and underline indicate the first and the second best performing algorithms.

Model	Ml1M			Amazon			Gowalla		
	NDCG	Recall		NDCG	Recall		NDCG	Recall	
	@20	@5	@20	@20	@5	@20	@20	@5	@20
BPR-MF	0.2602	0.1191	0.2756	0.0439	0.0377	0.0928	0.1021	0.0763	0.1721
LightGCN	0.2679	0.1254	0.2898	0.0446	0.0357	0.0956	0.1277	0.0980	0.2050
PureSVD	0.2621	0.1203	0.2755	0.0299	0.0229	0.0682	0.1162	0.0860	0.1836
EASE	**0.2969**	0.1415	0.3164	0.0509	0.0435	0.1028	0.1469	0.1114	0.2319
SGMC	0.2830	0.1369	0.3070	0.0528	0.0443	**0.1087**	0.1514	0.1167	0.2328
PSGE	0.2951	**0.1418**	**0.3230**	**0.0533**	**0.0458**	**0.1087**	**0.1641**	**0.1265**	**0.2519**
Statistics									
# users	5 949			9 279			29 858		
# items	2 810			6 065			40 988		
# inter	571 531			158 979			1 027 464		

1. $\tilde{P} = \tilde{R}\tilde{Q}\tilde{\Lambda}^{-1}$, and
2. $\tilde{P}\tilde{\Lambda}\tilde{Q}^T = \tilde{R}\tilde{Q}\tilde{Q}^T \approx \tilde{R}$.

Hence

$$\begin{aligned} D_U^{-\alpha} R D_I^{-\beta} \tilde{Q}\tilde{Q}^T &\approx D_U^{-\alpha} R D_I^{-\beta} && \text{(from (1) and (2) above)} \\ R D_I^{-\beta} \tilde{Q}\tilde{Q}^T &\approx R D_I^{-\beta} && \text{(dividing by } D_U^{-\alpha}\text{)} \\ R D_I^{-\beta} \tilde{Q}\tilde{Q}^T D_I^{\beta} \& &\approx R && \text{(multiplying by } D_I^{\beta}\text{)}. \end{aligned}$$

So, we set $\hat{R} = R D_I^{-\beta} \tilde{Q}\tilde{Q}^T D_I^{\beta}$, as the prediction matrix that directly approximates the observed interaction data, while being constructed in a manner that accounts for user and item propensity. It is worth noting that, although α does not appear explicitly in this formula, the eigenvectors in $\tilde{Q}$ depend on α, as they are computed from the user- and item-degree normalised matrix. In fact, using (1) we can equivalently write the prediction matrix as

$$\hat{R} = D_U^{\alpha} \tilde{P} \Lambda \tilde{Q}^T D_I^{\beta} . \tag{3}$$

We name this method *Pure Spectral Graph Embeddings* (PSGE).

6 Experiments

We conduct experiments on three real-world datasets: Movielens1M [6], Amazon Electronics [7] and Gowalla [11]. Following [8,20], we perform a k-core preprocessing step setting $k_{core} = 10$. We randomly split the interaction data of each user in train (80%), validation (10%) and test set (10%), we use the validation data to determine the best algorithm hyperparameters, subsequently, we assess their final performance on the test set by training the models with both train and validation data. We compare the proposed algorithm with BPR [13] and LightGCN [8] as well as the spectral methods PureSVD [4] and SGMC [1] and the linear model EASE [18]. The code used to produce the presented experiments is publicly available on github[2].

6.1 Recommendation Performance

To evaluate the algorithm's recommendation performance under the two different aspects of ranking and accuracy we report *NDCG@20* and *Recall@N* using two different cutoffs, $N = \{5, 20\}$ and present the results in Table 1. Except for the NDCG on Movielens1M, where it is the second best performer, PSGE gets the best results on the Recall and NDCG metrics for both cutoffs in all datasets studied, demonstrating its effectiveness in comparison to well-known, high-performing baselines from the graph convolution and spectral research domains. When compared to LightGCN, the model that inspired the study, PSGE consistently outperforms it in all datasets, with a minimum gain of 11% on the NDCG@20 on Movielens1M and a maximum increment of 29% on NDCG@20 on Gowalla. We conclude that in the context of implicit interaction data, we can mimic the effect of graph convolution without resorting to a costly gradient-based optimisation approach. PSGE corresponds to the SGMC algorithm with the setting $\alpha = \beta = 0.5$. We can see that in all the datasets and for all metrics and cutoffs, the introduction of the two tuneable parameters accounting for the propensity scoring of users and items, is capable of delivering substantial improvements over its hypergraph counterpart formulation in which the exponent is set to a fixed value.

6.2 Controlling Popularity Bias

PSGE reintroduces both user and item popularity to approximate the interaction matrix by rescaling the norm of the user and item embeddings by their respective degree (see Eq. 3). We note that rescaling on the users has no influence on the ranking at prediction time, but rescaling on the items increases the popularity on the recommendations. This enables us to control the popularity in the predictions by trading it off against recommendation performance- we achieve this by changing the value of β used to estimate $\hat{\mathrm{R}}$. To evaluate the algorithm's efficacy from this standpoint, in Fig. 2 we show the average popularity in the PSGE prediction against the performance when the exponent β (associated with the item degree rescaling) is varied. We also show a comparison to the behaviour of the baseline. The average popularity in the prediction is defined as the mean of the popularity of the items recommended, while the item popularity is defined as $pop_i = d_i/U$, where d_i indicates the item degree. For clarity, we refer to $\tilde{\beta}$ as the manipulated parameter while β refers to the value used in computing the normalised interaction matrix $\tilde{\mathrm{R}}$. We vary $\tilde{\beta}$ in the

[2] https://github.com/damicoedoardo/PSGE.

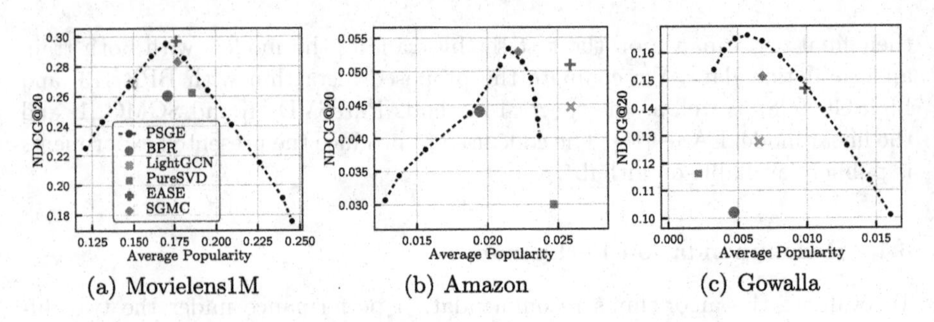

(a) Movielens1M (b) Amazon (c) Gowalla

Fig. 2. Manipulation of the hyperparameter β regulating the item degree norm rescaling. The tradeoff between accuracy and popularity in the predictions is reported plotting the NDCG@20 against the average popularity on the recommendations.

range $[0, 1]$ with a step size of 0.1. The mean popularity of the recommendation increases monotonically with $\tilde{\beta}$, while the NDCG peaks at a value of $\tilde{\beta}$ close to β. On Movielens1M and Amazon the peak is observed exactly at $\tilde{\beta} = \beta$, while on Gowalla we reach the best performance at $\tilde{\beta} = 0.3$ while $\beta = 0.4$. From the presented results we have empirically demonstrated how our algorithm can effectively trade off recommendation performance in favour of lowering popularity in the recommendations. It is worth mentioning that in all datasets, PSGE recommendations associated with peak performance have lower average popularity when compared to the second best performing algorithm, highlighting how the algorithm is capable of generating high quality predictions.

7 Conclusion

We presented a study on the graph convolution approach employed by LightGCN proving how the convolution acts as a fixed, high-pass filter in the spectral domain induced by the normalised adjacency matrix. We presented a detailed explanation of why this operation is beneficial to the top-N recommendation problem. Exploiting this spectral interpretation, we presented a scalable spectral algorithm based on the singular value decomposition of the propensity weighted interaction matrix. We empiracally showed how the presented model is able to emulate the behaviour of the light convolution by achieving better performance than LightGCN, requiring only a fraction of the training time and enabling the control of the tradeoff between accuracy and popularity on the set of provided recommendations.

Acknowledgments. This research was supported by Science Foundation Ireland (SFI) under Grant Number SFI/12/RC/2289_P2.

References

1. Chen, C., Li, D., Yan, J., Huang, H., Yang, X.: Scalable and explainable 1-bit matrix completion via graph signal learning. In: AAAI. AAAI Press (2021)
2. Chen, J., Dong, H., Wang, X., Feng, F., Wang, M., He, X.: Bias and debias in recommender system: A survey and future directions. CoRR abs/2010.03240 (2020)
3. Chen, L., Wu, L., Hong, R., Zhang, K., Wang, M.: Revisiting graph based collaborative filtering: A linear residual graph convolutional network approach. In: AAAI, pp. 27–34. AAAI Press (2020)
4. Cremonesi, P., Koren, Y., Turrin, R.: Performance of recommender algorithms on top-n recommendation tasks. In: RecSys, pp. 39–46. ACM (2010)
5. Garzón, E.M., García, I.: Parallel implementation of the lanczos method for sparse matrices: Analysis of data distributions. In: International Conference on Supercomputing, pp. 294–300. ACM (1996)
6. Harper, F.M., Konstan, J.A.: The movielens datasets: History and context. ACM Trans. Interact. Intell. Syst. **5**(4), 19:1–19:19 (2016)
7. He, R., McAuley, J.J.: Ups and downs: Modeling the visual evolution of fashion trends with one-class collaborative filtering. In: WWW, pp. 507–517. ACM (2016)
8. He, X., Deng, K., Wang, X., Li, Y., Zhang, Y., Wang, M.: Lightgcn: Simplifying and powering graph convolution network for recommendation. In: SIGIR, ACM (2020)
9. Kipf, T.N., Welling, M.: Semi-supervised classification with graph convolutional networks. In: ICLR (Poster). OpenReview.net (2017)
10. Kunegis, J.: Exploiting the structure of bipartite graphs for algebraic and spectral graph theory applications. Internet Math. **11**(3), 201–321 (2015)
11. Liang, D., Charlin, L., McInerney, J., Blei, D.M.: Modeling user exposure in recommendation. In: WWW, pp. 951–961. ACM (2016)
12. Magnus, J.R.: Handbook of matrices: H. lütkepohl, john wiley and sons, 1996. Econometric , pp. 379–380 (1998)
13. Rendle, S., Freudenthaler, C., Gantner, Z., Schmidt-Thieme, L.: BPR: bayesian personalized ranking from implicit feedback. In: UAI, AUAI Press (2009)
14. Schnabel, T., Swaminathan, A., Singh, A., Chandak, N., Joachims, T.: Recommendations as treatments: Debiasing learning and evaluation. In: ICML. JMLR Workshop and Conference Proceedings, vol. 48, pp. 1670–1679. JMLR.org (2016)
15. Shen, Y., et al.: How powerful is graph convolution for recommendation? In: CIKM, ACM (2021)
16. Shuman, D.I., Narang, S.K., Frossard, P., Ortega, A., Vandergheynst, P.: The emerging field of signal processing on graphs: Extending high-dimensional data analysis to networks and other irregular domains. IEEE Signal Process. (2013)
17. Steck, H.: Item popularity and recommendation accuracy. In: RecSys, pp. 125–132. ACM (2011)
18. Steck, H.: Embarrassingly shallow autoencoders for sparse data. In: WWW, pp. 3251–3257. ACM (2019)
19. Trendafilov, N.T.: P.-A. absil, r. mahony, and r. sepulchre. optimization algorithms on matrix manifolds. Found. Comput. Math. **10**(2), 241–244 (2010)
20. Wang, X., He, X., Wang, M., Feng, F., Chua, T.: Neural graph collaborative filtering. In: SIGIR, pp. 165–174. ACM (2019)
21. Zhu, Z., He, Y., Zhang, Y., Caverlee, J.: Unbiased implicit recommendation and propensity estimation via combinational joint learning. In: RecSys, pp. 551–556. ACM (2020)

Meta-learning Enhanced Next POI Recommendation by Leveraging Check-ins from Auxiliary Cities

Jinze Wang[1], Lu Zhang[2(✉)], Zhu Sun[3,4], and Yew-Soon Ong[3,5]

[1] Macquarie University, Balaclava Rd, Macquarie Park, Sydney, NSW 2109, Australia
[2] Chengdu University of Information Technology, Chengdu, China
zhang_lu010@outlook.com
[3] Centre for Frontier AI Research, A*STAR, Singapore, Singapore
[4] Institute of High Performance Computing, A*STAR, Singapore, Singapore
[5] School of Computer Science and Engineering, Nanyang Technological University, Singapore, Singapore

Abstract. Most existing point-of-interest (POI) recommenders aim to capture user preference by employing city-level user historical check-ins, thus facilitating users' exploration of the city. However, the scarcity of city-level user check-ins brings a significant challenge to user preference learning. Although prior studies attempt to mitigate this challenge by exploiting various context information, e.g., spatio-temporal information, they ignore to transfer the knowledge (i.e., common behavioral pattern) from other relevant cities (i.e., auxiliary cities). In this paper, we investigate the effect of knowledge distilled from auxiliary cities and thus propose a novel Meta-learning Enhanced next POI Recommendation framework (MERec). The MERec leverages the correlation of check-in behaviors among various cities into the meta-learning paradigm to help infer user preference in the target city, by holding the principle of "paying more attention to more correlated knowledge". Particularly, a city-level correlation strategy is devised to attentively capture common patterns among cities, so as to transfer more relevant knowledge from more correlated cities. Extensive experiments verify the superiority of the proposed MERec against state-of-the-art algorithms.

Keywords: Next POI Recommendation · Meta learning

1 Introduction

Next POI recommendation, which aims to recommend POIs for users that they are most likely to visit in the future, benefits both location-based social network services, e.g., Foursquare (`foursquare.com`), and individuals. As users' activities typically limit within a city, most existing studies exploit the city-level user check-in records to develop next POI recommenders. Table 1 shows the statistics of user-POI interactions for four cities on Foursquare, which are widely explored

H. Kashima et al. (Eds.): PAKDD 2023, LNAI 13937, pp. 322–334, 2023.
https://doi.org/10.1007/978-3-031-33380-4_25

Table 1. Statistics of four datasets from Foursquare.

	#Users	#POIs	#Check-ins	#Categories	Density
Calgary (CAL)	435	3,013	13,911	293	**1.06%**
Phoenix (PHO)	2,945	7,247	47,980	344	**0.22%**
Singapore (SIN)	8,648	33,712	355,337	398	**0.12%**
New York (NYC)	16,387	56,252	511,431	420	**0.05%**

in prior studies [20,21]. We can observe that CAL with relatively higher density being 1.06%, while the extremely lower density is 0.05% in NYC. Obviously, the sparsity of user-POI interactions in many cities severely hinders the capability of existing approaches for more accurate user preference learning.

To ease this issue, various context information, e.g., spatial and temporal contexts, has been widely exploited in existing next POI recommenders. Specifically, most current research devotes to capturing the spatio-temporal relations between users and POIs. They are built upon various techniques, ranging from matrix factorization [9,16], Markov chain models [2], to advanced deep learning frameworks, e.g., recurrent neural networks [20] and graph neural networks [12]. However, they are restricted by insufficient training data for more accurate user preference learning due to the sparse user-POI interactions within a city.

Intuitively, users' check-in behaviors among different cities may share common patterns. This motivates us to conduct an in-depth analysis of the check-in records across different cities (i.e., auxiliary cities), and transfer useful knowledge from such cities for assisting user preference inference within the target city. However, non-overlapping visited POIs between different cities bring challenges in knowledge transfer, that is, blindly leveraging check-in behaviors from auxiliary cities to augment the target city may result in harmful knowledge transfer. We thus seek to investigate two fundamental problems when transferring knowledge from auxiliary cities to the target city as follows.

(1) ***What to transfer?*** In e-commerce, overlapping items can be found on shopping sites in different regions. While in the city-level location recommendation scenario, non-overlapping visited POIs across different cities present a challenge to transferring common behavioral knowledge. Fortunately, mining users' check-in behavioral knowledge over the categorical context (i.e., common category-level patterns) helps address this challenge. For example, the category transition *Shop&Service→Food* are common to all four cities, which indicates that users in different cities are most likely heading to a restaurant after shopping. By contrast, the transition *Travel&Transport→Shop* is quite common only in SIN due to the developed public transportation. **(2)** ***How to transfer?*** Although the common category-level patterns captured from auxiliary cities may enhance the recommendation quality for the target city, this inevitably introduces noise if we ignore the cultural diversity and geographical property of such cities. Hence, determining what extent we can transfer knowledge from the auxiliary cities to the target city is of great significance.

Accordingly, we propose a novel Meta-learning Enhanced next POI Recommendation (MERec) framework, which delicately considers the correlation of category-level behavioral patterns among different cities into the meta-learning paradigm, that is, paying more attention to more correlated knowledge. Specifically, MERec mainly consists of two components: a *two-channel encoder* to capture the transition patterns of categories and POIs, whereby a city-correlation based strategy is devised to attentively capture common knowledge (i.e., patterns) from auxiliary cities via the meta-learning paradigm; and a *city-specific decoder* to aggregate the latent representations of the two channels to perform the next POI prediction on the target city.

Overall, our main contributions lie in three folds: (1) we are the first to study to what extent we can transfer knowledge from auxiliary cities to the target city via differentiating the correlation of category-level behavioral patterns; (2) we propose a novel meta-learning based framework – MERec, which exploits both the transferred knowledge and user behavioral contexts within the target city to alleviate the data sparsity issue; and (3) we conduct extensive experiments on four datasets to validate the superiority of MERec against state-of-the-arts.

2 Related Work

Next POI Recommendation. It predicts future POI visits for users based on their historical successive check-in behaviors. Early studies generally employ the property of Markov chain to model the sequential influence [2,5,18]. Recently, recurrent neural network (RNN) based methods show great capability in capturing long-term sequential dependencies. Existing studies based on RNN and its variants mainly tend to exploit users' sequential check-ins by incorporating various context information, such as ST-RNN [11], SERM [17], MCARNN [10] ATST-LSTM [8], and iMTL [20]. Despite the great success of these methods, most of them suffer from the issue of insufficient user check-ins in many cities, which heavily limits their performance improvements. In this sense, transferring knowledge from auxiliary cities to the target city brings the possibility to further enhance the user preference learning for the next POI recommendation.

Meta-learning for Next POI Recommendation. Transfer learning (TL) aims to transfer knowledge from source domains to the target domain, which has shown strong capability in resolving the sparsity issue. Existing TL-based approach [4] focuses on the cross-city POI recommendation task due to the lack of large amount of overlapping user-POI interactions across cities. Meta-learning (ML) is able to transfer the knowledge learned from multiple tasks to a new task and has been recently introduced in next POI recommendation. For example, Chen et al. [1] proposed CHAML by fusing hard sample mining and curriculum learning into a meta-learning framework. Sun et al. [13] devised MFNP to integrate user preference and region-dependent crowd preference tasks in a meta-learning paradigm. Cui et al. [3] designed Meta-SKR by using sequential, spatiotemporal, and social knowledge to recommend next POIs. Meanwhile, Tan et al. [15] developed the METAODE which models city-irrelevant and -specified information separately to achieve city-wide next POI recommendation. However, the aforementioned ML-based next POI recommenders ignore to attend the correlation of user behavioral patterns when transferring knowledge from auxiliary cities to the target city, i.e., paying more attention to more correlated knowledge.

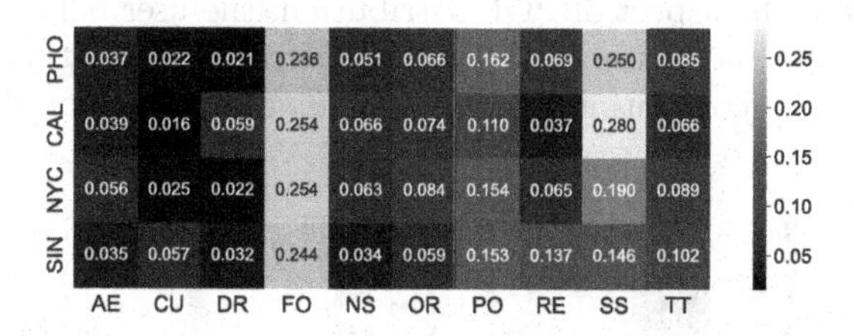

Fig. 1. The distribution of POIs at category level among four cities.

3 Data Analysis

There is a great necessity to analyze the correlation among different cities w.r.t. user check-in behaviors (see Table 1), so as to better guide the knowledge transfer from auxiliary cities to the target city. It is, however, non-trivial due to the non-overlapping visited POIs across cities. Fortunately, POIs in various cities share the same categories, which inspires us to study the POI distribution and user behavioral patterns at the category level to uncover the correlation among cities.

POI Distribution at Category Level. The number of POIs under each category varies a lot across cities due to different cultures and geography. Hence, we first study the nature of POI distributions among four cities to help explore the correlation of user behavioral patterns. Specifically, all POIs are characterized by ten first-level categories [14], including Arts & Entertainment (AE), College & University (CU), Drink (DR), Food (FO), Nightlife Spot (NS), Outdoor & Recreation (OR), Professional & Other Places (PO), Residence (RE), Shop & Service (SS), and Travel & Transport (TT). Figure 1 depicts the POI distribution at category level, where we note that cities exhibit high similarity in some categories while show dissimilarity in others. For example, the proportion of POIs under FO is relatively higher across the four cities, whereas the proportion of POIs under, e.g., AE, is lower than POIs under FO and SS. On the other hand, different cities show their unique characteristics, such as the higher proportion of CU-related POIs in SIN and the higher proportion of AE-related POIs in NYC.

Correlation of Cities w.r.t POI Distribution. The POI distribution of each city enables us to further explore the correlation between cities, i.e., measuring the similarity of cities from the aspect of POI distribution. Specifically, given any two cities, $A^{poi} = [A_1^{poi}, A_2^{poi} \cdots A_{|\mathcal{C}|}^{poi}]$ and $B^{poi} = [B_1^{poi}, B_2^{poi} \cdots B_{|\mathcal{C}|}^{poi}]$ denote the POI distributions among $|\mathcal{C}|$ categories within city A and city B, respectively. We thus derive their similarity $\gamma_{A,B}$ via the Pearson correlation coefficient, and the results are shown in Fig. 2(a). We find that NYC shows the highest similarity with PHO while the lowest similarity with SIN, implying that cities in the same country (i.e., USA) may have a higher correlation due to the similar property of culture. Besides, CAL (i.e., Canada) shows relatively higher similarity with NYC and PHO, which means that the geography property is also an important factor when measuring the correlation of cities. Although the correlation of cities can be measured from the aspect of POI distribution, the user behavioral transition pattern is a significant factor in the next POI recommendation task, we thus further explore such correlation from the angle of user sequential behaviors.

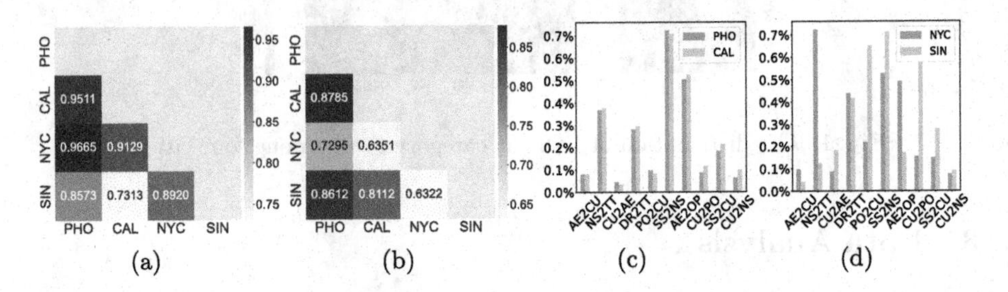

Fig. 2. (a-b) the correlation of four cities w.r.t POI distribution and behavioral patterns at category level; (c-d) two most correlated and least correlated cities.

Correlation of Cities w.r.t Behavioral Patterns. We examine the correlation of cities w.r.t. the categories of users' successive POI visits. In particular, given any two cities, $A^{cat} = [A_1^{cat}, A_2^{cat} ... A_{|S|}^{cat}]$ and $B^{cat} = [B_1^{cat}, B_2^{cat} ... B_{|S|}^{cat}]$ refer to the category transition distributions among S transition types, e.g., A_1^{cat} denotes the ratio of transition type $FO \rightarrow SS$ within city A. Analogously, the similarity among different cities can be calculated via the Pearson correlation coefficient, shown in Fig. 2(b). Interestingly, we observe that the correlation of cities w.r.t behavioral patterns is quite different from that w.r.t POI distribution. Specifically, PHO and CAL still keep higher similarity, whereas NYC shows comparably lower similarity with PHO and CAL. To further dig out how the four cities are correlated and different over the behavioral patterns, we compare the two most correlated cities (i.e., CAL and PHO) and the two least correlated cities (i.e., NYC and SIN). For ease of presentation, we select the 10 most frequent category transitions for comparison as shown in Fig. 2(c-d), where the x-axis denotes the category transitions, e.g., $AE \rightarrow CU$ (AE2CU), and the $y-$axis shows the proportion of such a transition within a city. We find that the more correlated cities possess consistent distributions over the frequent category transitions and *vice versa.* The above observations depict the various correlations between cities, which inspire us to differentiate their influence when transferring knowledge from auxiliary cities to the target city.

4 The Proposed MERec

This section presents the proposed MERec, which leverages the correlation of behavioral patterns when transferring knowledge from auxiliary cities to the target city, i.e., paying more attention to more correlated knowledge.

Problem Formulation. Each city has its unique user set $\mathcal{U}$ and POI set $\mathcal{P}$ without sharing any common users and POIs. For user u, all his check-in records, i.e., $r = (p, c, g, t)$, are ordered by timestamps as in [22], where p, c, g, t denote POI p, category c, coordinate g (i.e., longitude and latitude) and timestamp t. We then split his historical records into sequences by day and obtain two types of sequences: 1) the i-th category sequence denoted by a set of category tuples, i.e., $C^{u,i} = \{C^u_{t_1}, C^u_{t_2}, \cdots, C^u_{t_n}\}$, where $C^u_{t_k} = (c^u_{t_k}, t^u_k)$, and 2) the i-th POI sequence denoted by a set of POI tuples, i.e., $P^{u,i} = \{P^u_{t_1}, P^u_{t_2}, \cdots, P^u_{t_n}\}$, where $P^u_{t_k} = (p^u_{t_k}, d^u_{t_k}, t^u_k)$, and d_{t_k} is the distance between successive POIs calculated by their coordinates. Given $C^{u,i}$, $P^{u,i}$, *auxiliary cities* $\mathcal{Y}_\mathcal{A} = \{y^{(m)}_{aux} | m \in 1, 2, \cdots, M\}$ and the *target city* $\mathcal{Y}_\mathcal{T} = \{y_{tar}\}$, our goal is to predict user u's next POI $p_{t_{n+1}}$ at time t_{n+1} by transferring knowledge from the auxiliary cities to the target city.

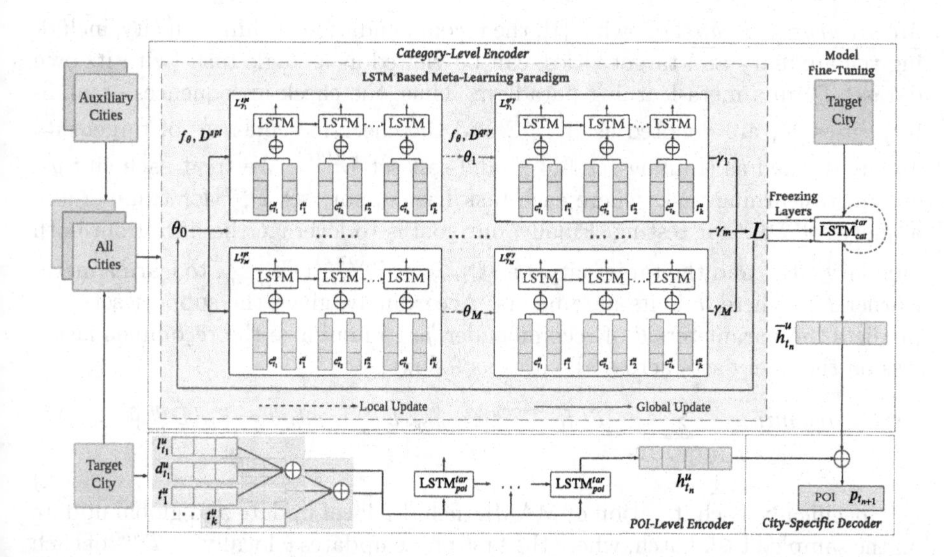

Fig. 3. The overall framework of our proposed MERec.

Overview of MERec. The overview of MERec is outlined in Fig. 3, mainly composed of a *two-channel encoder* (i.e., category- and POI-level encoders) with the embedding layer and a *city-specific decoder*. In particular, the category-level encoder exploits meta-learning to capture the common user check-in transition patterns at the category level in each city by holding the principle of "paying more attention to more correlated knowledge". The goal of the POI-level encoder is to learn the accurate POI transition patterns in the target city. Lastly, the city-specific decoder performs the next POI predictions by concatenating the hidden states of the above two encoders.

Embedding Layer. It maps each check-in record into an embedding vector. Specifically, in the category-level encoder, the embedding of a category tuple $\mathbf{e}^C \in \mathbb{R}^{2d}$ is the concatenation of the category embedding $\mathbf{e}^c \in \mathbb{R}^d$ and time embedding $\mathbf{e}^t \in \mathbb{R}^d$; thus the embedding of a category sequence $C^{u,i}$ is formed as $\mathbf{E}_{C^{u,i}} = [\mathbf{e}^C_{t_1}, \mathbf{e}^C_{t_2}, \cdots, \mathbf{e}^C_{t_n}]$. Analogously, in the POI-level encoder, the embedding of a POI sequence is denoted by $\mathbf{E}_{P^{u,i}} = [\mathbf{e}^P_{t_1}, \mathbf{e}^P_{t_2}, \cdots, \mathbf{e}^P_{t_n}]$, where $\mathbf{e}^P$ is the embedding of POI tuple represented by the concatenation of POI embedding $\mathbf{e}^p \in \mathbb{R}^d$, distance embedding $\mathbf{e}^{dist} \in \mathbb{R}^d$ and time embedding $\mathbf{e}^t \in \mathbb{R}^d$.

Cateogry-level Encoder. To distil knowledge from auxiliary cities and employ category-level user behavioral patterns, we extend model-agnostic meta-learning (MAML) [6] with LSTM as the framework for the meta-learning update. In particular, we devise a *correlation strategy* that can transfer knowledge based on the correlation of user behavioral patterns among cities. Meanwhile, *freezing layers and model fine-tuning* are exploited to obtain a generic model while better adapting to the data of the target city.

Meta-learning Setup. Following [1], the recommendation within each city, including the auxiliary and target cities, can be viewed as a single task (with its own dataset $\mathcal{D}$) in a meta-learning paradigm. Thus, the check-in sequences of auxiliary cities $\mathcal{Y}_A$ are denoted as $\mathbb{D}^{(aux)}_{meta}$, and the check-in sequences of target city $\mathcal{Y}_T$ are divided as training set $\mathbb{D}^{(tar)}_{train}$ and test set $\mathbb{D}^{(tar)}_{test}$. We treat each city y_m as a meta-learning task, where each task has support set $\mathcal{D}^{spt}_{y_m}$ for training and a query set $\mathcal{D}^{qry}_{y_m}$ for testing. Finally, our goal is to leverage the data from both auxiliary cities and the target city, i.e., $\mathbb{D}_{train} = \mathbb{D}^{(aux)}_{meta} \cup \mathbb{D}^{(tar)}_{train}$, to learn a meta-learner F_w, where w is its parameters. Accordingly, given the support sets, F_w predicts the parameters θ of recommender f_θ to minimize the recommendation loss on the query sets across all cities as follows,

$$w^* = \underset{w}{arg\,min} \sum_{y_m \in \{\mathcal{Y}_A \cup \mathcal{Y}_T\}} \mathcal{L}(f_\theta, \mathcal{D}^{qry}_{y_m} | \mathbb{D}_{train}, \mathcal{D}^{spt}_{y_m}),\ s.t.\ \theta = F_w(\mathcal{D}^{spt}_{y_m} | \mathbb{D}_{train}). \tag{1}$$

Specifically, each iteration of MAML includes local update and global update on the sampled task batch, where the first phase updates θ locally on $\mathcal{D}^{spt}$ of each task, and the second phase globally updates θ by gradient descent to minimize the sum of loss on $\mathcal{D}^{qry}$ of all tasks.

- *Local update:* we first sample a batch of cities, and then randomly sample N category sequences $\mathcal{D}^{spt}_{y_m}$ and $\mathcal{D}^{qry}_{y_m}$ for each sampled city. Thus, we calculate the training loss on $\mathcal{D}^{spt}_{y_m}$ and locally update θ by one step:

$$\theta'_{y_m} = \theta - \alpha \nabla_\theta \mathcal{L}_{y_m}(f_\theta, \mathcal{D}^{spt}_{y_m}), \tag{2}$$

where $\mathcal{L}$ is the cross-entropy loss; α is the local learning rate, and θ'_{y_m} is the locally updated parameters of recommender for each city.

- *Global update:* we calculate the testing loss on each $\mathcal{D}^{qry}_{y_m}$ with the corresponding θ'_{y_m} and then update the initialization θ by one gradient step on the sum of testing losses across all cities, where β is the global learning rate.

$$\theta = \theta - \beta \nabla_\theta \sum\nolimits_{y_m \in \{\mathcal{Y}_\mathcal{A} \cup \mathcal{Y}_\mathcal{T}\}} \mathcal{L}_{y_m}(f_{\theta'_{y_m}}, \mathcal{D}^{qry}_{y_m}). \tag{3}$$

Correlation Strategy. From the data analysis in Sect. 3, we observe that there exist various correlations w.r.t different aspects among different cities. Directly transferring user check-in behaviors from auxiliary cities to the target city may introduce noise thus hurting the recommendation performance. By holding the principle of "paying more attention to more correlated knowledge", we further consider the *correlation of behavioral patterns at category level* in different cities when conducting the global update. To be specific, we obtain the city-level correlation (e.g., γ_{cor}) based on behavioral patterns, and then attentively adapt the gradient across cities by employing their correlations. In other words, if the auxiliary city is more correlated to the target city, we adapt the gradient so that it updates faster in that direction. Therefore, Eq.(2) is reformulated as:

$$\theta'_{y_m} = \theta - \alpha \nabla_\theta [\mathcal{L}_{y_m}(f_\theta, \mathcal{D}^{spt}_{y_m}) \times \gamma_{cor}]. \tag{4}$$

Freezing Layers and Model Fine-Tuning. Inspired by [19], the network with freezing layers and fine-tuning is generalized better than the one trained directly on the target dataset. Therefore, after obtaining the well-trained category-level encoder for the target city (i.e., $LSTM^{tar}_{cat}$) by the meta-learning paradigm, we further consider fine-tuning it. In doing this, we can deliver a network that not only accommodates knowledge distilled from the auxiliary cities but also better adapts to the target city. Specifically, assuming $LSTM^{tar}_{cat}$ contains L layers, we freeze its first l $(1 \leq l \leq L)$ layers, while adding n layers after the l layers. The newly constructed model is denoted by $\overline{LSTM}^{tar}_{cat}$, which is further fine-tuned via category sequences from the target city, i.e., $\mathbb{D}^{(tar)}_{train}$. As such, the freezing-layers help generate a network that can better balance parameters between auxiliary cities and the target city after the fine-tuning. Accordingly, the hidden state $\overline{\boldsymbol{h}}^u_{t_k}$ of category at t_k is given by,

$$\overline{\boldsymbol{h}}^u_{t_k} = \overline{LSTM}^{tar}_{cat}(\mathbf{e}^C_{t_k}, \overline{\mathbf{h}}^u_{t_{k-1}}). \tag{5}$$

POI-level Encoder. It aims to model users' sequential check-in behaviors and the spatio-temporal context in the target city by using the LSTM model. As illustrated in the *Embedding Layer*, the embedding of a POI sequence is represented by $\mathbf{E}_{P^{u,i}} = [\mathbf{e}^P_{t_1}, \mathbf{e}^P_{t_2}, \cdots, \mathbf{e}^P_{t_n}]$, where each embedding $\mathbf{e}^P_{t_k}$ is feed into the $LSTM^{tar}_{poi}$ to infer the hidden state $\boldsymbol{h}^u_{t_k}$ of POI check-in at t_k, given by,

$$\boldsymbol{h}^u_{t_k} = LSTM^{tar}_{poi}(\mathbf{e}^P_{t_k}, \boldsymbol{h}^u_{t_{k-1}}). \tag{6}$$

City-specific Decoder. The city-specific decoder aims to perform the next POI prediction based on the last hidden states learned from the two-channel encoder (i.e., $\overline{\mathbf{h}}^u_{t_n}, \mathbf{h}^u_{t_n}$). Accordingly, the probability distribution on all candidate POIs is calculated by the softmax function, given by,

$$\hat{\boldsymbol{y}} = softmax(f(\overline{\mathbf{h}}^u_{t_n}; \mathbf{h}^u_{t_n})), \tag{7}$$

where f is a fully connected layer to transform $(\overline{\mathbf{h}}^u_{t_n}; \mathbf{h}^u_{t_n})$ into a $|\mathcal{P}|$-dimensional vector; and $|\mathcal{P}|$ is the number of POIs in the target city. Hence, the objective function for the next POI recommendation is defined by:

$$\mathcal{J} = -\sum_{i=1}^{|\mathcal{P}|} \boldsymbol{y}[i] \cdot log(\hat{\boldsymbol{y}}[i]), \tag{8}$$

where $\boldsymbol{y}$ is a one-hot embedding of the ground-truth POI. Algorithm 1 shows the training process of MERec, consisting of meta training (lines 3–9), freezing layers and model fine-tuning (lines 10–12), as well as next POI prediction (lines 13–14).

5 Experiments and Results

We conduct experiments to answer three research questions: **(RQ1)** does MERec outperform state-of-the-art baselines? **(RQ2)** how do different components of MERec affect its performance? **(RQ3)** how do essential hyper-parameters affect MERec? The code is available at https://github.com/oli-wang/MERec.

Datasets and Evaluation Metrics. The four datasets shown in Table 1 are used in our experiment, where we take one of the cities as the target city and the rest as auxiliary cities each time. Following [8], we chronologically divide the

---

Algorithm 1: The training process of MERec

Input: $\mathbb{D}_{train}, \mathcal{Y}_\mathcal{A}, \mathcal{Y}_\mathcal{T}, \alpha, \beta, Iter, N, l, n$
Output: A list of recommended next POIs
1 Randomly initialize parameters θ;
2 Calculate the correlation of behavioral patterns at category level;
3 **for** ($iter = 1; iter \leq Iter; iter++$) **do**
4   **for** *each city* $y_m \in \{\mathcal{Y}_\mathcal{A} \cup \mathcal{Y}_\mathcal{T}\}$ **do**
5   Sample N category sequences from $\mathbb{D}^{spt}_{y_m}$ as the adapt_batch;
6   Evaluate: $\nabla_\theta \mathcal{L}_{y_m}(f_\theta, \mathcal{D}^{spt}_{y_m})$ using the adapt_batch;
7   Calculate the gradient update of θ'_{y_m} by Eq.(4); `// local update`
8   Sample N category sequences from $\mathbb{D}^{qry}_{y_m}$ as the eval_batch;
9   Update θ using eval_batch by Eq.(3); `// global update`
10 Freeze the first l layers and add n layers as the new $\overline{LSTM}^{tar}_{cat}$ model;
11 Fine-tune $\overline{LSTM}^{tar}_{cat}$ via the training category sequences of the target city;
12 Get the last hidden states of the two-channel encoder shown in Eqs. (5-6);
13 Predict the next possible POI via Eq.(7);
14 Calculate the prediction loss for each check-in record via Eq.(8);

---

dataset of the target city into training, validation, and test sets with a ratio of 8:1:1. Note that we remove users and POIs with less than five and three check-ins, respectively. Two commonly-used metrics, i.e., $HR@K$ and $NDCG@K$ are adopted by following [1], where the former measures whether the ground-truth POI can be found in the top-K recommendation list, and the latter measures the ranking quality of the ground-truth POI in the recommendation list.

Compared Baselines. We compare the MERec with seven state-of-the-art approaches. (1) `MostPop` recommends the next POI based on the popularity of POIs; (2) `BPRMF` is a matrix factorization method optimized via Bayesian personalized ranking; (3) `NeuMF` [7] generalizes the matrix factorization by employing a multi-layer perceptron to model the user-item interactions; (4) `ATST-LSTM` [8] is an attention-based LSTM method by considering spatio-temporal contextual information; (5) `iMTL` [20] is a multi-task learning framework for next POI recommendation, which consists of a two-channel encoder and a task-specific decoder; (6) `MAML` [6] is a model-agnostic meta-learning for few-shot learning tasks; (7) `CHAML` [1] is a meta-learning based framework for next POI recommendation, which considers both city- and user-level hardness during meta training.

Hyper-parameter Settings. The optimal hyper-parameter settings for all methods are empirically found out based on the performance on the validation set. Specifically, the embedding size is searched from $\{32, 64, 128, 256\}$. For baselines (2–5), the learning rate is selected from $\{0.1, 0.05, 0.01, 0.005, 0.001, 0.0001\}$,

Table 2. Comparative results of all approaches on the four datasets, where 'H' refers to 'HR' and 'N' means 'NDCG'; the best results are highlighted in bold; the runner up is underlined; and the column 'Improve' indicates the improvements achieved by MERec relative to the runner up.

		Traditional		**Deep Learning**			**Meta Learning**			*Improve*
		MostPop	BPRMF	NeuMF	ASTA-LSTM	iMTL	MAML	CHAML	MERec	
CAL	H@5	0.0988	0.1304	0.1431	0.2924	0.2652	0.3987	0.3995	**0.4274**	6.98%
	H@10	0.1547	0.2349	0.2368	0.3705	0.3184	0.4618	0.4777	**0.5054**	5.80%
	N@5	0.0632	0.0928	0.0989	0.2134	0.1857	0.3178	0.3093	**0.3378**	6.29%
	N@10	0.0814	0.1672	0.1669	0.2383	0.2299	0.3362	0.3315	**0.3564**	6.01%
PHO	H@5	0.0682	0.1093	0.1316	0.2366	0.2410	0.3549	0.3660	**0.3928**	7.32%
	H@10	0.1068	0.1584	0.1852	0.3125	0.3370	0.4508	0.4419	**0.4531**	0.51%
	N@5	0.0419	0.0688	0.0869	0.1635	0.1753	0.2633	0.2648	**0.2796**	5.59%
	N@10	0.0547	0.0848	0.1042	0.1883	0.2065	0.2949	0.2891	**0.2993**	1.49%
SIN	H@5	0.0365	0.0848	0.1004	0.2165	0.2388	0.2991	0.3571	**0.3784**	5.96%
	H@10	0.0635	0.1450	0.1696	0.2879	0.3080	0.3816	0.4486	**0.4557**	1.58%
	N@5	0.0231	0.0452	0.0697	0.1532	0.1696	0.2188	0.2650	**0.2749**	3.73%
	N@10	0.0318	0.0648	0.0925	0.1760	0.1922	0.2451	0.2981	**0.3015**	1.14%
NYC	H@5	0.0214	0.0558	0.0959	0.1763	0.2187	0.2456	0.2745	**0.2991**	8.96%
	H@10	0.0336	0.0994	0.1495	0.2455	0.2879	0.3373	0.3526	**0.3995**	13.30%
	N@5	0.0134	0.0265	0.0595	0.1257	0.1484	0.1652	0.1865	**0.2107**	12.98%
	N@10	0.0173	0.0237	0.0770	0.1485	0.1705	0.2072	0.2118	**0.2436**	15.01%

and the batch size is set as 256. For meta-learning based baselines (6-7) and MERec, the learning rates α, β are searched from $\{0.5, 0.1, 0.01, 0.001, 0.0001\}$; and the batch size is set as 256 for a fair comparison. For MERec, the number of freezing layers l is searched in the range of $[1, 4]$ stepped by one, where the best setting is 3 for all cities; and $Iter = 500, N = 32, n = 2$ across all cities.

Performance Comparison (RQ1). The results are presented in Table 2. Across the four datasets, the traditional methods (MostPop, BPRMF) generally perform worse than deep learning methods (NeuMF, ATST-LSTM, iMTL) demonstrating the efficacy of neural networks on more accurate recommendation. RNN based methods (ATST-LSTM, iMTL) outperform NeuMF, which indicates the capability of RNN on modeling the sequential dependency. iMTL defeats ATST-LSTM, as it leverages multi-task learning (MTL) framework to jointly learn user preference on both categories and POIs, exhibiting the superiority of MTL on better next POI recommendation. Meta-learning based methods (MAML, CHAML, MERec) bring further enhancement compared with other methods, showcasing the efficacy of knowledge transfer in alleviating the data sparsity issue. Overall, our MERec consistently achieves the best performance across all the datasets, with an average lift of 6.3% and 6.53% w.r.t. HR and NDCG, respectively. This helps confirm the benefits of (1) leveraging check-ins of auxiliary cities to augment the target city, and (2) paying more attention to more correlated knowledge when transferring knowledge from auxiliary cities.

Ablation Study (RQ2). To check the impacts of various components in MERec, four variants are compared. (1) $\text{MERec}_{w/o\ cor}$ removes the correlation strategy from the meta-learner; (2) $\text{MERec}_{w/o\ frz}$ removes the freezing layers and fine-tuning from the category-level encoder; (3) $\text{MERec}_{w/o\ cor-frz}$ removes both correlation strategy, freezing layers and fine-tuning; and (4) $\text{MERec}_{w/o\ cat}$ removes the category-level encoder, but only retains the POI-level encoder. The results are shown in Fig. 4. We note that $\text{MERec}_{w/o\ cor-frz}$ performs worse than both $\text{MERec}_{w/o\ cor}$ and $\text{MERec}_{w/o\ frz}$, suggesting that both the correlation strategy, freezing layers, and fine-tuning operation indeed improve the recommendation performance. Generally, the performance decrease of $\text{MERec}_{w/o\ frz}$ far exceeds that of $\text{MERec}_{w/o\ cor}$, implying that the freezing layers and fine-tuning operation play more important roles than the correlation strategy. Besides, $\text{MERec}_{w/o\ cat}$ underperforms MERec, which helps verify the advantages of both the meta-learning paradigm with auxiliary check-ins and the correlation strategy.

Parameter Sensitivity Analysis (RQ3). We study the influence of two essential hyper-parameters, i.e., the number of local-update steps in Eq.(2) and the number of freezing layers. Figure 5 only reports the results on the CAL dataset and similar trends can be observed on the rest three datasets. Figures 5 (a-b) depict the model performance w.r.t. the number of local-update steps. We empirically find out that updating only one step is sufficient to obtain better recommendation accuracy, which also increases the model efficiency. Figures 5 (c-d) display the influence of the number of layers frozen on the model perfor-

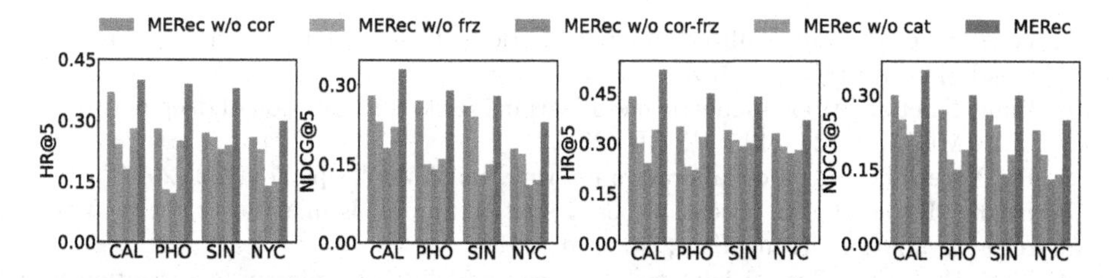

Fig. 4. Performance comparison for variants of MERec on the four datasets.

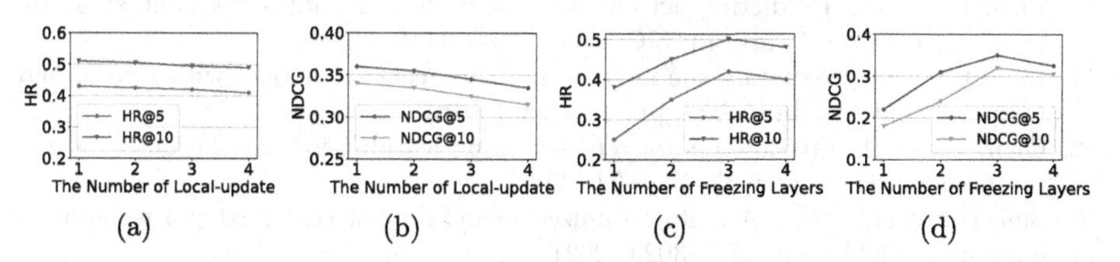

Fig. 5. Parameter sensitivity analysis on CAL.

mance. As observed, with the layer increasing, the performance first goes up and then drops slightly. The best setting for the number of freezing layers is 3 on the four datasets.

6 Conclusion

In this paper, we propose a Meta-learning Recommendation (MERec) framework for the next POI recommendation by leveraging check-ins from auxiliary cities to augment the target city, and holding the principle of "paying more attention to more correlated knowledge". In particular, we devise a two-channel encoder to capture the transition patterns of categories and POIs, whereby a city-correlation based strategy is devised to attentively capture common knowledge (i.e., patterns) from auxiliary cities via the meta-learning paradigm. The city-specific decoder then concatenates the latent representations of the two-channel encoder to perform the next POI prediction for the target city. Extensive experiments on four real-world datasets demonstrate the superiority of our proposed MERec.

References

1. Chen, Y., et al.: Curriculum meta-learning for next poi recommendation. In: SIGKDD, pp. 2692–2702 (2021)
2. Cheng, C., et al.: Where you like to go next: Successive point-of-interest recommendation. In: IJCAI (2013)
3. Cui, Y., et al.: Sequential-knowledge-aware next poi recommendation: a meta-learning approach. TOIS **40**(2), 1–22 (2021)
4. Ding, J., et al.: Learning from hometown and current city: cross-city poi recommendation via interest drift and transfer learning. IMWUT **3**(4), 1–28 (2019)

5. Feng, S., et al.: Personalized ranking metric embedding for next new poi recommendation. In: IJCAI (2015)
6. Finn, C., et al.: Model-agnostic meta-learning for fast adaptation of deep networks. In: ICML, pp. 1126–1135. PMLR (2017)
7. He, X., et al.: Neural collaborative filtering. In: WWW, pp. 173–182 (2017)
8. Huang, L., et al.: An attention-based spatiotemporal lstm network for next poi recommendation. Trans. Serv, Comput (2019)
9. Lian, D., et al.: Geomf: joint geographical modeling and matrix factorization for point-of-interest recommendation. In: SIGKDD, pp. 831–840 (2014)
10. Liao, D., et al.: Predicting activity and location with multi-task context aware recurrent neural network. In: IJCAI, pp. 3435–3441 (2018)
11. Liu, Q., et al.: Predicting the next location: A recurrent model with spatial and temporal contexts. In: AAAI, pp. 194–200 (2016)
12. Qian, T., et al.: Spatiotemporal representation learning for translation-based poi recommendation. TOIS **37**(2), 1–24 (2019)
13. Sun, H., et al.: Mfnp: A meta-optimized model for few-shot next poi recommendation. In: IJCAI, pp. 3017–3023 (2021)
14. Sun, Z., et al.: Point-of-interest recommendation for users-businesses with uncertain check-ins. TKDE (2021)
15. Tan, H., et al.: Meta-learning enhanced neural ode for citywide next poi recommendation. In: MDM, pp. 89–98. IEEE (2021)
16. Wang, J., et al.: The footprint of factorization models and their applications in collaborative filtering. TOIS **40**(4), 1–32 (2021)
17. Yao, D., et al.: Serm: A recurrent model for next location prediction in semantic trajectories. In: CIKM, pp. 2411–2414 (2017)
18. Ye, J., et al.: What's your next move: User activity prediction in location-based social networks. In: SDM, pp. 171–179. SIAM (2013)
19. Yosinski, J., et al.: How transferable are features in deep neural networks? In: NeurIPS, p. 27 (2014)
20. Zhang, L., et al.: An interactive multi-task learning framework for next poi recommendation with uncertain check-ins. In: IJCAI, pp. 3551–3557 (2021)
21. Zhang, L., et al.: Next point-of-interest recommendation with inferring multi-step future preferences. In: IJCAI (2022)
22. Zhao, S., et al.: Geo-teaser: Geo-temporal sequential embedding rank for point-of-interest recommendation. In: WWW, pp. 153–162 (2017)

Global-Aware External Attention Deep Model for Sequential Recommendation

Tianxing Wang and Can Wang(✉)

Griffith University, Parklands Dr, Southport, QLD 4222, Australia
can.wang@griffith.edu.au

Abstract. The sequential recommender plays a major role in contemporary recommendation systems, which shows the strong ability to model sequential patterns among the dataset. The classic sequential recommenders utilize the convolutional neural network, recurrent neural network, and self-attention mechanism to model the user's preferences of items. However, these existing sequential recommendation models face the "Filter Bubble" issue by putting too much attention on each user's own historical sequence, and they also ignore the feature-level item-item relationship. To address the existing challenges, we propose a novel global-aware external attention deep model (EDM) to learn both the global and local user preferences. The proposed EDM mainly contains a multi-embedding layer, an external attention layer, a feature-wise feed-forward network, and the candidate matching layer. Specifically, the external attention layer uses two external memory units shared across the entire input set to model the global interests of users. Then, by applying the feed-forward network to each feature dimension, the feature-wise feed-forward network is capable to learn the feature-level dependencies and properly model the local user preferences. In the experiments, three benchmark datasets are used with various validation metrics to show that our proposed EDM outperforms the state-of-the-art methods.

Keywords: Sequential Recommendation · External Attention · Global Interest

1 Introduction

With the continued development of mobile technology, many enterprises are focusing on expanding their online business, such as advertising, smart transport, and social media. The recommendation system is the core function of these online services [18]. By mining the user's previous interactions (e.g., purchases, and clicks), the recommender system can provide the personalized recommendation that the user may be interested in. The major benefits of the recommendation system are: (1) helping users to discover their potential needs for products, (2) increasing user loyalty for the company and creating more profit.

At present, owing to the revolutionary modeling ability of the neural network, sequential recommendation (SR) systems are becoming popular in both the research community and the business sector. In SR, the user's historical

H. Kashima et al. (Eds.): PAKDD 2023, LNAI 13937, pp. 335–347, 2023.
https://doi.org/10.1007/978-3-031-33380-4_26

interaction is treated as the chronological sequence determined by the timestamp. The purpose of SR is to capture the user's dynamic preferences and then predict their future preferred items. Recurrent neural network-based models use long short-term memory (LSTM) or gated recurrent units (GRU) to capture the user interests [8]. SasRec employs dot-product self-attention to retrieve the item-item representations for each user [9].

Although the existing recommendation models have gained the significant success in the area of SR, there are some issues that may affect the recommendation performance. On one hand, the "Filter Bubble" issue occurs in many previous SR models by focusing too much on the user's own historical sequence [1]. For example, suppose that there is a short video recommendation scenario, a user's historical watching sequence is mainly about surfing techniques. SR system might keep recommending him more surfing videos and fail to offer him some content following the current trend. The item preferences from other users may considered as well. On the other hand, the RNN or attention-based models ignore learning the feature-level dependencies between items. For instance, self-attention focuses on the instance-level correlation weights by addressing item-item product and ignores the item-item relationship in the latent feature-level locally [13].

To this end, we propose the Global-Aware **E**xternal Attention **D**eep **M**odel (EDM) to tackle the addressed issues without using the self-attention and other complex DNN structures (e.g., convolutional layer). Figure 1 depicts the structure of our proposed model, which consists of a Multi-Embedding Layer with Positional Embedding, an External Attention Layer, a Feature-Wise feed-forward network (FFN) layer, and the candidate matching layer. Through this design, our proposed method can effectively model user's global and local preferences.

Specifically, the external attention layer calculates the feature map by two external memory units (external *Key* and *Value*) independent of the individual user. Unlike self-attention, which uses the same input item representations from the user him/herself, the two external memories are shared across the whole dataset [5]. Optimized by the back-propagation, these two memory units are able to capture the global representations of the entire dataset. Through this design, our EDM can explicitly consider the global interest of the entire input set when modeling each individual user's interests. Additionally, to model users' own (local) preference from his/her historical sequence, the feature-wise FFN layer is used. We apply the FFN to each feature dimension to model the feature-level dependencies for the user. The appealing features can be highlighted under the feature-wise FFN layer. In the candidate matching layer, a point-wise FFN is employed to merge user's local and global feature representation. In the end, the model is optimized by the Bayesian personalized ranking objective function [14]. In the experiments, we evaluate our proposed EDM with many state-of-the-arts via various metrics including precision, MAP, and nDCG. The results demonstrate that our model has the advent performance against other baselines. The ablation analysis also indicates that EDM effectively models the users' global and local preferences.

The main contributions of this paper is specified as follows:

- We propose the global-aware external attention deep model (EDM) for sequential recommendation to learn both the users' global and local preferences to achieve comprehensive recommendations.
- We introduce the external attention to quantify the global interest correlations across the whole dataset.
- We employ the feature-wise feed-forward layer to capture the user's local preference by modeling the feature-level dependencies between items.
- The experiments under three benchmark datasets verify that our EDM outperforms the state-of-the-art models via four metrics.

2 Related Work

2.1 General Recommendation

The conventional recommendation system discovers users' general interests based on their historical interactions. A classical branch uses collaborative filtering (CF) to extract users' possible preferences from other strongly correlated users [16]. Matrix factorization (MF) is another typical approach that uses the shared latent space to represent the user and item, and optimize the inner product between user and item latent representations [10]. BPR, proposed by Rendle et al., introduces the Bayesian personalized ranking method to the matrix factorization to handle the implicit feedback data [14].

Recently, the research on the DNN-based recommendation model has gotten much attention. For example, NeuMF employs the multi-layer perception to improve the recommendation performance of MF [6]. xDeepFM uses a unified DNN structure to learn both the explicit and the implicit feedback [12]. Further more, many approaches consider the graph neural network (GNN) in the CF-based model. IGCN applies a GNN to learn the historical and temporal information to improve the embedding representations of user and item [23].

2.2 Sequential Recommendation

The objective of the sequential recommendation (SR) is to learn users' historical sequence to predict their next behaviors. The early approaches of SR use the Markov Chain to capture item-item transition patterns and predict the users' next items based on their last interactions [15]. After that, many SR models consider the DNN structures. Caser applies the horizontal and vertical convolution layer to capture item patterns [20]. GRU4Rec, proposed by Jannach et al., is a recurrent neural network-based recommender that uses the gated recurrent units to model the entire input session [8]. MANN fuses the user memory matrix into the user embedding [2]. On contrast, our proposed EDM consider the memory network on item embedding side. SASRec firstly adopts the self-attention structure to learn the users' long-term interests based on the historical user actions [9]. After the success of SASRec, many variants expanded the flexibility in this

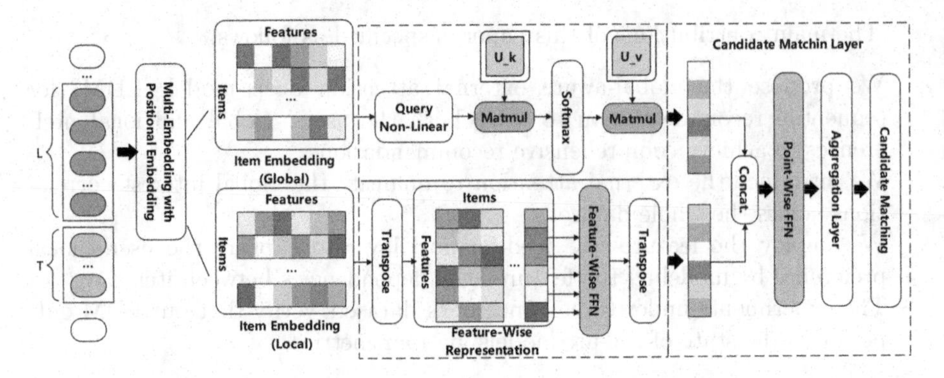

Fig. 1. The main structure of our proposed model: EDM.

direction [7,17,27]. For example, TiSASREC introduces the temporal information into self-attention that considers the transition time between two successive items [25]. SSE-PT proposes the self-attention structure with personalized user embedding [22]. STOSA models the user sequences with the stochastic Gaussian distribution that considers the uncertainty of input sequence [4].

3 Problem Formulation

The training data used in this paper is the users' implicit feedback on items. Firstly, the user and item set are defined as: $U = \{u_1, u_2, u_3, \ldots, u_M\}$, and $I = \{i_1, i_2, i_3, \ldots, i_N\}$, where M is the total number of users, and N is the total number of items. Then, for each user $u_i \in U$, the user-item interaction sequence is denoted as $\mathcal{S}^i = (\mathcal{S}^i_1, \mathcal{S}^i_2, \mathcal{S}^i_3, \cdots, \mathcal{S}^i_{|\mathcal{S}^i|})$, where $\mathcal{S}^i_t \in I$ is the item index that user i interacted with.

In order to model the sequential recommendation task, the problem can be specified as: given the earlier subsequence with length $|L|$, how likely all other N items will be interacted in future. In the training process, for each user, we use each $|L|$ successive item interactions $(\mathcal{S}^i_j, \ldots, \mathcal{S}^i_{j+|L|-1})$ as the input, and their next $|T|$ records as the targets to predict. A sliding window with size $|L+T|$ is applied to the user's sequence to extract the training objects. Each user i will generate $||\mathcal{S}^i| - (L+T) + 1|$ training objects with length $|L+T|$.

4 Our Proposed Method

In this section, we discuss our proposed global-aware external attention deep model for sequential recommendation (EDM). EDM utilizes the external attention mechanism to learn global user interests of items and the Feed-forward Network (FFN) Layer to capture local user preferences. Figure 1 shows the design of our proposed model. EDM is composed of four main components: (1) Multi-Embedding Layer, (2) External Attention Layer (middle upper side) for capturing the global user preferences on items, (3) Feature-Wise FFN Layer (middle bottom side) for modeling the local user interests, and (4) the Candidate Matching Layer to generate the recommendations. In the next pages, we will discuss the details of our EDM.

4.1 Multi-embedding Layer

The input of our proposed model is a subsequence with $|L|$ items from user i's historical interaction sequence, and each item is represented by a unique index. At the embedding layer, We propose a multi-embedding layer that allows the external attention layer and the feature-wise FFN layer to learn the separate item embeddings. The input item embeddings of user i are represented as: $E_{i,l}$ for the external attention layer, and $F_{i,l}$ for the feature-wise FFN layer, where $E_{i,l}, F_{i,l} \in R^{L \times d}$, d is the embedding dimension, and l means the embedding is the l-th subsequence of user.

Positional Embedding: Like the attention-based sequential recommender, the proposed EDM does not contain any recurrent or convolutional module, which means it is not able to memorize the positions of each item [21]. Hence, the positional embedding (PE) is adopted to represent the input item position:

$$PE_{(p,2i)} = sin(p/1000^{\frac{2i}{d}}),$$
$$PE_{(p,2i+1)} = cos(p/1000^{\frac{2i}{d}}), \quad (1)$$

where p is the item position, i s the i-th embedding dimension, and d is the size of the embedding dimension. After constructing the positional embedding matrix, the operation of addition is applied to add PE to two input item embeddings.

4.2 Learning User Preference Globally and Locally

External Attention: The external attention mechanism has reached success in the computer vision task, which employs two shared memory linear matrices (external *keys* and *values*) to capture the most critical information across the dataset [5]. Inspired by this research, we design the external attention module suitable for the sequential recommendation scenario, which learns the globally important item features in the sequential recommendation model. In the external attention, the weight map is generated by the multiplication between the input item embedding (*query*) and an external linear unit $U_k \in R^{L \times d}$. Another linear unit $U_v \in R^{L \times d}$ is applied to extract the attention score with the weight map. The structure of the external attention is on the middle-upper side of Fig. 1, and the equations can be formulated as follows:

$$Q = ReLU(E_{i,l} \cdot W_q),$$
$$G^{map} = softmax(Q \cdot U_k^T / \sqrt{d}), \quad (2)$$
$$G_{i,l}^{att} = G\_map \cdot U_v,$$

where $E_{i,l} \in R^{L \times d}$ is the input item embedding; $W_q \in R^{d \times d}$ is the trainable weight matrix; $ReLU(\cdot)$ is the rectified linear unit used to provide non-linearity; d is the embedding size; $[\cdot]$ represents the matrix multiplication operation; U_k and $U_v \in R^{L \times d}$ are the external *key* and *value*, they can be considered as two shared parameter matrices among the whole training dataset.

Feature-Wise Feed-Forward Layer: When considering the feature-level (local) information between items, existing researches applies the depthwise CNN to capture the local importance of each feature dimension from the input sequence [11,20]. Unlike the previous approaches, we propose a simple feature-wise FFN to capture user's local preferences on items to deal with the disadvantage (2) of the self-attention, which utilizes the feed-forward layer to every feature dimension (item embedding dimension). Each feature dimension has a shared set of parameters, which save the memory when the size of item embedding increases. Our proposed feature-wise FFN is on the middle-bottom side of Fig. 1, and the formula is as follows:

$$L_{t,*}^{FFN} = Z_{t,*} + W_{f2} \cdot ReLU(W_{f1} \cdot Z_{t,*}),\ for\ t = 1, 2, \cdots d, \tag{3}$$

where the $Z_{t,*} \in R^L$ is the t-th dimension of the input item embedding, which belongs to a transposed input item embedding $F_{i,l}^T \in R^{d \times L}$; $W_{f1} \in R^L$ and $W_{f2} \in R^L$ are two shared learnable parameters among all feature dimensions; $L_{t,*}^{FFN}$ denotes the t-th dimension of the output sequence embedding $L^{FFN} \in R^{d \times L}$. In the end, another transposing is used to extract the output:

$$L_{i,l}^{out} = (L^{FFN})^T. \tag{4}$$

4.3 Candidate Matching Layer

Point-Wise Feed-Forward Network. Through the above two branches of neural networks, we get two weighted item embeddings that represent the global and local user interests. In order to better refine the information in the two item representations, we apply a point-wise FFN with concatenating $G_{i,l}^{att}$ and $L_{i,l}^{out} \in R^{L \times d}$, via:

$$\begin{aligned} C_{i,l} &= [G_{i,l}^{att}\ L_{i,l}^{out}], \\ \hat{C_{i,l}} &= ReLU(C_{i,l} \cdot W_{c1} + b_{c1}) \cdot W_{c2} + b_{c2}, \end{aligned} \tag{5}$$

where $C_{i,l} \in R^{L \times 2d}$ is the concatenated matrix of $G_{i,l}^{att}$ and $L_{i,l}^{out}$; $W_{c1} \in R^{2d \times 2d}$ and $W_{c1} \in R^{2d \times d}$ are weight matrices; b_{c1} and b_{c2} are the bias vectors. After the point-wise FFN, $\hat{C_{i,l}} \in R^{L \times d}$ represents the output preference representation of user i's l-th input sequence.

Ranking Candidates. Two factors are considered when ranking item candidates: (a) the correlation between user i's output preference and the target item embedding; (b) the correlation between user i's embedding and the target item embedding. The correlation is measured by the euclidean distance: the smaller distance the two matrices has, the more related they are. Given user i's l-th output preference, the predicted ranking score of item j can be denoted as follows:

$$\hat{y_{i,j}} = (1-\alpha)\|mean(\hat{C_{i,l}}) - T_j\| + \alpha\|V_i - T_j\|, \tag{6}$$

where the T_j is the target embedding of item j; V_i is the user i's embedding; α denotes the weight factor; $mean(\cdot)$ is the mean operation that is used to aggregate the output weighted representation. The other aggregation operation, such as sum and max, can also be considered when ranking item candidates.

4.4 Model Learning

With the ranking score $\hat{y_{i,j}}$, we apply the Bayesian Personalized Ranking loss function to train the model [14], which is suitable for implicit feedback:

$$\arg\min_{\Theta} \sum_{j \in \mathcal{U}^+} \sum_{k \in \mathcal{U}^-} -\ln \sigma(\hat{y_{i,j}} - \hat{y_{i,k}}) + \lambda(\|\Theta\|^2), \tag{7}$$

where $\hat{y_{i,k}}$ and $\hat{y_{i,j}}$ represent the prediction score of the negative samples and the positive samples, respectively; $\sigma(x)$ denotes the sigmoid function $\frac{1}{1+e^x}$; the parameter Θ contains the set of parameters: $\{T, V, W_q, W_{c1}, W_{c2}, W_{f1}, W_{f2}\}$, which is learned by the loss function in Eq. (7).

Instead of using all items in the candidate set to train the network, $|T|$ item interactions following $|L|$ input items are used in the training session, which lowers the training cost. For each target item, we randomly select the same number of negative instances.

4.5 Computational Complexity

The computational complexity of our proposed EDM is based on three parts: the external attention, the feature-wise FFN, and the point-wise FFN, which is $O(3nd^2)$. Compared with SASRec, whose computational complexity is $O(n^2d + nd^2)$, our model has better scalability when the input length increases.

5 Experiments

In the experiments, we first evaluate the performance between our proposed EDM and the state-of-the-arts. Then, we further explore the effect of EDM's core components and hyperparameters.

5.1 Experiment Settings

Dataset: We conduct our experiments with three real-world datasets: MovieLens, Foursquare [24] and the Gowalla [3], which contain implicit feedbacks. For each dataset, we ignore the rating score of each data sample and transform it

Table 1. The Statistics of Three Datasets.

	User	Item	Interactions	Sparsity	Avg. user Ints.
Foursquare	1083	9989	227428	97.68%	210.00
Gowalla	2267	19358	284067	99.35%	125.3
MovieLens	6040	3416	999611	95.16%	165.5

into implicit feedback of 1. Meanwhile, we remove inactive users with fewer than 5 check-in records and inactive POIs that have interacted fewer than 5 times for MovieLens and Foursquare datasets. We set the inactive users and items threshold for Gowalla to 10 and 10. Table 1 shows the statistics of datasets.

Then, in the training procedure, the first 70% of the user interaction sequence is used to train the model, and the next 10% of interactions is used as the validation set for the hyperparameter searching. The last 20% of interactions is the test set for evaluating the model performance.

Metrics: To comprehensively evaluate the performance of the state-of-the-art methods and our proposed model, we select four metrics: precision (*Prec@K*), mean average precision (*MAP@K*), and normalized discounted cumulative gain (*nDCG@K*). For each user, *Prec@K* is what percentage of the K predicted items are in the test item set. *MAP@K* is the average of AP for all users, where AP is the average precision at all possible thresholds (from 1 to K). *nDCG@K* considers the ranking position of the predicted item with the normalization technique.

Baseline Models: We select the following state-of-the-art methods to compare with EDM. The matrix factorization-based model (**BPR** [14]), the RNN-based model (**GRU4Rec** [19]), the CNN-based model (**Caser** [20]), the attention-based model (**SasRec** [9], **AttRec** [26], and **LSAN** [11]).

5.2 Implementation Configuration

Our experiments are running by PyTorch[1] with a Nvidia GeForce GTX 3060. For our proposed EDM, we set $|L| = 5$ as the input sequence length of the model, and set $|T| = 3$ for model learning purpose. The model embedding size d is set to 128, 128, and 256 for Foursquare, Gowalla, and MovieLens, respectively. The learning rate and The L2 regularization λ are both 0.001. The batch size is 512, and the weight factor α is set to 0.1 in Eq. (6). The dropout rate for the point-wise feed-forward network is 0.5. All the above hyper-parameters are tuned by grid search.

For the baseline models, the following configuration is the same as our proposed model: the embedding dimension, the training batch size, the learning rate, and the λ. The models with CNN components (Caser and LSAN) use the kernel size of 3. The other hyperparameters are set by the original papers.

[1] https://pytorch.org/.

Table 2. Performance Comparison

Dataset	Metric	BPR	GRU4Rec	Caser	SasRec	AttRec	LSAN	EDM	Improv.
Foursquare	Prec@5	0.044	0.068	0.075	0.081	0.077	0.083	**0.089**	7.2%
	Prec@10	0.024	0.050	0.054	0.055	0.060	0.051	**0.061**	10.9%
	MAP@5	0.031	0.041	0.044	0.046	0.039	0.051	**0.058**	13.7%
	MAP@10	0.015	0.024	0.027	0.026	0.025	0.028	**0.033**	17.9%
	nDCG@5	0.048	0.076	0.081	0.088	0.078	0.092	**0.102**	10.9%
	nDCG@10	0.039	0.060	0.067	0.066	0.067	0.067	**0.078**	16.4%
Gowalla	Prec@5	0.034	0.061	0.058	0.072	0.070	0.066	**0.079**	9.7%
	Prec@10	0.028	0.050	0.049	0.058	0.055	0.053	**0.062**	6.9%
	MAP@5	0.021	0.039	0.041	0.050	0.044	0.047	**0.058**	16.0%
	MAP@10	0.015	0.031	0.029	0.037	0.036	0.035	**0.041**	10.8%
	nDCG@5	0.034	0.069	0.066	0.083	0.074	0.076	**0.093**	12.0%
	nDCG@10	0.033	0.066	0.068	0.076	0.072	0.070	**0.083**	9.2%
MovieLens	Prec@5	0.141	0.215	0.218	0.223	0.219	0.225	**0.237**	5.3%
	Prec@10	0.118	0.193	0.199	0.201	0.194	0.206	**0.210**	1.9%
	MAP@5	0.082	0.149	0.148	0.149	0.151	0.154	**0.162**	5.2%
	MAP@10	0.054	0.110	0.114	0.120	0.118	0.124	**0.129**	4.0%
	nDCG@5	0.147	0.231	0.224	0.229	0.228	0.233	**0.246**	5.6%
	nDCG@10	0.136	0.208	0.213	0.223	0.218	0.227	**0.234**	3.1%

5.3 Performance Comparison

The performance between our proposed EDM and the baseline models is summarized in Table 2. As the non-sequential recommenders, BPR has the worst performance among all baselines. Although BPR can effectively learn the long-term preference of the user, it fails to model the sequential information. Caser shows a promising result on Foursquare and MovieLens datasets in terms of MAP@K, but this CNN-based model has the second worst result on Gowalla. The possible reason is that Caser has a limitation on capturing users' general interest in a sparser dataset. GRU4Rec have a usual performance on three datasets. It uses the mini-batch strategy in the training session to improve their performance scalability. AttRec and SasRec are two self-attention approaches that show a strong ability for sequential behavior modeling. LSAN has the second-best performance on Foursquare and MovieLens, whose twin-attention structure can capture long-term and short-term users' preferences.

For our proposed model, the performance surpasses all baseline methods on three datasets, demonstrating EDM's effectiveness. The main reason is two-fold: (1) The external attention branch successfully models the global interests of the entire dataset and provides a positive supplement when predicting users' potential interests; (2) The feature-wise FFN considers the feature-level information that successfully learns the users' local preferences. The following Sect. 5.4 provides a detailed analysis of the components of EDM.

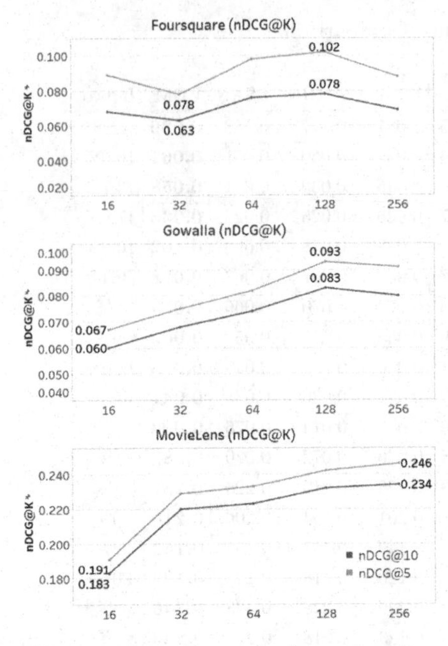

Fig. 2. The effect of the embedding dimension d

Table 3. Ablation Analysis (MA@5 and nD@5)

	Foursquare		Gowalla	
	MA@5	nD@5	MA@5	nD@5
1) EA (~~PE~~)	0.048	0.089	0.019	0.031
2) FW (~~PE~~)	0.049	0.089	0.047	0.079
3) EA + PE	0.051	0.093	0.023	0.038
4) FW + PE	0.053	0.094	0.051	0.082
5) EDM (~~PE~~)	0.053	0.094	0.056	0.091
6) **EDM**	**0.058**	**0.102**	**0.058**	**0.093**

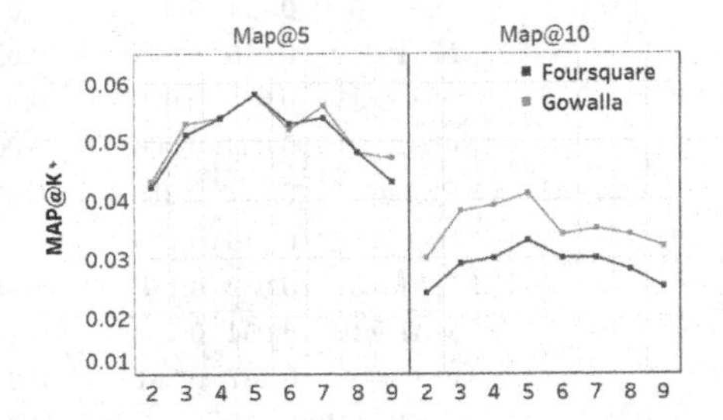

Fig. 3. The effect of item length L

5.4 Ablation Study

In this part, we perform the ablation study on the following three components: the external attention layer (EA), the feature-wise FFN layer (FW), and the positional embedding (PE). In order to test the stability, a denser dataset (Foursquare) and a sparser dataset (Gowalla) are used with the metrics *MAP@5* (*MA@5*) and *nDCG@5* (*nD@5*). The other parameters are set to the optimal model's configuration.

Table 3 shows the performance of EDM and two main components with and without the positional embedding (shown as ~~PE~~ in the table). According to the results in Table 3, the model performance decreases when we remove the positional embedding part. That is because both the external attention and the feature-wise FFN cannot encode the item position in the sequence. Such uncertainty of the positional relationship can lead to perturbations in the model performance. In addition, the external attention part performs similarly or slightly better on the denser dataset (Foursquare) but performs worse on a sparser dataset (Gowalla). The reason might be that the sparser dataset increases the uncertainty of user interests. Moreover, the large number of users might make it hard for the two linear units to memorize suitable global preferences for users (Fig. 2).

5.5 Influence of Hyperparameters

The Embedding Dimension d: We examine the size of the item embedding d from 16 to 256. Figure 3 depicts the result of the evaluation in terms of nDCG@K. The result indicates that a small dimension size is insufficient to preserve latent information of items. When the embedding size increases, the model performance grows steadily. The best performance comes when d equals 128, 128, and 256 for Foursquare, Gowalla, and MovieLens, respectively.

The Influence of the Input Sequence Length $|L|$: We test the effect of the input sequence length $|L|$ on Foursquare and Gowalla. Figure 3 shows the performance in terms of MAP@K. The trends for other metrics are similar. The result indicates that the $|L|$ successive items determine the users' future item interactions, and our model gets the highest MAP with a moderate value of $|L|$ ($|L| = 5$). The trend illustrates that the input item sequence with smaller the $|L|$ does not contain enough information. Meanwhile, the model does not get the extra performance with larger $|L|$ ($|L| > 5$). The longer input sequence may contain irrelevant and noisy information for a sparse dataset.

6 Conclusion

In this paper, we propose a novel global-aware external attention-based sequential recommender, EDM, for modeling both the global and local preferences of users. In our proposed model, we use the external attention layer with two shared linear memory units to capture the global interests of the entire dataset. At the same time, we also develop a feature-wise feed-forward network that considers the feature-level information dependency to capture the local preference of each user. The experiments demonstrate that EDM outperforms the state-of-the-art models in terms of four evaluation metrics. The ablation study indicates that external attention performs worse in the sparser dataset. Hence, we plan to explore the hierarchical structure and clustering method to deal with this potential drawback of external attention in future.

References

1. Chen, J., Dong, H., Wang, X., Feng, F., Wang, M., He, X.: Bias and debias in recommender system: a survey and future directions. arXiv preprint arXiv:2010.03240 (2020)
2. Chen, X., et al.: Sequential recommendation with user memory networks. In: Proceedings of the Eleventh ACM International Conference on Web Search and Data Mining, pp. 108–116 (2018)
3. Cho, E., Myers, S.A., Leskovec, J.: Friendship and mobility: user movement in location-based social networks. In: Proceedings of the 17th ACM SIGKDD International Conference on Knowledge Discovery and Data Mining, pp. 1082–1090 (2011)

4. Fan, Z., et al.: Sequential recommendation via stochastic self-attention. In: Proceedings of the ACM Web Conference 2022, pp. 2036–2047 (2022)
5. Guo, M.H., Liu, Z.N., Mu, T.J., Hu, S.M.: Beyond self-attention: external attention using two linear layers for visual tasks. arXiv preprint arXiv:2105.02358 (2021)
6. He, X., Liao, L., Zhang, H., Nie, L., Hu, X., Chua, T.S.: Neural collaborative filtering. In: Proceedings of the 26th International Conference on World Wide Web, pp. 173–182 (2017)
7. He, Z., Zhao, H., Lin, Z., Wang, Z., Kale, A., McAuley, J.: Locker: locally constrained self-attentive sequential recommendation. In: Proceedings of the 30th ACM International Conference on Information & Knowledge Management, pp. 3088–3092 (2021)
8. Jannach, D., Ludewig, M.: When recurrent neural networks meet the neighborhood for session-based recommendation. In: Proceedings of the Eleventh ACM Conference on Recommender Systems, pp. 306–310 (2017)
9. Kang, W.C., McAuley, J.: Self-attentive sequential recommendation. In: 2018 IEEE International Conference on Data Mining (ICDM), pp. 197–206. IEEE (2018)
10. Koren, Y., Bell, R., Volinsky, C.: Matrix factorization techniques for recommender systems. Computer **42**(8), 30–37 (2009)
11. Li, Y., Chen, T., Zhang, P.F., Yin, H.: Lightweight self-attentive sequential recommendation. In: Proceedings of the 30th ACM International Conference on Information & Knowledge Management, pp. 967–977 (2021)
12. Lian, J., Zhou, X., Zhang, F., Chen, Z., Xie, X., Sun, G.: xDeepFM: combining explicit and implicit feature interactions for recommender systems. In: Proceedings of the 24th ACM SIGKDD International Conference on Knowledge Discovery & Data Mining, pp. 1754–1763 (2018)
13. Ma, C., Kang, P., Liu, X.: Hierarchical gating networks for sequential recommendation. In: Proceedings of the 25th ACM SIGKDD International Conference on Knowledge Discovery & Data Mining, pp. 825–833 (2019)
14. Rendle, S., Freudenthaler, C., Gantner, Z., Schmidt-Thieme, L.: BPR: Bayesian personalized ranking from implicit feedback. In: Proceedings of UAI 2009, pp. 452–461. AUAI Press (2009)
15. Rendle, S., Freudenthaler, C., Schmidt-Thieme, L.: Factorizing personalized Markov chains for next-basket recommendation. In: Proceedings of the 19th International Conference on World Wide Web, pp. 811–820 (2010)
16. Schafer, J.B., Frankowski, D., Herlocker, J., Sen, S.: Collaborative filtering recommender systems. In: Brusilovsky, P., Kobsa, A., Nejdl, W. (eds.) The Adaptive Web. LNCS, vol. 4321, pp. 291–324. Springer, Heidelberg (2007). https://doi.org/10.1007/978-3-540-72079-9_9
17. Sun, F., et al.: Bert4rec: sequential recommendation with bidirectional encoder representations from transformer. In: Proceedings of the 28th ACM International Conference on Information and Knowledge Management, pp. 1441–1450 (2019)
18. Tan, Q., et al.: Sparse-interest network for sequential recommendation. In: Proceedings of the 14th ACM International Conference on Web Search and Data Mining, pp. 598–606 (2021)
19. Tan, Y.K., Xu, X., Liu, Y.: Improved recurrent neural networks for session-based recommendations. In: Proceedings of the 1st Workshop on Deep Learning for Recommender Systems, pp. 17–22 (2016)
20. Tang, J., Wang, K.: Personalized top-n sequential recommendation via convolutional sequence embedding. In: Proceedings of the Eleventh ACM International Conference on Web Search and Data Mining, pp. 565–573 (2018)

21. Vaswani, A., et al.: Attention is all you need. In: Advances in Neural Information Processing Systems, pp. 5998–6008 (2017)
22. Wu, L., Li, S., Hsieh, C.J., Sharpnack, J.: SSE-PT: sequential recommendation via personalized transformer. In: Fourteenth ACM Conference on Recommender Systems, pp. 328–337 (2020)
23. Xia, J., Li, D., Gu, H., Lu, T., Zhang, P., Gu, N.: Incremental graph convolutional network for collaborative filtering. In: Proceedings of the 30th ACM International Conference on Information & Knowledge Management, pp. 2170–2179 (2021)
24. Yang, D., Zhang, D., Zheng, V.W., Yu, Z.: Modeling user activity preference by leveraging user spatial temporal characteristics in LBSNs. IEEE Trans. Syst. Man Cybern. Syst. **45**(1), 129–142 (2014)
25. Ying, H., et al.: Time-aware metric embedding with asymmetric projection for successive poi recommendation. World Wide Web **22**(5), 2209–2224 (2019)
26. Zhang, S., Tay, Y., Yao, L., Sun, A.: Next item recommendation with self-attention. arXiv preprint arXiv:1808.06414 (2018)
27. Zhou, K., et al.: S3-Rec: self-supervised learning for sequential recommendation with mutual information maximization. In: Proceedings of the 29th ACM International Conference on Information & Knowledge Management, pp. 1893–1902 (2020)

Aggregately Diversified Bundle Recommendation via Popularity Debiasing and Configuration-Aware Reranking

Hyunsik Jeon, Jongjin Kim, Jaeri Lee, Jong-eun Lee, and U Kang(✉)

Seoul National University, Seoul, South Korea
{jeon185,j2kim99,jlunits2,kjayjay40,ukang}@snu.ac.kr

Abstract. How can we expose diverse items across all users while satisfying their needs in bundle recommendations? Diversified bundle recommendation is a crucial task since it leads to great benefits for both sellers and users. However, there have been no studies on aggregate diversity in bundle recommendation, while they have been intensively studied in item recommendation. Moreover, existing methods of aggregately diversified item recommendation are not fully suitable for bundle recommendation. In this paper, we propose PopCon (Popularity Debiasing and Configuration-aware Reranking), an accurate method for aggregately diversified bundle recommendation. PopCon mitigates the popularity bias of a recommendation model by a popularity-based negative sampling in training process, and maximizes accuracy and aggregate diversity by a configuration-aware reranking algorithm. We show that PopCon provides state-of-the-art performance on real-world datasets, achieving up to 60.5% higher Entropy@5 and 3.92× higher Coverage@5 with comparable accuracies compared to the best competitor.

Keywords: Bundle Recommendation · Aggregate Diversity · Popularity Debiasing · Configuration-aware Reranking

1 Introduction

How can we expose diverse items across all users as well as satisfying their needs in bundle recommendations? Recommender systems [9,10,13,16] have been indispensable techniques in online platforms providing customers with several relevant items from numerous ones [20]. Bundle recommendation aims to suggest sets of items instead of individual ones to users. It has been gaining attention in online platforms due to its advantage of providing items that customers need with one-stop convenience [14]. Furthermore, bundles are ubiquitous in real-world scenarios because they provide effectual marketing strategies (e.g., discount sales) which are appealing to customers [6]. However, traditional bundle recommendation models [3–6,8,14,18] have focused only on accuracy without paying attention to diversity. Figure 1 compares the traditional bundle recommendation and an aggregately diversified bundle recommendation. Note that

H. Kashima et al. (Eds.): PAKDD 2023, LNAI 13937, pp. 348–360, 2023.
https://doi.org/10.1007/978-3-031-33380-4_27

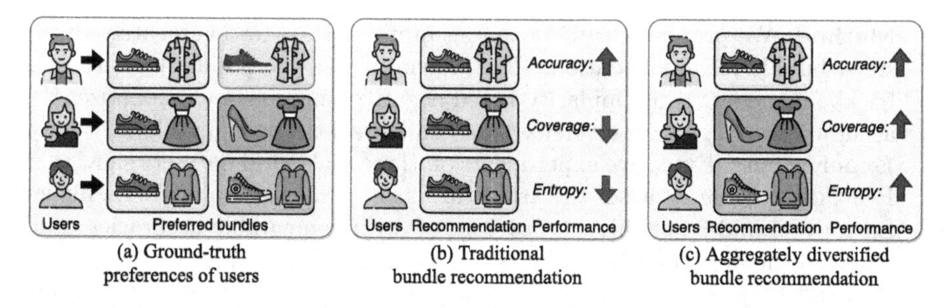

(a) Ground-truth preferences of users
(b) Traditional bundle recommendation
(c) Aggregately diversified bundle recommendation

Fig. 1. Illustrative comparison of (b) traditional bundle recommendation and (c) aggregately diversified bundle recommendation when the (a) ground-truth preferences of users are given.

aggregate diversity is measured by the degree of fair exposure of items (i.e., coverage and entropy) in recommendation results across all users. As shown in Fig. 1(b), the traditional bundle recommendation, despite achieving high accuracy, results in a low aggregate diversity by recommending bundles that contain a popular item (e.g., the red shoes). On the other side, as shown in Fig. 1(c), the *aggregately diversified bundle recommendation (our task)* further aims to achieve high aggregate diversity by exposing diverse items across all users.

In the last decade, there have been several studies for aggregate diversity in item recommendation. Reranking-based methods [2,7,11,15], which rerank the recommendation results of a trained model to achieve both high accuracy and high aggregate diversity, are the most prevailing approaches in aggregately diversified item recommendation owing to their effectiveness in handling aggregate diversity. However, they are not fully suitable for bundle recommendation due to the following two limitations. First, a bundle recommendation model used as a backbone is easily overfitted to some popular bundles, and thus relying on the backbone model's results inevitably results in sacrificing a lot of accuracies to increase aggregate diversity. Second, they do not consider the configuration of bundles which is pivotal information to address the diversity of item exposure in bundle recommendation.

We propose PopCon (Popularity Debiasing and Configuration-aware Reranking), an accurate method for aggregately diversified bundle recommendation. PopCon consists of two phases, model training and reranking. In the training phase, PopCon trains a bundle recommendation model as a backbone with a popularity-based negative sampling to mitigate the popularity bias of the model. In the reranking phase, PopCon reranks the recommendation result of the models to maximize both accuracy and aggregate diversity. PopCon exploits each bundle's configuration to effectively deal with the aggregate diversity in the reranking phase. The contributions of PopCon are summarized as follows.

- **Problem.** To the best of our knowledge, our work is the first study that focuses on aggregately diversified bundle recommendation, which is of large importance in real-world scenarios.

- **Method.** We propose PopCon, an accurate method for aggregately diversified bundle recommendation. PopCon mitigates the popularity bias of a backbone model via a popularity-based negative sampling and maximizes the accuracy and aggregate diversity by a configuration-aware reranking.
- **Experiments.** Extensive experiments on three real-world datasets show that PopCon provides state-of-the-art performance achieving up to 60.5% higher Entropy@5 and 3.92× higher Coverage@5 with comparable accuracies compared to the best competitor.

2 Problem Definition and Related Works

2.1 Problem Definition

Bundle recommendation aims to predict sets of items, instead of individual items, that users would prefer. In this work, we focus on aggregate diversity in the bundle recommendation. We give the formal definition of the problem, namely aggregately diversified bundle recommendation, as Problem 1.

Problem 1 (Aggregately diversified bundle recommendation). Let $\mathcal{U}$, $\mathcal{I}$, and $\mathcal{B}$ be the sets of users, items, and bundles, respectively. We have matrices of user-bundle interactions, user-item interactions, and bundle-item affiliations which are denoted as $\mathbf{X} = [x_{ub}] \in \mathbb{R}^{|\mathcal{U}|\times|\mathcal{B}|}$, $\mathbf{Y} = [y_{ui}] \in \mathbb{R}^{|\mathcal{U}|\times|\mathcal{I}|}$, and $\mathbf{Z} = [z_{bi}] \in \mathbb{R}^{|\mathcal{B}|\times|\mathcal{I}|}$, respectively. $x_{ub}, y_{ui}, z_{bi} \in \{0,1\}$ are binary values, indicating an observation or a non-observation of interaction or affiliation. Then, the problem is to recommend a list of k bundles to each user u as $\mathbf{r}_u(k) \subset \{b|b \in \mathcal{B}, x_{ub} = 0\}$, which have not been observed in the user-bundle interactions. The goal is to make $\mathbf{r}_u(k)$ accurate for each user u, and to make the overall recommendation results $\mathbf{R}(k) = (\mathbf{r}_1(k), \cdots, \mathbf{r}_{|\mathcal{U}|}(k))$ aggregately diverse.

The aggregate diversity is evaluated for the items in $\mathbf{R}(k)$ by two metrics.

- **Coverage** measures how many different items are contained in the results.

$$Coverage@k = \frac{1}{|\mathcal{I}|}\sum_{i\in\mathcal{I}} app(i, \mathbf{R}(k)), \tag{1}$$

where $app(i, \mathbf{R}(k)) = [i \in \bigcup_{b\in\mathbf{R}(k)} \Omega_b]$ indicates whether item i appears in $\mathbf{R}(k)$. $\Omega_b = \{i|i \in \mathcal{I}, z_{bi} = 1\}$ is the set of bundle b's constituent items. The Iverson bracket $[\cdot]$ returns 1 if the statement is true, 0 otherwise.
- **Entropy** measures how evenly all items appear in the results.

$$Entropy@k = -\sum_{i\in\mathcal{I}} p(i, \mathbf{R}(k)) \log p(i, \mathbf{R}(k)), \tag{2}$$

where $p(i, \mathbf{R}(k)) = \frac{Freq(i,\mathbf{R}(k))}{\sum_{j\in\mathcal{I}} Freq(j,\mathbf{R}(k))}$. $Freq(i, \mathbf{R}(k)) = \sum_{u\in\mathcal{U}} freq(i, \mathbf{r}_u(k))$ where $freq(i, \mathbf{r}_u(k)) = \sum_{b\in\mathbf{r}_u(k)} [i \in \Omega_b]$ indicates item i's frequency in user u's recommended bundles $\mathbf{r}_u(k)$.

2.2 Related Works

Bundle recommendation. Bundle recommendation aims to recommend a set of items instead of an individual one to users. Existing bundle recommendation methods are mainly divided into matrix factorization-based approaches [3,5,18] and graph learning-based approaches [4,6,14]. BR [18] and EFM [3] jointly factorize user-item and user-bundle interactions to predict unseen user-bundle interactions. DAM [5] further introduces an attention mechanism to effectively learn bundle embeddings. With the proliferation of graph learning approaches, several studies [4,6,14] formulate the bundle recommendation in a tripartite graph with nodes of users, items, and bundles. BundleNet [6] learns a graph convolutional network to predict interactions between the nodes, while BGCN [4] further decomposes user preferences into item-view and bundle-view to effectively predict the interactions. CrossCBR [14] captures cooperative association between the item-view and bundle-view by a contrastive learning method to improve performance. However, such previous works for bundle recommendation focus only on accuracy. In this work, we further address aggregate diversity which is of great importance but makes the problem more challenging.

Aggregately Diversified Recommendation. Aggregately diversified recommendation aims to increase diversity of recommendations across all users [2,12]. It is important to accomplish high aggregate diversity because it alleviates the long tail problems and maximizes the profit of the sales platform. Most existing methods for aggregately diversified recommendations modify the results of a backbone model to achieve high aggregate diversity since it is difficult to optimize the model both for accuracy and diversity. Kwon et al. [2] rerank the recommendation results of a backbone model based on item popularity and heuristic thresholds of scores. Karakaya et al. [11] replace recommended items with similar ones through a random walk on an item co-occurrence graph. FairMatch [15] finds high-quality but less frequently recommended items in a recommendation list by solving the maximum flow problem. UImatch [7] constrains the limit of each item and solves the matching problem with a greedy strategy. However, there has been no study of aggregate diversity for bundle recommendation, which is crucial in practical scenarios but more challenging to address.

3 Proposed Method

In this section, we propose PopCon (Popularity Debiasing and Configuration-aware Reranking) to address the aggregately diversified bundle recommendation.

3.1 Overview

We concentrate on the following challenges to achieve high aggregate diversity with comparable accuracy in bundle recommendation.

C1. **Mitigating popularity bias of a backbone model.** A bundle recommendation model easily overfits to some popular bundles. How can we mitigate the popularity bias of the backbone model?

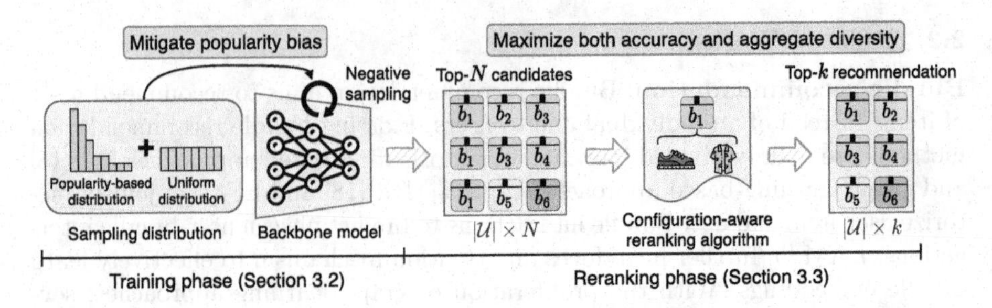

Fig. 2. Overview of PopCon which consists of training and reranking phases.

C2. **Fitting two opposite criteria, accuracy and diversity.** It is challenging to fit accuracy and diversity simultaneously since they are opposite criteria. How can we satisfy both opposite criteria?

C3. **Simultaneously considering how many items appear and how evenly items appear.** To achieve high aggregate diversity, we need to consider not only whether items appear or not, but whether items appear evenly. How can we consider both simultaneously?

The main ideas of PopCon are summarized as follows.

I1. **Popularity-based negative sampling.** It mitigates the popularity bias of a backbone model and enables us to effectively leverage the user-bundle relationship scores.

I2. **Accuracy-prioritized coupling.** It enables us to retain high-scored bundles in recommendation results and replace low-scored bundles with more diverse ones.

I3. **Maximizing the gains of coverage and entropy.** It encourages bundles that have not been recommended and that are less recommended to be recommended more.

Figure 2 shows the overall process of PopCon. PopCon consists of two phases, model training phase and reranking phase. In the training phase, PopCon trains a bundle recommendation model such as DAM [5] or CrossCBR [14] as a backbone while mitigating its popularity bias by a popularity-based negative sampling. In the reranking phase, PopCon selects candidate bundles for each user and reranks the candidates by a configuration-aware reranking algorithm to maximize both accuracy and aggregate diversity.

3.2 Training Phase with Popularity Debiasing

The objective of the training phase is to train a model $f(u, b)$ that accurately predicts the score between user u and bundle b. We first investigate the popularity bias of traditional models and propose a popularity-based negative sampling to mitigate the popularity bias of the models.

Real-world datasets for bundle recommendation commonly entail popularity bias because of various factors such as exposure mechanisms and public opinions. Accordingly, bundle recommendation models suffer from the popularity bias in their output [1]. Figure 3 shows the popularity bias of real-world datasets and that of trained models. We train DAM [5] and CrossCBR [14] which are state-of-the-art bundle recommendation models on real-world datasets. For each dataset, we split bundles into 50 groups in the order of their popularity, and sum up the number of incorrect recommendations for each group's bundles in the top-5 recommendation of the model. As shown in the figure, the real-world datasets entail the popularity bias (i.e., long-tail problem [17]) and the trained recommendation models emphasize popular items, showing their vulnerability to the popularity bias. The popularity bias of the model gives incorrect information about user-bundle relationships because popular bundles easily receive high scores regardless of user preferences, and makes it challenging to achieve high aggregate diversity when using the predicted scores in the reranking phase.

We propose a popularity-based negative sampling in training process to mitigate the popularity bias of a backbone model. Assume we have matrices of user-bundle interactions, user-item interactions, and bundle-item affiliations as $\mathbf{X}=[x_{ub}]\in\mathbb{R}^{|\mathcal{U}|\times|\mathcal{B}|}$, $\mathbf{Y}=[y_{ui}]\in\mathbb{R}^{|\mathcal{U}|\times|\mathcal{I}|}$, and $\mathbf{Z}=[z_{bi}]\in\mathbb{R}^{|\mathcal{B}|\times|\mathcal{I}|}$, respectively. $\mathcal{U}$, $\mathcal{B}$, and $\mathcal{I}$ are the sets of users, bundles, and items, respectively. Then, a bundle recommendation model f aims to predict the scores of user-bundle pairs. Specifically, the model f is defined as matrix factorization-based [3,5,18] or graph-based frameworks [4,6,14] to utilize $\mathbf{X}$, $\mathbf{Y}$, and $\mathbf{Z}$. Then, the model f is trained by minimizing the Bayesian Personalized Ranking (BPR) loss [19] as follows:

$$\sum_{(u,b,b')\in D} -\ln\sigma\left(f(u,b)-f(u,b')\right), \tag{3}$$

where $D=\{(u,b,b')|u\in\mathcal{U},b\in\mathcal{B},b'\in\mathcal{B},x_{ub}=1,x_{ub'}=0\}$, and $\sigma(\cdot)$ is the sigmoid function. In Equation (3), b is a positive sample which user u has interacted with, whereas b' is a negative sample which user u has not interacted with. However, the previous works [3–6,14,18] sample the negative bundles b' from the uniform distribution although popular bundles are more likely to be picked as positive samples. This makes the model overfit to some popular bundles and causes the popularity bias as in Fig. 3. To mitigate the popularity bias, we increase the probability that popular bundles are selected as negative samples. We propose the probability of sampling negative bundle b' as follows:

$$p(b')=\alpha\frac{freq(b')}{\sum_{j\in\mathcal{B}}freq(j)}+(1-\alpha)\frac{1}{|\mathcal{B}|}, \tag{4}$$

where $freq(j)$ is the number of bundle j's interactions (i.e., number of non-zeros in $\mathbf{X}$'s jth column), $\alpha\in[0,1]$ is a balancing hyper-parameter between the popularity-based distribution and the uniform distribution. If α is large, the sampling probability of a bundle is largely affected by its popularity, whereas if α is small, a bundle is selected almost uniformly regardless of its popularity.

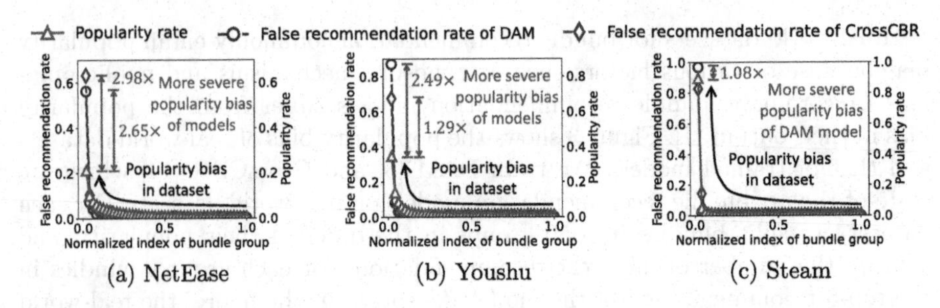

(a) NetEase (b) Youshu (c) Steam

Fig. 3. Popularity bias of real-world datasets (NetEase, Youshu, and Steam) and that of trained models on them. The datasets entail popularity biases, and the trained models have more severe ones.

3.3 Reranking Phase with Configuration-Awareness

The objective of the reranking phase is to maximize both accuracy and aggregate diversity using the trained backbone model f. We first select top-N candidate bundles for each user u using the scores $f(u, b) = \hat{x}_{ub} \in \mathbb{R}$ of all bundles $b \in \mathcal{B}$. Then, we rerank the candidate bundles to recommend k bundles ($N \gg k$) for each user. Specifically, we select the most suitable bundle among the candidates for each user and repeat it k times. The main challenge in the reranking phase is to measure which bundle is the best for user u at each time in terms both of accuracy and aggregate diversity.

It is straightforward to select the best bundle using a single criterion: accuracy or aggregate diversity. Assume we consider the candidate bundle b for user u currently. We compare $\sigma(\hat{x}_{ub})$ of each candidate to obtain the best accuracy because it measures how bundle b is appropriate for user u. To obtain the best aggregate diversity, we simultaneously measure the gains of coverage and entropy when recommending a bundle and select the one that maximizes it. Specifically, we propose to compare $DivGain(b, \hat{\mathbf{R}}(k)) \in \mathbb{R}$, which considers the appearance of new items and the fair appearance of items, as follows:

$$DivGain(b, \hat{\mathbf{R}}(k)) = \frac{1}{2}CovGain(b, \hat{\mathbf{R}}(k)) + \frac{1}{2}EntGain(b, \hat{\mathbf{R}}(k)), \tag{5}$$

where $DivGain(b, \hat{\mathbf{R}}(k)), CovGain(b, \hat{\mathbf{R}}(k))$, and $EntGain(b, \hat{\mathbf{R}}(k)) \in \mathbb{R}$ denote the gains of aggregate diversity, coverage, and entropy, respectively, and $\hat{\mathbf{R}}(k)$ is the current recommendation results for all users. $CovGain(b, \hat{\mathbf{R}}(k)) \in [0, 1]$ and $EntGain(b, \hat{\mathbf{R}}(k)) \in [-1, 1]$ are measured as the changes of Equations (1) and (2), respectively, when adding a bundle b to the current recommendation result $\hat{\mathbf{R}}(k)$; we obtain $EntGain(b, \hat{\mathbf{R}}(k))$ by dividing the original entropy gain by the maximum entropy so that the resulting value is in $[-1, 1]$.

However, the main difficulty of the reranking is to select the best bundle by measuring the accuracy and aggregate diversity simultaneously. For example, for user u, if $\sigma(\hat{x}_{ub}) > \sigma(\hat{x}_{ub'})$ and $DivGain(b, \hat{\mathbf{R}}(k)) < DivGain(b', \hat{\mathbf{R}}(k))$, it is difficult to decide which bundle should be recommended. It is essentially challenging because the accuracy and aggregate diversity are opposite in most cases. For instance, popular bundles usually provide high accuracy scores but less aggregate diversity scores.

Desired Properties. To deal with this conflict, we propose three desired properties for a measurement function $g(u, b, \hat{\mathbf{R}}(k))$, which is used to select the best bundle b for user u and the current recommendation results $\hat{\mathbf{R}}(k)$.

Property 1 (Increasing for accuracy). The function should satisfy $g(u, b, \hat{\mathbf{R}}(k)) \geq g(u, b', \hat{\mathbf{R}}(k))$ if $\sigma(\hat{x}_{ub}) > \sigma(\hat{x}_{ub'})$ and $DivGain(b, \hat{\mathbf{R}}(k)) = DivGain(b', \hat{\mathbf{R}}(k))$.

Property 2 (Increasing for diversity). The function should satisfy $g(u, b, \hat{\mathbf{R}}(k)) \geq g(u, b', \hat{\mathbf{R}}(k))$ if $\sigma(\hat{x}_{ub}) = \sigma(\hat{x}_{ub'})$ and $DivGain(b, \hat{\mathbf{R}}(k)) > DivGain(b', \hat{\mathbf{R}}(k))$.

Properties 1 and 2 are essential because they allow fair comparisons for accuracy and aggregate diversity when the other metrics are the same. One candidate measurement function to satisfy both Properties 1 and 2 are as follows.

$$g(u, b, \hat{\mathbf{R}}(k)) = (1 - \beta)\sigma(\hat{x}_{ub}) + \beta DivGain(b, \hat{\mathbf{R}}(k)), \tag{6}$$

where $\beta \in [0, 1]$ is a balancing hyper-parameter. Equation (6) is a weighted sum of the accuracy and aggregate diversity terms to measure two criteria together.

On the other hand, it is also necessary to ensure that bundles that users like a lot are recommended regardless of the gains of aggregate diversity to satisfy the users. This is challenging in our task because accuracy and aggregate diversity are opposite in most cases. Thus, we need to reduce the influence of the gain of aggregate diversity as the accuracy increases. In this regard, we propose a property of accuracy priority as follows.

Property 3 (Accuracy priority). The function should satisfy $\frac{\partial g(u,b,\hat{\mathbf{R}}(k))}{\partial DivGain(b,\hat{\mathbf{R}}(k))} < \frac{\partial g(u,b',\hat{\mathbf{R}}(k))}{\partial DivGain(b',\hat{\mathbf{R}}(k))}$ if $\sigma(\hat{x}_{ub}) > \sigma(\hat{x}_{ub'})$.

Accuracy-prioritized coupling. We propose a measurement function g that satisfies all the desired properties by prioritizing accuracy as follows.

$$g(u, b, \hat{\mathbf{R}}(k)) = \sigma(\hat{x}_{ub})^{\beta} + (1 - \sigma(\hat{x}_{ub})^{\beta}) DivGain(b, \hat{\mathbf{R}}(k)), \tag{7}$$

where $\beta \geq 1$ is a balancing hyper-parameter. If β is small, the recommendation result is highly dependent on accuracy, and if β is large, it is highly dependent on aggregate diversity because $\sigma(\hat{x}_{ub}) \in [0, 1]$. We show in Lemmas 1, 2, and 3 that Equation (7) satisfies all the desired properties. In the Lemmas, we denote $\sigma(\hat{x}_{ub})$ as $A(b)$, $DivGain(b, \hat{\mathbf{R}}(k))$ as $D(b)$, and $g(u, b, \hat{\mathbf{R}}(k))$ as $G(b)$ for brevity.

Lemma 1. *Equation* (7) *satisfies Property 1.*

Proof. If $A(b) > A(b')$ and $D(b) = D(b')$, then $G(b) - G(b') = (A(b)^{\beta} - A(b')^{\beta})(1 - D(b))$. Thus, $G(b) \geq G(b')$ because $A(b)^{\beta} > A(b')^{\beta}$ and $D(b) \leq 1$.

Table 1. Summary of bundle recommendation datasets. U, B, and I indicate users, bundles, and items, respectively.

Dataset	#U	#B	#I	#U-B (dens.)	#U-I (dens.)	#B-I (dens.)	Avg. B size
Steam[a]	29,634	615	2,819	87,565 (0.48%)	902,967 (1.08%)	3,541 (0.20%)	5.76
Youshu[b]	8,039	4,771	32,770	51,377 (0.13%)	138,515 (0.05%)	176,667 (0.11%)	37.03
NetEase[c]	18,528	22,864	123,628	302,303 (0.07%)	1,128,065 (0.05%)	1,778,838 (0.06%)	77.80

[a] https://github.com/technoapurva/Steam-Bundle-Recommendation
[b] https://github.com/yliuSYSU/DAM
[c] https://github.com/cjx0525/BGCN

Lemma 2. *Equation* (7) *satisfies Property 2.*

Proof. If $A(b) = A(b')$ and $D(b) > D(b')$, then $G(b) - G(b') = (1 - A(b)^\beta)(D(b) - D(b'))$. Thus, $G(b) \geq G(b')$ because $A(b)^\beta \leq 1$ and $D(b) > D(b')$.

Lemma 3. *Equation* (7) *satisfies Property 3.*

Proof. $\frac{\partial G(b)}{\partial D(b)} = 1 - A(b)^\beta$. Thus, $\frac{\partial G(b)}{\partial D(b)} < \frac{\partial G(b')}{\partial D(b')}$ if $A(b) > A(b')$.

Note that Equation (6) does not satisfy Property 3 because its $\frac{\partial G(b)}{\partial D(b)}$ is a constant value β, although it satisfies Properties 1 and 2.

Reranking Algorithm. We repeat recommending the most suitable bundle among the candidate bundles to each user, k times. Specifically, let the current recommendation results be $\hat{\mathbf{R}}(k) = (\hat{\mathbf{r}}_1(k), \hat{\mathbf{r}}_2(k), \cdots, \hat{\mathbf{r}}_{|\mathcal{U}|}(k))$, where $\hat{\mathbf{r}}_u(k)$ is the current recommendation result for user u; $\hat{\mathbf{r}}_u(k)$ for every $u \in \mathcal{U}$ is empty at the initial state. In random order of users $u \in \mathcal{U}$, we add $b' = \arg\max_b g(u, b, \hat{\mathbf{R}}(k))$ to $\hat{\mathbf{r}}_u(k)$ among u's candidate N bundles. We adopt a mini-batch technique that randomly selects m users in every step. We repeat this process k times, and finally obtain the recommendation results $\mathbf{R}(k)$.

4 Experiments

In this section, we perform experiments to answer the following questions.

Q1. **Performance Trade-off (Sect.** 4.2**).** Does PopCon provide the best trade-off between accuracy and aggregate diversity?
Q2. **Ablation Study (Sect.** 4.3**).** How do the main ideas in PopCon help improve the performance?
Q3. **Effects of number of candidates (Sect.** 4.4**).** How does the number N of candidates affect the performance of PopCon?

4.1 Experimental Setup

Datasets. We use three real-world datasets of bundle recommendation as summarized in Table 1. Steam [18] is constructed from Australian Steam community, a video game distribution platform. Youshu [5] is constructed from Youshu, a book review site. Netease [3] is constructed from Netease, a cloud music service.

Baselines. We compare PopCon with six baselines of aggregately diversified recommendation. Given a recommendation list of size $N(N > k)$ for each user, Reverse and Random pick bottom-k bundles and random-k bundles, respectively. Kwon [2] heuristically replaces the popular bundles of a recommendation list with unpopular ones. Karakaya [11] replaces bundles in a recommendation list with other bundles through random walk on an item co-occurrence network. Fairmatch [15] handles the maximum flow problem to replace bundles in a recommendation list with other bundles. UImatch [7] assigns capacity of each bundle to be recommended and generates a recommendation list in a greedy manner.

Backbone Models. We leverage two existing bundle recommendation models, DAM [5] and CrossCBR [14], as backbone models of PopCon and the baselines. DAM and CrossCBR are the state-of-the-art models among matrix factorization-based methods and graph learning-based methods, respectively.

Evaluation Metrics. We employ leave-one-out protocol [5] where one of each user's interactions is randomly selected for testing. We evaluate the performance in two criteria, accuracy and aggregate diversity. We use mean average precision (MAP@k) for the accuracy, and Coverage@k and Entropy@k for the aggregate diversity. MAP@k considers highly ranked bundles more importantly for accuracy. Coverage@k and Entropy@k are explained in Sect. 2.1. We investigate the trade-off curve between accuracy and aggregate diversity. We set the number k of bundles to 5, which is the most widely used setting.

Hyperparameters. We set the embedding dimensionality of DAM and CrossCBR to 20. We set the batch size m in the reranking phase to 10. For both DAM and CrossCBR, we set α to 0.1, 0.05, and 0.02 on Steam, Youshu, and NetEase, respectively. In Sects. 4.2 and 4.3, we set N to 100, 1,000, and 1,000 on Steam, Youshu, and NetEase, respectively. For each curve, β is not a fixed value but controls the trade-off between accuracy and aggregate diversity.

4.2 Performance Trade-Off (Q1)

We compare PopCon and baselines on real-world datasets in Fig. 4. As shown in the figure, PopCon outperforms the baselines noticeably, drawing better trade-off curves between accuracy and aggregate diversity than all baselines in most cases. Especially, PopCon using DAM backbone achieves up to 60.5% higher Entropy@5 with comparable MAP@5, and up to 56.3% higher MAP@5 with comparable Entropy@5 compared with the best competitor Karakaya on Steam dataset. Furthermore, PopCon using CrossCBR achieves 3.92× higher Coverage@5 than Karakaya with similar MAP@5 on Steam dataset.

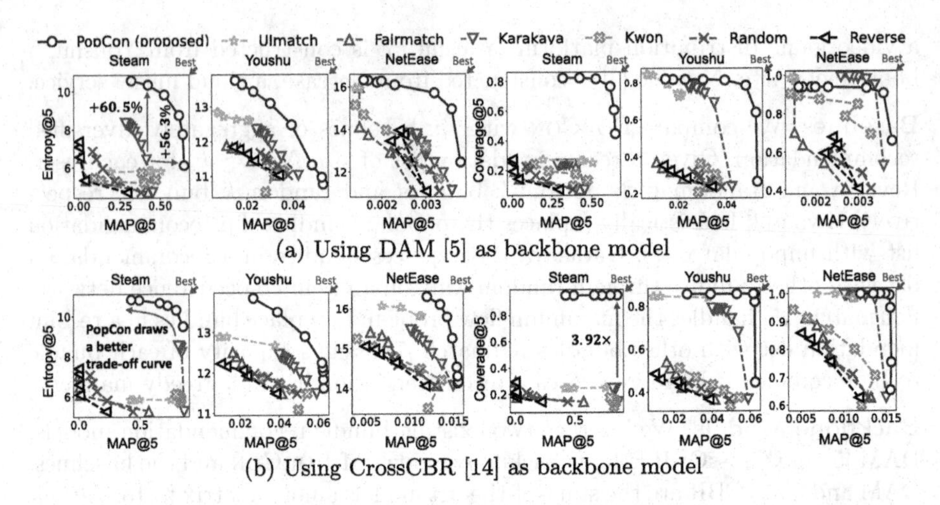

(a) Using DAM [5] as backbone model

(b) Using CrossCBR [14] as backbone model

Fig. 4. PopCon outperforms baselines in most cases using the two different backbone models (a) DAM [5] and (b) CrossCBR [14].

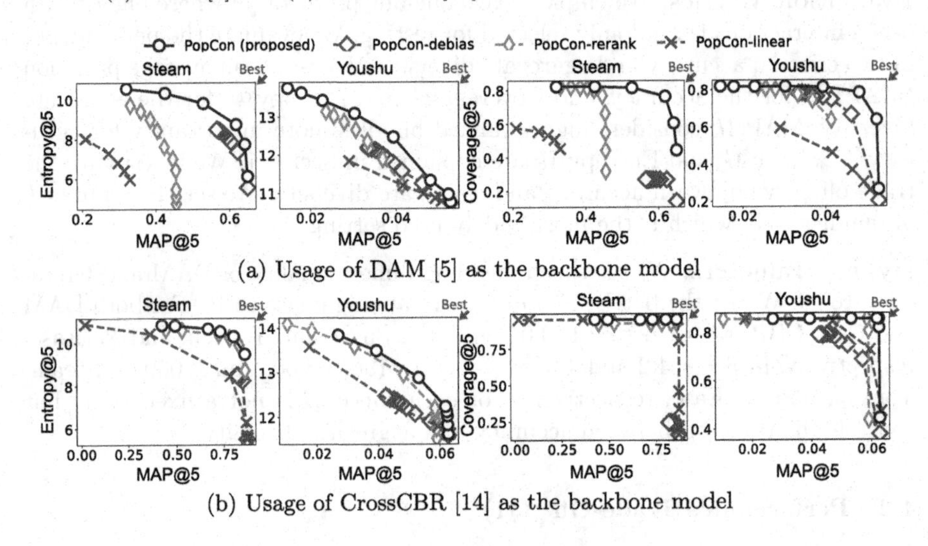

(a) Usage of DAM [5] as the backbone model

(b) Usage of CrossCBR [14] as the backbone model

Fig. 5. All the main ideas of PopCon help improve the performance.

4.3 Ablation Study (Q2)

Figure 5 provides an ablation study that compares PopCon with its three variants PopCon-debias, PopCon-rerank, and PopCon-linear on Steam and Youshu datasets. PopCon-debias adopts the proposed popularity debiasing in the training phase, but utilizes Karakaya in the reranking phase. PopCon-rerank does not adopt the popularity debiasing in the training phase while utilizing the proposed reranking algorithm in the reranking phase. PopCon-linear uses Equation (6) instead of Equation (7) in the reranking phase. As shown in the figure, PopCon outperforms all the variants, which verifies all the main ideas help improve the performance. Especially, PopCon-linear shows a severe performance drop compared with PopCon, justifying the importance of satisfying Property 3 (accuracy priority) in aggregately diversified bundle recommendation.

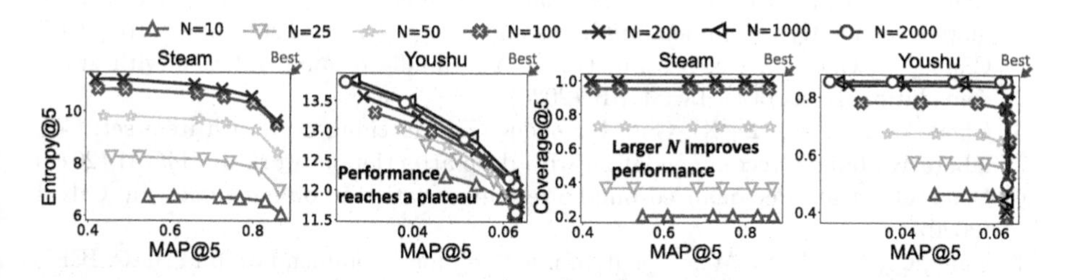

Fig. 6. The performance improves as N increased and reaches a plateau eventually. CrossCBR is used as the backbone of PopCon.

4.4 Effects of Number of Candidates (Q3)

Figure 6 shows the effects of the number N of candidates for the performance of PopCon using CrossCBR on Steam and Youshu datasets. We set N up to 200 on Steam dataset because Steam contains much fewer amount of bundles than Youshu. As shown in the figure, Entropy@5 and Coverage@5 are significantly improved as N increased, and finally reaches a plateau. Thus, we set N to 100 and 1,000 on Steam and Youshu, respectively, since they provide sufficient high performance despite being far lower than the total number of bundles.

5 Conclusion

In this paper, we propose PopCon, an accurate method for aggregately diversified bundle recommendation. PopCon mitigates the popularity bias of a backbone model using a popularity-based negative sampling, and reranks the recommendation results of the backbone model by a configuration-aware reranking algorithm to simultaneously maximize accuracy and aggregate diversity. PopCon provides the state-of-the-art performance in aggregately diversified bundle recommendation, achieving up to 60.5% higher Entropy@5 and 3.92× higher Coverage@5 with comparable accuracies compared to the best competitor.

Acknowledgments. This work was supported by Jung-Hun Foundation. The Institute of Engineering Research and ICT at Seoul National University provided research facilities for this work. U Kang is the corresponding author.

References

1. Abdollahpouri, H., Burke, R., Mobasher, B.: Controlling popularity bias in learning-to-rank recommendation. In: RecSys (2017)
2. Adomavicius, G., Kwon, Y.: Improving aggregate recommendation diversity using ranking-based techniques. IEEE Trans. Knowl. Data Eng. (2012)
3. Cao, D., Nie, L., He, X., Wei, X., Zhu, S., Chua, T.: Embedding factorization models for jointly recommending items and user generated lists. In: SIGIR (2017)
4. Chang, J., Gao, C., He, X., Jin, D., Li, Y.: Bundle recommendation with graph convolutional networks. In: SIGIR (2020)
5. Chen, L., Liu, Y., He, X., Gao, L., Zheng, Z.: Matching user with item set: Collaborative bundle recommendation with deep attention network. In: IJCAI (2019)
6. Deng, Q., et al.: Personalized bundle recommendation in online games. In: CIKM (2020)
7. Dong, Q., Xie, S., Li, W.: User-item matching for recommendation fairness. IEEE Access (2021)
8. Jeon, H., Jang, J.G., Kim, T., Kang, U.: Accurate bundle matching and generation via multitask learning with partially shared parameters. Plos one (2023)
9. Jeon, H., Kim, J., Yoon, H., Lee, J., Kang, U.: Accurate action recommendation for smart home via two-level encoders and commonsense knowledge. In: CIKM. ACM (2022)
10. Jeon, H., Koo, B., Kang, U.: Data context adaptation for accurate recommendation with additional information. In: BigData (2019)
11. Karakaya, M.Ö., Aytekin, T.: Effective methods for increasing aggregate diversity in recommender systems. Knowl. Inf. Syst. (2018)
12. Kim, J., Jeon, H., Lee, J., Kang, U.: Diversely regularized matrix factorization for accurate and aggregately diversified recommendation. In: PAKDD (2023)
13. Koo, B., Jeon, H., Kang, U.: Accurate news recommendation coalescing personal and global temporal preferences. In: PAKDD (2020)
14. Ma, Y., He, Y., Zhang, A., Wang, X., Chua, T.: Crosscbr: Cross-view contrastive learning for bundle recommendation. In: KDD (2022)
15. Mansoury, M., Abdollahpouri, H., Pechenizkiy, M., Mobasher, B., Burke, R.: Fairmatch: A graph-based approach for improving aggregate diversity in recommender systems. In: UMAP (2020)
16. Park, H., Jung, J., Kang, U.: A comparative study of matrix factorization and random walk with restart in recommender systems. In: BigData (2017)
17. Park, Y., Tuzhilin, A.: The long tail of recommender systems and how to leverage it. In: RecSys (2008)
18. Pathak, A., Gupta, K., McAuley, J.J.: Generating and personalizing bundle recommendations on Steam. In: SIGIR (2017)
19. Rendle, S., Freudenthaler, C., Gantner, Z., Schmidt-Thieme, L.: BPR: bayesian personalized ranking from implicit feedback. In: UAI (2009)
20. Zhang, S., Yao, L., Sun, A., Tay, Y.: Deep learning based recommender system: A survey and new perspectives. ACM Comput. Surv. (2019)

Diversely Regularized Matrix Factorization for Accurate and Aggregately Diversified Recommendation

Jongjin Kim, Hyunsik Jeon, Jaeri Lee, and U. Kang(✉)

Seoul National University, Seoul, South Korea
{j2kim99,jeon185,jlunits2,ukang}@snu.ac.kr

Abstract. When recommending personalized top-k items to users, how can we recommend them diversely while satisfying users' needs? Aggregately diversified recommender systems aim to recommend a variety of items across whole users without sacrificing the recommendation accuracy. They increase the exposure opportunities of various items, which in turn increase the potential revenue of sellers as well as user satisfaction. However, it is challenging to tackle aggregate-level diversity with matrix factorization (MF), one of the most common recommendation models, since skewed real-world data lead to the skewed recommendation results of MF.

In this work, we propose DivMF (Diversely Regularized Matrix Factorization), a novel matrix factorization method for aggregately diversified recommendation. DivMF exploits novel coverage regularizer and skewness regularizer which consider the top-k recommendation results of an MF model to aggregately diversify the recommendation results. We also propose a carefully designed training algorithm for effective training. Extensive experiments on real-world datasets show that DivMF gives the state-of-the-art performance, improving up to 34.7% aggregate-level diversity in the similar level of accuracy, and up to 27.6% accuracy in the similar level of aggregate-level diversity compared to the best competitors.

Keywords: Diversified Recommendation · Aggregate-level Diversity · Matrix Factorization

1 Introduction

When recommending personalized top-k items to users, how can we recommend them diversely while satisfying users' needs? Customers heavily rely on recommender systems [10,12,15] to choose items due to the flood of information nowadays. Thus, it is desired to expose as many items as possible to users to maximize the potential revenue of sales platforms [2] while improving users' experience [3]. Achieving aggregate-level diversity means fairly distributing items for the overall recommendation results. It requires that the results are of high coverage and

H. Kashima et al. (Eds.): PAKDD 2023, LNAI 13937, pp. 361–373, 2023.
https://doi.org/10.1007/978-3-031-33380-4_28

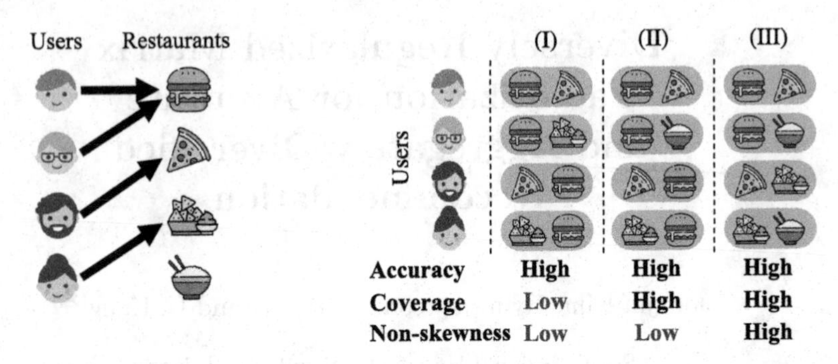

(a) Users' ground-truth preferences (b) Three different recommendation results

Fig. 1. Comparison of three different recommendation results. Note that all three results achieve high accuracy by recommending the ground-truth item to each user. However, the aggregate-level diversities (i.e., coverage and non-skewness) of the results (I), (II), and (III) are significantly different. Aggregately diversified recommendation aims to achieve high coverage and non-skewness while maintaining high accuracy as in the result (III).

low skewness; coverage indicates the proportion of recommended items among all items, and skewness indicates the degree of unfair frequencies of recommended items. Figure 1 demonstrates the coverage and non-skewness of three different recommendation results. Note that all three results achieve high accuracy but only the result (III) obtains high aggregate-level diversity by recommending every item twice. In other words, only the result (III) achieves a high aggregate-level diversity, recommending each item by the same amount, maximizing the potential revenue of sales platforms.

Matrix factorization (MF) [16] is the most widely used collaborative filtering method due to its powerful scalability and flexibility [13,19]. However, the traditional MF has a limitation in achieving high aggregate-level diversity on real-world data because it is vulnerable to the skewness of data [23]. To overcome this problem, previous works on aggregately diversified recommendation rerank the recommendation lists or recommendation scores of a given MF model [1,5,14,17]. However, these approaches do not give the best diversity since they focus only on post-processing the results of MF, which is already trained with skewed data. Thus, it is desired to deal with aggregate-level diversity in the training process of MF to achieve both high accuracy and diversity.

In this work, we propose Diversely Regularized Matrix Factorization (DivMF), a novel approach for aggregately diversified recommendation. DivMF regularizes a recommendation model in its training process so that more diverse items appear uniformly on top-k recommendations. DivMF effectively maximizes the coverage and non-skewness of the recommendation by utilizing two regularizers: coverage and skewness regularizers both of which consider the item occurrences in top-k recommendation list. This allows the model to achieve optimal aggregate-level diversity in the training process. We also propose a carefully designed training algorithm that first focuses on accuracy and then on diversity, and an unmasking mechanism for accurate and effective learning of DivMF.

Our contributions are summarized as follows:

- **Method.** We propose DivMF, a method for aggregately diversified recommendation. DivMF provides a new way to accurately and efficiently optimize an MF model to achieve both high accuracy and aggregate-level diversity for top-k recommendation.
- **Theory.** We theoretically prove that DivMF provides an optimal solution to maximize the aggregate-level diversity in top-k recommendation.
- **Experiments.** Extensive experiments show that DivMF achieves up to 34.7% higher aggregate-level diversity in the similar level of accuracy, and up to 27.6% higher accuracy in the similar level of aggregate-level diversity in personalized top-k recommendation compared to the best competitors, resulting in the state-of-the-art performance (see Fig. 2). The code and datasets are available at https://github.com/snudatalab/DivMF.

2 Aggregately Diversified Recommendation

In recent years, diversification has attracted increasing attention in recommendation research [11,25]. We focus on increasing diversity at the aggregate-level. Aggregate-level diversity considers the diversity in the overall recommendation results of all users to improve the potential profit of service platforms [2].

Aggregately diversified recommendation aims to improve two aspects of recommendation: coverage and non-skewness. Coverage is the total number of unique items recommended at least once. Non-skewness is the balance between frequencies of recommended items. The details of their evaluation are as follows.

- **Coverage.** Coverage measures how many different items a recommendation result contains from the whole items. It is defined as follows:

$$Coverage = \left|\cup_{u\in\mathbb{U}}\mathbb{L}(u)\right|/\left|\mathbb{I}\right|, \tag{1}$$

  where k is the number of items recommended, and $\mathbb{L}(u)$ is the set of recommended items for user u. $\mathbb{U}$ and $\mathbb{I}$ are sets of users and items, respectively. Coverage ranges from 0 to 1, and a higher value represents better coverage.
- **Gini index.** Gini index measures the inequality between item frequencies in recommendation results. It is defined as follows:

$$Gini = \frac{1}{|\mathbb{I}|-1}\sum_{j=1}^{|\mathbb{I}|}(2j-|\mathbb{I}|-1)p_j, \tag{2}$$

  where p_j is the j-th least value in $\{\frac{f(i)}{\sum_{j\in\mathbb{I}}f(j)}|i\in\mathbb{I}\}$ and $f(i)$ indicates the frequency of item i in the recommendation results for whole users. Gini index ranges from 0 to 1, and a lower value represents better non-skewness.

3 Proposed Method

In this section, we propose DivMF (Diversely Regularized Matrix Factorization), a matrix factorization method for accurate and aggregately diversified recommendation.

3.1 Overview

We address the following challenges to achieve high performance of aggregately diversified recommendation:

- **Coverage maximization.** Matrix factorization (MF) is prone to obtaining top-k recommendations with low coverage where only a few items are recommended. How can we train MF to recommend every item at least once?
- **Non-skewed frequency.** MF is liable to achieving skewed top-k recommendations. How can we train MF to recommend all items with similar frequencies?
- **Non-trivial optimization.** It is difficult to simultaneously handle both accuracy and diversity which are disparate criteria. How can we train MF to optimize both the accuracy and diversity?

The main ideas to address the challenges are as follows:

- **Coverage regularizer.** The coverage regularizer evenly balances the recommendation scores at the item-level, enabling us to recommend each item to at least one user.
- **Skewness regularizer.** The skewness regularizer equalizes all the recommendation scores to assist the coverage regularizer to make the model recommends all items by the same numbers of times.
- **Careful training.** We carefully design a training algorithm which first focuses on accuracy and then on diversity. This allows a model to be trained stably and efficiently, despite the conflict between accuracy and diversity. We also propose an unmasking mechanism for effective training.

3.2 Definition of Diversity Regularizer

Coverage Regularizer. We design a *coverage regularizer* to maximize the coverage. Focusing on the recommended items in the score matrix, we mask the scores of non-recommended items for each user to zero. After masking, a column filled with zeros corresponds to an item that is not recommended to any user. Hence, the coverage regularizer is required to distribute the remaining values in the masked matrix among all columns. In the following, we show how we construct the coverage regularizer from the fact that the equality condition of the arithmetic-geometric mean inequality states the equal distribution of values.

Assume that $\hat{\mathbf{R}} = [\hat{r}_{ui}] \in \mathbb{R}^{|\mathbb{U}|\times|\mathbb{I}|}$ is the recommendation score matrix where $\hat{r}_{ui}$ is a dot product between user u's embedding and item i's embedding. For $u \in \mathbb{U}$, consider $\mathbf{S} = [s_{ui}]$ where $\mathbf{S}_u = softmax(\hat{\mathbf{R}}_{\mathbf{u}})$, which means $(s_{u1}, ..., s_{u|\mathbb{I}|}) = softmax(\hat{r}_{u1}, ..., \hat{r}_{u|\mathbb{I}|})$. Then, we keep top-$k$ elements of each row in $\mathbf{S}$ while masking others to zero to construct a matrix $\mathbf{T} = [t_{ui}]$. Note that the nonzero t_{ui} implies that the top-k recommendation list of user u includes item i. Then, the coverage regularizer Reg_{cov} is defined as follows:

$$Reg_{cov} = -\log\left(\prod_{i\in\mathbb{I}}\sum_{u\in\mathbb{U}} t_{ui}\right) = -\sum_{i\in\mathbb{I}}\log\left(\sum_{u\in\mathbb{U}} t_{ui}\right).$$

This regularizer is useful to maximize coverage, as shown in Theorem 1.

Theorem 1. *If Reg_{cov} is minimized, then coverage is maximized.*

Proof. $\sum_{u\in\mathbb{U}} t_{ui} \leq \sum_{u\in\mathbb{U}} s_{ui}$ for all $i \in \mathbb{I}$ since $0 \leq t_{ui} \leq s_{ui}$ for all $u \in \mathbb{U}$ and $i \in \mathbb{I}$. Thus, using the fact that $\sum_{i\in\mathbb{I}} s_{ui} = 1$ for all $u \in \mathbb{U}$,

$$\sum_{i\in\mathbb{I}}\sum_{u\in\mathbb{U}} t_{ui} \leq \sum_{i\in\mathbb{I}}\sum_{u\in\mathbb{U}} s_{ui} = \sum_{u\in\mathbb{U}}\sum_{i\in\mathbb{I}} s_{ui} = |\mathbb{U}|.$$

We thus obtain

$$\exp(-Reg_{cov}) = \prod_{i\in\mathbb{I}}\sum_{u\in\mathbb{U}} t_{ui} \leq (\frac{|\mathbb{U}|}{|\mathbb{I}|})^{|\mathbb{I}|}, \tag{3}$$

from the arithmetic geometric mean inequality. Equality holds if and only if for all i, $\sum_{u\in\mathbb{U}} t_{ui} = |\mathbb{U}|/|\mathbb{I}|$. In this case, every column of $\mathbf{T}$ has at least one nonzero element. Thus, every item is included in at least one user's top-k recommendation list, so the coverage is 1. □

Skewness Regularizer. Although the condition to minimize the coverage regularizer guarantees the coverage of the model to be 1, this does not guarantee the non-skewness to be maximized. For example, assume that $(t_{11}, t_{21}, ..., t_{|\mathbb{U}|1}) = (\frac{1}{2}, \frac{1}{2}, 0, 0, ..., 0)$ and $(t_{12}, t_{22}, ..., t_{|\mathbb{U}|2}) = (\frac{1}{3}, \frac{1}{3}, \frac{1}{3}, 0, 0, ..., 0)$. In this case, $\sum_{u\in\mathbb{U}} t_{u1} = \sum_{u\in\mathbb{U}} t_{u2}$ but the item 1 is recommended twice while the item 2 is recommended three times. In other words, it is possible to meet the equality condition of Equation (3) even if the non-skewness of the model is not maximized, since the value of each nonzero element could vary.

To address this problem, we propose a *skewness regularizer*. Since the problem occurs because of the variance of nonzero elements, we design the skewness regularizer to equalize values of nonzero t_{ui}. After equalization, $\sum_{u\in\mathbb{U}} t_{ui}$ and $\sum_{u\in\mathbb{U}} t_{uj}$ would be equal if and only if items i and j are recommended for the same number of times, so the coverage regularizer would also optimize the non-skewness in recommendation lists.

Let $\mathbf{T}' = [t'_{ui}]$ be a row-normalized $\mathbf{T}$ which means $t'_{ui} = t_{ui}/\sum_{j\in\mathbb{I}} t_{uj}$. The skewness regularizer Reg_{skew} is defined as follows:

$$Reg_{skew} = \sum_{u\in\mathbb{U}}\sum_{i\in\mathbb{I}} t'_{ui}\log t'_{ui} = -\sum_{u\in\mathbb{U}} entropy(\mathbf{T}_u).$$

Since each entropy function is maximized if and only if nonzero elements of each $\mathbf{T}_u$ are equal, Reg_{skew} is minimized if and only if all nonzero elements of each row of $\mathbf{T}$ are equal.

Diversity Loss Function. Finally, we define the loss function for aggregate-level diversity in DivMF as $\mathcal{L}_{div}(\hat{\mathbf{R}}) = Reg_{cov} + Reg_{skew}$. This loss function satisfies the Theorem 2.

Theorem 2. *If $\mathcal{L}_{div}(\hat{\mathbf{R}})$ is minimized, then coverage and non-skewness are both maximized.*

Proof. The condition to minimize Reg_{cov} is $\sum_{u\in\mathbb{U}} t_{ui} = |\mathbb{U}|/|\mathbb{I}|$ for every item i, and the condition to minimize Reg_{skew} is that nonzero elements of $\mathbf{T}_u$ are equal for every user u. Thus, the condition to minimize $Reg_{cov} + Reg_{skew}$ is that each row of $\mathbf{T}$ contains k nonzero elements with value of $\frac{1}{k}$, and each column of $\mathbf{T}$ contains $\frac{|\mathbb{U}|k}{|\mathbb{I}|}$ nonzero elements. In this case, every item appears in the recommendation results with equal frequency. Therefore, both coverage and non-skewness are maximized if $\mathcal{L}_{div}(\hat{\mathbf{R}})$ is minimized. □

3.3 Model Training

Objective Function and Training Algorithm. In order to maximize accuracy and aggregate-level diversity of recommendation results simultaneously, we propose the following objective function.

$$\mathcal{L}_{total}(\theta; \mathbf{R}) = \mathcal{L}_{acc}(\hat{\mathbf{R}}) + \mathcal{L}_{div}(\hat{\mathbf{R}}),$$

where $\mathcal{L}_{total}(\cdot)$ is the total loss to be minimized, $\mathcal{L}_{acc}(\cdot)$ and $\mathcal{L}_{div}(\cdot)$ are losses for accuracy and aggregate-level diversity, respectively, $\mathbf{R}$ is the observed interaction matrix, $\hat{\mathbf{R}}$ is the recommendation score matrix, and θ is the parameter to be optimized. We use BPR loss function as an accuracy loss since it is known to show the best performance in top-k recommendation [21]. Thus,

$$\mathcal{L}_{acc}(\hat{\mathbf{R}}) = \sum_{u\in\mathbb{U},(i,j)\in\mathbb{Z}(u)} \log\Big(1 + \exp\Big(\hat{\mathbf{R}}_{uj} - \hat{\mathbf{R}}_{ui}\Big)\Big),$$

where $\mathbb{Z}(u) = \{(i,j)|\mathbf{R}_{ui} = 1, \mathbf{R}_{uj} = 0\}$.

A challenge in minimizing the loss $\mathcal{L}_{total}$ is that directly minimizing $\mathcal{L}_{total}$ or optimizing $\mathcal{L}_{acc}$ and $\mathcal{L}_{div}$ in an iterative, alternating fashion leads to poor performance (see Sect. 4.4). We presume that this problem happens because the gradients of accuracy loss and diversity regularizer cancel each other out. The accuracy loss tries to increase the gap between recommendation scores of high scored items and low scored items, while the diversity regularizer tries to decrease the gap. Thus, the net gradient is not large enough to prevent the model from being trapped in bad local optima.

Our idea to avoid this issue is to train DivMF model with only accuracy loss $\mathcal{L}_{acc}$ until the accuracy converges, and then train the model with the diversity regularizer $\mathcal{L}_{div}$. In this way, the gradients of accuracy loss and diversity regularizer do not cancel each other out since the optimizer minimizes only one loss at a time. To adjust the trade-off between accuracy and diversity, we control the number n_{ep} of epochs to optimize $\mathcal{L}_{div}$, since the model achieves higher diversity and lower accuracy as we increase n_{ep}.

Unmasking Mechanism. Gradients from $\mathcal{L}_{div}(\hat{\mathbf{R}})$ do not flow directly into unrecommended items since $\mathbf{T}$ masks $|\mathbb{I}| - k$ items with the lowest scores in $\mathbf{S}$ of each user. Thus, a straightforward gradient descent with $\mathcal{L}_{div}(\hat{\mathbf{R}})$ has limitation to find new items for diversity, optimizing only k scores of initially selected items.

We propose an unmasking mechanism to overcome this problem. The idea is to keep additional unmasked elements in each row of $\mathbf{S}$ when building $\mathbf{T}$. In this way, rarely recommended items have an opportunity to be unmasked. DivMF finds new rarely recommended items by a gradient descent with this unmasking mechanism. DivMF unmasks a fixed number of the highest-scored items other than already recommended items during each iteration of training, which is the best unmasking scheme as experimentally shown in Sect. 4.5.

4 Experiments

We perform experiments to answer the following questions:

Q1. **Diversity and accuracy (Section 4.2).** Does DivMF show high aggregate-level diversity without sacrificing the accuracy of recommendation?

Q2. **Regularizer (Section 4.3).** How do the diversity regularization terms Reg_{cov} and Reg_{skew} of DivMF help improve the diversity of DivMF?

Q3. **Training algorithm (Section 4.4).** Does the training algorithm of DivMF prevent the training from being trapped in bad local optima?

Q4. **Unmasking mechanism (Section 4.5).** How does the unmasking mechanism of DivMF affect the performance?

4.1 Experimental Setup

We introduce our experimental setup including datasets, evaluation protocol, baseline approaches, evaluation metrics, and the training process.

Datasets. We use five real-world rating datasets as summarized in Table 1. We preprocess extremely sparse datasets (Yelp, Gowalla, and Epinions) as core-15 following a previous work [17]. In other words, we make the datasets include only users and items that have at least 15 interactions. MovieLens-10M and MovieLens-1M datasets [9] contain movie ratings constructed by the GroupLens research group. Yelp-15 contains 15-core restaurant rating data collected from a restaurant review site with the same name. Epinions-15 [18] contains 15-core rating data of products constructed from a general consumer review site. Gowalla-15 [4] contains 15-core data of a friendship network of users constructed from a location-based social networking website. We remove the rating scores of datasets and obtain user-item interaction data which indicate whether the user has rated the item or not.

Evaluation Protocol. We employ *leave-one-out* protocol where one of each user's interaction instances is removed for testing. If the dataset includes timestamp, the latest instance of each user is removed, and if not, randomly sampled instances are removed.

Baselines. We compare DivMF with existing methods for aggregately diversified recommendation.

Table 1. Summary of datasets.

Dataset	Users	Items	Interactions	Density(%)
Yelp-15 [a]	69,853	43,671	2,807,606	0.0920
Gowalla-15 [b]	34,688	63,729	2,438,708	0.1111
Epinions-15 [c]	5,531	4,286	186,995	0.7888
MovieLens-10M [d]	69,878	10,677	10,000,054	1.3403
MovieLens-1M [e]	6,040	3,706	1,000,209	4.4684

[a] https://www.yelp.com/dataset.
[b] https://snap.stanford.edu/data/loc-gowalla.html.
[c] http://www.trustlet.org/downloaded_epinions.html.
[d] https://grouplens.org/datasets/movielens/10m/.
[e] https://grouplens.org/datasets/movielens/1m/.

- **Kwon.** Kwon et al. [1] adjust recommendation scores of items based on their frequencies to achieve aggregate level diversity.
- **Karakaya.** Karakaya et al. [14] replace items on recommendation lists with infrequently recommended similar items.
- **Fairmatch.** Fairmatch [17] utilizes a maximum flow problem to find important items.
- **UImatch.** UImatch [5] assigns recommendation capacity to each item and greedily constructs recommendation lists.

Evaluation Metrics. We evaluate the performance of the methods in two categories: accuracy and diversity. Accuracy metric checks whether a model recommends correct items or not, and diversity metrics evaluate aggregated diversity of the recommendation. For each experiment, a list of recommendation to each user is created and evaluated by the following metrics.

- **Accuracy.**
  - **nDCG@k.** nDCG@k measures the overall accuracy of the top-k recommendation. It ranges from 0 to 1, where the value 0 indicates the lowest accuracy and the value 1 represents the highest accuracy.
- **Diversity.**
  - **Coverage@k.** The coverage of the top-k recommendation.
  - **Negative Gini index@k.** The negative value of the Gini index of the top-k recommendation.

Training Details. We first train the MF model until convergence. Then, we apply each baseline and DivMF on the trained MF model. We min-max normalize the recommendation scores for Kwon and Karakaya since they need prediction ratings on a finite scale. We use reverse prediction scheme [1] and set $T_H = 0.8, T_R = 0.9$ for Kwon. We vary t in $\{30, 50, 75, 100\}$ and set $\alpha = 0.5$ for FairMatch. We unmask 50 items in Epinions-15 dataset, 100 items in ML-1M/ML-10M datasets, and 500 items in Gowalla-15/Yelp-15 datasets to apply

DivMF. All the models are trained with Adam optimizer with learning rate 0.001, l_2 regularization coefficient 0.0001, $\beta_1 = 0.9$, and $\beta_2 = 0.999$. We vary k in $\{5, 10\}$ for all datasets.

4.2 Diversity and Accuracy (Q1)

We show the change of accuracies and diversities of DivMF and the competitors on five real-world datasets in Fig. 2. For each method, we adjust hyperparameters to mark points on the plot and connect them to obtain the trade-off curve. We mark the point with the highest accuracy and the highest diversity in each plot as the 'best' point of the plot. Note that DivMF achieves the highest diversity while sacrificing the least accuracy compared to other baselines considering the balance of coverage and non-skewness.

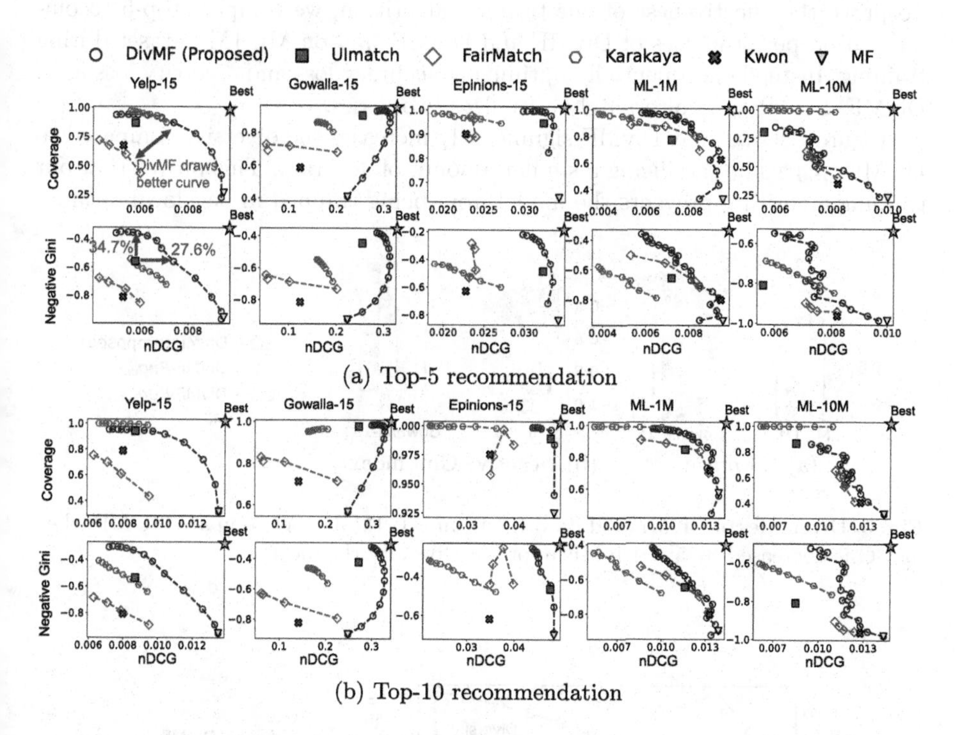

Fig. 2. Accuracy-diversity trade-off curves of top-5 and top-10 recommendations on five real-world datasets. DivMF achieves the highest aggregate-level diversity while sacrificing minimal accuracy.

4.3 Regularizer (Q2)

To verify the impact of coverage regularizer and skewness regularizer, we examine how much the diversity of top-5 recommendation results improves. We compare DivMF, DivMF-Reg_{skew}, and DivMF-Reg_{cov} on ML-1M and Gowalla-15 datasets; DivMF-Reg_{skew} and DivMF-Reg_{cov} are DivMF without the skewness regularizer and the coverage regularizer, respectively. For the fair comparison, we train each model until the nDCG is dropped by 5% compared to MF.

Figure 3 shows that DivMF increases both the coverage and the non-skewness the most, compared to other models. This verifies that both regularizers contribute to improving the aggregate-level diversity.

4.4 Training Algorithm (Q3)

To prove the effectiveness of our training algorithm, we compare top-5 recommendation performances of DivMF and DivMF $_{alter}$ on ML-1M dataset during training. Instead of sequentially optimizing accuracy loss and diversity loss as in DivMF, DivMF $_{alter}$ alternately optimizes two losses.

Figure 4 shows that DivMF significantly increases the diversity compared to DivMF $_{alter}$ while sacrificing a similar amount of accuracy. This proves that our training algorithm prevents the model from being trapped in bad local optima.

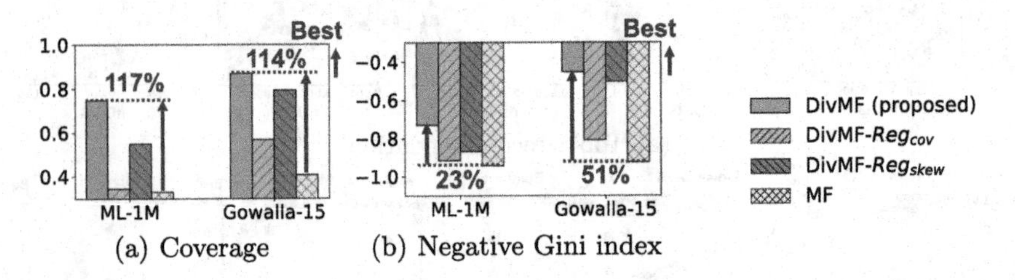

Fig. 3. Diversities of DivMF and its variants on ML-1M dataset compared to MF when nDCG is decreased by 5%. DivMF improves diversity the most.

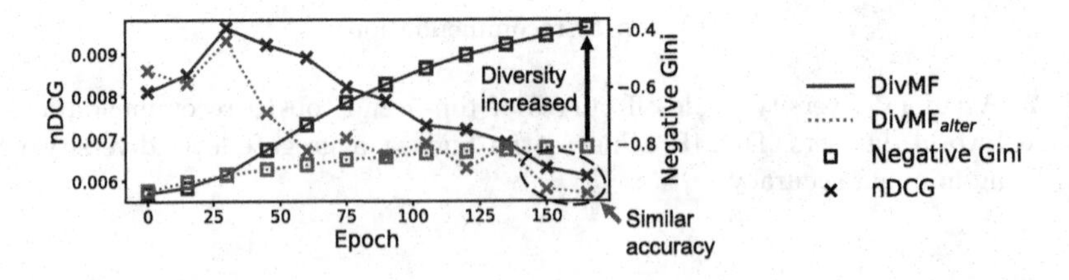

Fig. 4. Change of nDCG and Gini index of DivMF and DivMF $_{alter}$ during training on ML-1M dataset. DivMF improves diversity better by avoiding bad local optima.

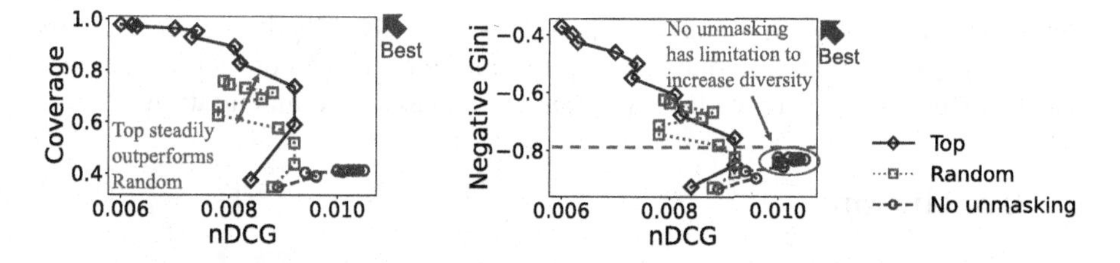

Fig. 5. Accuracy-coverage (left) and accuracy-negative Gini index (right) trade-off curves of different unmasking policies. '*Top*' shows the best overall performance.

4.5 Unmasking Mechanism (Q4)

To find the best unmasking policy for DivMF, we compare three policies: *No unmasking*, *Top*, and *Random* on ML-1M dataset. In addition to top-k items, *Top* unmasks n items with the highest prediction scores while *Random* unmasks random n items. We set $n = 100$ since it shows the best performance in both schemes. *No unmasking* does not unmask any item other than top-k items.

Figure 5 shows performances of the three policies in top-5 recommendation. We have two observations. First, *No unmasking* fails to increase aggregate-level diversity, while *Top* and *Random* further improve both coverage and non-skewness. Second, *Top* performs better than *Random* since it achieves higher coverage while non-skewnesses of the two schemes are comparable in the case. In summary, *Top* is the best unmasking scheme to achieve high aggregate-level diversity.

5 Related Works

Individually diversified recommendation. Individually diversified recommendation recommends diversified items to each user [25]. Maximizing individual diversity can maximize item novelty in each user's view, but it may recommend already known items in overall recommendation list for all users. Thus, maximizing individual-level diversity does not guarantee the improvement in aggregate-level diversity [1].

Fair Recommendation. Fair recommendation aims to design an algorithm that makes fair predictions devoid of discrimination [8]. Fairness in recommendation could be observed between different item groups [6] or between distinct items with similar attributes [20]. Aggregately diversified recommendation does not require any group or attribute of items, which is the main difference compared to the fair recommendation.

Popularity Debiased Recommendation. Popularity debiased recommendation aims to improve the quality of recommendation for long-tail items. Traditional recommender systems tend to show poor accuracy for infrequently appearing items because of the skewness in dataset [22]. There are researches to eliminate the popularity bias to achieve high accuracy in recommending long-tail items as well as popular items [7,24]. Aggregately diversified recommendation focuses on increasing the frequencies of long tail items instead of their accuracies, which is the main difference from popularity debiased recommendation.

6 Conclusion

We propose DivMF, a matrix factorization method which maximizes aggregate-level diversity while sacrificing minimal accuracy in top-k recommendation. DivMF exploits coverage regularizer and skewness regularizer for MF via a carefully designed training algorithm. Experiments on five real-world datasets show that DivMF achieves the state-of-the-art performance in aggregately diversified recommendation, outperforming the best competitor with up to 34.7% reduced Gini index in the similar level of accuracy and up to 27.6% higher nDCG in the similar level of diversity. Future works include extending DivMF for other recommendation models beyond the matrix factorization.

Acknowledgments. This work was supported by Jung-Hun Foundation. The Institute of Engineering Research and ICT at Seoul National University provided research facilities for this work. U Kang is the corresponding author.

References

1. Adomavicius, G., Kwon, Y.: Improving aggregate recommendation diversity using ranking-based techniques. IEEE TKDE (2012)
2. Brynjolfsson, E., Hu, Y., Simester, D.: Goodbye pareto principle, hello long tail: The effect of search costs on the concentration of product sales. Manag. Sci. (2011)
3. Brynjolfsson, E., Hu, Y., Smith, M.D.: Consumer surplus in the digital economy: Estimating the value of increased product variety at online booksellers. Manag. Sci. (2003)
4. Cho, E., Myers, S.A., Leskovec, J.: Friendship and mobility: User movement in location-based social networks. In: KDD. ACM (2011)
5. Dong, Q., Xie, S.S., Li, W.J.: User-item matching for recommendation fairness. IEEE Access (2021)
6. Ekstrand, M.D., Kluver, D.: Exploring author gender in book rating and recommendation. In: UMUAI (2021)
7. Ferraro, A.: Music cold-start and long-tail recommendation: bias in deep representations. In: RecSys (2019)
8. Gajane, P.: On formalizing fairness in prediction with machine learning. ArXiv (2017)
9. Harper, F.M., Konstan, J.A.: The movielens datasets: History and context. TiiS (2015)
10. Jeon, H., Jang, J.G., Kim, T., Kang, U.: Accurate bundle matching and generation via multitask learning with partially shared parameters. Plos one (2023)
11. Jeon, H., Kim, J., Lee, J., Lee, J., Kang, U.: Aggregately diversified bundle recommendation via popularity debiasing and configuration-aware reranking. In: PAKDD (2023)

12. Jeon, H., Kim, J., Yoon, H., Lee, J., Kang, U.: Accurate action recommendation for smart home via two-level encoders and commonsense knowledge. In: CIKM. ACM (2022)
13. Jeon, H., Koo, B., Kang, U.: Data context adaptation for accurate recommendation with additional information. In: BigData (2019)
14. Karakaya, M., Aytekin, T.: Effective methods for increasing aggregate diversity in recommender systems. In: KAIS (2018)
15. Koo, B., Jeon, H., Kang, U.: Accurate news recommendation coalescing personal and global temporal preferences. In: PAKDD (2020)
16. Koren, Y., Bell, R., Volinsky, C.: Matrix factorization techniques for recommender systems. Computer (2009)
17. Mansoury, M., Abdollahpouri, H., Pechenizkiy, M., Mobasher, B., Burke, R.: Fairmatch: A graph-based approach for improving aggregate diversity in recommender systems. In: UMAP. ACM (2020)
18. Massa, P., Souren, K., Salvetti, M., Tomasoni, D.: Trustlet, open research on trust metrics. In: SCPE (2008)
19. Park, H., Jung, J., Kang, U.: A comparative study of matrix factorization and random walk with restart in recommender systems. In: BigData (2017)
20. Patro, G.K., Biswas, A., Ganguly, N., Gummadi, K.P., Chakraborty, A.: Fairrec: Two-sided fairness for personalized recommendations in two-sided platforms. In: WebConf (2020)
21. Rendle, S., Freudenthaler, C., Gantner, Z., Schmidt-Thieme, L.: BPR: bayesian personalized ranking from implicit feedback. In: UAI. AUAI Press (2009)
22. Steck, H.: Item popularity and recommendation accuracy. In: RecSys (2011)
23. Wang, H., Ruan, B.: Matrec: Matrix factorization for highly skewed dataset. In: BigData (2020)
24. Wei, T., Feng, F., Chen, J., Wu, Z., Yi, J., He, X.: Model-agnostic counterfactual reasoning for eliminating popularity bias in recommender system. In: KDD (2021)
25. Wu, Q., Liu, Y., Miao, C., Zhao, Y., Guan, L., Tang, H.: Recent advances in diversified recommendation (2019)

kNN-Embed: Locally Smoothed Embedding Mixtures for Multi-interest Candidate Retrieval

Ahmed El-Kishky(✉), Thomas Markovich, Kenny Leung, Frank Portman, Aria Haghighi, and Ying Xiao

Twitter Cortex, San Francisco, CA 94103, USA
{aelkishky,tmarkovich,kennyleung,fportman,ahaghighi,yxiao}@twitter.com

Abstract. Candidate retrieval is the first stage in recommendation systems, where a light-weight system is used to retrieve potentially relevant items for an input user. These candidate items are then ranked and pruned in later stages of recommender systems using a more complex ranking model. As the top of the recommendation funnel, it is important to retrieve a high-recall candidate set to feed into downstream ranking models. A common approach is to leverage approximate nearest neighbor (ANN) search from a single dense query embedding; however, this approach this can yield a low-diversity result set with many near duplicates. As users often have multiple interests, candidate retrieval should ideally return a diverse set of candidates reflective of the user's multiple interests. To this end, we introduce kNN-Embed, a general approach to improving diversity in dense ANN-based retrieval. kNN-Embed represents each user as a smoothed mixture over learned item clusters that represent distinct "interests" of the user. By querying each of a user's mixture component in proportion to their mixture weights, we retrieve a high-diversity set of candidates reflecting elements from each of a user's interests. We experimentally compare kNN-Embed to standard ANN candidate retrieval, and show significant improvements in overall recall and improved diversity across three datasets. Accompanying this work, we open source a large Twitter follow-graph dataset (https://huggingface.co/datasets/Twitter/TwitterFollowGraph), to spur further research in graph-mining and representation learning for recommender systems.

Keywords: candidate retrieval · embedding · nearest neighbor, diversity

1 Introduction

Recommendation systems for online services such as e-commerce or social networks present users with suggestions in the form of ranked lists of items [5].

A. Haghighi and Y. Xiao—Equal contribution.

H. Kashima et al. (Eds.): PAKDD 2023, LNAI 13937, pp. 374–386, 2023.
https://doi.org/10.1007/978-3-031-33380-4_29

Often, these item lists are constructed through a two-step process: (1) candidate retrieval, which efficiently retrieves a manageable subset of potentially relevant items, and (2) ranking, which applies a computationally-expensive ranking model to score and select the top-k candidates to display to the user.

During candidate retrieval, we are primarily concerned with the *recall* of the system [11], as opposed to the ranking model which typically targets *precision*. Ensuring high recall for users with multiple interests is a challenging problem, which is exacerbated by the way we typically perform retrieval. The dominant paradigm for candidate retrieval is to embed users and items in the same vector space, and then use approximate nearest-neighbor (ANN) search to retrieve candidates close to the user [5,14]. However, ANN search will often return candidate pools that are highly intra-similar (e.g., all candidates pertain to one "topic" only) [27]. A side effect of training embeddings to place users close to relevant items, is that similar items are also placed close to each other. During ANN-based candidate retrieval, this unfortunately leads to similar candidates that may not reflect a user's diverse multi-topic interests, and hence low recall.

In this paper, we introduce kNN-Embed, a new strategy for retrieving a high-recall, diverse set of candidates reflecting a user's multiple interests. kNN-Embed captures multiple user interests by representing user preferences with a smoothed, mixture distribution. Our technique provides a turn-key way to increase recall and diversity while maintaining user relevance in any ANN-based candidate retrieval scheme. It does not require retraining the underlying user and item embeddings; instead, we build directly on top of pre-existing ANN systems. The underlying idea is to exploit the similarity of neighboring users to represent per-user interests as a mixture over learned high level clusters of item embeddings. Since user-item relevance signal is typically sparse, estimating the mixture weights introduces significance variance. Thus, we smooth the mixture weights with information from similar users. At retrieval time, we simply sample candidates from each cluster according to mixture weights. Within each cluster, we perform ANN search using a smoothed per-user per-cluster embedding.

Our contributions in this paper are (1) a principled method to retrieve a high-recall, diverse candidate set in ANN-based candidate retrieval systems and (2) a large open-source graph dataset for studying graph-mining and retrieval.

2 Related Works

Traditionally, techniques for candidate retrieval rely on fast, scalable approaches to search large collections for similar sparse vectors [1,3]. Approaches apply indexing and optimization strategies to scale sparse similarity search. One such strategy builds a static clustering of the entire collection of items; clusters are retrieved based on how well their centroids match the query [20,25]. These methods either (1) match the query against clusters of items and rank clusters based on similarity to query or (2) utilize clusters as a form of item smoothing.

For embedding-based recommender systems [28], large-scale dense similarity search has been applied for retrieval. Some approaches proposed utilize hashing-based techniques such as mapping input and targets to discrete partitions and selecting targets from the same partitions as inputs [26]. With the advent of fast approximate nearest-neighbor search [13,21], dense nearest neighbor has been applied by recommender systems for candidate retrieval [5].

When utilizing graph-based embeddings for recommender systems [8], some methods transform single-mode embeddings to multiple modes by clustering user actions [23]. Our method extends upon this idea by incorporating nearest neighbor smoothing to address the sparsity problem of generating mixtures of embeddings for users with few engagements.

Smoothing via k-nearest-neighbor search has been applied for better language modeling [16] and machine translation [15]. We smooth low-engagement user representations by leveraging engagements from similar users.

3 *k*NN-Embed

3.1 Preliminaries

Let $\mathcal{U} = \{u_1, u_2, \dots u_n\}$ be the set of source entities (i.e., users in a recommender system) and $\mathcal{I} = \{i_1, i_2, \dots i_m\}$ be the set of target entities (i.e., items in a recommender system). Let $\mathcal{G}$ constitute a bipartite graph representing the engagements between users ($\mathcal{U}$) and items ($\mathcal{I}$). For each user and item, we define a "relevance" variable in $\{0, 1\}$ indicating an item's relevance to a particular user. An item is considered relevant to a particular user if a user, presented with an item, will engage with said item. Based on the engagements in $\mathcal{G}$, each user, u_j, is associated with a d-dimensional embedding vector $\mathbf{u_j} \in \mathbb{R}^d$; similarly each target item i_k is associated with an embedding vector $\mathbf{i_k} \in \mathbb{R}^d$. We call these the *unimodal* embeddings, and assume that they model user-item relevance $p(\text{relevance}|u_j, i_k) = f(\mathbf{u_j}, \mathbf{i_k})$ for a suitable function f.

Given the input user-item engagement graph, our goal is to learn mixtures of embeddings representations of users that better capture the multiple interests of a user as evidenced by higher recall in a candidate retrieval task.

Unimodal User and Item Embeddings: While kNN-Embed presupposes a set of co-embedded user and item embeddings and is agnostic to the exact embedding technique used (the only constraint is that the embeddings must satisfy $p(i_k|u_j) = g(\mathbf{u_j}^T\mathbf{i_k})$ for monotone g), for completeness we describe a simple approach we applied to co-embed users and items into the same space. We form a bipartite graph $\mathcal{G}$ of users and items, where an edge represents relevance (e.g., user follows content producer). We seek to learn an embedding vector (i.e., vector of learnable parameters) for each user (u_j) and item (i_k) in this bipartite graph; we denote these learnable embeddings for users and items as $\mathbf{u_j}$ and $\mathbf{i_k}$ respectively. A user-item pair is scored with a scoring function of the form $f(\mathbf{u_j}, \mathbf{i_k})$. Our training objective seeks to learn $\mathbf{u}$ and $\mathbf{i}$ parameters that maximize a log-likelihood constructed from the scoring function for $(u, i) \in \mathcal{G}$ and minimize for $(u, i) \notin \mathcal{G}$. For simplicity, we apply a dot product comparison between user and item representations. For a user-item pair $e = (u_j, i)$, this is defined by:

$$f(e) = f(u_j, i_k) = \mathbf{u_j}^\intercal \mathbf{i_k} \tag{1}$$

As seen in Eq. 1, we co-embed users and items by scoring their respective embedded representations via dot product and perform edge (or link) prediction. We consume the input bipartite graph $\mathcal{G}$ as a set of user-item pairs of the form (u, i) which represent positive engagements between a user and item. The embedding training objective is to find user and item representations that are useful for predicting which users and items are linked via an engagement. While a softmax is a natural formulation to predict a user-item engagement, it is impractical due to the cost of computing the normalization over a large vocabulary of items. Following previous methods [10,22], negative sampling, a simplification of noise-contrastive estimation, can be used to learn the parameters $\mathbf{u}$ and $\mathbf{i}$. We maximize the following negative sampling objective:

$$\arg\max_{\mathbf{u},\mathbf{i}} \sum_{e \in \mathcal{G}} \left[\log \sigma(f(e)) + \sum_{e' \in N(e)} \log \sigma(-f(e')) \right] \tag{2}$$

where: $N(u, i) = \{(u, i') : i' \in \mathcal{I}\} \cup \{(u', i) : u' \in \mathcal{U}\}$. Equation 2 represents the log-likelihood of predicting a binary "real" (edges in the network) or "fake" (negatively sampled edges) label. To maximize the objective, we learn $\mathbf{u}$ and $\mathbf{i}$ parameters to differentiate positive edges from negative, unobserved edges. Negative edges are sampled by corrupting positive edges via replacing either the user or item in an edge pair with a negatively sampled user or item. Following previous approaches, negative sampling is performed both uniformly and proportional to node prevalence in the training graph [4,18].

3.2 Smoothed Mixture of Embeddings

To use embeddings for candidate retrieval, we need a method of selecting relevant items given the input user. Ideally, we would like to construct a full distribution over all items for each user $p(i_k|u_j)$ and draw samples from it. The sheer number of items makes this difficult to do efficiently, especially when candidate retrieval strategies are meant to be light-weight. In practice, the most common method is to greedily select the top few most relevant items using an ANN search with the unimodal user embedding as query. A significant weakness of this greedy selection is that, by its nature, ANN search will return items that are similar not only to the user embedding, but also to each other; this drastically reduces the *diversity* of the returned items. This reduction in diversity is a side-effect of the way embeddings are trained – typically, the goal of training embeddings is to put users and relevant items close in Euclidean space; however, this also places similar users close in space, as well as similar items. We will repeatedly exploit this "locality implies similarity" property of embeddings in this paper to resolve this diversity issue.

Clustering Items: Since neighboring items are similar in the embedding space, if we apply a distance-based clustering to items, we can arrive at groupings that represent individual user preferences well. As such, we first cluster items using spherical k-means [6] where cluster centroids are placed on a high-dimensional sphere with radius one. Given these item clusters, instead of immediately collapsing the distribution $p(i_k|u_j)$ to a few items as ANN search does, we can write the full distribution $p(i_k|u_j)$ as a mixture over item clusters:

$$p(i_k|u_j) = \sum_c p(c|u_j) \cdot p(i_k|u_j, c)$$

where in each cluster, we learn a separate distribution over the items in the cluster $p(i_k|u_j, c)$. Thus, we are modeling each user's higher level interests $p(c|u)$, and then within each interest c, we can apply an efficient ANN-search strategy as before. In effect, we are interpolating between sampling the full preference distribution $p(i_k|u_j)$ and greedily selecting a few items in an ANN.

Mixture of Embeddings via Cluster Engagements: After clustering target entities, we learn $p(c|u_j)$ through its maximum likelihood estimator (MLE):

$$p_{\text{mle}}(c|u_j) = \text{count}(u_i, c) / \sum_{c' \in \mathcal{M}_j} \text{count}(u_j, c') \tag{3}$$

where, $\text{count}(u_j, c)$ is the number of times u_j has a relevant item in cluster c. For computational efficiency, we take $\mathcal{M}_j$ to be u_j's top m most relevant clusters. We normalize these counts to obtain a proper cluster-relevance distribution.

Nearest Neighbor Smoothing: Unfortunately, we typically have few user-item engagements on a per-user basis; thus, while the MLE is unbiased and asymptotically efficient, it can also be high variance. To this end, we introduce a smoothing technique that once again exploits locality in the ANN search, this time for users.

Figure 1 illustrates identifying k nearest-neighbors ($\mathcal{K}_j$) to the query user u_j's, and leveraging the information from the neighbors' cluster engagements to augment the user's cluster relevance. We compute this distribution over item clusters by averaging the MLE probability for each nearest neighbor (item clusters that are not engaged with by a retrieved neighbor have zero probability).

$$p_{kNN}(c|u_j) = \frac{1}{|\mathcal{K}_j|} \sum_{u' \in \mathcal{K}_j} p_{\text{mle}}(c|u') \tag{4}$$

We apply Jelinik-Mercer smoothing to interpolate between a user's MLE distribution with the aggregated nearest neighbor distribution [12].

$$p_{\text{smoothed}}(c|u_j) = (1 - \lambda) p_{\text{mle}}(c|u_j) + \lambda p_{kNN}(c|u_j), \tag{5}$$

where $\lambda \in [0, 1]$ represents how much smoothing is applied. It can be manually set or tuned on a downstream extrinsic task.

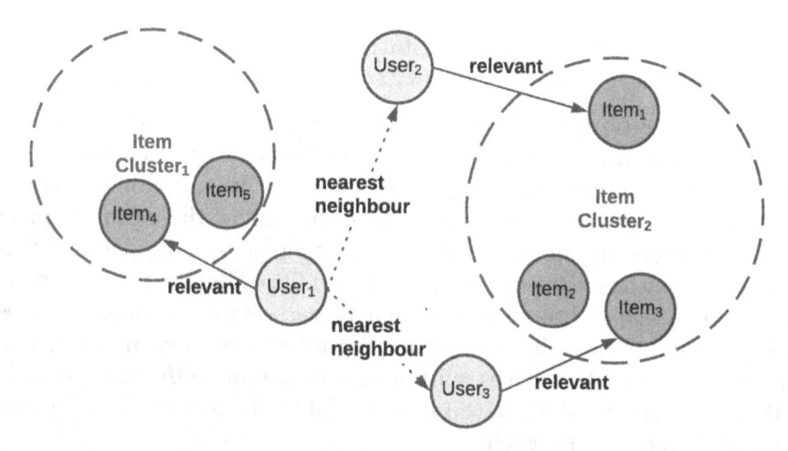

Fig. 1. Example of retrieving two candidates. In an ANN, items 4 and 5 would be deterministically returned for user 1. In our proposed kNN-Embed, even though the distances to cluster 2 are larger, smoothing means that we will sometimes return items from that cluster, yielding more diverse items. Note in this case, we don't even require that user 1 has previously relevant items in cluster 2.

Sampling within Clusters: Within each cluster there are many ways to retrieve items on a per user basis. A simple, but appealing, strategy is to represent each user as a normalized centroid of their relevant items in that cluster:

$$\text{centroid}(c, u_j) = \frac{\sum_{m \in R(c,u_j)} \mathbf{i_m}}{\| \sum_{m \in R(c,u_j)} \mathbf{i_m} \|}, \tag{6}$$

where $R(c, u_j)$ is the set of relevant items for user u_j in cluster c. However, since we are applying smoothing to the cluster probabilities $p(c|u_j)$, it may be case that u_j has zero relevant items in a given cluster. Hence, we smooth the user centroid using neighbor infomation to obtain the final user representation $\mathbf{u_j^c}$:

$$\mathbf{u_j^c} = (1 - \lambda)\,\text{centroid}(c, u_j) + \frac{\lambda}{|\mathcal{K}_j|} \sum_{u' \in \mathcal{K}_j} p_{\text{mle}}(c|u')\,\text{centroid}(c, u') \tag{7}$$

Equation 7 shows the **k**NN-smoothed user-specific embedding for cluster c. This embedding takes the user-specific cluster representations from Eq. 6, and performs a weighted averaging proportionate to each user's contribution to $p_{\text{smoothed}}(c|u_j)$. The final vector is once again normalized to unit norm.

4 Evaluation Datasets and Metrics

We evaluate on three datasets which we describe below:

HEP-TH Citation Graph: This paper citation network is collected from Arxiv preprints from the High Energy Physics category [9]. The dataset consists of: 34,546 papers and 421,578 citations.

DBLP Citation Graph: This paper citation network is collected from DBLP [24] and consists of 5,354,309 papers and 48,227,950 citation relationships.

Twitter Follow Graph: We curate Twitter user-follows-user (available via API) by first selecting a number of 'highly-followed' users that we refer to as 'content producers'; these content producers serve as 'items' in our recommender systems terminology. We then sampled users that follow these content producer accounts. All users are anonymized with no other personally identifiable information (e.g., demographic features) present. Additionally, the timestamp of each follow edge was mapped to an integer that respects date ordering, but does not provide any information about the date that follow occurred. In total, we have $261M$ edges and $15.5M$ vertices, with a max-degree of $900K$ and a min-degree of 5. We hope that this dataset will be of useful to the community as a test-bed for large-scale retrieval research.

Metrics: We evaluate kNN-Embed on three aspects: (1) the recall (2) diversity and (3) goodness of fit of retrieved candidates. Below, we formalize these metrics.

Recall@K: The most natural (and perhaps most important) metric for computing the efficacy of various candidate retrieval strategies is *Recall*@K. This metric is given by considering a fixed number of top candidates yield by a retrieval system (up to size K) and measuring what percent of these candidates are held-out relevant candidates. The purpose of most candidate retrieval systems is to collect a high-recall pool of items for further ranking, and thus recall is a relevant metric to consider. Additionally, recall provides an indirect way to measure diversity – to achieve high recall, one is obliged to return a large fraction of *all* relevant documents, which simple greedy ANN searches can struggle with.

Diversity: To evaluate the diversity among the retrieved candidates, we measure the spread in the embeddings of the retrieved candidates by calculating the average distance retrieved candidates are from their centroid. The underlying idea is that when 'locality implies similarity'; as a corollary, if candidates are *further* in Euclidean distance, then they are likely to be different. As such, for a given set of candidates $\mathcal{C}$, we compute diversity D as follows:

$$D(\mathcal{C}) = \frac{1}{|\mathcal{C}|} \sum_{i_k \in \mathcal{C}} \|\mathbf{i_k} - \hat{\mathbf{i}}\| \tag{8}$$

where $\mathcal{C}$ denotes the set of retrieved candidates and $\hat{\mathbf{i}} = \sum_{i_k \in \mathcal{C}} \mathbf{i_k}/|\mathcal{C}|$ is the mean of the unimodal embeddings of the retrieved candidates.

Goodness of Fit: In addition to diversity of retrieved items, we need to ensure that a user's mixture representation is an accurate model of their interests – that is the mixture of embeddings identifies points in the embedding space where relevant items lie. Thus, we compare held out relevant items to the user's mixture representation we use to query. We measure this "goodness of fit" by computing the Earth Mover's Distance (EMD) [19] between a uniform distribution over a user's relevant items and the user's mixture distribution. The EMD measures the distance between two probability distributions over a metric space [7,17].

We measure the distance between a user's cluster distribution (e.g., Eq. 3 and Eq. 4), to a uniform distribution over a held-out set of relevant items: $p(i|u_j)$ over a Euclidean space. We compute EMD by soft assigning all held-out relevant items of a user to clusters, minimizing the sum of item-cluster distances, with the constraint that the sum over soft assignments matches $p(c|u_j)$. As seen in Fig. 2, with standard unimodal representations, a single embedding vector is compared to the held-out items and the goodness of fit is the distance between the item embeddings and the singular user embedding. In comparison, for mixture representations (Fig. 2, each user multiple user embeddings who each have fractional probability mass that in total sums to 1. The goodness of fit is then the distance achieved by allocating the mass in each item to the closest user embedding cluster with available probability mass. Observing unimodal representations in Fig. 2, a single unimodal embedding is situated in the embedding space and compared to held-out relevant items. As shown, some held-out items are close to the unimodal embedding, while others are further away. In contrast, for mixture representations, each user has multiple user-embeddings and each of these embeddings lies close to a cluster of relevant items. The intuition is that if a user has multiple item clusters they are interested in, multiple user embeddings can better capture these interests.

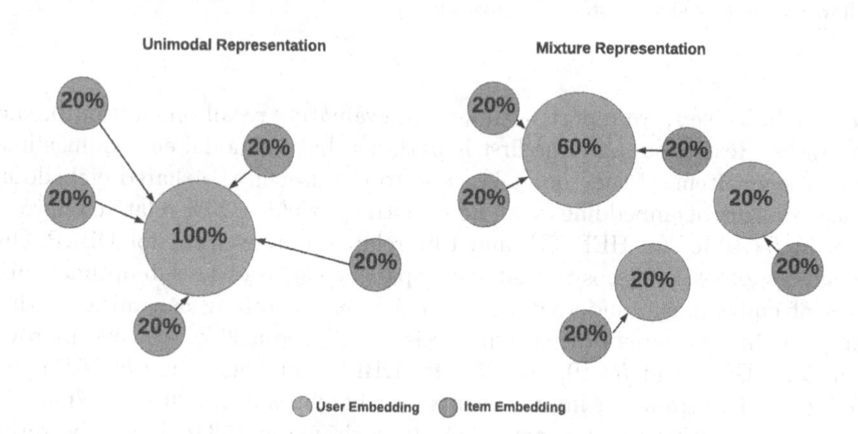

Fig. 2. Goodness of fit of unimodal representation vs mixture representation.

5 Experiments

Experimental Setup: For our underlying ANN-based candidate retrieval system, we start by creating a bipartite graph between source entities and target entities for each dataset, with each edge representing explicit relevance between items (e.g., citing paper *cites* cited paper or user *follows* content producer). We then learn unimodal 100-dimensional embeddings for users and items by training over 20 epochs and cluster them via spherical k-means over 20 epochs [2].

Evaluation Task: We evaluate three candidate retrieval strategies – baseline ANN with unimodal embeddings (which is how most ANN-based candidate retrieval systems work), mixture of embeddings with no smoothing [23], and mixture of embeddings with smoothing (i.e., kNN-Embed). For each strategy, we compute the *Recall@K*, diversity, and fit in a link prediction task.

Research Hypotheses: We explore two research hypotheses (as well as achieve some understanding of the hyperparameters): (1) Unimodal embeddings miss many relevant items due to the similarity of retrieved items. Mixtures yield more diverse and higher recall candidates. (2) Smoothing, by using information from neighboring users, further improves the recall of retrieved items.

Table 1. Recall of Retrieved Candidates

	HEP-TH			DBLP			Twitter-Follow		
Approach	R@10	R@20	R@50	R@10	R@20	R@50	R@10	R@20	R@50
Unimodal	20.0%	30.0%	45.7%	9.4%	13.9%	21.6%	0.58%	1.02%	2.06%
Mixture	22.7%	33.4%	49.3%	10.9%	16.1%	25.1%	3.70%	5.53%	8.79%
kNN-Embed	**25.8%**	**37.4%**	**52.5%**	**12.7%**	**18.8%**	**28.3%**	**4.13%**	**6.21%**	**9.77%**

Recall of unimodal vs mixture vs kNN-Embed- higher is better. HEP-TH ($\lambda = 0.8$, 2000 clusters, 5 embeddings). DBLP ($\lambda = 0.8$, 10000 clusters, 5 embeddings). Twitter-Follow ($\lambda = 0.8$, 40000 clusters, 5 embeddings).

Recall: In Table 1, we report results when evaluating recall on citation prediction tasks. Results support the first hypothesis that unimodal embeddings may miss relevant items if they don't lie close to the user in the shared embedding space. Mixture of embeddings with no smoothing, yields a 14% relative improvement in R@10 for for HEP-TH, and 16% relative improvement for DBLP. Our second hypothesis (2) posits that data sparsity can lead to sub-optimal mixtures of embeddings, and that nearest-neighbor smoothing can mitigate this. Our experiments support this hypothesis, as we see a 25% relative improvement for HEP-TH in R@10, and 35% for DBLP and when using kNN-Embed. We see similar significant improvements over baselines in R@20 and R@50. For Twitter-Follow, the improvements in recall are dramatic – 534% in relative terms going from unimodal embeddings to a mixture of embeddings in R@10. We suspect this significant improvement is because Twitter-Follow simultaneously has a much higher average degree than HEP-TH and DBLP and the number of unique nodes is much larger. It is a more difficult task to embed so many items, from many different interest clusters, in close proximity to a user. As such, we see a massive improvement by explicitly querying from each user's interest clusters. Applying smoothing provides an additional 74% in relative terms, and similar behaviours are observed in R@20 and R@50.

Diversity: We apply Eq. 8 to retrieved candidates and measure the spread of retrieved candidates' embedding vectors. As seen in Table 2, the candidates from unimodal retrieval are less diverse than candidates retrieved via multiple queries from mixture representations. This verifies our first research hypothesis that unimodal embeddings may retrieve many items that are clustered closely together as a by-product of ANN retrieval (i.e., diversity and recall is low). However, multiple queries from mixtures of embeddings broadens the search spatially; retrieved items are from different clusters, which are more spread out from each other. kNN-Embed (i.e., smooth mixture retrieval) results in slightly less diverse candidates than unsmoothed mixture retrieval. We posit that this is due to the high-variance of the maximum likelihood estimator of the $p_{\text{mle}}(c|u_j)$ multinomial (Eq. 3). While this high-variance may yield more diverse candidates, this yields less relevant candidates as seen in Table 1 where kNN-Embed consistently yields better recall than unsmoothed mixture retrieval. While high diversity is necessary for high recall, it is insufficient on its own.

Table 2. Diversity of Retrieved Candidates

	HEP-TH			DBLP			Twitter-Follow		
Approach	D@10	D@20	D@50	D@10	D@20	D@50	D@10	D@20	D@50
Unimodal	0.49	0.54	0.61	0.43	0.46	0.51	0.38	0.40	0.43
Mixture	**0.58**	**0.63**	**0.68**	**0.51**	**0.56**	**0.60**	**0.56**	**0.54**	**0.58**
kNN-Embed	0.54	0.60	0.66	0.46	0.52	0.57	0.47	0.52	0.55

Goodness of Fit: We evaluate how well unimodal, mixture, and smoothed mixture embeddings model a user's interests. The main idea is that the better fit a user representation is, the closer it will be to the distribution of held out relevant items for that user. As seen in Table 3, the results validate the idea that unimodal user embeddings do not model user interests as well as mixtures over multiple embeddings. Multiple embeddings yield a significant EMD improvement over a single embedding vector when evaluated on held-out items. Smoothing further decreases the EMD which we posit is due to the smoothed embedding mixtures being lower-variance estimates as they leverage engagement data from similar users in constructing the representations. These results suggest that the higher recall of smoothed mixtures is due to better user preferences modeling.

Table 3. Goodness of fit between user and held-out items as measured by earth mover's distance over a Euclidean embedding space. Lower EMD is better.

Approach	HEP-TH	DBLP	Twitter-Follow
Unimodal	0.897	0.889	1.018
Mixture	0.838	0.830	0.952
kNN-Embed	**0.811**	**0.808**	**0.940**

Hyper-parameter Sensitivity Analysis: We focus on recall as the *sine qua non* of candidate retrieval problems and analyze hyper-parameters on HEP-TH. In Fig. 3a, we vary the smoothing parameter λ (same parameter for both the mixture probabilities and the cluster centroids) and see heavy smoothing improves performance significantly. This likely stems from the sparsity of HEP-TH where most papers have only a few citations. In Fig. 3b, we vary the number of embeddings (i.e., the mixture size) and notice improved performance saturating at six mixture components. Out of all the hyperparameters, this seems to be the critical one in achieving high recall. In practice, latency constraints can be considered when selecting the number of embeddings per user, explicitly making the trade-off between diversity and latency. Finally, in Fig. 3c, we vary the number of k-means clusters; recall peaks at $k = 2500$ and then decreases. HEP-TH is a small dataset with only 34,546 items; it is likely that generating a very large number of clusters leads to excessively fine-grained and noisy sub-divisions of the items.

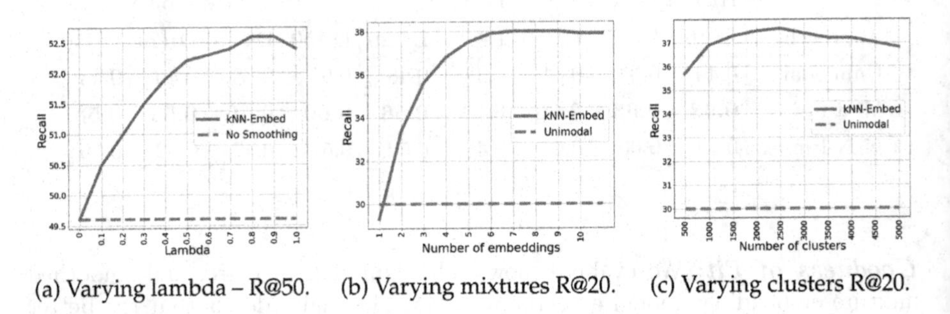

(a) Varying lambda – R@50. (b) Varying mixtures R@20. (c) Varying clusters R@20.

Fig. 3. We analyze the effect of three important hyper-parameters: (1) the λ smoothing (2) the number of embeddings in the mixture (3) the number of clusters for candidate retrieval in the HEP-TH dataset.

6 Conclusions

We present kNN-Embed, a method of transforming single user dense embeddings, into mixtures of embeddings, with the goal of better modeling user interests, increasing retrieval recall and diversity. This multi-embedding scheme represents a source entity with multiple distinct topical affinities by globally clustering items and aggregating the source entity's engagements with clusters. Recognizing that user-item engagements may often be sparse, we propose a nearest-neighbor smoothing to enrich these mixture representation. Our smoothed mixture representation better models user preferences retrieving a diverse set of candidate items reflective of a user's multiple interests. This significantly improves recall on candidate retrieval tasks on three datasets including Twitter-Follow, a dataset we curate and release to the community.

References

1. Andoni, A., Indyk, P.: Near-optimal hashing algorithms for approximate nearest neighbor in high dimensions. In: FOCS. IEEE (2006)
2. Arthur, D., Vassilvitskii, S.: k-means++: The advantages of careful seeding. Technical report, Stanford (2006)
3. Bayardo, R., Ma, Y., Srikant, R.: Scaling up all pairs similarity search. In: WWW (2007)
4. Bordes, A., Usunier, N., Garcia-Duran, A., Weston, J., Yakhnenko, O.: Translating embeddings for modeling multi-relational data. In: NeurIPS (2013)
5. Covington, P., Adams, J., Sargin, E.: Deep neural networks for youtube recommendations. In: RecSys (2016)
6. Dhillon, I., Modha, D.: Concept decompositions for large sparse text data using clustering. In: Machine learning (2001)
7. El-Kishky, A., Guzmán, F.: Massively multilingual document alignment with cross-lingual sentence-mover's distance (2020)
8. El-Kishky, A., Markovich, T., Park, S., et al.: Twhin: embedding the twitter heterogeneous information network for personalized recommendation. In: KDD (2022)
9. Gehrke, J., Ginsparg, P., Kleinberg, J.: Overview of the 2003 kdd cup. In: Sigkdd Explorations (2003)
10. Goldberg, Y., Levy, O.: word2vec explained: deriving mikolov et al'.s negative-sampling word-embedding method. arXiv preprint arXiv:1402.3722 (2014)
11. Huang, J., Sharma, A., Sun, S., Xia, L., et al.: Embedding-based retrieval in facebook search. In: SIGKDD (2020)
12. Jelinek, F.: Interpolated estimation of markov source parameters from sparse data. In: PRIP (1980)
13. Johnson, J., Douze, M., Jégou, H.: Billion-scale similarity search with gpus. In: BigData (2019)
14. Kang, W., McAuley, J.: Candidate generation with binary codes for large-scale top-n recommendation. In: CIKM (2019)
15. Khandelwal, U., Fan, A., Jurafsky, D., Zettlemoyer, L., Lewis, M.: Nearest neighbor machine translation. In: ICLR (2021)
16. Khandelwal, U., Levy, O., Jurafsky, D., Zettlemoyer, L., Lewis, M.: Generalization through memorization: nearest neighbor language models. In: ICLR (2020)
17. Kusner, M., Sun, Y., Kolkin, N., Weinberger, K.: From word embeddings to document distances. In: ICML (2015)
18. Lerer, A., et al.: Pytorch-biggraph: a large-scale graph embedding system. In: MLSys (2019)
19. Levina, E., Bickel, P.: The earth mover's distance is the mallows distance: some insights from statistics. In: ICCV. IEEE (2001)
20. Liu, X., Croft, B.: Cluster-based retrieval using language models. In: SIGIR (2004)
21. Malkov, Y., Yashunin, D.: Efficient and robust approximate nearest neighbor search using hierarchical navigable small world graphs. TPAMI **42**, 824–836 (2018)
22. n: Mikolov, T., Sutskever, I., Chen, K., Corrado, G., Dean, J.: Distributed representations of words and phrases and their compositionality. In: NeurIPS (2013)
23. Pal, A., Eksombatchai, C., Zhou, Y., et al.: Pinnersage: multi-modal user embedding framework for recommendations at pinterest. In: KDD (2020)
24. Tang, J., Zhang, J., Yao, L., Li, J., Zhang, L., Su, Z.: Arnetminer: extraction and mining of academic social networks. In: SIGKDD (2008)

25. Van Rijsbergen, C., Bruce, W.: Document clustering: an evaluation of some experiments with the cranfield 1400 collection. IPM **11**, 171–182 (1975)
26. Weston, J., Makadia, A., Yee, H.: Label partitioning for sublinear ranking. In: ICML
27. Wilhelm, M., Ramanathan, A., Bonomo, A.O.: Practical diversified recommendations on youtube with determinantal point processes. In: CIKM (2018)
28. Zhang, F., Yuan, N., Lian, D., Xie, X., Ma, W.: Collaborative knowledge base embedding for recommender systems. In: SIGKDD (2016)

Staying or Leaving: A Knowledge-Enhanced User Simulator for Reinforcement Learning Based Short Video Recommendation

Zhaoqi Yang and Hongyan Liu(✉)

School of Economics and Management, Tsinghua University, Beijing, China
yangzq21@mails.tsinghua.edu.cn, hyliu@tsinghua.edu.cn

Abstract. Reinforcement learning has been widely used in recommender systems in order to optimize users' long-term utilities. An accurate and explainable user simulator is crucial for reinforcement learning based recommendation, as an online interactive environment is often unavailable. On short video platforms, it is very important to keep users on the platform as long as possible in each session. Thus, session-based user utilities depend on two factors: how much users like every single video (video preference) and the number of videos watched (video views) in each session. To this end, the simulator should simultaneously model the user's degree of liking for each video and video views. However, most previous studies on the short video recommendation only paid attention to the former. In this work, we propose KESWA, a Knowledge-Enhanced Session-Wide Attention method for short video user simulation. KESWA fuses information foraging theory with a deep learning model for both video preference and video views modeling, providing an explainable prediction for users' staying and leaving behavior. Comparative experiments demonstrate that KESWA provides a better simulation of video views compared with existing models. Meanwhile, reinforcement learning agents can achieve higher session-based user utilities trained by KESWA than by other user simulators.

Keywords: Session-based recommendation · Reinforcement learning · Information foraging theory

1 Introduction

Reinforcement learning (RL) has been widely used in recommendation systems. RL algorithms aim to optimize the cumulative rewards, and thus can easily be used for maximizing users' long-term utilities. In RL, agents need to interact with the environment for learning. However, the online interaction environment is often unavailable for recommendation systems, because the immature RL agents may hurt user experience and thus reduce platform earnings.

H. Kashima et al. (Eds.): PAKDD 2023, LNAI 13937, pp. 387–399, 2023.
https://doi.org/10.1007/978-3-031-33380-4_30

There are two ways to address the lack of interaction environment. The first solution is offline reinforcement learning. Offline RL agents learn from logged data, which was collected by some unknown policies previously. The learning period of offline RL is extremely unstable because it highly depends on the quality of logged data. Data generated by sub-optimal policies will significantly affect the performance of offline RL [4]. The second solution is to design a simulator that can generate user behaviors in advance. Simulators serve as pseudo online users here, and RL agents are then trained by interacting with them. This approach has higher data efficiency and circumvents the influence of the data generation process. As a result, many recent works [3,12,15] adopted this solution. In this paper, we also adopt the second solution and focus on the design of user simulators.

The recent five years have witnessed a boom in short video platforms, such as TikTok and Kuaishou. Short video is a new way of life sharing. Video makers upload their works to the platform and video viewers watch them in an immersed mode: viewers rely largely on the recommendation system of the platform to push short videos to them. Viewers can decide at any time whether they want to skip the current video and directly watch the next video pushed by the platform. Meanwhile, viewers can give a like to short videos that attract them most. In this mode, users' preferences cannot be implied by common feedback like clicking and viewing. As a result, in this paper, we use one implicit factor, watching completion, and one explicit factor, "like", to fully reflect user preferences.

In a continuous period of time, the short videos watched by a user form a session. On short video platforms, it is very important to keep users on the platform as long as possible in each session. Thus, session-based user utilities (SUU) depend on two factors: how much users like every single video (video preference) and the number of videos watched (video views, VV) in each session. When designing RL-based recommendation systems, short video platforms are especially concerned about video views. This is because larger video views mean larger advertising audiences, which can attract more advertisers and help achieve higher profits. Therefore, before training RL agents for short video recommendations, it is of great importance to design user simulators to accurately model each user's video views. However, previous works paid little attention to this aspect. Most of them only focused on the prediction of users' preference on each video, overlooking the prediction of users' leaving behavior, and thus could not achieve a comprehensive understanding of session-based user utilities.

In this paper, we propose a novel Knowledge-Enhanced Session-Wide Attention (KESWA) framework for user simulation. By predicting the user's probability of leaving after watching each video, we find a way to model video views in each session. Due to the fact that users' leaving behaviors are rare in logged data, traditional deep learning methods tend to overfit them. As a result, we introduce information foraging theory (IFT) [9], a well-known theory about how people obtain information on the Internet, to help us overcome the overfitting problem and provide interpretability as well. Comparative experiments validate that KESWA provides a better simulation of video views compared with existing models. Meanwhile, RL agents trained by KESWA can achieve higher SUU than agents trained by other kinds of user simulators. To summarize, the main contributions of our work are as follows:

- We propose to measure session-based user utilities from both video preference and video views for session-based sequential short video recommendation.
- By fusing information foraging theory with a session-wide self-attention deep learning model, we propose a novel framework of user simulator for RL-based short video recommendation to improve utility estimation and enhance understanding of user behaviors on the platform.
- We conduct comparative experiments on a real-world data set and evaluate the performance of the proposed simulator through two different angles: the prediction performance of the simulator itself, and the RL recommendation performance based on the simulator. Experimental results demonstrate our proposed method's effectiveness.

2 Related Works

2.1 Short Video Recommendation

Many short video recommendation algorithms focused on static models [7,8]. This kind of modeling method does not consider the change in user preference over time, thus impossible to understand the user's real viewing intention. For RL-based sequential short video recommendation, Li et al. [5] proposed a multi-task prediction framework based on a multi-gate mixture of experts. However, this task can be fulfilled through a supervised learning paradigm, and treating it as a time-series decision-making problem lacks practical necessity. Cai et al. [2] aimed at achieving Pareto optimality for multiple optimization objectives on short video platforms by using a constrained RL method. However, RL agents were trained offline, causing instability in the training process. Meanwhile, the definition of cumulative reward in offline learning settings is unclear.

2.2 User Simulation for RL-Based Recommendation

There are multiple ways to simulate users' video preferences and video views for RL-based recommendations. Among them, behavioral cloning (BC) is the easiest and most effective way. What proposed by G. Zheng et al. [14] and L. Zou et al. [15] are typical BC methods. G. Zheng et al. analyzed user selection behavior during news browsing using a top-k ranking model. They assumed that news viewed first has the highest video preference. Meanwhile, they leveraged survival analysis to model user activeness and provided a probability for user leaving. This kind of analysis is equivalent to the constant leaving probability assumption, thus giving a video views prediction following geometric distribution. L. Zou et al. predicted users' video preferences through the attention mechanism and leveraged a multilayer perceptron (MLP) to predict video views. However, they did not conduct any experiments to validate the performance of MLP for video views simulation. Another way of user simulation is generative adversarial learning. J. Shi et al. [3] designed a simulator based on generative adversarial imitation learning. A similarity score between the simulated data and the logged data was used as the reward for video preference prediction. However, the adversarial learning process is extremely difficult for parameter adjustment. Besides, they also adopted an overly simplistic assumption that the number of video views is a constant value for all users. In conclusion, though video preference prediction has been widely discussed, no previous work could give a practical and user-specific video views prediction for different users.

3 Problem Statement

A user's watching behaviors on a short video platform can be divided into multiple sessions: $\tau_0, \tau_1, ..., \tau_N$. Each session includes a set of watching behaviors in a continuous period of time: $\tau_n = \{(a_i, f_i, l_i, e_i), i \in \mathbb{I}_n\}$, where a_i denotes the i-th short video recommended by the platform in session τ_n. The user response to video a_i contains three different parts: $f_i \in \{0, 1\}$ indicates whether user u finishes watching the entire video, $l_i \in \{0, 1\}$ indicates whether user u gives a like to a_i, and $e_i \in \{0, 1\}$ denotes the user leaving behavior, i.e.:

$$e_i = \begin{cases} 0, & \text{if user } u \text{ continue watching other videos after watching } a_i, \\ 1, & \text{if user } u \text{ leaves the platform after watching } a_i. \end{cases} \tag{1}$$

Based on the above notations, we give some important definitions used in this paper as follows.

Definition 1. *User's preference of short video a_i is the probability that user u finishes watching the entire video or gives a like to a_i:*

$$r_i = r(a_i) = \mathrm{P}((f_i | l_i) = 1). \tag{2}$$

Definition 2. *Video views (VV) of a given session τ_n is the number of (a_i, f_i, l_i, e_i) tuples in session τ_n:*

$$VV_n = |\tau_n| = |\mathbb{I}_n|. \tag{3}$$

Definition 3. *Session-based user utility (SUU) of a given session τ_n is the sum of user preferences regarding all videos in session τ_n:*

$$SUU_n = \sum_{i \in \mathbb{I}_n} r_i. \tag{4}$$

Definition 4. *Leaving probability (LP) after watching the i-th video is the probability that $e_i = 1$:*

$$LP_i = \mathrm{P}(e_i = 1). \tag{5}$$

As shown in Fig. 1, the purpose of the proposed framework KESWA is to provide a virtual online interactable environment for RL-based recommender systems. As a result, for each user u, the input of KESWA is a sequence of sessions $\tau_0, \tau_1, ..., \tau_N$, where τ_N is the current on-going session, and a candidate short video a_I for recommendation. Based on the input, KESWA is able to simulate user u's preference of the video and leaving probability LP_I at present, thus giving a better estimation of SUU and helping RL-based recommendation algorithms achieve higher SUU in applications.

4 Approach

4.1 Parameter Estimation for Information Foraging Theory

Information foraging theory (IFT) was proposed by Peter Pirolli and Stuart Card in the 1990s s to better understand how human users search for information [9]. IFT draws an analogy between the process by which humans acquire information and the process by which animals forage for food. To model and better simulate users' leaving or staying behavior on short video platforms, we adopt a variation of IFT [1]. The process of watching videos can be regarded as a process of information acquisition. IFT explains the information foraging process from two angles. The first angle is target-oriented (TO). People keeps viewing different videos until the utility cumulated during the watching process reaches the target set by themselves. The second angle is speed-sensitive (SS). When the speed of utility cumulation falls below the limit of viewers' tolerance, they stop watching anymore.

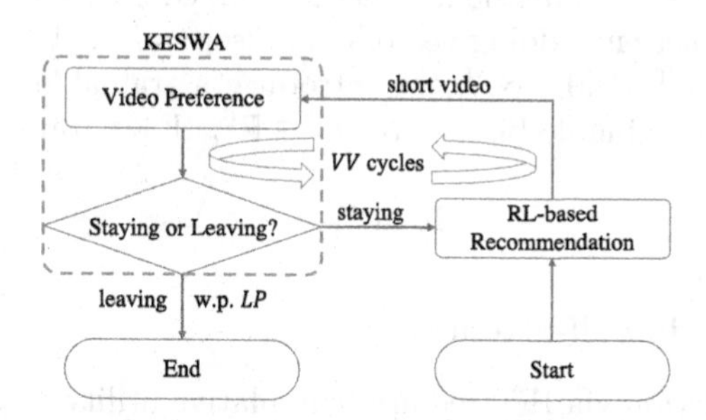

Fig. 1. A flow chart of how KESWA interacts with recommendation systems

As a result, we can use two equations to describe the leaving behavior of users when watching short videos. From the target-oriented angle, users set utility targets before they start to watch videos. The user's preference for each video reflects the utility gained by watching the video. Thus, we use the cumulation of user's preference on watched videos in the current session to represent user utility. With the cumulated utility acquired by users gradually approaching the target, the probability of staying on the platform decreases:

$$C_1 = 1 - (1 + \beta_1 \cdot e^{(\Gamma - \gamma)\alpha_1})^{-1}, \tag{6}$$

where C_1 represents the probability of staying, $\Gamma \in [0, +\infty)$ represents the utility target, γ represents the utility cumulated till now, $\beta_1 \in [0, +\infty)$ is a normalization parameter, and $\alpha_1 \in [0, +\infty)$ is a temperature coefficient, representing the sensitivity of the user response to the change of cumulated utilities. Higher α_1 represents higher sensitivity, and $\alpha_1 = 0$ implies that the cumulated utility has nothing to do with the user's probability of staying. β_1, Γ, and α_1 are trainable parameters for each user.

To model the speed-sensitive property, a utility-acquiring rate threshold is set before users start to watch videos. The probability of staying on the platform gradually decreases as the utility-acquiring rate falls behind the threshold:

$$C_2 = (1 + \beta_2 \cdot e^{(\Xi - \frac{\gamma}{\kappa})\alpha_2})^{-1}, \tag{7}$$

where C_2 represents the probability of staying, $\Xi \in [0, +\infty)$ represents the threshold, κ represents the number of videos viewed till now, $\frac{\gamma}{\kappa}$ represents the rate of utility cumulation, $\beta_2 \in [0, +\infty)$ is a normalization parameter, and $\alpha_2 \in [0, +\infty)$ is a temperature coefficient. Similarly, higher α_2 represents higher sensitivity, and $\alpha_2 = 0$ implies that the utility-acquiring rate has nothing to do with the user's probability of staying. β_2, Ξ, and α_2 are trainable parameters for each user.

Assuming that C_1 and C_2 are independent, the probability of the user leaving can be expressed as:

$$LP_i = 1 - C_1 C_2. \tag{8}$$

Suppose there are U users on the short video platform, then each of them can be represented by a U-dimensional one-hot embedding vector $\boldsymbol{u}$. In order to convert different user embedding vectors into user-specific IFT parameters $\theta_u = (\Gamma, \beta_1, \alpha_1, \Xi, \beta_2, \alpha_2)^{\mathrm{T}} \in [0, +\infty)^6$, we introduce a trainable embedding matrix $W_p \in \mathbb{R}^{6 \times U}$ and a trainable bias vector $b_p \in \mathbb{R}^6$. IFT parameters of user u are then decided by:

$$\theta_u = \mathrm{Relu}(W_p \boldsymbol{u} + b_p). \tag{9}$$

4.2 Session-Wide Self-attention

Video views prediction via IFT requires cumulative utility γ of all videos in the on-going session. User u's utility of a given video is related to his or her responses $f_i|l_i$ to all short videos that have already been watched. In order to calculate all video utilities in the current session, we adopt the self-attention framework [13]. We modify it to focus on the current session and the candidate video, leveraging history information to make predictions for them. Thus, we restrict the query items to be current session videos and the candidate video, regarding videos watched before each query video in the history to be the values and the keys. We call the modified self-attention method session-wide self-attention.

The input of this module contains two parts: candidate video a_I and user viewing history $\tau_{0:N} = \{\tau_0, \tau_1, ..., \tau_N\}$, where current session τ_N starts at short video a_{i_0}. The output is user's preference r_i for each video $a_i, i \in [i_0, I]$ and current leaving probability LP_I.

Suppose there are V short videos on the platform, then the candidate video and each video in the viewing history can all be represented by V-dimensional one-hot embedding vectors $\boldsymbol{a}_i$, $i \in [i_0, I]$ initially. They are then projected to a continuous space of size d_v by multiplying a trainable transformation matrix $W_v \in \mathbb{R}^{d_v \times V}$:

$$\phi_i = W_v \boldsymbol{a}_i. \tag{10}$$

To model user's response $f_i|l_i$ to each video a_i in the history (excluding the candidate video), we introduce two learnable $d_v \times d_v$ dimension user response matrices F_i, i.e., F_0 if $f_i|l_i = 0$; F_1 if $f_i|l_i = 1$.

Combining with user's response information, modified embedding φ_i of video a_i in the history is then obtained by:

$$\varphi_i = F_i \phi_i. \tag{11}$$

Matching scores ω_{ji} between any two different videos for $j, i \in [i_0, I]$ in the current session can be calculated:

$$\omega_{ji} = {\varphi_j}^{\mathrm{T}} \phi_i, \text{ if } j < i. \tag{12}$$

Then we can calculate our session-wide self-attention factors for all short videos in the current session:

$$\Omega_{ji} = \mathrm{Softmax}(\omega_{ji}). \tag{13}$$

Finally, viewing history feature $\eta_i \in \mathbb{R}^{d_v}$ $i \in [i_0, I]$ is:

$$\eta_i = \sum_{j<i} \Omega_{ji} \varphi_j. \tag{14}$$

Now, we could simulate user's preference for each video in the current session (including the candidate video) through an MLP with Softmax on the last layer :

$$r_i = \mathrm{Softmax}(\mathrm{MLP}(\eta_i)), \ i \in [i_0, I]. \tag{15}$$

To do the prediction of video views, we calculate the sum of video preferences cumulated in the session:

$$\gamma = \sum_{i=i_0}^{I} r_i. \tag{16}$$

The number of videos viewed in the current session is $\kappa = I - i_0 + 1$. Finally, by leveraging Eq. 6, 7, and 9, we output the prediction of LP_I based on Eq. 8, summarized below:

$$LP_I = f_{\theta_u}(\gamma, \kappa). \tag{17}$$

The overall framework of KESWA is shown in Fig. 2.

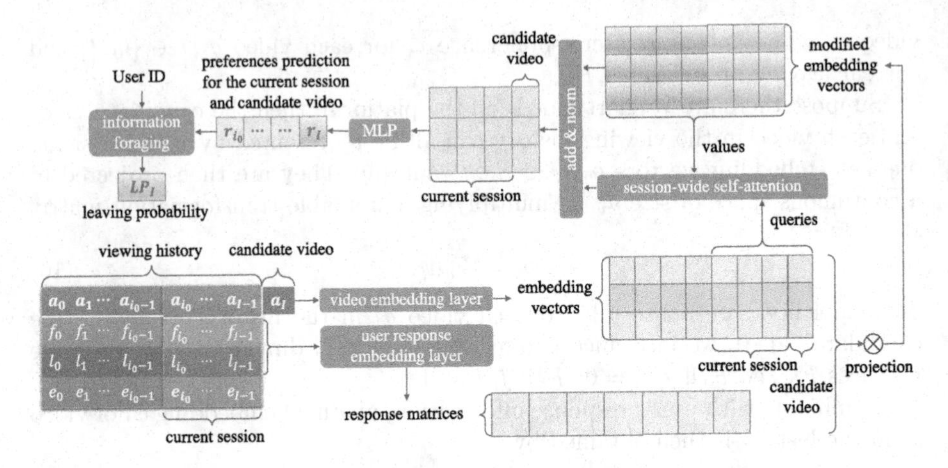

Fig. 2. KESWA: A Knowledge-Enhanced Session-Wide Attention user simulator

4.3 User Simulator Learning

The loss function of our model contains two parts: video preference loss L_v and user leaving loss L_u:

$$L = L_v + \rho L_u, \tag{18}$$

where ρ is a hyperparameter.

We take the binary cross entropy between model output r_i and real user response $f_i|l_i$ as L_v. There are two ways to compute L_v. One is focusing on candidate video a_I:

$$L_{v,I} = -(f_I|l_I)\log(r_I) - (1 - f_I|l_I)\log(1 - r_I). \tag{19}$$

The other takes into account all videos in current session τ_N:

$$L_v = \sum_{i \in \mathbb{I}_N} L_{v,i}. \tag{20}$$

Though Eq. 19 only considers the prediction loss about the candidate video, predictions of videos in the current session are used to predict the leaving probability, related to loss L_u.

L_u is the binary cross entropy between model output $LP_I = f_{\theta_u}(\gamma, \kappa)$ and real user leaving behavior e_I:

$$L_u = -e_I \log(f_{\theta_u}(\gamma, \kappa)) - (1 - e_I)\log(1 - f_{\theta_u}(\gamma, \kappa)). \tag{21}$$

We use gradient descent to update all of our trainable parameters ζ. When computing $\partial L_u / \partial \zeta$, we may partially cut the error back propagation process in order to improve computing efficiency, i.e., we may regard γ and κ with zero derivative to ζ.

5 Experiment

5.1 Data Set Pre-processing

The real-world data set comes from the biggest Chinese short video platform, Douyin, with more than 100 million watching behaviors. We first rule out the videos and the users whose appearance frequencies lie in the bottom 25%. Then each user's watching behaviors are divided into multiple sessions by setting the maximum interval between two videos within a session to be 10min. There are 38320729 behaviors, 55597 users, and 76023 videos in our final data set. All sessions have a mean length of 43. We use the first 60% of each user's watching behaviors sorted by time as the training set, and the last 40% as the test set.

5.2 Experimental Settings

The maximum length of user watching history inputted into the algorithm is 300. Dimension d_v of video embedding vectors ϕ_i and viewing history features η_i is 100. We set the batch size to be 64 and the learning rate to be 0.001.

In order to evaluate whether KESWA could help design better RL-based recommendation algorithms, we trained a PPO (Proximal Policy Optimization)-based [11] recommender agent by interacting with our user simulator. The input of the agent is the watching history of a given user. The output of the agent is the candidate short video. We train the PPO agent for 1000 epochs. In each epoch, we collect 256 viewing trajectories with 400 watching behaviors each by interacting with KESWA. Then we update our agent with a mini-batch sized 256. The learning rate of the PPO agent is 0.0001.

5.3 Experimental Results

Prediction Performance of KESWA. We adopt three metrics to evaluate the performance of KESWA against other baselines. The first is the area under the receiver operating characteristic curve (AUC) of preference prediction, the second is the AUC of leaving prediction, and the third is the estimation likelihood of video views, i.e., the estimation likelihood of our model to fit the real data.

We conduct comparative experiments between multiple baselines and different learning settings of KESWA. For video preference prediction, we compare our session-wide attention (SWA) with BPR-MF [10], a well-known Bayesian-based static recommendation algorithm, and NARM [6], a self-attention-based sequential recommendation algorithm similar to our model. For video views prediction, we choose all the existing methods to the best of our knowledge: survival analysis (SA) [14], and MLP [15] as our baselines.

Ablation studies are conducted according to the two angles of IFT: target-oriented (TO) and speed-sensitive (SS). We also try different learning settings of KESWA and consider the time (per mini-batch) and space consumption of them. Experiments are conducted on different settings of whether we only consider the candidate video (CV) in video preference loss L_v, and whether we conduct partial gradient cutting (PGC) in user leaving loss L_u. The results on the test set are shown in Table 1.

From the results we can see that, IFT with both TO and SS reaches the highest estimation likelihood of video views among all models, which proves the validity of IFT in this setting. Via IFT, our model can improve the AUC of leaving prediction by 5.61% compared with the strongest baseline. With CV and PGC, our model can achieve the highest AUC of preference prediction, meanwhile, save training time and space to a great extent. The advantage of CV may be that it allows the model to concentrate on the candidate video prediction.

RL Recommendation Performance Based on KESWA. We trained the same PPO-based recommender agent by interacting with KESWA and other kinds of user simulators in parallel. More specifically, we changed the method of video views prediction of simulators and see whether it affects the recommendation performance. When training RL-based agents, we compared two kinds of reward settings: one is to optimize the session-based user utility (SUU), and the other is to optimize the number of video views (VV). We also introduce an agent recommending videos randomly and an agent greedily recommending the video with the highest preference at each step as baselines.

Table 1. Prediction performance of KESWA.

Preference Prediction	Leaving Prediction	Preference AUC	Leaving AUC	Video Views Likelihood	Time	Space
BPR-MF	/	0.5549	/	/	/	/
NARM	/	0.6431	/	/	/	/
SWA+CV	SA	0.6434	0.5000	0.8776	/	/
SWA+CV	MLP	0.6427	0.5277	0.8779	/	/
SWA+CV	TO+PGC	0.6400	0.5475	0.8788	/	/
SWA+CV	SS+PGC	0.6435	0.5330	0.8787	/	/
SWA	IFT	0.6098	0.5471	0.8809	6′26″	11417MB
SWA+CV	IFT	0.6353	**0.5573**	**0.8822**	5′02″	11417MB
SWA+CV	IFT+PGC	**0.6451**	0.5519	0.8799	**1′52″**	**8011**MB

To avoid data breaches and ensure fair comparisons, following the instruction in [3,12], we first train different kinds of user simulators on the training set and then train RL-based agents with these simulators in parallel. For testing, we train another user simulator on the test set, and then all RL-based agents are tested with it. The metrics are average video preference (AVP), VV, and SUU. As shown in Table 2, KESWA can help the RL-based agent with rewards maximizing SUU achieve the highest AVP and SUU during the test period. The AVP and SUU of KESWA are significantly better than the strongest baseline (SWA+MPL) at a 99.9% confidence level.

Table 2. RL recommendation performance based on KESWA.

User Simulator	Reward Setting	AVP	VV	SUU
Random	/	0.261 ± 0.0014	30.893 ± 0.108	8.075 ± 0.007
Greedy	/	0.252 ± 0.0004	30.897 ± 0.083	7.778 ± 0.006
SWA+SA	max SUU	0.269 ± 0.0053	31.033 ± 0.872	8.348 ± 0.071
SWA+MLP	max SUU	0.269 ± 0.0057	$\mathbf{31.072 \pm 0.833}$	8.355 ± 0.048
KESWA	max VV	0.270 ± 0.0053	31.032 ± 0.851	8.388 ± 0.065
KESWA	max SUU	$\mathbf{0.273 \pm 0.0059}$	31.034 ± 0.871	$\mathbf{8.467 \pm 0.056}$

5.4 Case Study

KESWA can improve leaving probability prediction. As shown in Fig. 3, the left shows the leaving probability of a given user provided by model SWA+MLP, where vertical dotted lines distinguish different sessions. The results look like a random guess. The right is the results of the same user provided by our model. The leaving probability increases as the user keeps watching within a single session, with fluctuation caused by the change of preference-acquiring rate, which demonstrates the advantage of fusing IFT theory with the deep learning model.

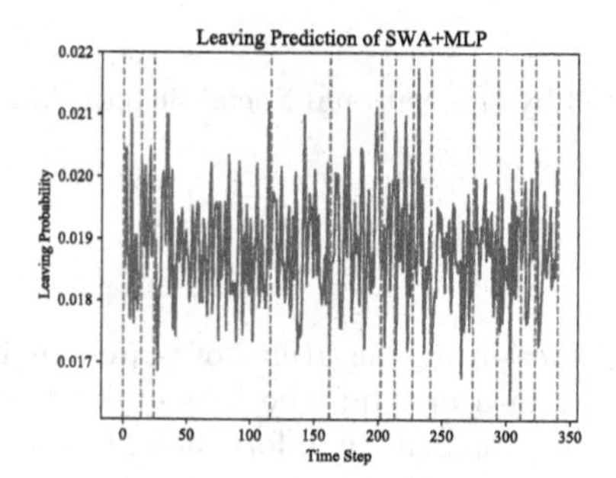

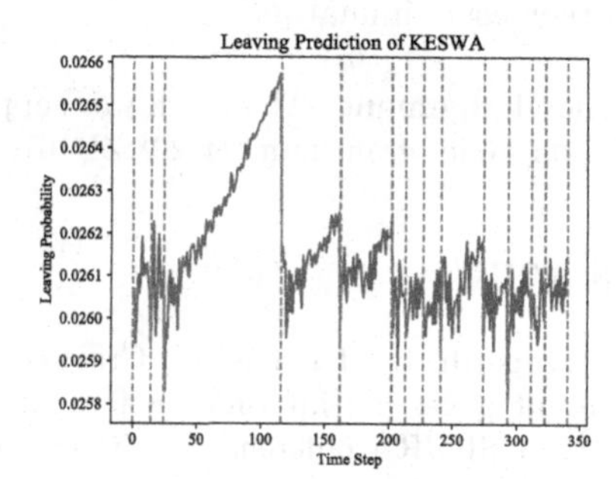

Fig. 3. Leaving Probability Prediction of KESWA.

KESWA can also identify user-specific IFT parameters, which demonstrates the interpretability of our model. As shown in Fig. 4, each node represents a user on the short video platform. On the left side, the horizontal axis represents user's utility target Γ, and the vertical axis represents temperature coefficient α_1. As a result, users in the upper part of the diagram are more sensitive to the cumulation of video preferences. Similarly, on the right side, the horizontal axis represents user's preference-acquiring speed threshold Ξ, and the vertical axis represents temperature coefficient α_2. Users in the upper part of the diagram are more sensitive to the change in preference-acquiring rate. Thus, with IFT, we can better understand user's behavior on the short video platform.

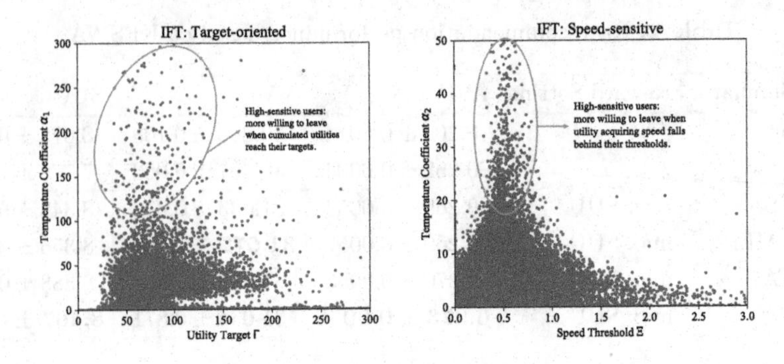

Fig. 4. User-specific IFT parameters.

6 Conclusion

In this paper, we propose KESWA, a knowledge-enhanced user simulator for RL-based short video recommendation. By fusing information foraging theory and session-wide self-attention, KESWA can provide better predictions for both video preferences and video views, thus giving a more comprehensive understanding of session-based user utilities. By taking advantage of this, RL-based recommender agents can achieve higher session-based user utilities trained by KESWA than by other user simulators.

Acknowledgement. This work was supported by the National Social Science Major Program under grant number 20&ZD161.

References

1. Azzopardi, L., Thomas, P., Craswell, N.: Measuring the utility of search engine result pages: an information foraging based measure. In: The 41st international ACM SIGIR Conference on Research & Development in Information Retrieval, pp. 605–614 (2018)
2. Cai, Q., et al.: Constrained reinforcement learning for short video recommendation. arXiv preprint arXiv:2205.13248 (2022)
3. Chen, X., Li, S., Li, H., Jiang, S., Qi, Y., Song, L.: Generative adversarial user model for reinforcement learning based recommendation system. In: International Conference on Machine Learning, pp. 1052–1061. PMLR (2019)
4. Fujimoto, S., Meger, D., Precup, D.: Off-policy deep reinforcement learning without exploration. In: International Conference on Machine Learning, pp. 2052–2062. PMLR (2019)
5. Li, D., Li, X., Wang, J., Li, P.: Video recommendation with multi-gate mixture of experts soft actor critic. In: Proceedings of the 43rd International ACM SIGIR Conference on Research and Development in Information Retrieval, pp. 1553–1556 (2020)
6. Li, J., Ren, P., Chen, Z., Ren, Z., Lian, T., Ma, J.: Neural attentive session-based recommendation. In: Proceedings of the 2017 ACM on Conference on Information and Knowledge Management, pp. 1419–1428 (2017)

7. Liu, S., Chen, Z., Liu, H., Hu, X.: User-video co-attention network for personalized micro-video recommendation. In: The World Wide Web Conference, pp. 3020–3026 (2019)
8. Liu, Y., Lyu, C., Liu, Z., Tao, D.: Building effective short video recommendation. In: 2019 IEEE International Conference on Multimedia & Expo Workshops (ICMEW), pp. 651–656. IEEE (2019)
9. Pirolli, P., Card, S.: Information foraging in information access environments. In: Proceedings of the SIGCHI Conference on Human Factors in Computing Systems, pp. 51–58 (1995)
10. Rendle, S., Freudenthaler, C., Gantner, Z., Schmidt-Thieme, L.: BPR: Bayesian personalized ranking from implicit feedback. arXiv preprint arXiv:1205.2618 (2012)
11. Schulman, J., Wolski, F., Dhariwal, P., Radford, A., Klimov, O.: Proximal policy optimization algorithms. arXiv preprint arXiv:1707.06347 (2017)
12. Shi, J.C., Yu, Y., Da, Q., Chen, S.Y., Zeng, A.X.: Virtual-Taobao: virtualizing real-world online retail environment for reinforcement learning. In: Proceedings of the AAAI Conference on Artificial Intelligence, vol. 33, pp. 4902–4909 (2019)
13. Vaswani, A., et al.: Attention is all you need. In: Advances in Neural Information Processing Systems, vol. 30 (2017)
14. Zheng, G., et al.: DRN: a deep reinforcement learning framework for news recommendation. In: Proceedings of the 2018 World Wide Web Conference, pp. 167–176 (2018)
15. Zou, L., et al.: Pseudo Dyna-Q: a reinforcement learning framework for interactive recommendation. In: Proceedings of the 13th International Conference on Web Search and Data Mining, pp. 816–824 (2020)

RLMixer: A Reinforcement Learning Approach for Integrated Ranking with Contrastive User Preference Modeling

Jing Wang[1], Mengchen Zhao[2], Wei Xia[2], Zhenhua Dong[2], Ruiming Tang[2], Rui Zhang[3], Jianye Hao[2,4], Guangyong Chen[5](✉), and Pheng-Ann Heng[1]

[1] The Chinese University of Hong Kong, Ma Liu Shui, Hong Kong
jing@link.cuhk.edu.hk, pheng@cse.cuhk.edu.hk

[2] Huawei Noah's Ark Lab, Quebec, Canada
{zhaomengchen,xiawei24,dongzhenhua,tangruiming,haojianye}@huawei.com

[3] Tsinghua University, Beijing, China
rayteam@yeah.net

[4] Tianjin University, Tianjin, China

[5] Zhejiang Lab, Zhejiang, China
gychen@zhejianglab.com

https://www.ruizhang.info

Abstract. There is a strong need for industrial recommender systems to output an integrated ranking of items from different categories, such as video and news, to maximize overall user satisfaction. Integrated ranking faces two critical challenges. First, there is no universal metric to evaluate the contribution of each item due to the huge discrepancies between items. Second, user's short-term preference may shift fast between diverse items during her interaction with the recommender system. To address the above challenges, we propose a reinforcement learning (RL) based framework called RLMixer to approach the sequential integrated ranking problem. Benefiting from the credit assignment mechanism, RLMixer can decompose the overall user satisfaction to items of different categories, so that they are comparable. To capture the user's short-term preference, RLMixer explicitly learns user interest vectors by a carefully designed contrastive loss. In addition, RLMixer is trained in a fully offline manner for the convenience in industrial applications. We show that RLMixer significantly outperforms various baselines on both public PRM datasets and industrial datasets collected from a widely used AppStore. We also conduct online A/B tests on millions of users through the AppStore. The results show that RLMixer brings over 4% significant revenue gain.

Keywords: Integrated ranking · Reinforcement Learning · Contrastive Learning

J. Wang and M. Zhao—The first two authors contributed equally to this work.

H. Kashima et al. (Eds.): PAKDD 2023, LNAI 13937, pp. 400–413, 2023.
https://doi.org/10.1007/978-3-031-33380-4_31

1 Introduction

Traditional ranking systems focus on ranking homogeneous items, such as a list of news, according to a specific metric like click-through rate. However, in practice, the final recommendation result presented to a user is usually a mixture of heterogeneous items. For example, in the news feeds scenarios, the recommendation list might consist of news, videos, advertisements and various form of cards. A straightforward way is to fix some slots for specific categories, which is commonly adopted in industrial recommender systems. However, this is clearly not the optimal strategy since the user's preferences towards each category evolve during interaction with the recommender system.

The optimal integrated ranking module is required to rank items of different categories in order to maximize the overall utility of the recommender system. The unique challenges of integrated ranking are two-fold. First, we lack a unified metric to evaluate the qualities of items from different categories. Thus they cannot be compared in one dimension directly. Second, users' preferences towards each category could be essentially different and shift during the interaction. For example, a user might find an interesting video while reading news and keep looking for similar videos. This personalized short-term preference shifting is hard to capture since the user feedbacks are usually implicit.

To address the first challenge, existing works try to use reinforcement learning to allocate categories to slots [11]. Unfortunately, their model focuses on inserting advertisements to news feed and does not generalize to integrated ranking with multiple categories of items. To address the second challenge, various methods have been proposed to learn representations of users' short-term interests [19,20]. However, they focus on implicitly mining knowledge from recent interacted items without explicitly modeling user preferences.

In this paper, we propose RLMixer for general integrated ranking problems. The novelties of RLMixer lie in the following three aspects. First, we propose a general and flexible MDP formulation that covers a broad range of integrated ranking problems. Second, we explicitly model the user's short-term preferences towards different categories and propose a carefully designed contrastive loss for learning them. Third, RLMixer can be trained fully offline, significantly saving online exploration costs and avoiding bias caused by simulation. Specifically, we implement RLMixer by conservative-q learning along with divergence penalty.

As far as we know, RLMixer is the first offline reinforcement learning approach to solve general integrated ranking problems. We successfully tackled the aforementioned issues with appropriate MDP modeling and a novel offline training framework. We compare RLMixer with several baselines on the public PRM datasets, as well as industrial datasets. We also deploy RLMixer on a widely used AppStore, where apps from different sources and categories are ranked together. Experimental results show that RLMixer significantly improves the original ranking quality and brings over 4% revenue gain.

2 Related Work

2.1 Integrated Ranking

Integrated ranking focuses on reranking items based on the roughly mixed heterogeneous items list while lacking a unified metric, while existing work mainly focuses on allocating advertisements to a list of organic items, which can be regarded as a particular case of integrated ranking. Koutsopoulos [7] defines ads allocation as a shortest-path problem on a weighted directed acyclic graph and apply the Bellman-Ford algorithm to solve it. Yan et al. [16] propose a uniform formula to rank advertisements and organic items together, considering the impact of interval between them. Zhao et al. [18] propose a novel deep Q-network to determine when and how to interpolate advertisements. Liao et al. [11] propose Cross-DQN to extract the crucial arrangement signal by crossing the embeddings of different items and modeling the crossed sequence by multi-channel attention. Unfortunately, existing works consider only advertisements and organic items, which limits their application in general integrated ranking problems with more than two categories of items. Moreover, their methods require online or off-policy training, which might incur huge online exploration costs.

2.2 Offline RL for Recommendation

Reinforcement learning for recommendation systems has attracted increasing interest in recent years. Zheng et al. [6] propose DRN for news recommendation, an off-policy framework with an online exploring network to balance exploration and exploitation. Chen et al. [2] propose a policy gradient method with various techniques to reduce the variance of policy gradients. Zhao et al. [17] propose a DDPG-based algorithm for learning optimal ranking weights to combat cheating sellers in e-commerce. However, existing work directly applies RL without considering how the user's preferences evolve during the interaction.

Inspired by the abundant historical interactions in recommendation scenarios, offline reinforcement learning is an emerging topic that aims to learn agent policy purely from dataset [4,8–10]. Offline RL strives for the issue of the overestimation of out-of-distribution actions, which introduces significant extrapolation error in policy learning. A popular method to address this is to use behavioral regularizations in RL training that compel the learned policy to stay close to the offline data. These regularizations consist of incorporating some divergence regularization into the critic [9], policy divergence penalties [4,14], and appropriate network initializations [12]. Regarding its application in recommendation, Xiao et al. [15] summarize several offline learning tricks and demonstrate their effectiveness in recommendation.

3 Integrated Ranking via Reinforcement Learning

In this section, we formally define the problem formulation for integrated ranking with the reinforcement learning settings.

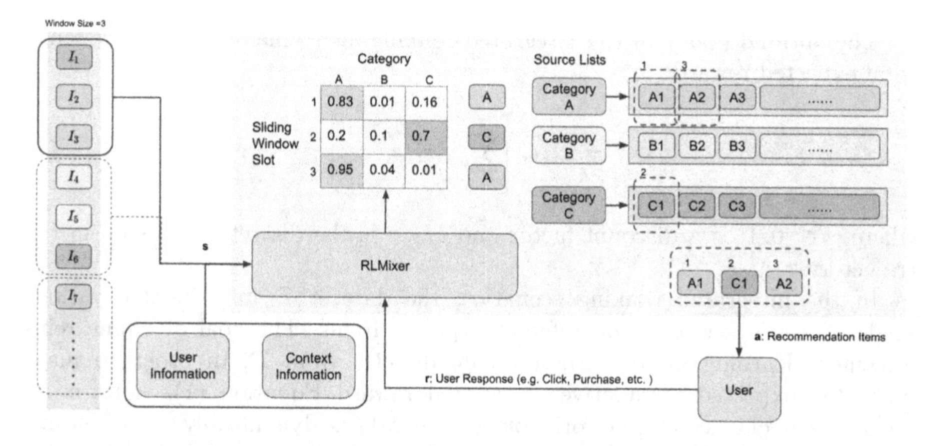

Fig. 1. Illustration of integrated ranking three categories of items with window size 3. The policy of RLMixer output a 3*3 matrix, where the rows represent the slots to be filled and the columns represent categories. During execution, the category with the highest score in each row is selected. Then the item that ranked highest in the corresponding category is fetched and filled in the slot.

Integrated ranking serves as a re-ranking module in the whole chain of the recommendation system, and aims to output a re-ranked list that maximizes the overall utility of the system. Figure 1 illustrates our decision making process with three categories of items. Due to the huge combinatorial action space of processing the whole list at the same time, we re-rank items within a sliding window among the original list step by step.

Integrated ranking is naturally a sequential decision making problem, we model the integrated ranking as a Markov Decision Process $< \mathcal{S}, \mathcal{A}, r, \mathcal{P}, \gamma >$:

State Space $\mathcal{S}$: $\mathcal{S}$ is the set of states describing the state space of the integrated ranking module. A state $s \in \mathcal{S}$ consists of the user information (e.g., age, gender, purchasing power), the originally ranked list, candidate items(i.e., items needed to be re-ranked at current step) and other contextual information.

Action Space $\mathcal{A}$: An action $a \in \mathcal{A}$ is a sequence of categories whose length is the size of sliding window. Assume there are C related categories in total. An action can be represented as a vector $a = (C_1, C_2, ..., C_W)$, where W is the size of the sliding window and $C_i \in \{1, 2, ..., C\}$ indicates its category.

Rewards: The reward is calculated based on the system's overall utility, which is the accumulated revenue(e.g., price) of cliked items in our case.

Transitions: $P(s_{t+1}|s_t, a_t)$ is the state transition function that indicates the state transferring from current state s_t to next state s_{t+1} after taking action a_t. Note that such updates on the raw states do not actually reflect the change of user preference. This motivates us to learn a mapping from raw states to explicit user preference representation. Please refer to Sect. 4.3 for details.

The optimal policy of the integrated ranking agent maximizes the system's total expected reward:

$$J = \mathbb{E}_{\tau \sim \pi}\left[\sum_{t=0}^{\mathcal{T}} \gamma^t r_t(s_t, a_t)\right], \tag{1}$$

where $\gamma \in [0,1]$ is a discount factor and $t \in \tau$ is the discrete time step in the trajectory τ.

In the integrated ranking scenarios, the dataset $\mathcal{D}$ may include various kinds of user's feedbacks of different types of items. The goal of offline reinforcement learning is to learn a policy directly from $\mathcal{D}$, in order to maximize the expected cumulative discounted reward Equation (1). Actor-critic scheme is a classical framework for solving MDPs dynamically. It maintains a parametric Q-function, Q_θ, and a parametric policy, $\pi_\omega(a|s)$. It alternates between policy evaluation, computing the Q^π that iterating the Bellman operator $\mathcal{B}^\pi Q = r + \gamma \mathbb{E}_{\mathbf{s}' \sim P(\mathbf{s}'|\mathbf{s},\mathbf{a}), \mathbf{a}' \sim \pi(\mathbf{a}'|\mathbf{s}')}\left[Q\left(\mathbf{s}', \mathbf{a}'\right)\right]$, and policy improvement, improving the policy $\pi(a|s)$ by updating it towards actions that maximize the expected Q-value. In this paper, we incorporate behavior regularization into the actor-critic framework via a critic penalty and policy regularization to address the overestimation and distribution shift. Details of training are illustrated in Sect. 4.5.

4 The Framework of RLMixer

4.1 Overview

The architecture of the policy network in RLMixer is presented in the Fig. 2. In order to capture local information for decision-making at every single step, we propose to maintain a sliding window that contains the current items to be reranked, which can also be interpreted as the user's current attention. And then, the global context extraction (GCE) module is expected to extract the context information from the original ranking list, the real-time preference capturing (RPC) module is utilized to learn the real-time user preferences on candidate items respectively. Finally, the user's preference on candidate items is concatenated together and fed into the RL module for policy execution.

In the following sections, we will take a sliding window size of 3 as an example to elaborate on implementing the aforementioned modules and networks.

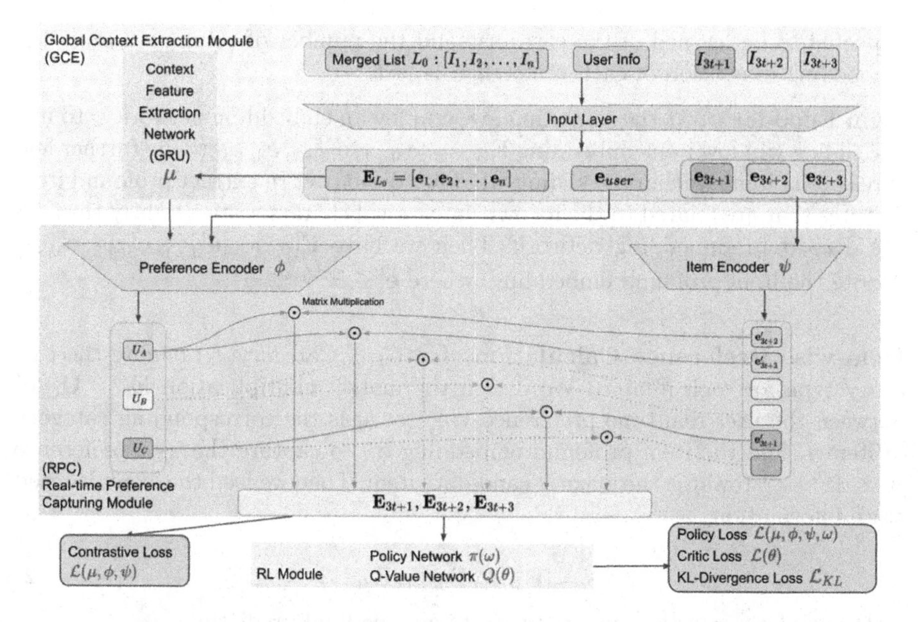

Fig. 2. Overall architecture of the policy network in RLMixer.

4.2 Global Context Extraction Module

In the training stage, we take the original ranking list $L_O = [I_1, I_2, ..., I_n]$ as part of the state information, where n is the total number of items in source lists. Let $W_t = [I_{3t+1}, I_{3t+2}, I_{3t+3}]$ denote the candidate items inside the sliding window W_t at time step t. As introduced in the Sect. 3, the full state input consists of the user information(e.g., age, gender, purchasing power, etc.), the original ranking list L_O, and candidate items W_t. Initially, we employ the input layer to map the user information features to the user embedding vector $\mathbf{e}_{user}$, and obtain the item embedding $\mathbf{e}_i$ for each item I_i. Then we adopt the traditional GRU cells to extract the contextual information from the original ranking list through context feature extraction network μ, $\mathbf{h}_i = GRU(\mathbf{e}_i)$, where $GRU(\cdot)$ denotes the traditional GRU cell, $\mathbf{h}_i$ denote the hidden state about item I_i. After feature extraction, we construct the overall embedded state information $\hat{\mathbf{s}} = [\mathbf{e}_{user}, \mathbf{e}_{3t+1}, \mathbf{e}_{3t+2}, \mathbf{e}_{3t+3}, \mathbf{h}_1, \mathbf{h}_2, ..., \mathbf{h}_n]$, and then feed it into the RPC module to model the user's real-time preferences toward different categories.

4.3 Real-time Preference Capturing Module

Preference Encoder ϕ. The preference encoder employs the embedded state information $\hat{\mathbf{s}}$ to model the user preference embedding matrices. Assume there are C (i.e., 3 categories in our scenario) categorical lists that need to be reranked, we have the user real-time preference overall matrix $\mathbf{U} = [\mathbf{U}_1, \mathbf{U}_2, ..., \mathbf{U}_c]$ learned by the network ϕ, where $\mathbf{U}_c \in \mathbb{R}^{d_{\text{pref}} \times d_{\text{emb}}}$ is the User Preference Matrix related to Category **c** respectively. The number of rows d_{pref} is the preference depth expected to be learned of the category, and the number of columns d_{emb} is the embedding dimension of each preference aspect.

Item Encoder ψ. At the same time, we employ an embedding network ψ to utilize sliding window item embedding $\mathbf{E}_{W_t} = [\mathbf{e}_{3t+1}, \mathbf{e}_{3t+2}, \mathbf{e}_{3t+3}]$ to dig further features of the items within the sliding window. We expect to extract profound item features that are related to the its own categorical characteristics tightly through the deep item embedding network. Then we have $\mathbf{E}'_{W_t} = [\mathbf{e}'_{3t+1}, \mathbf{e}'_{3t+2}, \mathbf{e}'_{3t+3}]$ denote the item profound embedding, where $\mathbf{e}'_i \in \mathbb{R}^{d_{emb}}$.

Item-wise Preference Calculation. Let C_i = Category(I_i) denote the category type for each item I_i. We design the matrix multiplication $\mathbf{E}_i = \mathbf{U}_{C_i}\mathbf{e}'_i$ between the user real-time preference $\mathbf{U}_{C_i}$ towards the corresponding category of item I_i and the item profound embedding $\mathbf{e}'_i$, to capture the user preference $\mathbf{E}_i \in \mathbb{R}^{d_{\text{pref}} \times 1}$ towards the exactly candidate item. Then we feed the exact learned state information

$$\mathbf{s} = [\mathbf{E}_{3t+1}, \mathbf{E}_{3t+1}, \mathbf{E}_{3t+3}] \tag{2}$$

to the reinforcement learning network to get the final prediction.

In conclusion, the real-time preference capturing module help to unify a comparable embedding scheme for different category items to benefit further development in the reinforcement learning module.

4.4 Contrastive User Preference Modeling

To provide a supervision signal for strengthening the learning of user preferences, we propose an auxiliary contrastive user preference loss. Inspired by the fact that user has common interests among different items implicitly, we believe that items clicked by the same user has common interests factor. Hence, we divide items in the sliding window to two sets $S_{clicked}$ and $S_{unclicked}$. We expect the similarity between the user-item interests embedding of user clicked items and unclicked items to be far away as much as possible, and of the same set items to be closed. Then we optimize the contrastive loss:

$$\mathcal{L}_C(\mu, \phi, \psi) = -\frac{\sum_{i \in S_{clicked}} \sum_{j \in S_{unclicked}} e^{d(\mathbf{E}_i, \mathbf{E}_j)}}{\sum_{i,j \in S_{clicked}} e^{d(\mathbf{E}_i, \mathbf{E}_j)} + \sum_{i,j \in S_{unclicked}} e^{d(\mathbf{E}_i, \mathbf{E}_j)}}, \tag{3}$$

$\forall \mathbf{E}_i, \mathbf{E}_j \in \{\mathbf{E}_{3t+1}, \mathbf{E}_{3t+2}, \mathbf{E}_{3t+3}\}$, where d is the similarity calculation function such as the Euclidean distance or cosine similarity distance. We adopt the square of Euclidean distance in our work.

With the help of the auxiliary contrastive user-item preference, we suppose to learn representations of the user-item interests regardless of the category.

4.5 Conservative Offline Reinforcement Learning

We adopt the actor-critic framework in our RL module, which includes a policy network $\pi(\omega)$ and a Q-function network $Q(\theta)$. Both networks inherit the output of the RPC module as input, the learned state information $\mathbf{s}$ from Eq.(2). In order to address the overestimation of the OOD actions, we extend the conservative Q-learning [9] to our scenario with further policy divergence regularization.

Conservative Policy Evaluation: In consistent with the CQL, we penalize the Q function at states in the dataset for actions not observed in the dataset. Then the Q function associated with the current policy π is conservatively updated by the following optimization function:

$$\begin{aligned} \mathcal{L}_Q(\theta) = \lambda \left(\mathbb{E}_{\mathbf{s}\sim\mathcal{D},\mathbf{a}\sim\pi(\cdot|\mathbf{s})}[Q(\mathbf{s},\mathbf{a})] - \mathbb{E}_{\mathbf{s},\mathbf{a}\sim\mathcal{D}}[Q(\mathbf{s},\mathbf{a})]\right) \\ +\frac{1}{2}\mathbb{E}_{\mathbf{s},\mathbf{a},\mathbf{s}'\sim\mathcal{D}}\left[\left(Q(\mathbf{s},\mathbf{a}) - \widehat{\mathcal{B}}^{\pi}Q(\mathbf{s},\mathbf{a})\right)^2\right]. \end{aligned} \tag{4}$$

The $\widehat{\mathcal{B}}^{\pi}Q(\mathbf{s},\mathbf{a}) := r(\mathbf{s},\mathbf{a}) + \gamma Q'(\mathbf{s}',\mathbf{a}')$ is the empirical bellman operator that only backs up a single sample, where $(\mathbf{s},\mathbf{a},\mathbf{s}')$ is a single transition from the given dataset, $\mathbf{a}' \sim \pi(\cdot|\mathbf{s}')$, Q'_θ is the target Q Network which has the same structure of Q_θ and is substituted by Q_θ network periodically. The second term in the Eq.(4) is the conventional loss that minimizes the squared error of the target Q value and prediction Q value. Significantly, the first term in the Eq.(4) enables a conservative estimation of the value function for learned policy to mitigate the overestimation bias.

Conservative Policy Improvement with Divergence Penalty: The goal of policy learning is to give prediction towards action that maximizes the expected Q value. With the help of a conservative critic Q, the policy network ω is improved by the optimization function:

$$\mathcal{L}_\pi(\mu,\phi,\psi,\omega) = -\mathbb{E}_{\mathbf{s}\sim\mathcal{D},\mathbf{a}\sim\pi(\cdot|\mathbf{s})}\left[Q(\mathbf{s},\mathbf{a})\right]. \tag{5}$$

In order to address the distributional shift challenge in the offline setting, we utilize the following KL-Divergence loss as regularization.

$$\mathcal{L}_{KL} = \mathbb{E}_{(\mathbf{s},\mathbf{a},r,\mathbf{s}')}D_{KL}(\pi(\cdot|\mathbf{s})||\pi_\beta(\cdot|\mathbf{s})). \tag{6}$$

This regularization aims to constrain the bound of the state distributional shift between the learned policy $\pi(\cdot|\mathbf{s})$ and the behavior policy $\pi_\beta(\cdot|\mathbf{s})$. However, the behavior policy $\pi_\beta(\cdot|\mathbf{s})$ is often a mixture of multiple policies due to the complex online logic. Note that in our scenario, the policy actually outputs a distribution of categories at each time step. With this observation, we could recover an approximate behavior policy by simply calculating the distribution of categories in the logged trajectories. Specifically, we calculate the overall category distribution from the real dataset $q = [q_1, q_2, ..., q_C]$. We denote the category distribution inferred from the training policy by $p = [p_1, p_2, ..., p_C]$. Then we can reformulate the KL-regularizert as Eq. 7. This reformulated regularizer implicitly pushes our learned policy $\pi(\cdot|\mathbf{s})$ to be close to the behavior policy $\pi_\beta(\cdot|\mathbf{s})$.

$$\mathcal{L}_{KL} = D_{KL}(p||q) \tag{7}$$

Implementation Details. During the training stage, the Q-function and policy network are updated separately. We train the policy network with the GCE and RPC module networks together through the gradient passed by the learned state information $\mathbf{s}$ while we stop the gradient of the input $\mathbf{s}$ for the Q-function network training. In summary, we alternately train networks between the loss function Eq.(4) and Eq.(8).

$$\mathcal{L} = \mathcal{L}_{\pi}(\mu, \phi, \psi, \omega) + \alpha \mathcal{L}_C(\mu, \phi, \psi) + \beta \mathcal{L}_{KL}, \tag{8}$$

where the α and β are the hyper-parameters to adjust the weight of each loss.

5 Experiments

We present both offline and online evaluation results of RLMixer. In the offline evaluation, we compare RLMixer with existing baselines on the public PRM datasets and industrial datasets collected from an industrial AppStore. In the online experiments, we deploy RLMixer to provide re-ranking services to the industrial AppStore and conduct online A/B testing.

5.1 Offline Experiment Setting

Datasets. We give a detailed description of the two datasets as follows.

PRM Dataset. We adopt the public PRM dataset released by [13], which is a large-scale dataset (E-commerce Re-ranking dataset) built from a real-world E-commerce recommender system. The dataset includes a huge number of sessions that record interactions between the recommender system and users. For each session, features (e.g., category, identity, price, etc.) of a recommendation list items recommended to a user and the corresponding user click-through response are stored. To avoid significant variance, we keep the interactions between the user and three main (i.e., most frequently presented) category items recommended by the system with primal orders presented in the recommendation list, which also matches the integrated ranking application scenario.

Industrial Dataset. We collected a real-world dataset from an industrial AppStore platform for 15 consecutive days. It contains similar user and item features as the PRM dataset, with two additional high-level categories.

The statistics of two offline datasets are presented in the Table 1. We split each dataset into training and test sets with a ratio of 4:1.

Table 1. Statistics of two offline datasets

	Sessions	Users	Category A	Category B	Category C
PRM dataset	7,919,659	605,668	1,411,185	291,629	311,364
Industrial dataset	194,233	184,443	9,018	4,771	-

Baselines. We compare RLMixer with the following representative methods.

Original. The primal recommendation list is presented in the original dataset.

MMR. [1]. Maximal Marginal Relevance(MMR) is a ranking algorithm that allows controlling the diversity and the relevance of provided information.

LinkedIn-Det. [5]. LinkedIn-Det proposes several deterministic algorithms for fair re-ranking of top-K results based on desired proportions over one or more protected attributes.

DHCRS. [3]. Deep Hierarchical Category-based Recommender System utilizes a high-level DQN to select a category and then a low-level DQN to choose an item in this category. Due to the order preservation constraints in our integrated ranking scenario, we implement the category-level structure of DHCRS, and add the KL-Divergence loss to make it more suitable to the offline setting.

Evaluation Metrics. The aim of an optimal integrated ranking system is to maximize the revenue for the platform. Accordingly, we adopt utility and $\alpha-$utility metrics to evaluate the performance of our method instead of NDCG. Furthermore, as mentioned in Sect. 5.1, the diversity or we call the ratio of each category information, is also important to the integrated problem. We introduce the ratio metric to make all algorithms fit into similar comparable levels.

NDCG@K (Normalized Discounted Cumulative Gain) is a measure of ranking quality in information retrieval area. It evaluates the quality of the recommendation list by calculating the fraction of Discounted Cumulative Gain (e.g., the click signal in our scenario) over the Ideal Discounted Cumulative Gain.

utility@K is the average utility of a session with regard to the top-K recommended items. The utility of a single item is related to the Sect. 3, we then formulate the calculation of *utility@K* as the Eq.(9).

$$utility@K = \frac{1}{|S|}\sum_{s \in S}\sum_{i=1}^{K} price(I_{s,i}) * Click(I_{s,i}), \tag{9}$$

where S is the set of sessions, $I_{s,i}$ is the i-th recommended item in the session s, $Click(I_{s,i})$ is the click signal indicating whether the item $I_{s,i}$ is clicked.

$\alpha-$utility@K is a metric that considers category information and evaluates whether the utility is balanced among different categories,

$$\alpha - utility@K = \frac{1}{|S|}\sum_{s \in S}\sum_{i=1}^{K} \alpha^{N_{category(I_{s,i})}} price(I_{s,i}) * Click(I_{s,i}), \tag{10}$$

where α is the discounted factor, and $N_{(category(I_{s,i}))}$ is the total counts of category $category(I_{s,i})$ appears in the session s.

ratio@K is used to evaluate the distribution of each category among all items. The ratio among two categories C_i and C_j only is defined as follows:

$$\underset{i,j}{ratio}@K = \frac{\sum_{s \in S}\sum_{k=1}^{K} \mathbb{I}(category(I_{s,k}) = C_i)}{\sum_{s \in S}\sum_{k=1}^{K} \mathbb{I}(category(I_{s,k}) = C_j)}, \tag{11}$$

where $\mathbb{I}$ is the indicator function.

To compute the evaluate the distribution of category when the number of categories larger than two, we elaborate the rotate ratio to roughly represent the average distribution in the following:

$$ratio@K = \frac{\sum_{i=1}^{|C|-1} ratio_{i,i+1}@K + ratio_{|C|,1}@K}{|C|}, \tag{12}$$

where $C = C_1, C_2, ..., C_n$ is the set of n categories supposed to be constrained.

5.2 Offline Results and Analysis

Results on PRM Dataset. We conduct experiments with our model and other baseline models on the PRM dataset and focus on the top-10 performance of the recommendation list since the average session length is 30. We first compute the basic statistics of the testing set, which is presented in the Table 2 as original baseline. To evaluate variant methods fairly, we set the target desired ratio 2.3 during model training, which is the approximation of the ratio metric in original recommendation among dataset. And then we select the best model according to the utility performance of top-10 when the ratio of model predictions within the range of 2.3 ± %10. The performance results are presented in the Table 2.

Table 2. Evaluation on PRM dataset.

Top-10	ratio 2.3 ± 10%			
	NDCG	utility	α−utility	ratio
original	0.5114	0.0339	0.0106	2.3124
MMR	0.4088	0.0386	0.0144	2.1436
LInkedINn-Det	0.4213	0.0354	0.0130	2.2883
DHCRS	0.4158	0.0385	0.0096	2.1522
RLMixer	0.4003	**0.0471**	**0.0203**	2.4239

- Compared our RLMixer with other baselines, RLMixer outperforms them in both utility and α−utility metrics, which is at least 22% and 44% higher than others respectively.
- Considering both MMR and Linkedin-Det are ranking algorithm related to balancing the diversity and utility, our RLMixer outperforms these two algorithms even though we were bound into the same level ratio. It shows that our algorithm can be applied to the scenario to earn profit much higher while fulfilling the desired distribution requirements.
- We achieve better performance than DHCRS even though DHCRS is a RL-based method and giving the category prediction first as well. This indicates that our real-time user preference capturing module and corresponding auxiliary contrastive loss design might contribute a lot to the final prediction. We will discuss this later in the ablation study.
- The NDCG value of the original list is the highest, but it brings the lowest utility. The reason is that the computation of NDCG uses only click signals, ignoring the real values of each click. Therefore, compared with NDCG, the utility-related metrics better align with the online performance. This gives us an intuition that pursuing the most clicks may not be the best strategy.

Results on Industrial Dataset. Since users are presented with at least 7 items at a time, we compare top-20 results of RLMixer with the original rank on the industrial dataset. Table 3 shows their performance comparisons. Similar to the PRM dataset, our RLMixer leads to a lower NDCG value but is compensated by a 1.7% utility gain.

Table 3. Evaluation on Industrial dataset.

Top-20	ratio 0.5 ± 10%			
	NDCG	utility	α−utility	ratio
Original	0.3659	3.056	1.152	0.5284
RLMixer	0.3455	**3.108**	**1.341**	0.5405

Ablation Study. To verify the impact of Real-time Preference Capturing(RPC) module and the auxiliary contrastive loss, we conduct two sets of experiments on the public dataset PRM. The two sets of experiments train RLMixer without the entire RPC module or auxiliary contrastive loss, respectively. We still constraint the desired ratio within the range of 2.3 ± %10 during the model training, and we select the best model based on top-10 performance. Then we evaluate top-3, top-5, and top-10 performance of the best model of each RLMixer variant. As shown in the Table 4, the full RLMixer comprehensively presents higher performance in utility and α−utility metrics of all top-k levels, while compared with the variants without RPC module or auxiliary contrastive loss. Especially, it can be observed that the cooperation of RPC module and auxiliary loss brings key improvements to the primal algorithm over other baseline algorithms.

Table 4. Ablation study of contrastive user preference modeling in RLMixers on PRM dataset.

Method	Top-K	NDCG	utility	α−utility	ratio
RLMixer	3	0.2575	**0.0246**	**0.0216**	2.4061
	5	0.3140	**0.0302**	**0.0211**	2.0444
	10	0.4003	**0.0471**	**0.0203**	2.4239
RLMixer w/o RPC	3	0.2585	0.0192	0.0159	2.3125
	5	0.3144	0.0288	0.0166	2.5253
	10	0.4010	0.0446	0.0151	2.4495
RLMixer w/o Contrastive Loss	3	0.2564	0.0214	0.0171	2.1858
	5	0.3123	0.0296	0.0165	1.9008
	10	0.3997	0.0437	0.0150	2.1805

5.3 Online A/B Test

The online setting is follow that of offline experiments, and the quantified criteria of the A/B experiment is to compare the revenue(i.e., utility) with the baseline in three days. We deploy RLMixer online and compare its performance with the fine-tuned rule-based model that is currently deployed online. Noted that the goal of our method is to adjust the ratio to a certain target value while maximizing the utility. Table 5 shows the regularized average utility obtained during three consecutive days. We find that RLMixer achieves 4.05% utility gain compared with the current model, which demonstrated the effectiveness of our method.

Table 5. The results of A/B experiments.

Policy	Baseline	RLMixer	Impr
utility	0.0074	0.0077	**4.05%**

6 Conclusion

We propose a general offline RL framework with contrastive user preference modeling called RLMixer for integrated ranking problems. With the aid of the Global Context Extraction (GCE) module and Real-time Preference Capturing (RPC) Module, RLMixer is able to synthesize values of items from different categories and capture the user's short-term preference shifting. Furthermore, it incorporates behavior regularization into the actor-critic framework to address the distribution shift problem that exists in the offline setting. We compare RLMixer with existing baselines on the public PRM datasets and datasets collected from an industrial AppStore. We also deploy RLMixer to provide re-ranking services to an industrial AppStore and conduct an online A/B test, which shows that RLMixer brings 4.05% utility gain.

References

1. Carbonell, J., Goldstein, J.: The use of mmr, diversity-based reranking for reordering documents and producing summaries. In: Proceedings of the 21st Annual International ACM SIGIR Conference on Research and Development in Information Retrieval, pp. 335–336 (1998)
2. Chen, M., Beutel, A., Covington, P., Jain, S., Belletti, F., Chi, E.H.: Top-k off-policy correction for a reinforce recommender system. In: Proceedings of the Twelfth ACM International Conference on Web Search and Data Mining. pp. 456–464 (2019)
3. Fu, M., Agrawal, A., Irissappane, A.A., Zhang, J., Huang, L., Qu, H.: Deep reinforcement learning framework for category-based item recommendation. IEEE Transactions on Cybernetics (2021)
4. Fujimoto, S., Meger, D., Precup, D.: Off-policy deep reinforcement learning without exploration. In: International Conference on Machine Learning, pp. 2052–2062 (2019)
5. Geyik, S.C., Ambler, S., Kenthapadi, K.: Fairness-aware ranking in search & recommendation systems with application to linkedin talent search. In: Proceedings of the 25th acm sigkdd international conference on knowledge discovery & data mining, pp. 2221–2231 (2019)
6. Guanjie, Z., et al.: Drn: A deep reinforcement learning framework for news recommendation. In: Proceedings of the 2018 World Wide Web Conference, pp. 167–176 (2018)
7. Koutsopoulos, I.: Optimal advertisement allocation in online social media feeds. In: Proceedings of the 8th ACM International Workshop on Hot Topics in Planet-Scale Mobile Computing and Online Social Networking, pp. 43–48 (2016)
8. Kumar, A., Fu, J., Soh, M., Tucker, G., Levine, S.: Stabilizing off-policy q-learning via bootstrapping error reduction. In: Advances in Neural Information Processing Systems, pp. 11761–11771 (2019)
9. Kumar, A., Zhou, A., Tucker, G., Levine, S.: Conservative q-learning for offline reinforcement learning. In: Advances in Neural Information Processing Systems, pp. 1179–1191 (2020)
10. Levine, S., Kumar, A., Tucker, G., Fu, J.: Offline reinforcement learning: Tutorial, review, and perspectives on open problems. arXiv preprint arXiv:2005.01643 (2020)

11. Liao, G., et al.: Cross dqn: Cross deep q network for ads allocation in feed. arXiv preprint arXiv:2109.04353 (2021)
12. Matsushima, T., Furuta, H., Matsuo, Y., Nachum, O., Gu, S.S.: Deployment-efficient reinforcement learning via model-based offline optimization. In: International Conference on Learning Representations (2021)
13. Pei, C., et al.: Personalized re-ranking for recommendation. In: Proceedings of the 13th ACM Conference on Recommender Systems, pp. 3–11 (2019)
14. Wu, Y., Tucker, G., Nachum, O.: Behavior regularized offline reinforcement learning. arXiv preprint arXiv:1911.11361 (2019)
15. Xiao, T., Wang, D.: A general offline reinforcement learning framework for interactive recommendation. In: Proceedings of the 35th AAAI Conference on Artificial Intelligence (2021)
16. Yan, J., Xu, Z., Tiwana, B., Chatterjee, S.: Ads allocation in feed via constrained optimization. In: Proceedings of the 26th ACM SIGKDD International Conference on Knowledge Discovery & Data Mining, pp. 3386–3394 (2020)
17. Zhao, M., Li, Z., Bo, A., Haifeng, L., Yifan, Y., Chen, C.: Impression allocation for combating fraud in e-commerce via deep reinforcement learning with action norm penalty. In: Proceedings of the 27th International Joint Conference on Artificial Intelligence, pp. 3940–3946 (2018)
18. Zhao, X., Gu, C., Zhang, H., Yang, X., Liu, X., Tang, J., Liu, H.: Dear: Deep reinforcement learning for online advertising impression in recommender systems. In: Proceedings of the 35th AAAI Conference on Artificial Intelligence, pp. 750–758 (2021)
19. Zhou, G., et al.: Deep interest evolution network for click-through rate prediction. In: Proceedings of the AAAI Conference on Artificial Intelligence, pp. 5941–5948 (2019)
20. Zhou, G., et al.: Deep interest network for click-through rate prediction. In: Proceedings of the 24th ACM SIGKDD International Conference on Knowledge Discovery & Data Mining, pp. 1059–1068 (2018)

Author Index

H. Kashima et al. (Eds.): PAKDD 2023, LNAI 13937, pp. 415–417, 2023.
https://doi.org/10.1007/978-3-031-33380-4

Printed in the United States
by Baker & Taylor Publisher Services

Printed in the United States
by Baker & Taylor Publisher Services